CONFORMAL
FIELD
THEORY

CONFORMAL
FIELD
THEORY

Yavuz Nutku,
Cihan Saclioglu,
and Teoman Turgut,
Editors

Advanced Book Program

CRC Press
Taylor & Francis Group
Boca Raton London New York

CRC Press is an imprint of the
Taylor & Francis Group, an **informa** business

First published 2000 by Westview Press

Published 2018 by CRC Press
Taylor & Francis Group
6000 Broken Sound Parkway NW, Suite 300
Boca Raton, FL 33487-2742

CRC Press is an imprint of Taylor & Francis Group, an Informa business

Visit the Taylor & Francis Web site at
http://www.taylorandfrancis.com

and the CRC Press Web site at
http://www.crcpress.com

A Cataloging-in-Publication data record for this book is available from the Library of Congress.

ISBN 13: 978-0-8133-4214-6 (pbk)

Frontiers in Physics
David Pines, Editor

Volumes of the Series published from 1961 to 1973 are not officially numbered. The parenthetical numbers shown are designed to aid librarians and bibliographers to check the completeness of their holdings.

Titles published in this series prior to 1987 appear under either the W. A. Benjamin or the Benjamin/Cummings imprint; titles published since 1986 appear under the Westview Press imprint.

Volumes published from 1974 onward are being numbered as an integral part of the bibliography.

Contents

Editor's Foreword

The problem of communicating in a coherent fashion recent developments in the most exciting and active fields of physics continues to be with us. The enormous growth in the number of physicists has tended to make the familiar channels of communication considerably less effective. It has become increasingly difficult for experts in a given field to keep up with the current literature; the novice can only be confused. What is needed is both a consistent account of a field and the presentation of a definite "point of view" concerning it. Formal monographs cannot meet such a need in a rapidly developing field, while the review article seems to have fallen into disfavor. Indeed, it would seem that the people who are most actively engaged in developing a given field are the people least likely to write at length about it.

Frontiers in Physics was conceived in 1961 in an effort to improve the situation in several ways. Leading physicists frequently give a series of lectures, a graduate seminar, or a graduate course in their special fields of interest. Such lectures serve to summarize the present status of a rapidly developing field and may well constitute the only coherent account available at the time. One of the principal purposes of the *Frontiers in Physics* series is to make notes on such lectures available to the wider physics community.

As *Frontier in Physics* has evolved, a second category of book, the informal text/monograph, an intermediate step between lecture notes and formal text or monographs, has played an increasingly important role in the series. In an informal text or monograph an author has reworked his or her lecture notes to the point at which the manuscript represents a coherent summation of a newly developed field, complete with reference and problems, suitable for either classroom teaching or individual study.

The present lecture note volume, *Conformal Field Theory*, edited by Yavuz Nutku, Cihan Saçlioglu, and Teoman Turgut, represents something of a departure from the usual Frontiers volume, in that it contains notes on a related series of lectures given during the 1998 Summer Research Semester on Conformal Field Theories, Matrix Models, and Dualities at the Feza Gursey Institute in Istanbul. The topics of these lectures are central to our emerging understanding of conformal field theory and its importance to both statistical mechanics and string theory. The lecturers, all experts in their respective subfields, have done an excellent job of introducing advanced graduate students and experienced researchers alike to this rapidly developing subfield of both physics and mathematics, and it is a pleasure to welcome them to the ranks of authors in *Frontiers in Physics*.

David Pines
Tesuque, NM
December, 1999

Preface

Quantum Field Theory has been with us for nearly three quarters of a century, but it is roughly only in the last quarter that physicists and mathematicians have jointly ventured out to explore its realms beyond the reach of perturbation theory, to the great benefit of both disciplines. The lectures in the present volume well exemplify some of these non-perturbative approaches and the way in which they lead not only to the Frontiers of Physics, but also of Mathematics.

Conformal Field Theory (CFT), to which four of the six lectures are devoted, naturally lends itself to non-perturbative treatment by virtue of the infinite-dimensional conformal symmetry present in two dimensions. It is of interest in physics, among other things, for describing the long distance behavior of certain two-dimensional Statistical Mechanical systems and String Theory, and also in mathematics, through its connections to Affine Lie Algebras, The Monster group and modular invariance. Mark Walton's lectures introduce Wess-Zumino-Novokov-Witten (WZNW) models and their current algebras, the affine Kac-Moody algebras. This is followed by representation theory, leading to the computation of 3-point functions and fusion rules. The latter can be regarded as a generalization, albeit a very non-trivial one, one of the familiar reduction of the tensor product of two representations using Clebsch-Gordan coefficients. Krzysztof Gawedzki uses the WZNW model as a prototype of general (rational) CFT models. The theory is formulated on a general Riemann surface to make it adaptable to string perturbation theory applications. In the later sections, the WZNW model with local boundary conditions is introduced, making contact with D-branes. The lecture notes of Matthias Gaberdiel and Peter Goddard (delivered by M. Gaberdiel in Istanbul) describe their axiomatic formulation of Meromorphic CFT (MCFT), in which the vacuum expectation values of fields are taken as the starting point for the reconstruction first of a topological vector space of states, followed by vertex operators corresponding to the states. They show that although the theory has only Möbius symmetry at the outset, this can be extended to full conformal symmetry. The general formalism is then applied to a U(1) theory, Virasoro and affine Lie algebras, lattice theories, and even to a case which does not have conformal structure. Terry Gannon provides an introduction to Monstrous Moonshine, an area of String theory-related mathematics for which Borcherds was awarded one of the 1998 Fields Medals, and to the classification of Rational CFTs. Along the way, he discusses the basic ideas and motivations behind mathematical subjects unfamiliar to most physicists, such as the finite simple groups classification, Galois theory, Fermat's Last Theorem and Category Theory.

Philip Argyres' lectures introduce methods that, starting with the ideas of Sen and Seiberg and Witten about five years ago, have developed to the point of providing exact information about the strong coupling limit of a class of four dimensional field theories. A variant of the Seiberg-Witten approach also provides a dramatic simplification in the study of four-manifolds; thus, just as with CFTs, these non-perturbative methods also shed light on important current problems in Mathematics. In fact, Selman Akbulut delivered a set of lectures of Seiberg-Witten-Donaldson 4-manifold theory in parallel with those of Argyres; an extended version of these will appear elsewhere. In the final set of lectures, Sarada Rajeev obtains the parton structure functions for the proton by first reducing QCD to an effectively two-dimensional theory in the Bjorken limit, and then showing its equivalence to a new theory of hadrons, whose phase space is the Grassmanian of complex Hilbert space. The symplectic form and the Hamiltonian lead to a dynamical system equivalent to QCD with an infinite number of colors; 't Hooft's planar limit is the linearized approximation of the resulting theory. The proton emerges as a topological soliton with numerically calculable structure functions which agree with experiment.

The lectures, varying from 8 to 16 hours, were aimed at advanced graduate students attending the 1998 Summer Research Semester on Conformal Field Theories, M(atrix) Models and Dualities held at Bogaziçi University-TÜBITAK Feza Gürsey Institute in Istanbul. We thank TÜBITAK for its support, which made the meeting possible.

Yavuz Nutku, Cihan Saçlioglu, Teoman Turgut

Istanbul, 1999

Conformal field theory: a case study

Krzysztof Gawędzki

C.N.R.S., I.H.E.S., 91440 Bures-sur-Yvette, France

Abstract

This is a set of introductory lecture notes devoted to the Wess-Zumino-Witten model of two-dimensional conformal field theory. We review the construction of the exact solution of the model from the functional integral point of view. The boundary version of the theory is also briefly discussed.

1 Introduction

Quantum field theory is a structure at the root of our understanding of physical world from the subnuclear scales to the astrophysical and the cosmological ones. The concept of a quantum field is very rich and still poorly understood although much progress have been achieved over some 70 years of its history. The main problem is that, among various formulations of quantum field theory it is still the original Lagrangian approach which is by far the most insightful, but it is also the least precise way to talk about quantum fields. The strong point of the Lagrangian approach is that it is rooted in the classical theory. As such, it permits a perturbative analysis of the field theory in powers of the Planck constant and also captures some semi-classical non-perturbative effects (solitons, instantons). On the other hand, however, it masks genuinely non-perturbative effects. In the quest for a deeper understanding of quantum field theory an important role has been played by two dimensional models. Much of what we have learned about nonperturbative phenomena in quantum field theory has its origin in such models. One could cite the Thirring model with its anomalous dimensions and the fermion-boson equivalence to the sine-Gordon model, the Schwinger model with the confinement of electric charge, the non-linear sigma model with the non-perturbative mass generation, and so on.

The two-dimensional models exhibiting conformal invariance have played a specially important role. On one side, they are not without direct physical importance, describing, in their Euclidean versions, the long-distance behavior of the two-dimensional

1

statistical-mechanical systems, like the Ising or the Potts models, at the second order phase transitions. On the other hand, the (quantum) conformal field theory (CFT) models constitute the essential building blocks of the classical vacua of string theory, a candidate "theory of everything", including quantum gravity. The two-dimensional space-time plays simply the role of a string world sheet parametrizing the string evolution, similarly as the one-dimensional time axis plays the role of a world line of point particles. The recent developments seem to indicate that string theory, or what will eventually emerge from it, provides the appropriate language to talk about general quantum fields, whence the central place of two-dimensional CFT in the quantum field theory edifice.

Due to the infinite-dimensional nature of the conformal symmetry in two space-time dimensions, the two-dimensional models of CFT lend themselves to a genuinely non-perturbative approach based on the infinite symmetries and the concept of the operator product expansion [1]. In the present lectures, we shall discuss a specific model of two-dimensional CFT, the so called Wess-Zumino-(Novikov)-Witten model (WZW) [2][3][4]. It is an example of a non-linear sigma model with the classical fields on the space-time taking values in a manifold which for the WZW model is taken as a group manifold of a compact Lie group G. We shall root our treatment in the Lagrangian approach and will work slowly our way towards a non-perturbative formulation. This will, hopefully, provide a better understanding of the emerging structure which, to a large extent, is common to all CFT models. In fact, the WZW theory is a prototype of general (rational) CFT models which may be obtained from the WZW one by different variants of the so called coset construction. In view of the stringy applications, where the perturbation expansion is built by considering two-dimensional conformal theories on surfaces of arbitrary topology, we shall define and study the WZW model on a general Riemann surface.

A word of warning is due to a more advanced audience. The purpose of these notes is not to present a complete up-to-date account of the WZW theory, even less of CFT. That would largely overpass the scope of a summer-school lecture notes. As a result, we limit ourselves to the simplest version of the model leaving completely aside the ramifications involving models with non-simply connected groups, orbifolds, etc, as well as applications to string theory. We profit, however, from this simple example to introduce on the way some of the main concepts of two-dimensional CFT. Much of the material presented is not new, even old, by the time-scale standard of the subject, with the possible exception of the last section devoted to the boundary WZW models. The author still hopes that the following exposition, which he failed to present at the 1998 Istanbul summer school, may be useful to a young reader starting in the field.

The notes are organized as follows. In Sect. 2, we discuss a simple quantum-mechanical version of the WZW model: the quantum particle on a group manifold. This simple model, exactly solvable by harmonic analysis on the group, permits to describe many structures similar to the ones present in the two-dimensional theory and to understand better the origin of those. Sect. 3 is devoted to the definition of the action functional of the WZW model. The action contains a topological term, which requires

2

a special treatment. We discuss separately the case of the surfaces without and with boundary, in the latter case postponing the discussion of local boundary conditions to the last section. In Sect. 4, we introduce the basic objects of the (Euclidean) quantum WZW theory: the quantum amplitudes taking values in the spaces of states of the theory and the correlation functions. We state the infinite-dimensional symmetry properties of the theory related to the chiral gauge transformations and to the conformal transformations. The symmetries give rise to the action of the two copies of the current and Virasoro algebras in the Hilbert space of states of the theory constructed with the use of the representation theory of those algebras. We discuss briefly the operator product expansions which encode the symmetry properties of the correlation functions. Sect. 5 is devoted to the relation between the WZW theory and the Schrödinger picture quantum states of the topological three-dimensional Chern-Simons theory. The relation is established via the Ward identities expressing the behavior of the WZW correlation functions, coupled to external gauge field, under the chiral gauge transformations. We discuss the structure of the spaces of the Chern-Simons states, the fusion ring giving rise to the Verlinde formula for their dimensions and their Hilbert-space scalar product, as well as the Knizhnik-Zamolodchikov connection which permits to compare the states for different complex structures. In particular, we explain how the knowledge of the scalar product of the Chern-Simons states permits to obtain exact expressions for the correlation functions of the WZW theory. In Sect. 6 we give a brief account of the coset construction of a large family of CFT models which may be solved exactly, given the exact solution of the WZW model. Finally, Sect. 7 is devoted to the WZW theory with local boundary conditions. Again, for the sake of simplicity, we restrict ourselves to a simple family of the conditions that do not break the infinite-dimensional symmetries of the theory. We discuss how to define the action functional of the model in the presence of such boundary conditions and what are the elementary properties of the corresponding spaces of states, quantum amplitudes and correlation functions.

2 Quantum mechanics of a particle on a group

2.1 The geodesic flow on a group

Non-linear sigma models describe field theories with fields taking values in manifolds. These lectures will be devoted to a special type of sigma models, known under the name of Wess-Zumino(-Novikov)-Witten (or WZW) models. They are prototypes of conformal field theories in two-dimensional space-time. As such they play a role in string theory whose classical solutions are built out of two-dimensional quantum conformal field theory models by a cohomological construction. Before we plunge, however, into the details of the WZW theory, we shall discuss a simpler but largely parallel model in one dimension, i.e. in the domain of mechanics rather than of field theory.

One-dimensional sigma models describe the geodesic flows on manifolds M endowed with a Riemannian or a pseudo-Riemannian metric γ. The classical action for the trajectory $[0, T] \ni t \mapsto x(t) \in M$ of such a system is

$$S(x) = \frac{1}{2} \int_0^T \gamma_{\mu\nu}(x) \frac{dx^\mu}{dt} \frac{dx^\nu}{dt} \, dt \qquad (2.1)$$

and the classical solutions $\delta S = 0$ correspond to the geodesic curves in M parametrized by a rescaled length. If $M = \mathbf{R}^n$, for example, and $\gamma_{\mu\nu} = m\,\delta_{\mu\nu}$, we obtain the action of the free, non-relativistic particle of mass m undergoing linear classical motions $x(t) = x_0 + \frac{p}{m}t$. The action (2.1) is not parametrization-invariant but it may be viewed as a gauged-fixed version of the action

$$S_p(x) = \frac{1}{2} \int_0^T \gamma_{\mu\nu}(x) \frac{dx^\mu}{dt} \frac{dx^\nu}{dt} \eta^{-\frac{1}{2}} \, dt + \frac{1}{2} \int_0^T \eta^{\frac{1}{2}} \, dt \,,$$

where the reparametrization invariance is restored by coupling the system *à la* Polyakov to the metric $\eta(t)(dt)^2$ on the word-line of the particle. The $\eta = 1$ gauge reproduces then the action (2.1), whereas extremizing over η, one obtains the relativistic action

$$S_r(x) = \int_0^T \left(\gamma_{\mu\nu}(x) \frac{dx^\mu}{dt} \frac{dx^\nu}{dt} \right)^{\frac{1}{2}} dt$$

given by the geodesic length of the trajectory.

The exact solvability of the geodesic equations can be achieved in sufficiently symmetric situations. In particular, we shall be interested in the case when M is a manifold of a compact Lie group G and when γ is a left-right invariant metric on G given by $\frac{k}{2}$ times a positive bilinear ad-invariant form $\mathrm{tr}(XY)$ on the Lie algebra[1] $\mathbf{g}$. For matrix algebras, as the algebra $su(N)$ of the hermitian $n \times n$ traceless matrices, the form is given by the matrix trace in the defining representation, hence the notation. The positive constant k will play the role of a coupling constant. The action (2.1) may be then rewritten as

$$S(g) = -\frac{k}{4} \int_0^T \mathrm{tr} \, (g^{-1} \frac{d}{dt} g)^2 dt.$$

The variation of the action under the infinitesimal change of g vanishing on the boundary is

$$\delta S(g) = \frac{k}{2} \int_0^T \mathrm{tr} \, (g^{-1} \delta g) \frac{d}{dt} (g^{-1} \frac{d}{dt} g)) \, dt \,.$$

Consequently, the classical trajectories are solutions of the equations

$$\frac{d}{dt} (g^{-1} \frac{d}{dt} g) = 0. \qquad (2.2)$$

The case $G = \mathbf{T}^n$, where $\mathbf{T}^n = \mathbf{R}^n / \mathbf{Z}^n$ is the n-dimensional torus, is the prototype of an integrable system whose trajectories are periodic or quasiperiodic motions with the

---

[1] we use the physicists' convention in which the exponential map between the Lie algebra and the group is $X \mapsto e^{iX}$

angles evolving linearly in time. The case $G = SO(3)$ corresponds to the symmetric top whose positions are parametrized by rigid rotations. The classical trajectories solving Eq. (2.2) have a simple form:

$$g(t) = g_\ell \, e^{it\lambda/k} g_r^{-1}, \qquad (2.3)$$

where $g_{\ell,r}$ are fixed elements in G and λ may be taken in the Cartan subalgebra $\mathbf{t} \subset \mathbf{g}$. For the later convenience, we have introduced the factor $\frac{1}{k}$ in the exponential.

The space $\mathcal{P}$ of classical solutions forms the phase space of the system. It may be parametrized by the initial data $(g(0), p(0))$ where the momentum[2] $p(t) = \frac{k}{2i}\frac{d}{dt}g$. As usually, the phase space $\mathcal{P}$ comes equipped with the symplectic form

$$\Omega = \frac{1}{i}d\,\mathrm{tr}\,(p\,g^{-1}dg), \qquad (2.4)$$

where the right hand side may be calculated at any instance of time giving a result independent of time. The symplectic structure on $\mathcal{P}$ allows to associate the vector fields $\mathcal{X}_f$ to functions f on $\mathcal{P}$ by the relation $-df = \iota_{\mathcal{X}_f}\Omega$, where $\iota_{\mathcal{X}}\alpha$ denotes the contraction of a vector field $\mathcal{X}$ with a differential form α. These are the Hamiltonian vector fields that preserve the symplectic form: $\mathcal{L}_{\mathcal{X}_f}\Omega = 0$, where $\mathcal{L}_{\mathcal{X}}$ is the Lie derivative that acts on differential forms by $\mathcal{L}_{\mathcal{X}}\alpha = \iota_{\mathcal{X}}d\alpha + d\iota_{\mathcal{X}}\alpha$. The Poisson bracket of functions is defined by: $\{f, g\} = \mathcal{X}_f(g)$. In particular, the time evolution is induced by the vector field associated with the classical Hamiltonian

$$h = \frac{1}{k}\,\mathrm{tr}\,p^2 = -\frac{k}{4}\,\mathrm{tr}\,(g^{-1}\tfrac{d}{dt}g)^2$$

which stays constant during the evolution. In the alternative way (2.3) to parametrize the solutions, $h = \frac{1}{4k}\mathrm{tr}\,\lambda^2$ and the symplectic structure splits:

$$\Omega = \Omega_\ell - \Omega_r, \qquad \text{where} \qquad \Omega_\ell = \frac{1}{2}\,\mathrm{tr}\,[\lambda(g_\ell^{-1}dg_\ell)^2 - d\lambda\,g_\ell^{-1}dg_\ell] \qquad (2.5)$$

and Ω_r is given by the same formula with the subscript ℓ replaced by r.

There are two commuting actions of the group G on $\mathcal{P}$: from the left $g(t) \mapsto g_0 g(t)$ and from the right $g(t) \mapsto g(t)g_0^{-1}$. Both preserve the symplectic structure and the Hamiltonian h. The vector fields corresponding to the left and right actions of the infinitesimal generators $t^a \in \mathbf{g}$ are induced by the functions

$$j^a = \frac{k_1}{2}\,\mathrm{tr}\,(t^a g\tfrac{d}{dt}g^{-1}) = \frac{1}{2}\,\mathrm{tr}\,(t^a g_\ell \lambda g_\ell^{-1})$$

$$\tilde{j}^a = \frac{k_1}{2}\,\mathrm{tr}\,(t^a g^{-1}\tfrac{d}{dt}g) = -\frac{1}{2}\,\mathrm{tr}\,(t^a g_r \lambda g_r^{-1}),$$

respectively. Note that, if we normalize t^a's so that $\mathrm{tr}\,(t^a t^b) = \frac{1}{2}\delta^{ab}$, then

$$h = \frac{2}{k}\,j^a j^a = \frac{2}{k}\,\tilde{j}^a \tilde{j}^a$$

(summation convention!). The symplectic form Ω_ℓ gives, for λ fixed, the canonical symplectic form on the (co)adjoint orbit $\{g_\ell \lambda g_\ell^{-1} \,|\, g_\ell \in G\}$ passing through λ. The left action of the group is $g_\ell \mapsto g_0 g_\ell$ so that it coincides with the (co)adjoint action on the orbit. As is well known, upon geometric quantization of the coadjoint orbits for appropriate λ, this action gives rise to irreducible representations of G.

---

[2] we identify $\mathbf{g}$ with its dual using the bilinear form $\mathrm{tr}(XY)$

2.2   The quantization

The geodesic motion on a group is easy to quantize. As the Hilbert space $\mathcal{H}$ one takes the space $L^2(G, dg)$ of functions on G square integrable with respect to the normalized Haar measure dg. The two commuting actions of G in $\mathcal{H}$:

$$f \mapsto {}^h f = f(h^{-1} \cdot), \qquad f \mapsto f^h = f(\cdot h),$$

give rise to the actions

$$J^a f = \left.\frac{1}{\imath} \frac{d}{d\epsilon}\right|_{\epsilon=0} {}^{e^{\imath \epsilon t^a}} f, \qquad \tilde{J}^a f = \left.\frac{1}{\imath} \frac{d}{d\epsilon}\right|_{\epsilon=0} f^{e^{\imath \epsilon t^a}},$$

of the infinitesimal generators t^a of **g**. The commutation relations

$$[J^a, J^b] = \imath f^{abc} J^c \qquad [\tilde{J}^a, \tilde{J}^b] = \imath f^{abc} \tilde{J}^c,$$

reflect the relation $[t^a, t^b] = \imath f^{abc} t^c$ in the Lie algebra **g**. The quantum Hamiltonian

$$H = \frac{2}{k} J^a J^a = \frac{2}{k} \tilde{J}^a \tilde{J}^a$$

coincides with $-\frac{2}{k}$ times the Laplace-Beltrami operator on G and is a positive self-adjoint operator.

   The irreducible representations R of the compact Lie group G are finite dimensional and are necessarily unitarizable so that we may assume that they act in finite-dimensional vector spaces V_R preserving their scalar product. We shall denote by g_R and X_R the endomorphisms of V_R representing $g \in G$ and $X \in$ **g**. Up to isomorphism, the irreducible representation of G may be characterized by their **highest weights**. Let us recall what this means. The complexified Lie algebra may be decomposed into the eigenspaces of the adjoint action of its Cartan subalgebra **t** as

$$\mathbf{g}^C = \mathbf{t}^C \oplus (\bigoplus_{\alpha} \mathbf{C} e_\alpha) \tag{2.6}$$

where $[X, e_\alpha] = \operatorname{tr}(\alpha X) e_\alpha$ for all $X \in$ **t**. The set of the roots $\alpha \in$ **t** may be divided into the positive roots and their negatives. We shall normalize the invariant form tr on **g** so that the long roots have the length squared 2 (this agrees with the normalization of the matrix trace for **g** $= su(N)$). The "step generators" $e_{\pm\alpha}$ may be chosen so that $[e_\alpha, e_{-\alpha}]$ is equal to the coroot $\alpha^\vee \equiv \frac{2\alpha}{\operatorname{tr}\alpha^2}$ corresponding to α. The elements $\lambda \in$ **t** such that $\operatorname{tr}(\alpha^\vee \lambda)$ is integer for all roots are called weights. A non-zero vector $v \in V_R$ (unique up to normalization) is called a highest weight (HW) vector if it is an eigenvector of the action of the Cartan algebra: $X_R v = \operatorname{tr}(\lambda_R X) v$ and if $(e_\alpha)_R v = 0$ for all positive roots α. The element λ_R of the Cartan algebra, a weight, is called the highest weight (HW) of the representation R and it determines completely R. All weights $\lambda \in$ **t** such that $\operatorname{tr}(\alpha^\vee \lambda)$ is a non-negative integer for each positive α appear as HW's of irreducible representations[3] of G. The representations R may be obtained by the geometric quantization of the (co)adjoint orbit passing through λ_R. For the $su(2)$ Lie

---

[3] such λ are usually called dominant weights

algebra spanned by the Pauli matrices σ_i, one usually takes σ_3 as the positive root and the matrices $\sigma_\pm = \frac{1}{2}(\sigma_1 \pm i\sigma_2)$ as the corresponding step generators. The HW's are of the form $j\sigma_3$ with $j = 0, \frac{1}{2}, 1, \ldots$ called the **spin** of the representation.

With respect to the left-right action of $G \times G$, the Hilbert space $L^2(G, dg)$ decomposes as

$$\mathcal{H} \cong \bigoplus_R V_R \otimes V_{\overline{R}}, \tag{2.7}$$

where the (infinite) sum is over the (equivalence classes of) irreducible representations of G and $\overline{R}$ denotes the representation complex-conjugate to R, i.e. $V_{\overline{R}} = \overline{V}_R$ and $g_{\overline{R}} = \overline{g}_R$. Recall that the complex conjugate vector space $\overline{V}$ is composed of the vectors $v \in V$, denoted for distinction by $\overline{v}$, with the multiplication by scalars defined by $\mu\overline{v} = \overline{\overline{\mu}v}$. A linear transformation A of V, when viewed as a transformation $\overline{A}$ of $\overline{V}$, is still linear. The above factorization of the Hilbert space reflects the classical splitting (2.5). Let (g_R^{ij}) be the (unitary) matrix of the endomorphism g_R with respect to a fixed orthonormal bases (e_R^i) in V_R. The decomposition (2.7) is given by the assignment

$$V_R \otimes V_{\overline{R}} \ni e_R^i \otimes \overline{e_R^j} \quad \mapsto \quad d_R^{\frac{1}{2}} \, g_R^{ij} \in L^2(G, dg). \tag{2.8}$$

The Schur orthogonality relations

$$\int_G \overline{g_{R'}^{ij}} \, g_R^{rs} \, dg = \frac{1}{d_R} \delta_{R'R} \, \delta^{ir} \delta^{js}$$

assure that this assignment preserves the scalar product. The matrix elements g_R^{ij} span a dense subspace in $L^2(G, dg)$.

The function on G invariant under the adjoint action $g \mapsto Ad_{g_0}(g) \equiv g_0 \, g \, g_0^{-1}$ are called class functions. They are constant on the conjugacy classes

$$C_\lambda = \{ g_0 \, e^{2\pi i \lambda / k} g_0^{-1} \,|\, g_0 \in G \} \tag{2.9}$$

with λ in the Cartan algebra $\mathbf{t}$. The characters $\chi_R(g) = \mathrm{tr}_{V_R} \, g_R$ of the irreducible representations R are class functions. The Schur relations imply that

$$\int_G \overline{\chi_{R'}(g)} \, \chi_R(g) \, dg = \delta_{R'R}.$$

The class functions in $L^2(G, dg)$ form a closed subspace and the characters χ_R form an orthonormal bases of it. Note that under the isomorphism (2.7),

$$\overline{\chi_R} \cong d_R^{-\frac{1}{2}} e_R^i \otimes \overline{e_R^i} \tag{2.10}$$

(sum over i!).

The Hamiltonian H becomes diagonal in the decomposition (2.7) of the Hilbert space. It acts on $V_R \otimes V_{\overline{R}}$ as the multiplication by $\frac{2}{k} c_R$ where c_R is the value of the quadratic Casimir $c = t^a t^a$ in the representation R. In terms of the HW's, $c_R =$

$\frac{1}{2}\operatorname{tr}(\lambda_R(\lambda_R + 2\rho))$, where ρ, the **Weyl vector**, is equal to half the sum of the positive roots. The Hamiltonian generates a 1-parameter family of unitary transformation e^{itH} describing the time evolution of the quantum system. In the Euclidean spirit, we shall be more interested, however, in the semigroup of the thermal density matrices $e^{-\beta H}$ obtained by the Wick rotation of time $\beta = it \geq 0$. Their (heat) kernels are given by:

$$e^{-\beta H}(g_0, g_1) = \sum_R d_R\, e^{-\frac{2}{k}\beta c_R}\, \chi_R(g_0 g_1^{-1}).$$

In particular, at $\beta = 0$, we obtain a representation for the delta-function concentrated at an element $g_0 \in G$:

$$\delta_{g_0}(g_1) = \sum_R d_R\, \chi_R(g_0 g_1^{-1}).$$

We shall also need below the delta-functions concentrated on the conjugacy classes $\mathcal{C}_\lambda$. They may be obtained by smearing the delta-function δ_g over $\mathcal{C}_\lambda$:

$$\delta_{\mathcal{C}_\lambda}(g_1) = \int_G \delta_{g_0 e^{2\pi i\lambda/k} g_0^{-1}}(g_1)\, dg_0 = \sum_R d_R \int_G \chi_R(g_0\, e^{2\pi i\lambda/k} g_0^{-1} g_1^{-1})\, dg_0$$
$$= \sum_R d_R \int_G (g_0)_R^{ij}\, (e^{2\pi i\lambda/k})_R^{jl}\, \overline{(g_0)_R^{nl}}\, \overline{(g_1)_R^{in}}\, dg = \sum_R \chi_R(e^{2\pi i\lambda/k})\, \overline{\chi_R(g_1)},$$

where we have used the Schur relations. It follows from the correspondence (2.10) that in the language of the isomorphism (2.7),

$$\delta_{\mathcal{C}_\lambda} \cong \sum_R \chi_R(1)^{-\frac{1}{2}}\, \chi_R(e^{2\pi i\lambda/k}) \Upsilon\, e_R^i \otimes \overline{e_R^i}. \tag{2.11}$$

More exactly, $\delta_{\mathcal{C}_\lambda}$ is not a normalizable state in $\mathcal{H}$ but it defines an antilinear functional on a dense subspace in $\mathcal{H}$, e.g. the one of vectors with a finite number of components in the decomposition (2.7).

The delta-functions $\delta_{\mathcal{C}_\lambda}$ may be used to disintegrate the Haar measure dg into the measures along the conjugacy classes and over the set of different conjugacy classes:

$$dg = \frac{1}{|T|}\, |\Pi(e^{2\pi i\lambda/k})|^2\, \delta_{\mathcal{C}_\lambda}(g)\, d\lambda\, dg. \tag{2.12}$$

We shall choose the measure $d\lambda$ such that it corresponds to the normalized Haar measure on the Cartan group $T \subset G$ under the exponential map $\lambda \mapsto e^{2\pi i\lambda/k}$. Then

$$\Pi(e^{2\pi i\lambda/k}) = \prod_{\alpha > 0} \left(e^{\pi i\operatorname{tr}(\alpha\lambda)/k} - e^{-\pi i\operatorname{tr}(\alpha\lambda)/k} \right) \tag{2.13}$$

is the so called **Weyl denominator**. In particular, for class functions constant on the conjugacy classes one obtains the Weyl formula:

$$\int_G f\, dg = \int f(e^{2\pi i\lambda/k})\, |\Pi(e^{2\pi i\lambda/k})|^2\, d\lambda, \tag{2.14}$$

where each conjugacy class should be represented ones. We shall employ this representation of the integral of class functions later.

The Feynman-Kac formula allows to express the heat kernel on the group as a path integral:

$$e^{-\beta H}(g_0, g_1) = \int\limits_{\substack{g:[0,\beta] \to G \\ g(0)=g_0, \ g(\beta)=g_1}} e^{-S(g)} \, Dg \, ,$$

where Dg stands for the product of the Haar measures $dg(t)$. The integral on the right hand side may be given a rigorous meaning as the one with respect to the Brownian bridge measure $dW_{g_0,g_1}(g)$ supported by continuous paths in G. The path integral may be also used to define the thermal **correlation function**

$$< \prod_n g_{R_n}^{i_n j_n}(t_n) >_\beta \ \equiv \ \frac{\int \prod\limits_{n=1}^{N} g_{R_n}^{i_n j_n}(t_n) \ e^{-S(g)} \, Dg}{\int e^{-S(g)} \, Dg} \, , \tag{2.15}$$

where on the right hand side one integrates over the periodic paths $g : [0,\beta] \to G$. Upon ordering the (Euclidean) times $t_1 \leq \ldots \leq t_N$, the above path integral may be expressed in the operator language:

$$\begin{aligned}
&< \prod_n g_{R_n}^{i_n j_n}(t_n) >_\beta \\
&= \frac{\mathrm{tr}_{\mathcal{H}} \left(e^{-t_1 \mathcal{H}} g_{R_1}^{i_1 j_1} e^{-(t_2 - t_1)\mathcal{H}} g_{R_2}^{i_2 j_2} \cdots e^{-(t_N - t_{N-1})\mathcal{H}} g_{R_N}^{i_N j_N} e^{-(\beta - t_N)\mathcal{H}} \right)}{\mathrm{tr}_{\mathcal{H}} \ e^{-\beta H}} \, .
\end{aligned} \tag{2.16}$$

where the functions g_R^{ij} on G are viewed as the multiplication operators in $L^2(G, dg)$.

The right hand side of Eq. (2.16) may be calculated using harmonic analysis on G. Indeed, what is really needed for such a computation are the matrix elements

$$(e_{R_1}^{i_1} \otimes \overline{e_{R_1}^{j_1}}, \ g_R^{ij} \ e_{R_2}^{i_2} \otimes \overline{e_{R_2}^{j_2}}) \ = \ d_{R_1}^{\frac{1}{2}} d_{R_2}^{\frac{1}{2}} \int_G \overline{g_{R_2}^{i_2 j_2}} \ g_{R_1}^{i_1 j_1} \ g_R^{ij} \, dg \tag{2.17}$$

encoding the decomposition of the tensor product of the irreducible representations

$$V_R \otimes V_{R_1} \ \cong \ \bigoplus_{R_2} M_{R_1 R}^{R_2} \otimes V_{R_2} \, . \tag{2.18}$$

In particular, the dimensions $N_{R_1 R}^{R_2}$ of the multiplicity spaces $M_{R_1 R}^{R_2}$ may be obtained from the traces of the above matrix elements:

$$N_{R_1 R}^{R_2} \ = \ \int_G \overline{\chi_{R_2}(g)} \ \chi_{R_1}(g) \ \chi_R(g) \, dg \, .$$

The finite combinations with integer coefficients $\sum n_i \chi_{R_i}$ of characters of irreducible representations form a subring $\mathcal{R}_G$ in the commutative ring of class functions. The identity

$$\chi_R \, \chi_{R_1} \ = \ \sum_{R_2} N_{R R_1}^{R_2} \ \chi_{R_2} \tag{2.19}$$

shows that the integers $N_{R_1 R}^{R_2} = N_{R R_1}^{R_2}$ play the role of structure constants of this ring.

9

One can define a version of the correlation functions by replacing the integral over the periodic paths $g : [0, \beta] \to G$ in Eq. (2.15) by the one over the paths constraint to fixed conjugacy classes on the boundary of the interval:

$$\frac{\int\limits_{g:[0,\beta]\to G} \prod\limits_{n=1}^{N} g_{R_n}^{\iota_n \jmath_n}(t_n) \, \delta_{c_{\lambda_1}}(g(0)) \, \delta_{c_{\lambda_2}}(g(\beta)) \, e^{-S(g)} \, Dg}{\int\limits_{g:[0,\beta]\to G} \delta_{c_{\lambda_1}}(g(0)) \, \delta_{c_{\lambda_2}}(g(\beta)) \, e^{-S(g)} \, Dg} \equiv \, < \prod\limits_{n} g^{\iota_n \jmath_n}(t_n) >_{\beta, \lambda_1 \lambda_2}. \quad (2.20)$$

For $G = SU(2) = \{x_0 + ix_i\sigma_i \,|\, x_0^2 + x_i^2 = 1\} \cong S^3$, the conjugacy classes are the 2-spheres with fixed x_0 so that one integrates over the paths as in Fig. 1.

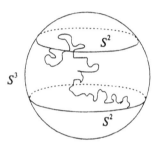

Fig. 1

The above functional integral may be rewritten in the operator language as

$$< \prod\limits_{n} g^{\iota_n \jmath_n}(t_n) >_{\beta, \lambda_1 \lambda_2} = \frac{\left(\delta_{c_{\lambda_1}}, \, e^{-t_1 \mathcal{H}} g_{R_1}^{\iota_1 \jmath_1} \cdots e^{-(t_N - t_{N-1})\mathcal{H}} g_{R_N}^{\iota_N \jmath_N} e^{-(\beta - t_N)\mathcal{H}} \delta_{c_{\lambda_2}} \right)}{\left(\delta_{c_{\lambda_1}}, \, e^{-\beta \mathcal{H}} \delta_{c_{\lambda_2}} \right)}. \quad (2.21)$$

Although δ_{c_λ} are generalized functions rather than normalizable states in $L^2(G, dg)$, the matrix elements on the right hand side are finite and may again be computed by harmonic analysis on G.

We shall encounter field-theoretical generalization of the above quantum-mechanical constructions below.

3 The WZW action

3.1 Two-dimensional sigma models

The two-dimensional sigma models describe field theories with fields mapping a surface Σ to a target manifold M, both equipped with metric structures. Such field configurations represent evolution of a string in the target M with Σ being the string world sheet. The (Euclidean) action functional of the field configuration $X : \Sigma \to M$ is

$$S^\gamma(X) = \frac{1}{4\pi} \int_\Sigma \gamma_{\mu\nu}(X) \, \partial_\alpha X^\mu \partial_\beta X^\nu \eta^{\alpha\beta} \sqrt{\eta} , \qquad (3.1)$$

where $\gamma_{\mu\nu}$ is the Riemannian metric on M, $\eta_{\alpha\beta}$ the one on Σ and $\sqrt{\eta} \equiv \sqrt{\det \eta_{\alpha\beta}}$ is the Riemannian volume density on Σ. In particular, if $M = \mathbf{R}^n$ with the standard metric, we obtain the quadratic action of the free field on a two-dimensional surface leading to linear classical equations. The general case, however, results in a non-linear classical theory.

The term S^γ does not change under the local rescalings $\eta_{\alpha\beta} \mapsto e^{2\sigma} \eta_{\alpha\beta}$ of the metric on Σ i.e. it possesses two-dimensional conformal invariance. For oriented Σ, conformal classes of the metric are in one to one correspondence with complex structures on Σ such that $\eta_{zz} = \eta_{\bar{z}\bar{z}} = 0$ in the holomorphic coordinates and that the latter preserve the orientation. The action S^γ may be written using explicitly only the complex structure of Σ:

$$S^\gamma(X) = \frac{1}{2\pi} \int_\Sigma \gamma_{\mu\nu}(X) \, \partial X^\mu \bar{\partial} X^\nu , \qquad (3.2)$$

where $\partial = dz \partial_z$ and $\bar{\partial} = d\bar{z} \partial_{\bar{z}}$. One-dimensional complex manifolds are called Riemann surfaces. It follows that the action $S^\gamma(X)$ may be defined on such surfaces.

To the S^γ term, one may add the expression

$$S^\beta(X) = \frac{1}{4\pi i} \int_\Sigma \beta_{\mu\nu}(X) \, \partial_\alpha X^\mu \partial_\beta X^\nu \epsilon^{\alpha\beta} \qquad (3.3)$$

where $\beta_{\mu\nu} = -\beta_{\nu\mu}$ are the coefficients of a 2-form β on M. Geometrically, S^β is proportional to the integral of the pullback of β by X:

$$S^\beta(X) = \frac{1}{4\pi i} \int_\Sigma X^* \beta . \qquad (3.4)$$

The imaginary coefficient is required by the unitarity of the theory after the Wick rotation to the Minkowski signature. The term S^β does not use the metric on Σ but only the orientation and is often called a topological term. Hence the classical two-dimensional conformal invariance of the model with the action $S = S^\gamma + S^\beta$.

On the quantum level, the sigma model requires a renormalization which often imposes the addition to the action of further terms

$$S^{tach}(X) = \frac{1}{2\pi} \int_\Sigma \mathcal{T}(X) \sqrt{\eta} \quad \text{and} \quad S^{dil}(X) = \frac{1}{2\pi} \int_\Sigma \mathcal{D}(X) \, r \sqrt{\eta} , \qquad (3.5)$$

where where $\mathcal{T}$ and $\mathcal{D}$ are functions on M called tachyonic and dilatonic potentials, respectively, and r is the scalar curvature of the metric $\eta_{\alpha\beta}$. The renormalization breaks the conformal invariance (note that S^{tach} and S^{dil} are not conformal invariant). We shall be interested, however, in the case of the WZW model [2], an example of a CFT, where the classical conformal invariance is (almost) not broken on the quantum level.

The WZW model is the two-dimensional counterpart of the particle on a group and may be thought of as describing the movement of a string on a group manifold $M = G$ equipped with the invariant metric γ described before. We then have for $g : \Sigma \to G$,

$$S^\gamma(g) \;=\; \frac{k}{4\pi i} \int_\Sigma \mathrm{tr}\,(g^{-1}\partial g)(g^{-1}\bar\partial g), \qquad (3.6)$$

where k is a positive constant. The quantization of a model with such an action leads, however, to a theory without conformal invariance. To restore the latter, one adds to the S^γ term, following Witten [2], a topological term, the so called Wess-Zumino (WZ) term S^{WZ}. In the first approximation, $S^{WZ} = kS^\beta$ where β is a 2-form on G such that $d\beta$ is equal to the canonical 3-form $\chi \equiv \frac{1}{3}\mathrm{tr}\,(g^{-1}dg)^3$ on G. If the group G is abelian, such a description is indeed possible and the overall action is a simple version of the free field one. In the non-abelian case, however, the difficulty comes from the fact that the 3-form χ is closed but not globally exact so that the forms β exist only locally and are defined only up to closed 2-forms. Hence the definition of the WZ term of the action requires a more refined discussion.

3.2   Particle in the field of a magnetic monopole

It may be useful to recall a simpler situation where one is confronted with a similar problem. Suppose that we want to define the contribution $S^{Dir}(x)$ to the action of a mechanical particle of the term

$$e \int_0^T A_\nu(x)\frac{dx^\nu}{dt}\,dt \;=\; e \int x^* A$$

describing the coupling to the abelian gauge field $A = A_\nu dx^\nu$ with the field strength $F_{\nu\lambda} = \frac{1}{2}(\partial_\nu A_\lambda - \partial_\lambda A_\nu)$, or in the language of differential forms, with $F = dA$, where $F = F_{\nu\lambda}dx^\nu dx^\lambda$. The constant e stands for the electric charge of the particle. For concreteness, suppose that $F_{\mu\nu}$ corresponds to the magnetic field of a monopole of magnetic charge μ placed at the origin of $\mathbf{R}^3$:

$$F_{\nu\lambda} \;=\; \frac{1}{2}\mu\,\epsilon_{\nu\lambda\kappa}\,\frac{x^\kappa}{|x|^3}\,.$$

There is no global 1-form A on $\mathbf{R}^3$ without the origin such that $dA = F$. For a closed trajectory $t \mapsto x(t)$, however, we may pose

$$S^{Dir}(x) \;=\; e \int_D \tilde x^* F\,, \qquad (3.7)$$

where $\tilde x$ is a map of a disc D into $\mathbf{R}^3 \setminus \{0\}$ coinciding on the boundary of the disc with x. For two different extensions $\tilde x$, however, the above prescription may give different results. Their difference may be written as the integral

$$e \int_{S^2} \tilde x^* F \qquad (3.8)$$

over the 2-sphere S^2 obtained by gluing the two disc D, one with the inverted orientation, along the boundary and for the map $\tilde{x} : S^2 \to \mathbf{R}^3 \setminus \{0\}$ glued from the two extensions of x to the respective discs, see Fig. 2.

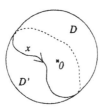

Fig. 2

The ambiguities (3.8) are the periods of the closed form F over the cycles of the 2nd integer homology $H_2(\mathbf{R}^3 \setminus \{0\}) = \mathbf{Z}$. They take discrete values which are multiples of $4\pi e\mu$. The latter value is obtained for the unit sphere in $\mathbf{R}^3$, a generator of $H_2(\mathbf{R}^3 \setminus \{0\})$. The discrete ambiguities are acceptable in classical mechanics where one studies the extrema of the action. In quantum mechanics, however, we have to give sense to the Feynman amplitudes $e^{i S^{Dir}(x)}$, hence only the ambiguities in the action with values in $2\pi\mathbf{Z}$ are admissible. Demanding that the quantum-mechanical amplitudes be unambiguously defined reproduces this way the Dirac quantization condition $e\mu \in \frac{1}{2}\mathbf{Z}$.

For open trajectories $[0, T] \ni t \mapsto x(t)$, the amplitudes $e^{ie \int x^* A}$ may not, in general, be unambiguously assigned numerical values. They may be only defined as maps between the fibers $\mathcal{L}_{x(0)}$ and $\mathcal{L}_{x(T)}$ of a line bundle $\mathcal{L}$. Geometrically, they give the parallel transport in the bundle corresponding to a $U(1)$-connection with the curvature form F. We shall recover the analogous situation below when discussing how to give meaning to the WZ term in the action of the WZW model.

3.3 Wess-Zumino action on surfaces without boundary

Let us first consider the case of compact Riemann surfaces without boundary.

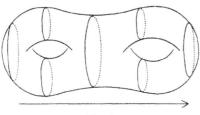

Fig. 3

Topologically, such surfaces are characterized by the genus g_Σ equal to the number of handles of the surface. They may be viewed as world sheets of a closed string created from the vacuum, undergoing in the evolution g_Σ splittings and recombinations and finally disappearing into the vacuum, see Fig. 3 where a surface of genus 2 was represented. At genus zero, there is only one (up to diffeomorphisms) Riemann surface, the Riemann sphere $CP^1 = C \cup \{\infty\}$, see Fig. 4(a).

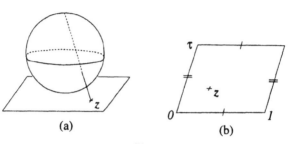

(a)   (b)

Fig. 4

At genus one, there is a complex one-parameter family of Riemann surfaces: the complex tori $T_\tau = C/(Z + \tau Z)$ with τ in the upper half plane H^+ of the complex numbers such that $\mathrm{Im}\,\tau > 0$, see Fig. 4(b). The tori T_τ and $T_{\tau'}$, where $\tau' = \frac{a+b\tau}{c+d\tau}$ for $\left(\begin{smallmatrix} a & b \\ c & d \end{smallmatrix}\right)$ in the modular group $SL(2, Z)$, may be identified by the map $z \mapsto z' = (c\tau + d)^{-1} z$. The space of the diffeomorphism classes (i.e. the **moduli space**) of genus one Riemann surfaces is equal to $H^+/SL(2, Z)$ and has complex dimension 1. For higher genera, the moduli spaces of Riemann surfaces have complex dimension $3(g_\Sigma - 1)$.

Let us return to the discussion of the action of the WZW model. Assume that G is connected and simply connected and that Σ is a compact Riemann surface without boundary. Following [2] and mimicking the trick used for a particle in a monopole field, one may extend the field $g : \Sigma \to G$ to a map $\tilde{g} : B \to G$ of a 3-manifold B such that $\partial B = \Sigma$ and set:

$$S^{WZ}(g) = \frac{k}{4\pi i} \int_B \tilde{g}^* \chi. \tag{3.9}$$

By the Stokes formula, this expression coincides with $k S^\beta(g)$ whenever the image of $\tilde{g}$ is contained in the domain of definition of a 2-form β such that $d\beta = \chi$, but it makes sense in the general case. The price is that the result depends on the extension $\tilde{g}$ of the field g. The ambiguities have the form of the integrals

$$\frac{k}{4\pi i} \int_{\tilde{B}} \tilde{g}^* \chi \tag{3.10}$$

over 3-manifolds $\tilde{B}$ without boundary with $\tilde{g} : \tilde{B} \to G$, see Fig. 5.

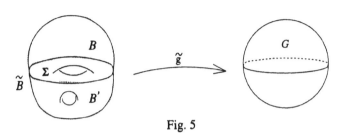

Fig. 5

They are proportional to the periods of the 3-form χ over the integer homology $H_3(G)$. Such discrete contributions do not effect the classical equations of motion $\delta S = 0$. In quantum mechanics, however, where we deal with the Feynman amplitudes[4] $e^{-S(g)}$, only ambiguities in $2\pi i \mathbf{Z}$ are allowed. Hence we have to find conditions under which the periods (3.10) lie in $2\pi i \mathbf{Z}$.

Recall that we normalized the invariant form tr on the Lie algebra $\mathbf{g}$ of G so that the long roots in $\mathbf{t}$ have length squared 2. When $G = SU(2) \cong \{x \in \mathbf{R}^4 \,|\, |x|^2 = 1\}$, the 3-form χ equals then to 4 times the volume form of the unit 3-sphere. Since the volume of the latter is equal to $2\pi^2$, we infer that

$$\frac{1}{4\pi i} \int_{SU(2)} \chi = -2\pi i. \tag{3.11}$$

For the other simple, simply connected groups, the roots α determine the sub-algebras $su(2)_\alpha \subset \mathbf{g}$, obtained by identifying the corresponding coroot $\alpha^\vee = \frac{2\alpha}{\operatorname{tr}\alpha^2}$ and the step generators $e_{\pm\alpha}$ with the Pauli matrices σ_3 and $\sigma_\pm$, respectively. By exponentiation, we obtain the $SU(2)_\alpha$ subgroups of G. Clearly,

$$\frac{1}{4\pi i} \int_{SU(2)_\alpha} \chi = -\frac{4\pi i}{\operatorname{tr}\alpha^2}. \tag{3.12}$$

The ratio $\frac{2}{\operatorname{tr}\alpha^2}$ is equal to 1 for long roots and is a positive integer for the others. It appears that any of the subgroups $SU(2)_\alpha \cong S^3$ for α a long root generates $H_3(G) = \mathbf{Z}$. Thus the unambiguous definition of the amplitudes $e^{-S^{WZ}(g)}$ requires that the coupling constant k, called the **level** of the model, be a (positive) integer, in the analogy to the Dirac quantization of the magnetic charge.

It is easy to see that, although the action $S^{WZ}(g)$ cannot be expressed, in general, as a local integral over Σ, the variation of S^{WZ} has such a form:

$$\delta S^{WZ}(g) = \frac{k}{4\pi i} \int_\Sigma \operatorname{tr}(g^{-1}\delta g)(g^{-1}dg)^2. \tag{3.13}$$

The above formula is a special case of the general, very useful, geometric identity: $\delta \int f^*\alpha = \int \mathcal{L}_{\delta f}\alpha$, where $\mathcal{L}_X$ is the Lie derivative. Applied to $f = g$ and $\alpha = \chi$ it gives,

---

[4] there is no i in front of S since we work with the Euclidean action

15

in conjunction with the Stokes formula, the above relation. It is also important to note the behavior of S^{wz} under the point-wise multiplication of fields, a basic property of the WZ term:

$$S^{wz}(g_1 g_2) = S^{wz}(g_1) + S^{wz}(g_2) + W(g_1, g_2), \tag{3.14}$$

where

$$W(g_1, g_2) = \frac{k}{4\pi i} \int_{\Sigma} \mathrm{tr}\, (g_1^{-1} dg_1)(g_2 dg_2^{-1}). \tag{3.15}$$

The relation follows easily from the definition (3.9), again by applying the Stokes formula.

The complete action of the WZW model on a closed Riemann surface Σ is the sum of the γ- and the WZ-terms with the same coupling constant k: $S(g) = S^{\gamma}(g) + S^{wz}(g)$. Since S^{γ} is unambiguous, it has the same ambiguities as S^{wz}, requiring that k be a (positive) integer. The relations (3.13) and (3.14) get also contributions from S^{γ} and become:

$$\delta S(g) = -\frac{k}{2\pi i} \int_{\Sigma} \mathrm{tr}\, (g^{-1} \delta g)\partial(g^{-1} \bar{\partial} g), \tag{3.16}$$

$$S(g_1 g_2) = S(g_1) + S(g_2) + \frac{k}{2\pi i} \int_{\Sigma} \mathrm{tr}\, (g_1^{-1} \bar{\partial} g_1)(g_2 \partial g_2^{-1}). \tag{3.17}$$

The last relation is often called the Polyakov-Wiegmann formula. From Eq. (3.16), we obtain the classical equations of motion

$$\partial(g^{-1} \bar{\partial} g) = 0 \qquad \text{or, equivalently,} \qquad \bar{\partial}(g \partial g^{-1}) = 0. \tag{3.18}$$

They have few solutions with values in G (this would not be the case if we considered Σ with a Minkowski metric). In all the above formulae, however, we could have taken fields g with values in the complexified group $G^{\mathbf{C}}$. For such fields, the general local solutions of Eqs. (3.18) have the form

$$g(z, \bar{z}) = g_\ell(z)\, g_r(\bar{z})^{-1} \tag{3.19}$$

where g_ℓ (g_r) are local holomorphic (anti-holomorphic) maps with values in $G^{\mathbf{C}}$. Thus Eqs. (3.18) constitute a non-linear generalization of the Laplace equation in two dimensions whose solutions are harmonic functions which are, locally, sums of holomorphic and anti-holomorphic ones. In particular, multiplying a solution (3.19) by a holomorphic map into $G^{\mathbf{C}}$ on the left and by an anti-holomorphic one on the right we obtain another solution. Similarly, composing a solution with a local holomorphic map or inverting it in $G^{\mathbf{C}}$ after composition with a local anti-holomorphic map of Σ one produces new solutions. Hence a rich symmetry structure of the classical theory. This structure will be preserved by the quantization leading to the current and Virasoro algebra symmetries of the quantum WZW model.

3.4   Wess-Zumino action on surfaces with boundary

What if the surface Σ has a boundary? Of course only the WZ term in the action causes problems due to its non-local character. The term S^γ is defined unambiguously for any compact surface. It will be convenient to represent Σ as $\Sigma' \setminus (\underset{n}{\cup} \mathring{D}_n)$ where D_n are disjoint unit discs $\{ z \,|\, |z| \leq 1 \}$ embedded in a closed surface Σ' without boundary, see Fig.6.

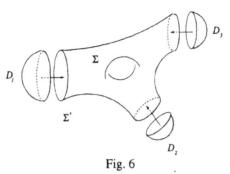

Fig. 6

Note that the boundaries of Σ are then naturally parametrized by the unit circles. One way to proceed in the presence of boundaries is to extend the field $g : \Sigma \to G$ to a map $g' : \Sigma' \to G$ and to consider the action $S^{WZ}_{\Sigma'}(g')$ pertaining to the surface Σ', as stressed by a subscript. We are again confronted with the question as to how the action on Σ' depends on the extension of the field. The answer is easy to work out. If $g'' = g'h$ is another extension of g then, by Eq. (3.14),

$$S^{WZ}_{\Sigma'}(g'') \;=\; S^{WZ}_{\Sigma'}(g') \,+\, S^{WZ}_{\Sigma'}(h) \,+\, W_{\Sigma'}(g', h)\,. \tag{3.20}$$

It will be convenient to localize the changes in the discs D_n by rewriting the last formula as

$$S^{WZ}_{\Sigma'}(g'') \;=\; S^{WZ}_{\Sigma'}(g') \,+\, \sum_n \left(S^{WZ}_{S^2}(h_n) \,+\, W_{D_n}(g', h) \right), \tag{3.21}$$

where h_n, mapping spheres (compactified planes) S^2_n to G, extend the maps $h|_{D_n}$ by unity and W_{D_n} are as in Eq. (3.15) but with the integration restricted to D_n. To account for the change (3.21), we shall define the following equivalence relation between the pairs (g', z) where $g' : D \to G$ and $z \in \mathbf{C}$:

$$(g',\, z) \;\sim\; (g'h,\, z\, e^{-S^{WZ}_{S^2}(h) - W_D(g',h)})$$

for $h : S^2 \to G$ equal to unity outside the unit disc $D \subset S^2$. The set of equivalence classes forms a complex line bundle $\mathcal{L}$ over the loop group LG of the boundary values of the maps g'. Comparing to Eq. (3.21), we infer that for $g : \Sigma \to G$ with Σ as above, the amplitude $e^{-S^{WZ}_\Sigma(g)}$ makes sense as the element of a tensor product of the line bundles $\mathcal{L}$, one for each boundary component of Σ,

$$e^{-S^{WZ}_\Sigma(g)} \;\in\; \underset{n}{\otimes}\, \mathcal{L}_{g|_{\partial D_n}}\,,$$

where $\mathcal{L}_h$ denotes the fiber of $\mathcal{L}$ over the loop $h \in LG$. Hence the WZ amplitudes $e^{-S_\Sigma^{WZ}(g)}$ take values in line bundles instead of having numerical values, exactly as for the amplitudes giving the parallel transport in a $U(1)$-gauge field mentioned before.

The line bundle $\mathcal{L}$ is an interesting object. It carries a hermitian structure given by the absolute value of z. The fibers of $\mathcal{L}$ over g and $\breve{g}$ where $\breve{g}$ ia a reversed loop, $\breve{g}(e^{i\varphi}) = g(e^{-i\varphi})$, may be naturally paired so that, for $g : \tilde{\Sigma} \to G$, where $\tilde{\Sigma}$ is obtained from two surfaces Σ and Σ' by gluing them along some boundary components, see Fig. 7,

$$\langle e^{-S_\Sigma^{WZ}(g|_\Sigma)}, e^{-S_{\Sigma'}^{WZ}(g|_{\Sigma'})} \rangle = e^{-S_{\tilde{\Sigma}}(g)}. \tag{3.22}$$

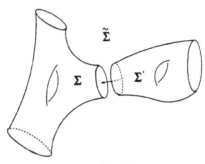

Fig. 7

$\mathcal{L}$ may be also equipped with a product structure such that

$$e^{-S_\Sigma(g_1)} \cdot e^{-S_\Sigma(g_2)} = e^{-S_\Sigma(g_1 g_2)} e^{W_\Sigma(g_1, g_2)}, \tag{3.23}$$

compare to Eq. (3.14). Under the product, the elements of unit length in $\mathcal{L}$ form a group $\hat{G}$ which is a **central extension** of the loop group LG by the circle group $U(1)$:

$$1 \to U(1) \to \hat{G} \to LG \to 1. \tag{3.24}$$

The second arrow sends $e^{i\varphi}$ to the equivalence class of $(1, e^{i\varphi})$. The extensions for $k > 1$ are powers of the universal one corresponding to $k = 1$. On the infinitesimal level, one obtains the central extensions of the loop algebra Lg of the maps of the circle S^1 to the Lie algebra g by the real line:

$$0 \to \mathbf{R} \to \hat{g} \to Lg \to 0. \tag{3.25}$$

The Lie algebra $\hat{g}$, called the **current** or the affine Kac-Moody **algebra**, may be described explicitly in terms of the complexified generators t_n^a corresponding to the loops $t^a e^{in\varphi}$ in $Lg^{\mathbf{C}}$ and the central element K satisfying the commutation relations:

$$[t_n^a, t_m^b] = i f^{abc} t_{n+m}^c + \frac{1}{2} K n \, \delta^{ab} \delta_{n+m,0}. \tag{3.26}$$

The algebra $\hat{\mathbf{g}}$ is the same for all levels k but the central element $K \in \hat{\mathbf{g}}$ is the image of $k \in \mathbf{R}$ under the second arrow in the exact sequences (3.25). Note that the generators t_0^a span a subalgebra $\mathbf{g} \subset \hat{\mathbf{g}}$. As we shall see, the group $\hat{G}$ and the algebra $\hat{\mathbf{g}}$ play in the WZW theory a similar role to that of G and $\mathbf{g}$ for the particle on the group.

3.5 Coupling to gauge field

We may couple the WZW model to the gauge field $iA \equiv i(A^{10} + A^{01})$, a 1-form with values in the Lie algebra $\mathbf{g}$ (or, more generally, $\mathbf{g}^{\mathbf{C}}$), where $A^{10} = t^a A_z^a dz$ and $A^{01} = t^a A_{\bar{z}}^a d\bar{z}$ are, respectively, a 1,0- and a 0,1-form (the chiral components of the gauge field). In most what follows, we shall treat the gauge field as external, i.e. non-dynamical. Nevertheless, the coupling will allow to test the variation of the quantum system under the changes of the gauge field background and, finally, will facilitate the exact solution of the model. For a surface without boundary, we define

$$S(g, A) = S(g) + \frac{ik}{2\pi} \int_\Sigma \operatorname{tr} \left[A^{10}(g^{-1}\bar{\partial}g) + (g\partial g^{-1}) A^{01} + g A^{10} g^{-1} A^{01} \right]. \quad (3.27)$$

Under the local gauge transformations $h : \Sigma \to G$, the gauge fields transform in the standard way:

$$A^{10} \mapsto {}^h A^{10} = h A^{10} h^{-1} + h \partial h^{-1}, \qquad A^{01} \mapsto {}^h A^{01} = h A^{01} h^{-1} + h \bar{\partial} h^{-1}.$$

The reaction of the action to the chiral changes of the gauge is encoded in the identity

$$S(g, A) = S(h_1 g h_2^{-1}, {}^{h_2}A^{10} + {}^{h_1}A^{01}) + S(h_2, A^{10}) + S(h_1^{-1}, A^{01}) \quad (3.28)$$

which follows in a direct manner from the Polyakov-Wiegmann formula (3.17). For the later convenience, we have chosen a modified way of coupling to the gauge field, as compared to the more standard way with the addition of the term $-A^{10}A^{01}$ in the brackets on the right hand side of Eq. (3.27). The latter way would render the action invariant with respect to the diagonal (i.e. non-chiral) gauge transformations with $h_1 = h_2$.

For surfaces with boundary, we define $e^{-S(g,A)}$ by the same prescription but on the level of the amplitudes with values in the product of line bundles. The definition (3.23) of the product implies then the transformation rule:

$$e^{S(g, A)} = e^{-S(h_1^{-1}, A^{01})} \cdot e^{-S(h_1 g h_2^{-1}, {}^{h_2}A^{10} + {}^{h_1}A^{01})} \cdot e^{-S(h_2, A^{10})} \quad (3.29)$$

which extends the property (3.28) to the case with boundaries.

4   Quantization of the WZW model

4.1   Quantum amplitudes

The Feynman quantization prescription instructs us that in the quantum WZW model we should sum the amplitudes of different classical configurations. This leads to formal functional integrals such as, for example,

$$\mathcal{A}_\Sigma = \int e^{-S_\Sigma(g)} \, Dg \, ,$$

where one integrates over the maps $g : \Sigma \to G$ and Dg stands for the local product $\prod_\xi dg(\xi)$ of the Haar measures. If Σ is closed, then the above integral should take a numerical value $\mathcal{Z}_\Sigma$ called the **partition function** (because of the statistical physics analogy). For Σ with boundary, it should define, instead, a Hilbert space state. Let $\Gamma(\mathcal{L})$ denote the space of sections of the line bundle $\mathcal{L}$ over the loop group LG. $\Gamma(\mathcal{L})$ plays the role of the space of states of the quantized theory. If Σ has a boundary, we should fix in the functional integration the boundary values $\underline{g} = (g_n)$ of fields $g : \Sigma \to G$:

$$\mathcal{A}_\Sigma(\underline{g}) = \int\limits_{g|_{\partial D_n} = g_n} e^{-S_\Sigma(g)} \, Dg \, .$$

The result, in its dependence on $\underline{g}$, should give an element of the tensor product $\otimes_n \Gamma(\mathcal{L})$ of the state spaces: the **quantum amplitude** corresponding to the surface Σ. More generally, we shall consider the quantum amplitudes in the presence of external gauge field:

$$\mathcal{A}_\Sigma(\underline{g}; A) \equiv \; = \int\limits_{g|_{\partial D_n} = g_n} e^{-S_\Sigma(g, A)} \, Dg \tag{4.1}$$

again with the values in $\otimes_n \Gamma(\mathcal{L})$. We would like to give a rigorous meaning to such objects. In general, the functional integrals require complicated renormalization procedures which, besides, work only in some cases (of the so called renormalizable theories) and even then, in most instances, have been implemented only on the level of formal perturbation series. The WZW models are perturbatively renormalizable. In this case, however, one may follow a shortcut by exploiting formal symmetry properties of the functional integrals and showing that they fix uniquely the quantum amplitudes. This will be the line of thought adopted below, although we shall only describe the essential points of the argument and make detours to introduce other important notions.

Let us start by discussing the formal structure of the space of states $\Gamma(\mathcal{L})$. The scalar product and the bilinear form

$$(\psi, \psi') = \int_{LG} (\psi(g), \psi'(g)) \, Dg \, , \qquad \langle \psi, \psi' \rangle = \int_{LG} \langle \psi(g), \psi'(\check{g}) \rangle \, Dg \, , \tag{4.2}$$

which employ the hermitian structure and the duality (3.22) on the line bundle $\mathcal{L}$, should turn $\Gamma(\mathcal{L})$ into a Hilbert space $\mathcal{H}$ and should allow the identification of $\mathcal{H}$ with its dual. The space $\Gamma(\mathcal{L})$ carries also two commuting actions of the group $\hat{G}$: $\psi \mapsto {}^{\hat{h}}\psi$ and $\psi \mapsto \psi^{\hat{h}}$. They are defined by:

$$^{\hat{h}}\psi(g) = \hat{h} \cdot \psi(h^{-1}g) \qquad \psi^{\hat{h}}(g) = \psi(gh) \cdot \hat{h}^{-1} \, , \tag{4.3}$$

where g and h are elements of the loup group LG and h is the projection of $\hat{h} \in \hat{G}$. Formally, these actions preserve the scalar product and the bilinear form in $\Gamma(\mathcal{L})$. On infinitesimal level, they give rise to two commuting actions of the current algebra $\hat{\mathbf{g}}$ in $\Gamma(\mathcal{L})$. We shall denote by J_n^a and $\tilde{J}_n^a$ the operators in $\Gamma(\mathcal{L})$ corresponding to the left and right action of the generators t_n^a of $\hat{\mathbf{g}}$. The central generator K acts in $\Gamma(\mathcal{L})$ as multiplication by k. Of course, J_n^a and $\tilde{J}_n^a$ satisfy the commutation relation (3.26).

As stressed by Segal in [5], there are two important properties of the quantum amplitudes $\mathcal{A}_\Sigma$ which are crucial for their rigorous construction. The first one, is the gluing property

$$\mathcal{A}_{\tilde{\Sigma}}(g_n, g_{n'}) = \int \langle \mathcal{A}_\Sigma(g_n, g_{n_0}), \mathcal{A}_{\Sigma'}(\check{g}_{n_0}, g_{n'}) \rangle \, Dg_{n_0} \qquad (4.4)$$

which states that for a surface $\tilde{\Sigma}$ glued along boundary components of two pieces Σ and Σ', as in Fig. 7, the functional integral may be computed iteratively, by first keeping the values of g fixed on the gluing circle and integrating over them only after the integration over the fields on Σ and Σ', see Eq. (3.22). Using the bilinear form on $\Gamma(\mathcal{L})$ applied to the glued channel, we may write this relation as the identity

$$\mathcal{A}_{\tilde{\Sigma}} = \langle \mathcal{A}_\Sigma, \mathcal{A}_{\Sigma'} \rangle. \qquad (4.5)$$

It is often more convenient to view the quantum amplitudes $\mathcal{A}_\Sigma$ as operators from the tensor product of some of the boundary spaces $\mathcal{H}$ into the others. This is always possible because of the linear isomorphism between $\mathcal{H}$ and its dual. Then Eq. (4.5) may be simply rewritten with the use of the product of operators:

$$\mathcal{A}_{\tilde{\Sigma}} = \mathcal{A}_\Sigma \mathcal{A}_{\Sigma'}. \qquad (4.6)$$

One may also glue two boundary components in a single connected surface Σ. The amplitude for the resulting surface $\tilde{\Sigma}$ is then obtained from that of Σ by pairing the two corresponding factors in the product of the Hilbert spaces or, in the operator interpretation, by the partial trace applied to the glued channel:

$$\mathcal{A}_{\tilde{\Sigma}} = \operatorname{tr}_{\mathcal{H}} \mathcal{A}_\Sigma. \qquad (4.7)$$

In fact, as pointed out in [5], the last relation encompasses also the previous one if one introduces the amplitudes for the disconnected Riemann surfaces defining them as the tensor product of the amplitudes of the components. Clearly, similar gluing relation should hold for the amplitudes in external gauge field.

The second important property of the quantum amplitudes follows formally from the transformation property (3.29) of the classical amplitudes under the chiral gauge transformations $h_{1,2} : \Sigma \to G$. It reads:

$$\widehat{h_1} \mathcal{A}_\Sigma(A) \widehat{h_2} = \mathcal{A}_\Sigma(^{h_2}A^{10} + {}^{h_1}A^{01}) \qquad (4.8)$$

for $\hat{h}_1^{-1} = e^{-S_\Sigma(h_1^{-1}, A^{01})}$ and $\hat{h}_2 = e^{-S_\Sigma(h_2, A^{10})}$. The identity (4.8) expresses the covariance of the quantum amplitudes under the chiral gauge transformations. It is at the basis of the rich symmetry structure of the quantized WZW theory.

4.2 The spectrum

To give a rigorous construction of the Hilbert space $\mathcal{H}$ of the WZW model, whose vectors represent quantum states of a closed string moving on the group manifold, one may resort to the representation theory of the current algebras. The algebra $\hat{\mathbf{g}}$ possesses a distinguished family of irreducible unitary representations labeled by pairs $\hat{R} = (k, R)$, where k, a non-negative integer called the level, is the value taken in the representation by the central generator K of $\hat{\mathbf{g}}$ and where R is an irreducible representation of G (and of $\mathbf{g}$). The irreducible unitary representations $\hat{R}$ act in spaces $V_{\hat{R}}$ possessing a (unique) subspace $V_R \subset V_{\hat{R}}$ annihilated by all the generators t_n^a with $n > 0$ and carrying the irreducible representation R of the subalgebra $\mathbf{g} \subset \hat{\mathbf{g}}$ generated by t_0^a. They are characterized by this property. Not all irreducible representations R of G appear for the fixed level k but only the ones corresponding to the the the so called integrable HW's which satisfy the condition

$$\operatorname{tr}\left(\phi^\vee \lambda_R\right) \leq k, \tag{4.9}$$

where $\phi = \phi^\vee$ is the highest root of $\mathbf{g}$, i.e. such a root that $\phi + \alpha$ is not a root for any positive root α. Given k, there is only a finite number of integrable HW's. For $\widehat{su(2)}$, the integrable HW's correspond to spins $j = 0, \frac{1}{2}, 1, \ldots, \frac{k}{2}$. If λ_R satisfies the condition (4.9) then so does $\lambda_{\overline{R}}$ and the space $V_{\hat{R}}$ is canonically isomorphic to $\overline{V_{\hat{R}}}$. The scalar product on $V_{\hat{R}}$ may then be viewed as a bilinear pairing between $V_{\hat{R}}$ and $V_{\hat{\overline{R}}}$.

The rigorous definition of the Hilbert space of states $\mathcal{H}$ for the WZW model of level k makes the two notions of the level coincide:

$$\mathcal{H} = \bigoplus_{\hat{R} \text{ of level } k} \left(V_{\hat{R}} \otimes V_{\hat{\overline{R}}}\right)^-, \tag{4.10}$$

where the symbol $(\ldots)^-$ stands for the Hilbert space completion. This is the loop group analogue of the decomposition (2.7) of $L^2(G, dg)$. The operators J_n^a and $\tilde{J}_n^a$ representing the action of the generators t_n^a in, respectively, $V_{\hat{R}}$ and $V_{\hat{\overline{R}}}$ satisfy the unitarity conditions $J_n^{a\dagger} = J_{-n}^a$, $\tilde{J}_n^{a\dagger} = \tilde{J}_{-n}^a$. It is not difficult to motivate the above choice of the Hilbert space. One may, indeed, realize the space (4.10) as a space of sections of $\mathcal{L}$. Formally, this may be done by the assignment (compare to the relation (2.8))

$$V_{\hat{R}} \otimes V_{\hat{\overline{R}}} \supset V_R \otimes V_{\overline{R}} \ni e_R^i \otimes \overline{e_R^j} \mapsto \zeta \int g_R^{ij}(0) \, e^{-S_D(g)} \, Dg, \tag{4.11}$$

where the normalization constant ζ will be fixed later. The functional integral on the right hand side, as a function of $g|_D = h$ is the corresponding section of $\mathcal{L}$. One may argue that the above integral is given, up to normalization, by its semi-classical value,

$$\overline{(g_{cl})_R^{ij}(0)} \, e^{-S_D(g_{cl})} \tag{4.12}$$

where $g_{cl} : D \to G^{\mathbf{C}}$ is the solution of the classical equations (3.18) with the boundary condition $g_{cl}|_{\partial D} = h$. As shown in [6], the expression (4.12) defines a non-singular section of $\mathcal{L}$ only if the HW of R is integrable at level k (recall that the action $S(g_{cl})$ is proportional to k). One obtains then a rigorous embedding of the space $\oplus_{\widehat{R}} V_R \otimes V_{\overline{R}}$ into $\Gamma(\mathcal{L})$, and, identifying the actions of the current algebra $\widehat{\mathbf{g}}$ in $\Gamma(\mathcal{L})$ and in $\oplus_{\widehat{R}} V_{\widehat{R}} \otimes V_{\widehat{\overline{R}}}$, also of the latter space. The formal scalar product and the formal bilinear form (4.2) on $\Gamma(\mathcal{L})$ correspond to the scalar product and the bilinear form on $\mathcal{H}$ induced by the scalar product on the representation spaces $V_{\widehat{R}}$ and the bilinear pairing between $V_{\widehat{R}}$ and $V_{\widehat{\overline{R}}}$ (the latter induces the pairing between the $\widehat{R}$ and $\overline{\widehat{R}}$ summands in $\mathcal{H}$).

The action of the pair of the current algebras in $\mathcal{H}$ leads to the (projective) action in $\mathcal{H}$ of the algebra of conformal symmetries. Let us discuss how this occurs. The spaces $V_{\widehat{R}}$ of the irreducible unitary representations $\widehat{R}$ of $\widehat{\mathbf{g}}$ appear to carry also the unitary representations of the **Virasoro algebra** Vir, the central extension of the algebra of the vector fields $Vect(S^1)$ on the circle,

$$0 \to \mathbf{R} \to Vir \to Vect(S^1) \to 0. \tag{4.13}$$

The complex generators ℓ_n of Vir corresponding to the vector fields $ie^{in\varphi}\partial_\varphi$ satisfy the commutation relations

$$[\ell_n, \ell_m] = (n-m)\ell_{n+m} + \frac{1}{12}C(n^3 - n)\delta_{n+m,0}, \tag{4.14}$$

where C is the central generator, the image of 1 under the second arrow in the exact sequence (4.13). The action of the generators ℓ_n in the spaces of the representations $\widehat{R}$ of $\widehat{\mathbf{g}}$ gives rise to the set of operators L_n and $\widetilde{L}_n$ acting in the space of states $\oplus_{\widehat{R}} V_{\widehat{R}} \otimes V_{\widehat{\overline{R}}}$. They implement a projective action of $Vect(S^1) \oplus Vect(S^1)$, the Lie algebra of Minkowskian conformal transformations. $Vect(S^1) \oplus Vect(S^1)$ is, indeed, the Lie algebra of the infinitesimal transformations preserving the conformal class of the Minkowski metric $dx^2 - dt^2 = dx^+ dx^-$, where $x^\pm = x \pm t$ are the light-cone coordinates on the cylinder with periodic space-coordinate x.

Explicitly, the operators L_n's and $\widetilde{L}_n$'s are given in terms of the operators J_n^a and $\widetilde{J}_n^a$, generating the actions of $\widehat{\mathbf{g}}$, by the so called Sugawara construction:

$$L_n = \frac{1}{k+h^\vee}\sum_{m=-\infty}^{\infty} J_{n-m}^a J_m^a \quad \text{for } n \neq 0, \qquad L_0 = \frac{2}{k+h^\vee}\sum_{m=0}^{\infty} J_{-n}^a J_n^a \tag{4.15}$$

and similarly for $\widetilde{L}_n$. Above, $h^\vee$ (the **dual Coxeter number**) stands for the value of the quadratic Casimir in the adjoint representation of $\mathbf{g}$ and is equal to N for the $SU(N)$ group. The operators L_n and $\widetilde{L}_n$ satisfy the relations (4.14) with C acting as the multiplication by $c = \frac{k\dim(G)}{k+h^\vee}$, the value of the **Virasoro central charge** of the WZW theory. Besides,

$$[L_n, J_m^a] = -mJ_{n+m}^a$$

and similarly for $[\widetilde{L}_n, \widetilde{J}_m^a]$. The operators L_n (and $\widetilde{L}_n$) satisfy the unitarity conditions $L_n^\dagger = L_{-n}$. In particular L_0 is a self-adjoint operator, bounded below on $V_{\widehat{R}}$ by the **conformal dimensions** $\Delta_R = \frac{c_R}{k+h^\vee}$, the eigenvalue of L_0 on the subspace $V_R \subset V_{\widehat{R}}$.

23

The latter subspace is annihilated by all L_n with $n > 0$. The Hamiltonian of the WZW theory is $H = L_0 + \tilde{L}_0 - \frac{c}{12}$ whereas $L_0 - \tilde{L}_0$ defines the momentum operator P. The tensor product of the HW vectors in the subspace $V_{\hat{1}} \otimes \overline{V}_{\hat{1}} \subset \mathcal{H}$ corresponding to the trivial representation $R = 1$ gives the **vacuum state** Ω of the theory annihilated by L_0 and $\tilde{L}_0$.

A certain role in what follows will be played by the characters of the representations $V_{\hat{R}}$ defined as traces of loop group operators acting in $V_{\hat{R}}$. To avoid domain problems, one often considers only the endomorphisms $g_{\hat{R}}$ of $V_{\hat{R}}$ representing the action of the elements $g \in G$ (or in $G^{\mathbf{C}}$) and obtained by the integration of the action of the generators t_0^a. One then defines

$$\chi_{\hat{R}}(\tau, g) \;=\; \mathrm{tr}_{V_{\hat{R}}} e^{2\pi i \tau (L_0 - \frac{c}{24})} \, g_{\hat{R}} \,, \tag{4.16}$$

where τ is a complex number in the upper half plane: $\mathrm{Im}\,\tau > 0$. The presence of the factor $e^{2\pi i \tau L_0}$ renders the trace finite. The characters $\chi_{\hat{R}}(\tau, g)$ are class functions of g and may be explicitly computed. Their decomposition into characters of G encodes the multiplicities of the eigenvalues of the Virasoro generator L_0 in the subspaces of $V_{\hat{R}}$ transforming according to a given representation of G.

The central charge c entering the commutation relations of the Virasoro generators is an important characteristic of a conformal field theory. It appears also in the rigorous version of the quantum amplitudes $\mathcal{A}_\Sigma$. It enters into them in a somewhat subtle way, measuring their change under the local rescalings $\eta \mapsto e^{2\sigma}\eta$ of the metric of Σ (recall that the amplitudes of the classical configurations $e^{-S(g)}$ were invariant under such rescalings). Under the change $\eta \mapsto e^{2\sigma}\eta$ with σ vanishing around the boundary,

$$\mathcal{A}_\Sigma \;\mapsto\; e^{\frac{c}{12\pi} \int_\Sigma [\frac{1}{2} \partial_\alpha \sigma \partial_\beta \sigma \eta^{\alpha\beta} + \sigma r + \mu(e^{2\sigma}-1)]\sqrt{\eta}} \, \mathcal{A}_\Sigma \tag{4.17}$$

due to renormalization effects, with μ depending on the renormalization prescription. We shall use the prescription corresponding to $\mu = 0$. The same transformation rule is obeyed by the amplitudes $\mathcal{A}_\Sigma(A)$ in external gauge field. Hence, the quantum amplitudes are only projectively invariant under the conformal rescalings of the metric η. This is an example of the standard effect leading to projective actions of symmetries in quantum theory. Due to this effect, some care will have to be taken when making sense out of formal properties of the quantum amplitudes, like the gluing property (4.4). We shall always assume that the metric η of Σ, which, together with the orientation of Σ, defines its complex structure, is of the special form around the boundary. Namely, that, in terms of the complex coordinate of the unit discs D_n holomorphically embedded into the surface Σ' without boundary such that $\Sigma = \Sigma' \backslash (\underset{n}{\cup} \mathring{D}_n)$, it is equal to the cylindrical metric $|z|^{-2}|dz|^2$. Upon gluing of surfaces along boundary components, such metrics will automatically give smooth metrics on the resulting surfaces.

In particular, the metrics on the unit discs D will have the form $e^{2\sigma}|dz|^2$ with $\sigma = -\ln|z|$ around the boundary of D. Unless otherwise stated, we shall also assume that $\sigma = 0$ around the center of D. Consider the Riemann sphere $\mathbf{C}P^1 = \mathbf{C} \cup \{\infty\}$

composed from the two copies of the disc D glued along the boundary. The choice

$$\zeta = \mathcal{Z}_{\mathbf{C}P^1}^{-\frac{1}{2}} \tag{4.18}$$

for the normalizing constant will make precise the assignment (4.11). This choice guarantees that the change of ζ under the rescalings of the metric on D will cancel the change of the functional integral $\int \overline{g_R^{ij}} \, e^{-S(g)} \, Dg$.

4.3 Correlation functions

The formalism of Green functions encoding the action of field operators constitutes a traditional tool in quantum field theory. In the Minkowski space, the Green functions allow to express easily the scattering matrix elements (at least for the massive theories, via the LSZ formalism) whereas in the Euclidean space they coincide with correlation functions of continuum statistical models, providing a bridge between quantum field theory and statistical mechanics. In the context of CFT, the correlation functions defined on a general Riemann surface Σ without boundary constitute a somewhat easier objects to deal with than the quantum amplitudes $\mathcal{A}_\Sigma$ for surfaces with boundary. Besides, even considered on the simplest Riemann surface, the Riemann sphere $\mathbf{C}P^1$, they already contain the full information about the model. Formally, the correlation functions of the WZW model are given by the functional integrals

$$< \prod_n g_{R_n}^{i_n j_n}(\xi_n) >_\Sigma (A) = \mathcal{Z}_\Sigma(A)^{-1} \int \prod_n g_{R_n}^{i_n j_n} \, e^{-S_\Sigma(g,A)} \, Dg \,, \tag{4.19}$$

where ξ_n are disjoint points in a Riemann surface Σ without boundary. For the Riemann surface without boundary Σ' obtained by gluing unit discs D_n to a surface Σ with boundary[5], see Fig. 6, and for the points ξ_n placed at the centers of the discs D_n, the correlation functions without the gauge field may be expressed, with the use of the assignment (4.11) and of the gluing property (4.5), by the scalar products of the quantum amplitudes $\mathcal{A}_\Sigma$ with special vectors in the Hilbert space of states:

$$< \prod_n g_{R_n}^{i_n j_n}(\xi_n) >_{\Sigma'} = \mathcal{Z}_\Sigma^{-1} \left(\bigotimes_n (e_{R_n}^{i_n} \otimes \overline{e_{R_n}^{i_n}}), \, \mathcal{A}_\Sigma \right).$$

The normalization factor is given by the partition function of the surface with boundary Σ defined by

$$\mathcal{Z}_\Sigma = \mathcal{Z}_{\Sigma'} \prod_n \zeta_n \tag{4.20}$$

with ζ_n as in Eq. (4.18). The combination of the partition functions on the right hand side does not change under local rescalings of the metric inside the discs D_n.

On the level of correlation functions, the symmetry properties of the theory are encoded in the so called Ward identities. For example, the behavior (3.28) of the action

---

[5]the metric η on Σ' is assumed to come from metrics on Σ and on the discs D_n of the type described above

under the chiral gauge transformations with for $h_1 = h$ and $h_2 = 1$ implies formally that

$$\mathcal{Z}_\Sigma(A) < \bigotimes_n g_{R_n}(\xi_n) >_\Sigma(A)$$

$$= e^{-S(h^{-1}, A^{01})} \bigotimes_n h_{R_n}^{-1}(\xi_n) \, \mathcal{Z}_\Sigma(A^{10} + {}^h A^{01}) < \bigotimes_n g_{R_n}(\xi_n) >_\Sigma(A^{10} + {}^h A^{01}), \quad (4.21)$$

where we view $\otimes g_{R_n}$ as taking value in $\bigotimes_n End(V_{R_n})$ and collecting all the matrix elements $\prod g_{R_n}^{i_n j_n}$. Similarly, for $h_1 = 1$ and $h_2 = h$, we obtain the mirror relation:

$$\mathcal{Z}_\Sigma(A) < \bigotimes_n g_{R_n}(\xi_n) >_\Sigma(A)$$

$$= e^{-S(h, A^{10})} \, \mathcal{Z}_\Sigma({}^h A^{10} + A^{01}) < \bigotimes_n g_{R_n}(\xi_n) >_\Sigma({}^h A^{10} + A^{01}) \bigotimes_n h_{R_n}(\xi_n). \quad (4.22)$$

These are the Ward identities expressing the symmetry of the correlation functions under the chiral gauge transformations.

It is useful and customary to introduce more general correlation functions with insertions of **currents** testing the reaction of the functions (4.19) to infinitesimal changes of the gauge fields. On the surface Σ' they are defined by

$$\mathcal{Z}_{\Sigma'} < J^a(z_m) \prod_n g_{R_n}^{i_n j_n}(\xi_n) >_{\Sigma'} = -\pi \left. \frac{\delta}{\delta A_{\bar{z}}^a(z_m)} \right|_{A=0} \mathcal{Z}_{\Sigma'}(A) < \prod_n g_{R_n}^{i_n j_n}(\xi_n) >_{\Sigma'}(A), \quad (4.23)$$

where z_m is the complex coordinate of the disc D_m, or by

$$\mathcal{Z}_{\Sigma'} < \tilde{J}^a(\bar{z}_m) \prod_n \cdot g_{R_n}^{i_n j_n}(\xi_n) >_{\Sigma'} = -\pi \left. \frac{\delta}{\delta A_z^a(z_m)} \right|_{A=0} \mathcal{Z}_{\Sigma'}(A) < \prod_n g_{R_n}^{i_n j_n}(\xi_n) >_{\Sigma'}(A). \quad (4.24)$$

Its is not very difficult to show, expanding the Ward identities (4.21) and (4.22) to the first order in h around 1, that the insertions of $J^a(z)$ $(\tilde{J}^a(\bar{z}))$ are analytic (anti-analytic) in $z \neq 0$ but that they have simple poles at $z = 0$, the location point of one of the insertions $g_R(\xi)$, with the behavior

$$J^a(z) \, g_R(\xi) = -\frac{1}{z} t_R^a \, g_R(\xi) + \cdots, \qquad \tilde{J}^a(\bar{z}) \, g_R(\xi) = \frac{1}{\bar{z}} \, g_R(\xi) \, t_R^a + \cdots, \quad (4.25)$$

where the dots denote non-singular terms. The latter are related to the action of the current algebra generators in the space of states by the following relations involving the contour integrals[6]:

$$\frac{1}{2\pi i} \int_{|z_m|=\rho} < J^a(z_m) \prod_n g_{R_n}^{i_n j_n}(\xi_n) >_{\Sigma'} z_m^p \, dz_m = -\mathcal{Z}_\Sigma^{-1} \left(\bigotimes_n (J_p^{a\#} e_{R_m}^{i_n} \otimes \overline{e_{R_n}^{i_n}}), A_\Sigma \right), \quad (4.26)$$

$$-\frac{1}{2\pi i} \int_{|z_m|=\rho} < \tilde{J}^a(\bar{z}_m) \prod_n g_{R_n}^{i_n j_n}(\xi_n) >_{\Sigma'} \bar{z}_m^p \, d\bar{z}_m = -\mathcal{Z}_\Sigma^{-1} \left(\bigotimes_n (e_{R_n}^{i_n} \otimes \tilde{J}_p^{a\#} \overline{e_{R_n}^{i_n}}), A_\Sigma \right) \quad (4.27)$$

---

[6] oriented counter-clockwise

with $\rho < 1$ and the superscript # indicating that the operator appears only for $n = m$. Eqs. (4.25) are examples of the **operator product expansions**, in this case, the ones stating that $g_R(\xi)$ are **primary fields** of the current algebra, in the CFT jargon.

Multiple current insertions, see Fig. 8, integrated over contours of increasing radii lead to multiple insertions of the current algebra generators. For example:

$$\frac{1}{2\pi i} \int_{|z_m|=\rho_1} \frac{1}{2\pi i} \int_{|w_m|=\rho_2} < J^a(z_m)\, J^b(w_m) \prod_n g_{R_n}^{i_n j_n}(\xi_n) >_{\Sigma'} z_m^p\, w_m^q\, dz_m\, dw_m$$

$$= \mathcal{Z}_\Sigma^{-1}\left(\otimes_n (J_p^{a\#}\, J_q^{b\#}\, e_{R_n}^{i_n} \otimes \overline{e_{R_n}^{i_n}}),\, \mathcal{A}_\Sigma\right)$$

for $\rho_1 > \rho_2$ whereas for $\rho_1 < \rho_2$ the order of $J_p^a J_q^b$ should be reversed. In particular, the commutator $[J_p^a, J_q^b]$ corresponds to the difference of the two double contour integrals. It follows, that general matrix elements of the quantum amplitudes $\mathcal{A}_\Sigma$ may be read of the correlation function in the external gauge. It is then enough to find the latter to describe the complete theory.

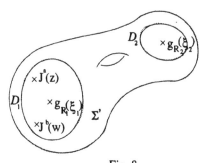

Fig. 8

The action of the Virasoro generators may be interpreted similarly in terms of the insertions of the **energy-momentum** tensor into the correlation functions which test their variation under the changes of the metric on the surface:

$$\mathcal{Z}_{\Sigma'} < T(z_m) \prod_n g_{R_n}^{i_n j_n}(\xi_n) >_{\Sigma'} = 4\pi \frac{\delta}{\delta\eta^{zz}(z_m)} \mathcal{Z}_{\Sigma'} < \prod_n g_{R_n}^{i_n j_n}(\xi_n) >_{\Sigma'},$$

$$\mathcal{Z}_{\Sigma'} < \tilde{T}(\bar{z}_m) \prod_n g_{R_n}^{i_n j_n}(\xi_n) >_{\Sigma'} = 4\pi \frac{\delta}{\delta\eta^{\bar{z}\bar{z}}(z_m)} \mathcal{Z}_{\Sigma'} < \prod_n g_{R_n}^{i_n j_n}(\xi_n) >_{\Sigma'}. \qquad (4.28)$$

Under the local rescaling of the metric $\eta \mapsto e^{2\sigma}\eta$ with σ vanishing around the insertion points ξ_n, the correlation functions (4.19) are invariant. This is not any more the case for general σ due to (the "wave function") renormalization of the insertions. For general σ, the correlation functions pick up the product of local factors equal to $e^{-2\Delta_{R_n}\sigma(\xi_n)}$, where the **conformal dimension** Δ_R of the fields $g_R(\xi)$ coincide with the lowest eigenvalues of the Virasoro generator L_0 in the HW representations of the current algebra discussed before. The partition functions transform under the metric rescaling according to the rule (4.17). The infinitesimal versions of the above transformation properties together with the covariance of the whole scheme under infinitesimal diffeomorphisms of the surface Σ may be shown to imply that the insertions of $T(z_m)$ $(\tilde{T}(\bar{z}_m))$ are analytic (anti-analytic)[7] in z_m for $z_m \neq 0$ with the singular part given by the operator product expansion

$$T(z)\, g_R(\xi) \;=\; \frac{1}{z^2} \Delta_R\, g_R(\xi) \;+\; \frac{1}{z} \partial_z g_R(\xi) \;+\; \dots \,,$$

$$\tilde{T}(\bar{z})\, g_R(\xi) \;=\; \frac{1}{\bar{z}^2} \Delta_R\, g_R(\xi) \;+\; \frac{1}{\bar{z}} \partial_{\bar{z}} g_R(\xi) \;+\; \dots \,.$$

The latter expansions state that $g_R(\xi)$ are primary fields of the Virasoro algebra. The insertions of the energy momentum tensor encode the action of the Virasoro algebra generators in the space of states:

$$\frac{1}{2\pi i} \int_{|z_m|=\rho} < T(z_m) \prod_n g_{R_n}^{i_n j_n}(\xi_n) >_{\Sigma'} z_m^{p+1}\, dz_m \;=\; \mathcal{Z}_{\Sigma}^{-1} \left(\otimes_n (L_p^\# e_{R_n}^{i_n} \otimes \overline{e_{R_n}^{i_n}}), \mathcal{A}_\Sigma \right), \qquad (4.29)$$

$$-\frac{1}{2\pi i} \int_{|z_m|=\rho} < \tilde{T}(\bar{z}_m) \prod_n g_{R_n}^{i_n j_n}(\xi_n) >_{\Sigma'} \bar{z}_m^{p+1}\, d\bar{z}_m \;=\; \mathcal{Z}_{\Sigma}^{-1} \left(\otimes_n (e_{R_n}^{i_n} \otimes \tilde{L}_p^\# \overline{e_{R_n}^{i_n}}), \mathcal{A}_\Sigma \right). \qquad (4.30)$$

The Ward identities of the chiral gauge symmetry together with the transformation properties under the local rescalings of the metric and under diffeomorphisms of the surface, expanded to the second order in the symmetry generators, yield the operator product expansions

$$J^a(z)\, J^b(w) \;=\; \frac{k\delta^{ab}}{2(z-w)^2} + \frac{if^{abc}}{z-w} J^c(w) + \dots \,,$$

$$T(z)\, T(w) \;=\; \frac{c}{2(z-w)^4} + \frac{2}{(z-w)^2} T(w) + \frac{1}{z-w} \partial_w T(w) + \dots \,,$$

$$T(z)\, J^a(w) \;=\; \frac{1}{(z-w)^2} J^a(w) + \frac{1}{z-w} \partial_w J^a(w) + \dots \,,$$

$$\tag{4.31}$$

$$\tilde{J}^a(\bar{z})\, \tilde{J}^b(\bar{w}) \;=\; \frac{k\delta^{ab}}{2(\bar{z}-\bar{w})^2} + \frac{if^{abc}}{\bar{z}-\bar{w}} \tilde{J}^c(\bar{w}) + \dots \,,$$

$$\tilde{T}(\bar{z})\, \tilde{T}(w) \;=\; \frac{c}{2(\bar{z}-\bar{w})^4} + \frac{2}{(\bar{z}-\bar{w})^2} \tilde{T}(\bar{w}) + \frac{1}{\bar{z}-\bar{w}} \partial_{\bar{w}} \tilde{T}(\bar{w}) + \dots \,,$$

$$\tilde{T}(\bar{z})\, \tilde{J}^a(\bar{w}) \;=\; \frac{1}{(\bar{z}-\bar{w})^2} \tilde{J}^a(\bar{w}) + \frac{1}{\bar{z}-\bar{w}} \partial_{\bar{w}} \tilde{J}^a(\bar{w}) + \dots \,,$$

where the dots denote the non-singular terms analytic (anti-analytic) in z around w. The above expansions hold when inserted into the correlation functions as above with z and w corresponding to the values of the same local coordinate for two different points

---

[7] in the standard metric $|dz|^2$ around the insertion point

and in the standard metric. They encode through the relations (4.26), (4.27), (4.29) and (4.30) the commutation relations of the current and Virasoro generators J_n^a, $\tilde{J}_n^a$, L_n, $\tilde{L}_n$ obtained from the above expansions by the deformation of the integration contours

$$\int_{|z|=\rho+\epsilon} dz \int_{|w|=\rho} dw - \int_{|z|=\rho-\epsilon} dz \int_{|w|=\rho} dw = \int_{|w|=\rho} dw \int_{|z-w|=\epsilon} dz, \qquad (4.32)$$

see Fig. 9, and the use of the residue theorem.

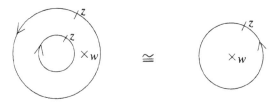

Fig. 9

The operator expansion algebra of the insertions into the correlation functions substitutes then for the operator commutation relations but allows to encode also more complicated algebraic relations between the CFT operators, see the last section. It is the basic technique of two-dimensional CFT.

As we have discussed before, the action of the Virasoro generators in the space of states of the WZW model may be expressed in terms of the current algebra action, see Eq. (4.15). This relation may be translated into the language of the insertions into the correlation functions, giving rise to the Sugawara construction of the energy-momentum tensor:

$$T(w) = \lim_{z \to w} \tfrac{1}{k+h^\vee} \left(J^a(z)\, J^a(w) - \tfrac{k\,\dim(G)}{2(z-w)^2} \right)$$

and similarly for $\tilde{T}(\bar{w})$.

5 Chiral WZW theory and the Chern-Simons states

As we have seen, the whole information about the quantum amplitudes of the WZW theory resides in the correlation functions (4.19) in an external gauge field. We shall look now more closely into the gauge-field dependence of these functions. By (formal) analytic continuation, the chiral Ward identities (4.21) and (4.22) should also hold for the complexified gauge fields A with values in $\mathbf{g}^{\mathbf{C}}$ and for the complexified gauge transformations h with values in $G^{\mathbf{C}}$. As we shall see, they give a powerful tool for analysis of the correlation functions.

Let us consider first the Ward identity (4.21). The holomorphic maps Ψ on the space $\mathcal{A}^{01}$ of $\mathbf{g}^{\mathbf{C}}$-valued 0,1-gauge fields A^{01} with values in $\underset{n}{\otimes} V_{R_n}$ satisfying the equation

$$\Psi(A^{01}) = e^{-S(h^{-1}, A^{01})} \underset{n}{\otimes} h_{R_n}^{-1}(\xi_n) \; \Psi(^h A^{01}) \tag{5.1}$$

for h in the group $\mathcal{G}^{\mathbf{C}}$ of $G^{\mathbf{C}}$-valued gauge transformations have an interesting geometric interpretation. On one side, they may be viewed as holomorphic sections of a vector bundle W with typical fiber $\underset{n}{\otimes} V_{R_n}$ over the orbit space[8] $\mathcal{N} = \mathcal{A}^{01}/\mathcal{G}^{\mathbf{C}}$. Mathematically, the orbit space $\mathcal{N}$ is the moduli space of the holomorphic $G^{\mathbf{C}}$-bundles and the mathematicians like to view Ψ's as non-abelian generalizations of the classical theta functions. Indeed, the latter are holomorphic sections of a line bundle over the moduli space (the Jacobian) of the holomorhic $\mathbf{C}^*$-bundles over a Riemann surface.

5.1 States of the Chern-Simons theory

Physically, the holomorphic maps Ψ satisfying the Ward identity (5.1) may be identified as the quantum states of the three-dimensional Chern-Simons (CS) gauge theory [7]. The classical phase space of the CS theory on the 3-manifold $\Sigma \times \mathbf{R}$ is composed of the flat $\mathbf{g}$-valued gauge fields iA on Σ modulo G-valued gauge transformations. The flatness condition is

$$F(A) \equiv dA + A^2 = 0. \tag{5.2}$$

In the holomorphic quantization à la Bargmann, the quantum states of the theory are described as holomorphic functionals Ψ on the space $\mathcal{A}^{01}$ with the condition (5.2) imposed as a quantum constraint:

$$F(A) \; \Psi(A^{01}) = 0, \tag{5.3}$$

with $F(A)$ as in Eq. (5.2) but with A^{01} acting as the multiplication operator and A^{10} as the differentiation: $A_{\bar{z}}^a = -\frac{2\pi}{k} \frac{\delta}{\delta A_{\bar{z}}^a}$. The constraint (5.3) is closely related to the infinitesimal Ward identity:

$$\left(F(A) + \frac{4\pi i}{k} \sum_n \delta_{\xi_n} \, t^a t_{R_n}^a \right) \Psi(A) = 0 \tag{5.4}$$

obtained by expanding the global Ward identity (5.1) to the first order in h around 1. The infinitesimal identity (5.4) is equivalent to its global version (5.1). In the absence of insertions, it coincides with Eq. (5.3). The modifications involving the insertions correspond in the CS gauge theory language to the insertions of the Wilson lines $\{\xi_n\} \times \mathbf{R}$ in representations R_n.

It is a crucial fact that the (k-dependent) spaces $\mathbf{W}_\Sigma(\underline{\xi}, \underline{R})$ of the holomorphic maps Ψ's satisfying the Ward identities (5.1) or (5.4) are finite-dimensional, with the dimension given by the celebrated Verlinde formula [8]. In particular, only representations

---

[8]this space requires a careful definition with a special treatment of bad orbits

R_n with HW's λ_{R_n} integrable at level k may give rise to non-trivial spaces $\mathbf{W}_\Sigma(\underline{\xi}, \underline{R})$. It is instructive to look more carefully into the case of the Riemann sphere $\mathbf{C}P^1$. On $\mathbf{C}P^1$ all gauge fields A^{01} may be written in the form $h^{-1}\bar{\partial}h$ or may be approximated by the fields of this form. In other words, the gauge orbit of $A^{01} = 0$ is dense in $\mathcal{A}^{01}$. But by Eq. (5.1),

$$\Psi(h^{-1}\bar{\partial}h) \ = \ e^{S(h)} \bigotimes_n h_{R_n}^{-1}(\xi_n) \, \Psi(0) \, , \qquad (5.5)$$

where $\Psi(0) \in \bigotimes_n V_{R_n}$ is an element of a finite-dimensional space. Hence $\Psi(0)$ determines Ψ on a dense set of gauge fields A, so everywhere. In fact, $\Psi(0)$ belongs to the subspace $\left(\bigotimes_n V_{R_n}\right)^G$ of tensors invariant under the diagonal action of G, as is easy to see by taking constant h in Eq. (5.5). We obtain then a natural embedding

$$\mathbf{W}_{\mathbf{C}P^1}(\underline{\xi}, \underline{R}) \ \subset \ \left(\bigotimes_n V_{R_n}\right)^G . \qquad (5.6)$$

The images of $\mathbf{W}_{\mathbf{C}P^1}(\underline{\xi}, \underline{R})$ in the spaces of invariant tensors are, in general, proper subspaces of $\left(\bigotimes_n V_{R_n}\right)^G$ depending on k. The reason is that the Ψ's defined by Eq. (5.5) on $A^{01} = h^{-1}\bar{\partial}h$ do not extend holomorphically to all of $\mathcal{A}^{01}$ for all invariant tensors $\Psi(0)$. In particular, the image of $\mathbf{W}_{\mathbf{C}P^1}(\underline{\xi}, \underline{R})$, which is zero if some of HW's λ_{R_n} are not integrable at level k, becomes the whole space of invariant tensors for sufficiently large k.

For genus one, i.e. on the complex tori $T_\tau = \mathbf{C}/(\mathbf{Z} + \tau\mathbf{Z})$, a dense set of gauge fields is formed by the gauge orbits of the fields

$$A_u^{01} \ = \ \frac{\pi}{\mathrm{Im}\,\tau} u \, d\bar{z}$$

with u in the complexified Cartan algebra $\mathbf{t}^\mathbf{C} \subset \mathbf{g}^\mathbf{C}$. It is then enough to know the CS states Ψ only on the gauge fields A_u^{01}. In particular, in the case with no insertions, the holomorphic functions ψ defined by

$$\psi(u) \ = \ e^{-\frac{\pi k}{2\mathrm{Im}\,\tau}\,\mathrm{tr}\,u^2} \, \Psi(A_u^{01})$$

characterize completely the CS states Ψ.

It appears that the functions $\psi(u)$ are arbitrary combinations of the characters $\chi_{\widehat{R}}(\tau, e^{2\pi\imath u})$ of the HW representations of the current algebra $\widehat{\mathbf{g}}$, see Eq. (4.16). This fact implies an important property of the latter. Recall that the tori T_τ and $T_{\tau'}$ for $\tau' = -\frac{1}{\tau}$ may be identified by the map $z \mapsto z' = -z/\tau$. Under this identification, $A_{u'}^{01} \mapsto A_u^{01}$ if $u' = -u/\tau$. It follows then that the characters $\chi_{\widehat{R}'}(\tau', e^{2\pi\imath u'})$ of the current algebra are combinations of the characters $\chi_{\widehat{R}}(\tau, e^{2\pi\imath u})$:

$$\chi_{\widehat{R}'}(\tau', e^{2\pi\imath u'}) \ = \ \sum_{\widehat{R}} S_{R'}^R \, \chi_{\widehat{R}}(\tau, e^{2\pi\imath u}) \, .$$

Hence the modular transformation $\tau \mapsto -1/\tau$ (and more generally, the transformations of $SL(2, \mathbf{Z})$) may be implemented on the characters of the current algebra. The symmetric unitary matrices $(S_{R'}^R)$ representing the action of the transformation $\tau \mapsto -1/\tau$

may be expressed explicitly by the characters χ_R of the group G and by the Weyl denominator of Eq. (2.13),

$$S^R_{R'} = 1^{\frac{1}{4}} |T|^{-\frac{1}{2}} \chi_{R'}(e^{2\pi i \widehat{\lambda}_R/\widehat{k}}) \, \Pi(e^{2\pi i \widehat{\lambda}_R/\widehat{k}}) = S^{R'}_R = \overline{S^{\overline{R}}_{R'}} \qquad (5.7)$$

in the notation: $\widehat{\lambda} \equiv \lambda + \rho$ and $\widehat{k} \equiv k + h^\vee$ with ρ the Weyl vector and $h^\vee$ the dual Coxeter number. The normalizing factor $|\widehat{T}|$ is the number of the Cartan group elements of the form $e^{2\pi i \widehat{\lambda}/\widehat{k}}$ with λ a weight. $1^{\frac{1}{4}}$ is a fourth root of unity. For the $SU(2)$ group, the above formula reduces to

$$S^j_{j'} = \left(\frac{2}{k+2}\right)^{\frac{1}{2}} \sin \frac{\pi(2j+1)(2j'+1)}{k+2} . \qquad (5.8)$$

5.2 Verlinde dimensions and the fusion ring

The dimensions $\widehat{N}_{\underline{R}}$ of the spaces $\mathbf{W}_\Sigma(\underline{\xi}, \underline{R})$ are independent of the complex structure of Σ and the locations $\underline{\xi}$ of the insertion points (but dependent on the level k of the theory suppressed in the notation). They are given by the Verlinde formula which, in the present context, is a natural generalization of the classical formula for the dimensions $N_{\underline{R}}$ of the spaces $(\otimes V_{R_n})^G$ of group G invariant tensors. The dimensions $N_{\underline{R}}$ may be computed from the characters of the representations R_n:

$$N_{\underline{R}} = \int_G \prod_n \chi_{R_n}(g) \, dg = \int \prod_n \chi_{R_n}(e^{2\pi i \lambda/k}) \, |\Pi(e^{2\pi i \lambda/k})|^2 \, d\lambda , \qquad (5.9)$$

where we have used the relation (2.14). For simple, simply connected groups, the last integral may be taken over the symplex

$$\Delta_k = \{ \lambda \in \mathbf{t} \mid \operatorname{tr}(\alpha^\vee \lambda) \geq 0 \text{ for } \alpha > 0, \ \operatorname{tr}(\phi^\vee \lambda) \leq k \} \qquad (5.10)$$

whose elements label the conjugacy classes classes $\mathcal{C}_\lambda$ in a one to one way. Note that the weights in the symplex Δ_k are exactly the HW's integrable at level k, see the definition (4.9). The numbers $N_{R \, R_1 \overline{R}_2}$ coincide with the dimensions $N^{R_2}_{R \, R_1}$ of the multiplicity spaces in the decomposition (2.18) of the tensor product of representations, i.e. with the structure constants of the character ring $\mathcal{R}_G$ of the group G, see Eq. (2.19). For example for the $SU(2)$ group, $N^{j_2}_{j \, j_1} = 1$ if $|j - j_1| \leq j_2 \leq j + j_1$ and $j + j_1 + j_2$ is an integer and $N^{j_2}_{j \, j_1} = 0$ otherwise. The ring $\mathcal{R}_G$ comes with an additive $\mathbf{Z}$-valued form ω given by the integral over G. ω assigns to the combination $\sum n_i \chi_{R_i}$ of characters the coefficient of the character $\chi_1 = 1$ of the trivial representation $R = 1$. The dimensions $N_{\underline{R}}$ are the values of ω on the product of the characters χ_{R_n} in $\mathcal{R}_G$.

The dimensions $\widehat{N}_{\Sigma, \underline{R}}$ of the spaces $\mathbf{W}_\Sigma(\underline{\xi}, \underline{R})$ are given by the formula

$$\widehat{N}_{\Sigma, \underline{R}} = \frac{1}{|\widehat{T}|} \sum_{\substack{\text{weights} \\ \lambda \in \Delta_k}} \prod_n \chi_{R_n}(e^{2\pi i \widehat{\lambda}/\widehat{k}}) \, |\Pi(e^{2\pi i \widehat{\lambda}/\widehat{k}})|^{2-g_\Sigma} , \qquad (5.11)$$

in the notations from the end of the last subsection and with g_Σ denoting the genus of the surface Σ. The above equation is a rewrite of the original Verlinde formula:

$$\widehat{N}_{\Sigma,\underline{\xi}} = \sum_{\widehat{R}} \prod_n (S^R_{R_n}/S^R_1) \, (S^R_1)^{2-g_\Sigma} \tag{5.12}$$

which may be easily obtained from Eq. (5.11) with the use of the explicit expression (5.7) for the modular matrix $(S^R_{R'})$. For the particular case of $\Sigma = \mathbf{C}P^1$, Eq. (5.11) is clearly a deformation of Eq. (5.9). More exactly, the sum in Eq. (5.11) is a Riemann sum approximation of the integral in Eq. (5.9). The genus zero 3-point dimensions $\widehat{N}_{R\,R_1\bar{R}_2} \equiv \widehat{N}^{R_2}_{R\,R_1}$ give the structure constants of a commutative ring $\widehat{\mathcal{R}}_G$ which is additively generated by the representations R with the HW's integrable at level k. The (k-dependent) ring $\widehat{\mathcal{R}}_G$ is called the **fusion ring** of the WZW model. For the $SU(2)$ group and all spins $\leq \frac{k}{2}$, $\widehat{N}^{j_2}_{j\,j_1} = 1$ if $|j - j_1| \leq j_2 \leq j + j_1$ and $j + j_1 + j_2$ is an integer $\leq k$ and $\widehat{N}^{j_2}_{j\,j_1} = 0$ otherwise. The fusion ring is a deformation of the character ring $\mathcal{R}_G$. More exactly,

$$\widehat{\mathcal{R}}_G \cong \mathcal{R}_G/\widehat{\mathcal{I}}, \tag{5.13}$$

where $\widehat{\mathcal{I}}$ is the (k-dependent) ideal in $\mathcal{R}_G$ composed of the functions vanishing on the Cartan group elements $e^{2\pi i\widehat{\lambda}/k}$ for weights $\lambda \in \Delta_k$. The isomorphism identifies the image of χ_R in $\mathcal{R}_G/\widehat{\mathcal{I}}$ with the generator of $\widehat{\mathcal{R}}_G$ corresponding to R for representations R with integrable HW's. The coefficient at the generator corresponding to the trivial representation defines an additive $\mathbf{Z}$-valued form $\widehat{\omega}$ on $\widehat{\mathcal{R}}_G$. For all R_n with integrable HW's, the genus zero Verlinde dimensions $\widehat{N}_R$ are given by the values of $\widehat{\omega}$ on the image in the fusion ring of the product of the characters χ_{R_n}. For fixed representations, $\widehat{N}^{R_2}_{R\,R_1} = N^{R_2}_{R\,R_1}$ for sufficiently high k. The fusion ring may be also identified as the character ring of the quantum deformation $\mathcal{U}_q(\mathbf{g})$ of the enveloping algebra of $\mathbf{g}$ for $q = e^{\pi i/(k+h^\vee)}$, an example of the intricate relations between the WZW model and the quantum groups.

5.3   Holomorphic factorization

Consider now the Ward identity (4.22) for the mirror chiral gauge transformations. The anti-holomorphic maps Φ of the space $\mathcal{A}^{10}$ of the $\mathbf{g}^\mathbf{C}$-valued 1,0-gauge fields A^{10} with values in $\underset{n}{\otimes} V_{\bar{R}_n}$ such that

$$\Phi(A^{10}) = e^{-S(h,\,A^{10})} \underset{n}{\otimes} h^t_{\bar{R}_n}(\xi_n) \, \Phi(^h A^{10})$$

are the complex conjugates of the holomorphic maps Ψ satisfying the relation (5.1):

$$\Phi(A^{10}) = \overline{\Psi(-(A^{10})^*)}, \tag{5.14}$$

where the star denotes the anti-linear involution of the complexified Lie algebra $\mathbf{g}^\mathbf{C}$ leaving $\mathbf{g}$ invariant (it coincides with the hermitian conjugation for $\mathbf{g} = su(N)$). It follows that the correlation functions, in their dependence of the external gauge field, are sesqui-linear combinations of the elements of the space $\mathbf{W}_\Sigma(\underline{\xi}, \underline{R})$ of holomorphic solutions of Eq. (5.1):

$$\mathcal{Z}_\Sigma(A) < \otimes_n g_{R_n}(\xi_n) >_\Sigma (A) = H^{\alpha\beta} \Psi_\alpha(A^{01}) \otimes \overline{\Psi_\beta(-(A^{01})^*)}, \tag{5.15}$$

where the states Ψ_α form a basis of $\mathbf{W}_\Sigma(\underline{\xi}, \underline{R})$ and the right hand side should be summed over α and β. The equality involves the natural identification of the vector spaces $(\otimes_n V_{R_n}) \otimes (\otimes_n V_{\overline{R}_n}) \cong \otimes_n End(V_{R_n})$. The partition function $\mathcal{Z}(A)$ has to be given by similar expressions pertaining to the case without insertions. In particular, on the complex torus T_τ, and in the vanishing gauge field,

$$\mathcal{Z}_{T_\tau} = \sum_{\widehat{R}, \widehat{R}'} H^{\widehat{R}\widehat{R}'} \operatorname{ch}_{\widehat{R}}(\tau, 1) \overline{\operatorname{ch}_{\widehat{R}'}(\tau, 1)}. \tag{5.16}$$

The matrices $(H^{\alpha\beta})$ should be specified for a given choice of the basis (Ψ_α) for each complex structure on the surface and for each configuration of the insertions points so that, if we did not have means to compute them, the above formulae would mean little progress towards the solution of the WZW theory. Fortunately, there exist effective ways to determine the coefficients $H^{\alpha\beta}$.

5.4   Scalar product of the Chern-Simons states

It was argued in [9], see also [10] for a formal functional integral argument, that the matrices $(H^{\alpha\beta})$ appearing in Eq. (5.15) are inverse to the matrices $(H_{\alpha\beta})$ with matrix elements

$$H_{\alpha\beta} = (\Psi_\alpha, \Psi_\beta)$$

given by the scalar product of the CS states. According to the rules of holomorphic quantization, the latter is given by the functional integral

$$(\Psi, \Psi') = \int_{A^{01}} (\Psi(A^{01}), \Psi'(A^{01}))_{\otimes V_{R_n}} e^{-\frac{k}{2\pi} \|A\|^2} DA \tag{5.17}$$

over the **g**-valued gauge fields $iA = i(-(A^{01})^* + A^{01})$ with $\|A\|^2 \equiv i \int_\Sigma \operatorname{tr}(A^{01})^* A^{01}$. This is again a formal expression. The point is, however, that the DA-integral may be calculated exactly by reducing it to doable Gaussian (i.e. free field) functional integrals. Ones this is done, the exact solution for the correlation functions follows by Eq. (5.15). Note that the above solution for coefficients $H^{\alpha\beta}$ guarantees that the right hand side of Eq. (5.15) is independent of the choice of a basis of the CS states. Let us briefly sketch how one achieves the reduction of the integral (5.17) to the free field ones.

In the first step, the integral (5.17) may be rewritten by a trick resembling the Faddeev-Popov treatment of gauge theory functional integrals. The reparametrization of the gauge fields

$$A^{01} = {}^{h^{-1}} A^{01}(n) \tag{5.18}$$

by the chiral $G^{\mathbf{C}}$-valued gauge transforms of a (local) slice $n \mapsto A^{01}(n)$ in the space $\mathcal{A}^{01}$ cutting each gauge orbit in one point, see Fig. 10,

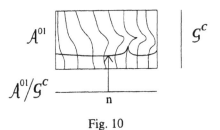

$$\mathcal{A}^{01} \qquad\qquad \mathcal{G}^C$$

$$\mathcal{A}^{01}\!/\mathcal{G}^C \underline{\hspace{5cm}}$$

$$n$$

Fig. 10

permits to rewrite the functional integral expression for the norm squared of a CS state in the new variables as

$$\|\Psi\|^2 = \int \big(\Psi(A^{01}(n)), \otimes(hh^*)_{R_n}^{-1} \Psi(A^{01}(n)) \big)_{V_{R_n}} e^{(k+2h^\vee)S(hh^*,A^{01}(n))} \, D(hh^*) \, d\mu(n). \quad (5.19)$$

To obtain the expression on the right hand side, we have used Eq. (5.1). The term $2h^\vee S(hh^*)$ in the action comes from the Jacobian of the change of variables (5.18) contributing also to the measure $d\mu(n)$ on the local slice in $\mathcal{A}^{01}$.

Unlike in the standard Faddeev-Popov setup, the integral over the group of gauge transformations did not drop out since the integrand in Eq. (5.17) is invariant only under the G-valued gauge transformations. Instead, we are left with a functional integral (5.19) similar to the one for the original correlation functions, see Eq. (4.19), except that it is over fields hh^*. These fields may be considered as taking values in the contractible hyperbolic space $G^{\mathbf{C}}/G$. $D(hh^*)$ is the formal product of the $G^{\mathbf{C}}$-invariant measures on $G^{\mathbf{C}}/G$. The gain is that the functional integral (5.19) for the hyperbolic WZW model correlation functions is doable. For example for $G = SU(2)$ and for $\Sigma = \mathbf{C}P^1$, where we may set $A^{01}(n) = 0$ (in this case the gauge orbit of $A^{01} = 0$ gives already a dense open subset $\mathcal{A}^{01}$),

$$S(hh^*) = -\tfrac{i}{2\pi} \int \partial\phi \wedge \bar\partial\phi - \tfrac{1}{2\pi} \int (\partial + \partial\phi)\bar v \wedge (\bar\partial + \bar\partial\phi)v$$

in the Iwasawa parametrization $h = \left(\begin{smallmatrix} e^{\phi/2} & 0 \\ 0 & e^{-\phi/2} \end{smallmatrix}\right) \left(\begin{smallmatrix} 1 & v \\ 0 & 1 \end{smallmatrix}\right) u$ of the 3-dimensional hyperboloid $SL(2, \mathbf{C})/SU(2)$ by $\phi \in \mathbf{R}$ and $v \in \mathbf{C}$, with $u \in SU(2)$. Although the action is not quadratic in the fields, it is quadratic in the complex field v so that the v integral can be explicitly computed. Somewhat miraculously, the resulting integral appears to depend on the remaining field ϕ again in a Gaussian way so that the integration may be carried out further. The same happens for more complicated groups and on surfaces with handles, except that the procedure requires more steps. At the end, one obtains explicit finite-dimensional integrals. Hence, the integral (5.19) belongs to the class of functional integrals that may be explicitly evaluated. The Gaussian functional integrals encountered in its computation require mild renormalizations (the zeta-function or similar regularization of determinants, Wick ordering of insertions) but these are well understood. They are responsible for the mild non-invariance of the WZW correlation functions under the local rescalings of the metric, leading to the values of the Virasoro central charge and of the conformal dimensions discussed above.

On the complex tori T_τ with no insertions and in the constant metric $|dz|^2$, the scalar product of the CS states takes a particularly simple form: the current algebra characters $\chi_{\widehat{R}}(\tau, \cdot)$ appear to give an orthonormal basis of the space of states. It follows that the toroidal partition function in the constant metric is

$$\mathcal{Z}(\tau) \;=\; \sum_{\widehat{R}} |\chi_{\widehat{R}}(\tau, 1)|^2, \tag{5.20}$$

see Eq. (5.16). The exact normalization of the constant metric on T_τ is not important since at genus one constant rescalings of the metric, exceptionally, do not effect the partition functions. The latter fact has an important consequence. It implies that the partition function $\mathcal{Z}(\tau)$ has to be a modular invariant:

$$\mathcal{Z}(\tau) \;=\; \mathcal{Z}(\tfrac{a\tau+b}{c\tau+d})$$

for $\left(\begin{smallmatrix} a & b \\ c & d \end{smallmatrix}\right) \in SL(2, \mathbf{Z})$. This is indeed the property of the right hand side of Eq. (5.20) since the matrices implementing the modular transformations on the characters of the current algebra are unitary.

Explicit finite-dimensional integral formulae for the scalar product (5.17) have been obtained for general groups at genus zero and one and, for $G = SU(2)$, for higher genera. It is clear that the case of general group and genus >1 could be treated along the same lines, but the explicit calculation has not been done. It should be also said that the general proof of the convergence of the resulting finite-dimensional integrals is also missing, although several special cases have been settled completely.

5.5   Case of $G = SU(2)$ at genus zero

To give a feeling about the form of the explicit expressions for the scalar product of the CS states, let as describe the result for $G = SU(2)$ and $\Sigma = \mathbf{CP}^1$ with insertions at points z_n in the standard complex coordinate z. In this case, as we have discussed above, the CS states correspond to invariant tensors v in $\left(\bigotimes_n V_{j_n}\right)^{SU(2)}$, where we label the irreducible representations of $SU(2)$ by spins. The spin j representation acts in the space V_j spanned by the vectors $(\sigma_-)^\ell_j v_j$ with $\ell = 0, 1, \ldots, 2j$, where v_j is the normalized HW vector annihilated by $(\sigma_+)_j$, with σ_i denoting the Pauli matrices. For the scalar product of the CS states, the procedure described in the previous subsection gives the following integral expression:

$$\|v\|^2 \;=\; f(\sigma, \underline{z}, \underline{j}, k) \int_{\mathbf{C}^J} \left| (v, \omega(\underline{z}, \underline{y}))_{\otimes V_{R_n}} \; e^{-\frac{2}{k+2} U(\underline{z}, \underline{y})} \right|^2 \prod_{a=1}^J d^2 y_a. \tag{5.21}$$

Let us explain the terms on the right hand side. First,

$$f(\sigma, z, j, k) = e^{-\frac{1}{2\pi i(k+2)}\int \partial\sigma\bar{\partial}\sigma} \left(\frac{\det'(-\Delta)}{area}\right)^{3/2} \prod_n e^{2\frac{2n(jn+1)}{k+2}\sigma(z_n)} \tag{5.22}$$

carries the dependence on the metric $e^{2\sigma}|dz|^2$ on $\mathbf{CP}^1$, with $\det'(-\Delta)$ denoting the zeta-function regularized determinant of the (minus) Laplacian on $\mathbf{CP}^1$ with omission of the zero eigenvalue. The σ-dependence of $(\det'(-\Delta)/area)$ is given by a term $e^{\frac{1}{4\pi}\int \partial\sigma\bar{\partial}\sigma}$ leading altogether to the value $\frac{3k}{k+2}$ of the Virasoro central charge of the WZW theory (recall that this is the inverse of the scalar product that enters the WZW correlation functions). Similarly, the conformal weight Δ_j of the fields $g_j(\xi)$ of the $SU(2)$ WZW theory may be read from Eq. (5.22) to be $\frac{j(j+1)}{k+2}$. Next, $\omega(z, y)$ is a meromorphic $\bigotimes_n V_{j_n}$-valued function of z and $y = (y_1, \ldots, y_J)$, where $J = \sum_n j_n$:

$$\omega(z, y) = \prod_{r=1}^{J} \left(\sum_n \frac{1}{y_r - z_n}(\sigma_-)_{jn}\right) \bigotimes_n v_{J_n}. \tag{5.23}$$

Finally, $U(z, y)$ is a multivalued function

$$U(z, y) = \sum_{n<m} j_n j_m \ln(z_n - z_m) - \sum_{n,r} j_n \ln(z_n - y_r) + \sum_{r<s} \ln(y_r - y_s). \tag{5.24}$$

The integral (5.21) is over a positive density with singularities at the coinciding y_r's and the question arises as to whether it converges. A natural conjecture states that the integral is convergent if and only if the invariant tensor v is in the image of the space of states $\mathbf{W}(\mathbf{CP}^1, z, j)$ explicitly described as the set of $v \in (\bigotimes_n V_{jn})^{SU_2}$ such that

$$(\bigotimes v_{j_n}, \prod_n (\sigma_+)_{jn}^{p_n} e^{z_n (\sigma_+)_{jn}} v)_{\bigotimes V_{J_n}} = 0 \quad \text{for} \quad \sum_n p_n \leq J - k - 1. \tag{5.25}$$

In particular, for two or three points, the image does not depend on the location of the insertions and gives the whole space of invariant tensors if $\sum_n j_n \leq k$ and zero otherwise. In this case, the integrals in Eq. (5.21) may indeed be computed in a closed form confirming the conjecture. Numerous other special cases have been checked. In general, the "only if" part of the conjecture is easy but the "if" part remains to be verified.

5.6 Knizhnik-Zamolodchikov connection

There is another way to construct the matrices $(H^{\alpha\beta})$ entering the formula (5.15) for the correlation functions. Let us describe it briefly.

The spaces $\mathbf{W}_{\Sigma}(\xi, R)$ of the CS states depend on the complex structure of the surface Σ and on the insertion points. They form, in a natural way, a holomorphic vector bundle $\mathcal{W}$ whose base is the space of complex structures on a given smooth surface Σ and of non-coinciding insertions ξ (modulo diffeomorphisms). The scalar product of the CS states equips this bundle with a hermitian structure. In turn, a hermitian structure on a holomorphic vector bundle determines a unique connection such that the covariant derivatives of the structure and of the holomorphic sections vanish. Although the scalar product of the CS states has been rigorously defined in the general case only modulo the convergence of finite-dimensional integrals (see the end of the last subsection), the connection on the bundle of the states may be easily constructed with full mathematical rigor. It appears to be projectively flat (i.e. with a curvature that is a scalar 2-form on the base space), a crucial fact. In other words, the parallel transport of a CS state around a closed loop in the space of complex structures and insertions at most changes its normalization.

For the genus zero case, there is only one complex structure modulo diffeomorphisms. If we fix it as the standard complex structure on $\mathbf{C}P^1$, then we are only left with the freedom to move the insertion points $\underline{z}$. The bundle $\mathcal{W}$ is in this case a subbundle of the trivial bundle with the fiber formed by the invariant group tensors $(\underset{n}{\otimes} V_{R_n})^G$ and the connection may be extended to this trivial bundle. The covariant derivatives of the sections of the latter are given explicitly by the equations:

$$\nabla_{\bar{z}_n} v = \partial_{\bar{z}_n} v, \qquad \nabla_{z_n} v = \partial_{z_n} v + \frac{2}{k+h^\vee} \sum_{m \neq n} \frac{t^a_{R_m} t^a_{R_n}}{z_m - z_n} v \qquad (5.26)$$

which go back to the work [3] of Knizhnik-Zamolodchikov on the WZW theory. In fact the above connection is flat as long as the insertion points stay away from infinity and the article [3] studied their horizontal sections such that $\nabla v = 0$. The higher genus generalizations of these equations were first considered by Bernard [11][12] and we shall call the connection on the bundle $\mathcal{W}$ the Knizhnik-Zamolodchikov (KZ) connection for genus zero or the Knizhnik-Zamolodchikov-Bernard (KZB) one for higher genera.

In general, the KZB connection can be made flat by some choices (as in the case of genus zero, where the curvature has been concentrated at infinity). For a flat connection, one may choose locally a basis (Ψ_α) of horizontal sections. The gain from such a choice of the basis of the CS states is that the coefficients $H^{\alpha\beta}$ in Eq. (5.15) become then independent of the complex structure or the positions of the insertions. Indeed, since $H_{\alpha\beta} = (\Psi_\alpha, \Psi_\beta)$ and the KZB connection preserves the scalar product of the states, the above scalar products are constant for horizontal Ψ_α. Since the horizontal sections are, in particular, holomorphic, Eq. (5.15) gives then a holomorphic factorization of the correlation functions into sesqui-linear combinations of the **conformal blocks** holomorphic in their dependence on the complex structure and positions of insertions. Such a finite factorization is the characteristic feature of **rational** CFT's. As we see, the conformal blocks of the WZW theory are given by the horizontal sections Ψ_α of the bundle $\mathcal{W}$ of the CS states. For example, at genus zero with no insertions, the conformal blocks are formed by the characters of the current algebra and Eq. (5.20) provides a particular realization of the holomorphic factorization.

Since the KZB connection, although flat, has nevertheless a non-trivial holonomy, the conformal blocks are, in general, multivalued. The coefficients $H^{\alpha\beta}$ in Eq. (5.15) may then be fixed, up to normalization, by demanding that the correlation functions be uni-valued. The overall normalization may be fixed, in turn, by considering the limits

when the insertion points coincide. This was the strategy used in the original work
[3] to compute the 4-point correlation function of the spin $\frac{1}{2}$ field $g_{\frac{1}{2}}(\xi)$ of the $SU(2)$
WZW model on the Riemann sphere. The horizontality relations for the conformal
blocks reduce in this case to the hypergeometric equation and the calculations of the
conformal blocks and of their monodromy are easy to perform. For general genus-zero
conformal blocks, one obtains generalizations of the hypergeometric equation whose
solutions may be expressed by contour integrals [13][14]. The latter are, essentially, the
holomorphic versions of the integrals (5.21) so that the two strategies to obtain explicit
solutions for the correlation functions, one based on the study of the monodromy of the
conformal blocks and the other one involving a calculation of the scalar product of the
CS states, are closely related.

There appears to be a very rich structure behind the connection (5.26) and its generalizations. It is closely related to the integrable systems of mechanics and statistical
mechanics, see e.g. [21][22]. The holonomy of the connection gives representations of the
braid groups which played an important role in the construction of the Jones-Witten
invariants of knots and 3-manifold invariants [7]. The perturbative solutions of the
horizontality equations enter the Vasiliev invariants of knots [23]. The KZ connection
is also closely connected to quantum groups [24] and to Drinfel'd's quasi-Hopf algebras
[25], to mention only some interrelated topics.

6   Coset theories

There is a rich family of CFT's which may be obtained from the WZW models by a
simple procedure known under the name of a **coset construction** [15][16]. On the
functional integral level, the procedure consists of coupling the G-group WZW theory
to a gauge field $iB = i(B^{10} + B^{01})$ with values in a subalgebra $\mathbf{h} \subset \mathbf{g}$. The field B is
then integrated over with gauge-invariant insertions [17][18][19][20]. Let $H \subset G$ be the
subgroup of G corresponding to $\mathbf{h}$. Choose elements t_n in the space $(End(V_{R_n}, V_{r_n}))^H$
of the intertwiners of the action of H in the irreducible G- and H-representation spaces.
The simplest correlation functions of the G/H coset theory take the form

$$< \prod_{i=1}^{n} \mathrm{tr}_{V_{r_n}}(t_n g_{R_n}(x_n) t_n^{\dagger}) >_{\Sigma}$$

$$= \frac{1}{Z_{\Sigma}^{G/H}} \int \prod_n \mathrm{tr}_{V_{r_n}}(t_n g_{R_n}(x_n) t_n^{\dagger}) \, e^{-kS(g,B)-\frac{k}{2\pi}\|B\|^2} \, Dg \, DB \,, \qquad (6.1)$$

where $Z_{\Sigma}^{G/H} = \int e^{-kS(g,B)-\frac{k}{2\pi}\|B\|^2} Dg \, DB$ is the partition function of the G/H theory.
Note that the g-field integrals are the ones of the WZW theory and are given by Eq.
(5.15). Consequently,

$$Z_{\Sigma}^{G/H} < \prod_n \mathrm{tr}_{V_{r_n}}(t_n g_{R_n}(x_n) t_n^{\dagger}) >_{\Sigma}$$

$$= H^{\alpha\beta} \int (\otimes t_n \Psi_{\beta}(B^{01}), \otimes t_n \Psi_{\alpha}(B^{01}))_{\otimes V_{r_n}} \, e^{-\frac{k}{2\pi}\|B\|^2} \, DB \,. \qquad (6.2)$$

The composition with $\otimes t_n$ defines a linear map T between the spaces of the group G and the group H CS states, i.e. $T : \mathbf{W}_\Sigma(\underline{\xi}, \underline{R}) \to \mathbf{W}_\Sigma(\underline{\xi}, \underline{r})$ with

$$(T\Psi)(B^{01}) = \otimes_n t_n \Psi(B^{01}). \tag{6.3}$$

Indeed, it is straightforward to check that the right hand side satisfies the the group H version of the Ward identity (5.1). Eq. (6.2) may be rewritten with the use of the map T as

$$\mathcal{Z}_\Sigma^{G/H} < \prod_n \mathrm{tr}_{V_{r_n}}(t_n g_{R_n}(x_n) t_n^\dagger) >_\Sigma = H^{\alpha\beta} (T\Psi_\beta, T\Psi_\alpha) = \mathrm{tr}_{\mathbf{W}_\Sigma(\underline{\xi},\underline{R})} T^\dagger T. \tag{6.4}$$

Let (T_α^μ) denote the ("branching") matrix of the linear map T in the bases (Ψ_α) and (ψ_μ) of, respectively, $\mathbf{W}_\Sigma(\underline{\xi}, \underline{R})$ and $\mathbf{W}_\Sigma(\underline{\xi}, \underline{r})$, i.e. $T\Psi_\alpha = T_\alpha^\mu \psi_\mu$. Then

$$\mathcal{Z}_\Sigma^{G/H} < \prod_n \mathrm{tr}_{V_{r_n}}(t_n g_{R_n}(x_n) t_n^\dagger) >_\Sigma = H^{\alpha\beta} \overline{T_\beta^\mu} h_{\mu\nu} T_\alpha^\nu,$$

where $h_{\mu\nu} = (\psi_\mu, \psi_\nu)$. Since the above relations hold also for the partition function, it follows that the calculation of the coset theory correlation functions (6.1) reduces to that of the scalar products of group G and group H CS states, both given by explicit, finite-dimensional integrals.

Among the simplest examples of the coset theories is the case with the group $G = SU(2) \times SU(2)$ with level $(k, 1)$ (for product groups, the levels may be taken independently for each group) and with H being the diagonal $SU(2)$ subgroup. The resulting theories coincide with the unitary **minimal series** of CFT's with the Virasoro central charges $c = 1 - \frac{6}{(k+2)(k+3)}$, first considered by Belavin-Polyakov-Zamolodchikov [1]. The Hilbert spaces of these theories are built from irreducible unitary representations of the Virasoro algebras with $0 < c < 1$. The simpliest one of them with $k = 1$ and $c = \frac{1}{2}$ describes the continuum limit of the Ising model at critical temperature or the scaling limit of the massless ϕ_2^4 theory. In particular, in the continuum limit the spins in the critical Ising model are represented by fields $\mathrm{tr}\, g_{\frac{1}{2}}(\xi)$ where g takes values in the first $SU(2)$. The corresponding correlation functions may be computed as above. One obtains this way for the 4-point function on the complex plane (or the Riemann sphere) an explicit expression in terms of the hypergeometric function.

The G/H coset theory with $H = G$ is a prototype of a two-dimensional topological field theory. As follows from Eq. (6.4), the correlation functions of the fields $\mathrm{tr}\, g_R(\xi)$ are equal to the dimension of the spaces $\mathbf{W}_\Sigma(\underline{\xi}, \underline{R})$, normalized by the dimension of $\mathbf{W}_\Sigma(\emptyset, \emptyset)$ and are given by the Verlinde formula (5.11). In particular, they do not depend on the position of the insertion points, a characteristic feature of the correlation functions in topological field theories.

7 Boundary conditions in the WZW theory

Discussing in Sects. 3.4 and 4.1 above the classical and the quantum amplitudes for the WZW model on surfaces with boundary, we have admitted an arbitrary behavior of the fields on the boundary. In physical situations, one often has to constrain this behavior by imposing the boundary conditions (BC) on the fields. The simplest example is provided by the Dirichlet or Neumann BC's for the free fields which fix to zero, respectively, the tangent or the normal derivative of the field (the absorbing versus the reflecting condition). Such conditions leave unbroken an infinite-dimensional set of symmetries of the free field theory. We shall be interested in BC's in the WZW model with a similar property.

The theory of boundary CFT's was pioneered by Cardy [26] and Cardy-Lewellen [27]. It found its applications e.g. in the theory of isolated impurities in condensed matter physics (in the so called Kondo problem, a traditional playground for theoretical ideas) [28]. In string theory, the use of the Neumann BC for free open strings has a long tradition [29]. The realization that one should also consider free open strings with the Dirichlet BC came much later and gave rise to a theory of Dirichlet- or D-branes [30]: the end of an open string, some of whose coordinates are restricted by the ·Dirichlet BC, moves on a surface (brane) of a lower dimension. The D-branes provide the basic tool in the analysis of the non-perturbative effects in string theory: of the stringy solitons and of the strong-weak coupling dualities [30]. The general theory of boundary CFT's is slowly becoming an important technique of string theory (see e.g. [31]). Here, for the sake of illustration, we shall discuss a particular class of BC's for the WZW theory. These conditions constrain the group G valued field g to stay over the boundary components in fixed conjugacy classes of G. Such BC's were discussed in [33], see also [32]. They clearly generalize the Dirichlet BC of the free fields, contrary to the claim in [33] (based on the conventions of reference [34]) associating them to the Neumann BC. Our presentation, along similar lines, clarifies, hopefully, some of the discussions of the above papers.

7.1   The action

As before, we shall represent a Riemann surface Σ with boundary as $\Sigma' \setminus (\underset{m}{\cup} \overset{\circ}{D}_m)$, where D_m are disjoint unit discs embedded in a closed surface Σ' without boundary, see Fig. 6. We have seen in Sect. 3.4 that the classical amplitudes $e^{-S(g)}$ of the fields $g : \Sigma \to G$ of the WZW model take values in a line bundle $\mathcal{L}$ rather than being numbers. The line bundle $\mathcal{L}$ over the loop group LG is not trivial but it may be trivialized over certain subsets of LG, for example the ones formed by the loops taking values in special conjugacy classes $\mathcal{C}_\lambda$. It will then be possible to give numerical values to the amplitudes $e^{-S_\Sigma(g)}$ for $g : \Sigma \to G$ satisfying the BC's

$$g(\partial D_m) \subset \mathcal{C}_{\lambda_m} . \tag{7.1}$$

In order to achieve this goal, we shall fix the 2-forms

$$\omega_\lambda = \text{tr}\,(g^{-1}dg)(1 - Ad_g)^{-1}(g^{-1}dg) = \text{tr}\,(g_0^{-1}dg_0)e^{2\pi i\lambda/k}(g_0^{-1}dg_0)e^{-2\pi i\lambda/k} \tag{7.2}$$

on the conjugacy classes $\mathcal{C}_\lambda$ composed of the elements $g = g_0\,e^{2\pi i\lambda/k}g_0^{-1}$ (the operator $(1 - Ad_g)$ is invertible on the vectors tangent to $\mathcal{C}_\lambda$). It is easy to check by a direct calculation that $d\omega_\lambda$ coincides with the restriction of the 3-form $\chi = \frac{1}{3}\text{tr}\,(g^{-1}dg)^3$ to C_λ.

Since the conjugacy classes in a simply connected group G are simply connected, any field $g : \Sigma \to G$ satisfying the BC's (7.1) may be extended to a field $g' : \Sigma' \to G$ in such a way that $g'(D_m) \subset C_{\lambda_m}$ and then to a field $\tilde{g}$ on a 3-manifold B such that $\partial B = \Sigma'$, see Fig. 11.

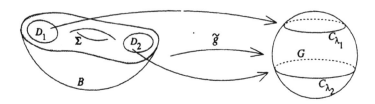

Fig. 11

Having done this, we define the Wess-Zumino part of the action as

$$S_\Sigma^{WZ}(g) \;=\; \frac{k}{4\pi i}\int_B \tilde{g}^*\chi \;-\; \frac{k}{4\pi i}\sum_m\int_{D_m} \tilde{g}\big|_{D_m}^*\, \omega_{\lambda_m}\,.$$

The ambiguities in this definition are the values of the integrals

$$\frac{k}{4\pi i}\int_{\tilde{B}} \tilde{g}^*\chi \;-\; \frac{k}{4\pi i}\sum_m\int_{S_m^2} \tilde{g}\big|_{S_m^2}^*\, \omega_{\lambda_m} \tag{7.3}$$

for 3-manifolds $\tilde{B}$ with $\partial\tilde{B} = \bigcup\limits_m S_m^2$ and for maps $\tilde{g} : \tilde{B} \to G$ such that $\tilde{g}(S_m^2) \subset C_{\lambda_m}$, see Fig. 12.

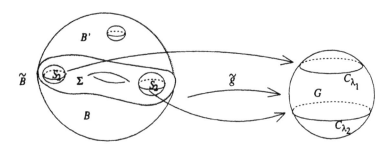

Fig. 12

In other words, they are proportional to the periods of (χ, ω) over the cycles of the relative integer homology $H_3(G, \bigcup\limits_m C_{\lambda_m})$, as noticed in [32].

42

It is not difficult to get a hold on these ambiguities. Let us glue the unit 3-balls B_m to $\bar{B}$ along the boundary spheres S_m^2 to obtain a 3-manifold $\bar{B}'$ without boundary and let us extend $\tilde{g}$ to a map $\tilde{g}' : \bar{B}' \to G$. The expression (7.3) may be now rewritten as

$$\frac{k}{4\pi i} \int_{\bar{B}'} \tilde{g}'^* \chi \; - \; \frac{k}{4\pi i} \sum_m \left(\int_{B_m} \tilde{g}_m'^* \chi \; - \; \int_{\partial B_m} \tilde{g}_m'^* \omega_{\lambda_m} \right),$$

where $\tilde{g}_m'$ are the restrictions of $\tilde{g}'$ to B_m and they satisfy $\tilde{g}_m'(\partial B_m) \subset C_{\lambda_m}$. As we have discussed in Sect. 3.3, the first term, involving the integral over the 3-manifold without boundary $\bar{B}'$, takes values in $2\pi i \mathbf{Z}$ as long as k is an integer.

Let us consider the other terms. For $G = SU(2) = \{x_0 + ix_i\sigma_i \,|\, x_0^2 + x_i^2 = 1\} \cong S^3$, the conjugacy classes corresponding to $\lambda = j\sigma_3$, with $0 \le 2j \le k$, are the 2-spheres with $x_0 = \cos\frac{2\pi j}{k}$ fixed (except for $j = 0$ or $\frac{k}{2}$ corresponding to the center elements). They are boundaries of two 3-balls B_j and B_j' with $x_0 \ge \cos\frac{2\pi j}{k}$ and $x_0 \le \cos\frac{2\pi j}{k}$, see Fig. 13.

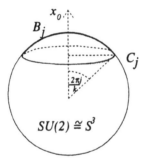

Fig. 13

A direct calculation shows that

$$\frac{k}{4\pi i}\left(\int_{B_j} \chi \; - \; \int_{\partial B_j} \omega \right) \; = \; -4\pi i j. \tag{7.4}$$

If we used B_j' instead of B_j, the result would be $4\pi i(\frac{k}{2} - j)$. We infer that j, between 0 and $\frac{k}{2}$, must be an integer or a half-integer for the ambiguity to belong to $2\pi i \mathbf{Z}$. This result has already been stated in [33].

For the other groups, the restrictions come from the the 2-spheres in C_λ of the form

$$\{ g_0 \, e^{2\pi i \lambda/k} g_0^{-1} \;|\; g_0 \in SU(2)_\alpha \}, \tag{7.5}$$

where $SU(2)_\alpha$ is the $SU(2)$ subgroup of G corresponding to a root α, see Sect. 3.3. Decomposing $\lambda = (\lambda - \frac{1}{2}\alpha^\vee \operatorname{tr}(\alpha\lambda)) + \frac{1}{2}\alpha^\vee \operatorname{tr}(\alpha\lambda)$, we observe that the first term commutes with the generators $\alpha^\vee$, $e_{\pm\alpha}$ of the Lie algebra of $SU(2)_\alpha$ and plays a spectator role. The calculation of the ambiguity is now essentially the same as for $G = SU(2)$ with j replaced by $\frac{1}{2} \operatorname{tr}(\alpha \lambda)$ and an overall factor $\frac{2}{\operatorname{tr} \alpha^2}$ due to the different normalization of $\operatorname{tr} \alpha^2$. We infer the condition

$$\frac{2}{\operatorname{tr} \alpha^2} \operatorname{tr}(\alpha \lambda) = \operatorname{tr}(\alpha^\vee \lambda) \in \mathbf{Z}.$$

Since the conjugacy classes $\mathcal{C}_\lambda$ are in one to one correspondence with λ in the symplex (5.10), the admissible conjugacy classes are in one to one correspondence with the HW's λ integrable at level k, see the definition (4.9).

The full action $S_\Sigma(g)$ of the boundary WZW model is still obtained by adding to the WZ action S^{WZ} the S^γ term of Eq. (3.6). The coupling to the gauge field is given again by Eq. (3.27). The behavior of the complete action under the chiral gauge transformation may be shown to obey the following BC version of the Eq. (3.28):

$$S(g, A) = S(h_1 g h_2^{-1}, {}^{h_2}A^{01} + {}^{h_1}A^{01}) + S(h_1^{-1} h_2, A) - \frac{ik}{2\pi} \int_\Sigma \operatorname{tr}({}^{h_2}A^{10}\, {}^{h_1}A^{01}), \quad (7.6)$$

provided that g satisfies the BC's (7.1) and that $h_1|_{\partial D_m} = h_2|_{\partial D_m}$. Note that under this conditions, the field $h_1 g h_2^{-1}$ is constrained on the boundary to the same conjugacy classes as g and $h_1 h_2^{-1}$ to the trivial one so that the actions on the right hand side make sense. The above relation will be employed below to infer the chiral gauge symmetry Ward identities for the boundary WZW theory. It may be used as the principle that selects the BC's (7.1).

Summarizing: if the field $g : \Sigma \to G$ satisfies the BC's (7.1) with integrable weights λ_m, then the amplitude $e^{-S_\Sigma^{WZ}(g)}$, and consequently also $e^{-S_\Sigma(g)}$ and $e^{-S_\Sigma(g,A)}$, may be well defined as complex numbers. Of course, the mixed case, where the BC's (7.1) with integrable λ_m are satisfied only on some boundary components of Σ and no conditions are imposed on the other ("free") boundary components can be treated in the same way. It results in the amplitudes

$$e^{-S_\Sigma(g,A)} \in \bigotimes_{n \text{ free}} \mathcal{L}_{g|_{\partial D_n}}.$$

7.2 Quantum amplitudes and correlation functions

The functional integral definition (4.1) of the quantum amplitudes of the WZW model may be naturally generalized to the case where on some boundary components of Σ we impose the BC's (7.1) with integrable weights λ_m. The resulting amplitudes $\mathcal{A}_{\Sigma,\underline{\lambda}}(A)$ will be now elements of $\bigotimes_{n \text{ free}} \Gamma(\mathcal{L})$. They may be represented as (partial) contractions of the amplitudes $\mathcal{A}_\Sigma(A)$ with all boundaries free with appropriate states:

$$\mathcal{A}_{\Sigma,\underline{\lambda}}(A) = (\bigotimes_m \hat{\delta}_{\lambda_m}, \mathcal{A}_\Sigma(A)).$$

The non-normalizable states[9] $\hat{\delta}_\lambda$ are given by Cardy's formula [26]:

$$\hat{\delta}_\lambda = \sum_{\hat{R}} (S_R^1)^{-\frac{1}{2}} S_R^{R_\lambda}\, e_{\hat{R}}^{\hat{\imath}} \otimes \overline{e_{\hat{R}}^{\hat{\imath}}}, \qquad (7.7)$$

---

[9]technically, antilinear forms on a dense subspace of the Hilbert space $\mathcal{H}$

where R_λ denotes the representation of G with the HW λ, the vectors $e^{\hat{i}}_{\hat{R}}$ form an orthonormal basis of the space $V_{\hat{R}}$ and the sum over $\hat{i}$ is understood. Note the analogy with Eq. (2.11) for the delta-function supported by a conjugacy class C_λ. The matrix $(S^{R_\lambda}_R)$ has replaced the one with the elements $(\chi_R(e^{2\pi i\lambda/k}))$ and the representation spaces $V_{\hat{R}}$ of the current algebras those of the finite-dimensional group. The state $\hat{\delta}_\lambda$ should be interpreted as a delta-function concentrated on the loops in LG contained in the conjugacy class C_λ. The non-normalizable states $e^{\hat{i}}_{\hat{R}} \otimes \overline{e^{\hat{i}}_{\hat{R}}}$ are called the Ishibashi states [35] and generalize the (properly normalized) characters of the group, see Eq. (2.10).

The correlation functions $< \otimes g_{R_n}(\xi_n) >_{\Sigma,\Delta}(A)$ in the presence of the boundaries with fields constrained by the the BC's (7.1) may be again defined by the functional integral (4.19) taking numerical values. The transformation property (7.6) of the action implies now that

$$Z_{\Sigma,\Delta}(A) < \bigotimes_n g_{R_n}(\xi_n) >_{\Sigma,\Delta}(A) = e^{-S(h_1^{-1}h_2, A) + \frac{ik}{2\pi}\int_\Sigma \mathrm{tr}\,(^{h_2}A^{10}\,^{h_1}A^{01})} Z_{\Sigma,\Delta}(^{h_2}A^{10} + ^{h_1}A^{01})$$

$$\cdot \bigotimes_n (h_1)^{-1}_{R_n}(\xi_n) < \bigotimes_n g_{R_n}(\xi_n) >_{\Sigma,\Delta}(^{h_2}A^{01} + ^{h_1}A^{01}) \bigotimes_n (h_2)_{R_n}(\xi_n). \qquad (7.8)$$

This is the chiral gauge symmetry Ward identity for the boundary WZW correlation functions, a variant of the identities (4.21) and (4.22) in presence of the BC's. Note, however, that the identity (7.8) may be factorized as the latter ones only if we assume that h_1 and h_2 are equal to 1 on the boundary. The general case where on the boundary $h_1 = h_2$ requires the presence of both gauge transformations h_1 and h_2.

It is illuminating to rewrite the Ward identity (7.8) in a different way. To this end, let us define a "doubled" Riemann surface without boundary $\tilde{\Sigma}$ by gluing Σ to its complex conjugate $\bar{\Sigma}$ along the boundary components, see Fig. 14.

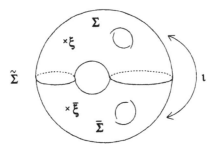

Fig. 14

We shall denote by ι the anti-holomorphic involution of $\tilde{\Sigma}$ exchanging Σ with its complex conjugate: $\iota(\xi) = \bar{\xi}$. Each chiral gauge field $\tilde{A}^{01}$ on the Riemann surface $\tilde{\Sigma}$

defines a complexified gauge field $A = (\iota^* A^{01} + A^{01})|_\Sigma$ on the surface Σ. Let us define a holomorphic functional

$$\Psi_\lambda : \tilde{A}^{01} \longrightarrow (\underset{n}{\otimes} V_{R_n}) \otimes (\underset{n}{\otimes} V_{\bar{R}_n})$$

of the chiral gauge fields on the doubled surface by

$$\Psi_\lambda(\tilde{A}^{01}) = Z_{\Sigma,\lambda}(A) < \underset{n}{\otimes} g_{R_n}(\xi_n) >_{\Sigma,\lambda}(A),$$

where we identify the space $(\otimes V_{R_n}) \otimes (\otimes V_{\bar{R}_n})$ with $\otimes End(V_{R_n})$. Let $\tilde{h}$ be a $G^{\mathbf{C}}$-valued gauge transformation on $\tilde{\Sigma}$. We shall pose $h_1 = \tilde{h}|_\Sigma$ and $h_2 = \iota^* \tilde{h}|_\Sigma$. Note that $h_1 = h_2$ on the boundary of Σ. It is not difficult to prove that

$$e^{-S_\Sigma(h_1^{-1} h_2, A) + \frac{ik}{4\pi} \int_\Sigma \mathrm{tr}\,(^{h_2} A^{10}\, ^{h_1} A^{01})} = e^{-S_{\tilde{\Sigma}}(\tilde{h}^{-1}, \tilde{A}^{01})}. \tag{7.9}$$

The Ward identity (7.8) implies then that

$$\Psi_\lambda(\tilde{A}^{01}) = = e^{-S_{\tilde{\Sigma}}(\tilde{h}^{-1}, \tilde{A}^{01})} (\underset{n}{\otimes} \tilde{h}_{R_n}^{-1}(\xi_n)) \otimes (\underset{n}{\otimes} \tilde{h}_{\bar{R}_n}^{-1}(\bar{\xi}_n)) \Psi_\lambda(^{\tilde{h}}\tilde{A}^{01}),$$

i.e. that Ψ_λ is a CS state on the doubled surface $\tilde{\Sigma}$ with the doubled insertions at points ξ_n and at their complex conjugates $\bar{\xi}_n$, associated, respectively, to the representations R_n and to the complex conjugate representations $\bar{R}_n$.

We infer that the correlations functions of the boundary WZW theory on a surface Σ may be viewed, in their gauge field dependence, as the special states Ψ_λ belonging to the space $\mathbf{W}_{\tilde{\Sigma}}(\underline{\xi}, \underline{\bar{\xi}}, \underline{R}, \underline{\bar{R}})$ of the CS states on the doubled surface $\tilde{\Sigma}$. The states Ψ_λ may be shown, moreover, to be preserved by the parallel transport with respect to the KZB connection on the restriction of the bundle $\widetilde{\mathcal{W}}$ of the CS states on the doubled surface to the subspace of the "doubled" complex structures and insertions. These properties are often summarized by saying that the boundary CFT is chiral since its correlation functions are given by special conformal blocks of the WZW theory on $\tilde{\Sigma}$. It would be desirable to characterize geometrically the CS states Ψ_λ. Some of their special properties are easy to find. For example, they are preserved by the antilinear involution $\Psi \mapsto {}^\iota\Psi$ of $\mathbf{W}_{\tilde{\Sigma}}(\underline{\xi}, \underline{\bar{\xi}}, \underline{R}, \underline{\bar{R}})$ induced by the involution ι of the doubled surface $\tilde{\Sigma}$ and defined by

$$^\iota\Psi(\tilde{A}^{01}) = \overline{\Psi(-(\iota^* \tilde{A}^{01})^*)}. \tag{7.10}$$

A complete geometric characterization of the states Ψ_λ for different choices λ of the BC's seems, however, still missing.

7.3 Piece-wise boundary conditions

Up to now, we have imposed the boundary conditions forcing the fields to take values in the special conjugacy classes uniformly on the component circles of $\partial\Sigma$. Since the conditions are local, it should be also possible to do this locally on the pieces of the boundary. Suppose that the boundary $\partial\Sigma$ is divided into intervals I_r (the entire boundary circles are also admitted). We shall associate integrable weight labels λ_r to some of these intervals in such a way that two labeled intervals in the same boundary component are separated by an unlabeled ("free") one, see Fig. 15.

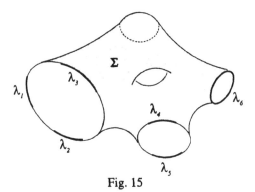

Fig. 15

We shall now consider the fields g on Σ which on the labeled intervals take values in the corresponding conjugacy classes $\mathcal{C}_{\lambda_r}$ and are not restricted on the free intervals. One may still define the classical amplitudes $e^{-S^{WZ}(g)}$ for such fields although this requires a more local (Čech cohomology type) technique than the one developed above [36]. Let us sketch how this is done.

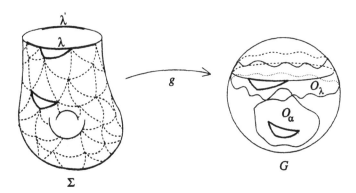

Fig. 16

Recall from the end of Sect. 3.1 that $S^{WZ}(g)$ in the first approximation is equal to $\frac{k}{4\pi i}\int g^{*}\beta$ where $d\beta = \chi$ is the canonical closed 3-form on G. The problem stemmed

47

from the fact that such 2-forms β exist only locally. However, on the sets of a sufficiently fine open covering $(\mathcal{O}_\alpha)$ of G, we may choose 2-forms β_α such that $d\beta_\alpha = \chi$. Choose a triangulation $\mathcal{T}$ of Σ with the triangles t, edges e and vertices v. If $\mathcal{T}$ is fine enough then each of the simplices s of $\mathcal{T}$ is mapped by g into an open set, say $\mathcal{O}_{\alpha_s}$, see Fig. 16. The main contribution to the amplitude $e^{-S_\Sigma^{WZ}(g)}$ will come from $\exp[-\frac{k}{4\pi i}\sum_t \int_t g^* \beta_{\alpha_t}]$. The above expression depends, however, on the choice of the forms β_α and of the triangulation. The idea is to compensate this dependence by contributions from simplices of lower dimension. Let $\eta_{\alpha_0\alpha_1} = -\eta_{\alpha_1\alpha_0}$ be 1-forms defined on the non-empty intersections $\mathcal{O}_{\alpha_0\alpha_1} \equiv \mathcal{O}_{\alpha_0} \cap \mathcal{O}_{\alpha_1}$ such that

$$d\eta_{\alpha_0\alpha_1} = \beta_{\alpha_1} - \beta_{\alpha_0}$$

and let $f_{\alpha_0\alpha_1\alpha_2}$ be functions on the triple intersections $\mathcal{O}_{\alpha_0\alpha_1\alpha_2}$, antisymmetric in the indices, satisfying

$$df_{\alpha_0\alpha_1\alpha_2} = \eta_{\alpha_1\alpha_2} - \eta_{\alpha_0\alpha_2} + \eta_{\alpha_0\alpha_1}$$

and such that on the four-fold intersections

$$f_{\alpha_1\alpha_2\alpha_3} - f_{\alpha_0\alpha_2\alpha_3} + f_{\alpha_0\alpha_1\alpha_3} - f_{\alpha_0\alpha_1\alpha_2} \in 8\pi^2 \mathbf{Z}.$$

Such data may, indeed, be chosen. We define then

$$e^{-S_\Sigma^{WZ}(g)} = \exp\left[-\frac{k}{4\pi i}\left(\sum_t \int_t g^*\beta_{\alpha_t} - \sum_{e \subset t} \int_e g^*\eta_{\alpha_e\alpha_t} + \sum_{v \in e \subset t} (\pm) f_{\alpha_v\alpha_e\alpha_t}(g(v))\right)\right], \quad (7.11)$$

where in the last sum the sign is taken according to the orientation of the vertices v inherited from the triangles t via the edges e. One may show that for the surface without boundary, the above expression does not depend on the choices involved and coincides with the definition given in Sect. 3.3.

In the presence of boundary circles with unconstrained fields, the above expression may be used to define the amplitudes with values in a line bundle and it provides an alternative construction of the bundle $\mathcal{L}$ over the loop group [36]. In the presence of the boundary conditions on the intervals I_r we shall still employ the same definition, but with some care. Namely, we include neighborhoods $\mathcal{O}_\lambda$ of the conjugacy classes $\mathcal{C}_\lambda$ into the open covering $(\mathcal{O}_\alpha)$ of G. We choose 2-forms β_λ on $\mathcal{O}_\lambda$ coinciding with ω_λ of Eq. (7.2) when restricted to $\mathcal{C}_\lambda$. The triangulations used in Eq. (7.11) are required to be compatible with the splitting of the boundary. To the simplices in the labeled intervals I_r we assign the open sets $\mathcal{O}_{\lambda_r}$, see Fig. 16. The amplitudes resulting from Eq. (7.11) coincide then with those defined in the previous section for the special case when the labeled intervals fill entire circles. In the general case,

$$e^{-S_\Sigma^{WZ}(g)} \in \prod_{\text{free } I_r} (\mathcal{L}_{I_r})_{g|_{I_r}} \quad (7.12)$$

where $\mathcal{L}_{I_r}$ is a line bundle over the space of maps from an interval I_r to G taking on the boundary of I_r the values in the conjugacy classes $\mathcal{C}_{\lambda_r}$ and $\mathcal{C}_{\lambda'_r}$ specified by the labels of the neighboring intervals[10].

---

[10]if I_r is a full circle, $\mathcal{L}_{I_r} = \mathcal{L}$

The space of sections $\Gamma(\mathcal{L}_{I_r})$ plays, as before, the role of the space of states of the WZW theory but, this time, on the interval and with boundary conditions specified by the conjugacy classes $\mathcal{C}_{\lambda_r}$ and $\mathcal{C}_{\lambda'_r}$. In the string language, these are states of the open string moving on the group with the ends on the **branes** $\mathcal{C}_{\lambda_r}$ and $\mathcal{C}_{\lambda'_r}$, see Fig. 17.

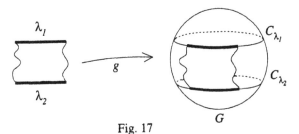

Fig. 17

One may still define an action of the central extension of the loop group in the spaces $\Gamma(\mathcal{L}_{I_r})$ (a single one) and base on its analysis a rigorous construction of the open-string Hilbert spaces of states $\mathcal{H}_{\lambda\lambda'}$, as we did in Sect. 4.1 for the closed-string states, see Eq. (4.10). One obtains

$$\mathcal{H}_{\lambda\lambda'} = \bigoplus_{\widehat{R}} M_{\lambda\lambda'}^{R} \otimes V_{\widehat{R}} . \tag{7.13}$$

The multiplicity spaces may be naturally identified with the spaces of the genus zero CS states $\mathcal{W}(\mathbf{C}P^1, \underline{\xi}, \underline{R})$ with three insertion points in representations $\overline{R}_\lambda$, $R_{\lambda'}$ and R. In particular, the dimension of the multiplicity spaces is given by the fusion ring structure constants $\widetilde{N}_{\overline{R}_\lambda R_{\lambda'}}^{R}$. The spaces $\mathcal{H}_{\lambda\lambda'}$ carry the obvious action of the current algebra $\widehat{\mathbf{g}}$ and of the Virasoro algebra, the latter obtained by the Sugawara construction (4.15). The generator $L_0 - \frac{c}{24}$ gives the Hamiltonian of the open string sectors. The spaces $\mathcal{H}_{\lambda\lambda}$ with the same BC on both sides contain the vacua Ω_λ, i.e. the states annihilated by L_0 (unique up to normalization).

7.4 Elementary quantum amplitudes

The quantum amplitudes with the general boundary conditions are given now by the formal functional integrals.

$$\mathcal{A}_{\Sigma, \underline{L}, \underline{\lambda}}(A) = \int_{g(I_r) \subset \mathcal{C}_{\lambda_r}} e^{-S_\Sigma(g, A)} Dg \tag{7.14}$$

and, should take values in the space $\underset{\text{free } I_r}{\otimes} \mathcal{H}_{\lambda_r \lambda'_r}$. They should possess a gluing property along free boundary intervals with opposite boundary weight assignment, generalizing the gluing properties (4.6) or (4.7). As discussed in detail by Segal in [5] for the closed string sector, the general amplitudes may be constructed by gluing from the elementary ones for the geometries listed on Fig. 18.

49

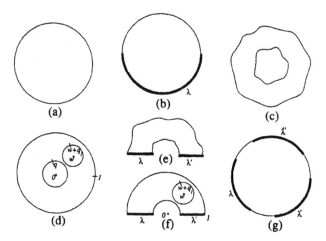

Fig. 18

The elementary amplitudes (a) and (b) represent, respectively, the vacuum state Ω in the closed string space $\mathcal{H}$, and the vacua Ω_λ in the open string spaces $\mathcal{H}_{\lambda\lambda}$. The amplitudes (c) for arbitrary annuli encode the action of the pair of Virasoro algebras in $\mathcal{H}$. In particular, for a complex number $q \neq 0$ inside the unit disc one may consider the annular regions $A_q = \{\, z \mid |q| \leq |z| \leq 1 \,\}$, see Fig. 19, obtained from $\mathbf{CP}^1$ by taking out the unit discs embedded by the maps $z \mapsto qz$ and $z \mapsto z^{-1}$.

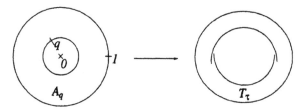

Fig. 19

Viewing the amplitude of A_q as an operator from the space $\mathcal{H}$ associated to the first boundary component to $\mathcal{H}$ associated to the second boundary, one has:

$$\mathcal{Z}_{A_q}^{-1}\, \mathcal{A}_{A_q} \;=\; q^{L_0}\, \tilde{q}^{\tilde{L}_0}\,.$$

The gluing of the two boundary circles of A_q leads to the complex torus T_τ where

$q = e^{2\pi i \tau}$. According to the gluing relation (4.7), this produces the toroidal partition function

$$\mathcal{Z}(T_\tau) \;=\; \mathcal{Z}_{A_q} \; \mathrm{tr}_{\mathcal{H}} \left(q^{L_0} \bar{q}^{\tilde{L}_0} \right). \tag{7.15}$$

Upon choosing a flat metric on T_τ and working out the partition function[11] $\mathcal{Z}_{A_q}$, one obtains finally

$$\mathcal{Z}(\tau) \;=\; \mathrm{tr}_{\mathcal{H}} \; q^{L_0 - \frac{c}{24}} \, \bar{q}^{\tilde{L}_0 - \frac{c}{24}} \tag{7.16}$$

which is nothing else but Eq. (5.20).

The amplitude for a disc P_{w,q,q_1} with two round holes, as in Fig. 18(d), gives rise to a 3-linear form on $\mathcal{H}$ which may be also viewed as an operator from the space $\mathcal{H} \otimes \mathcal{H}$ associated to the inner discs to $\mathcal{H}$ corresponding to the outer one. It is customary in CFT to rewrite this amplitude as an operator $\Phi(e; w)$ in $\mathcal{H}$ labeled by the vectors e in (a dense subspace of) $\mathcal{H}$ and the point w inside the unit disc:

$$\Phi(e; w)\, e' \;=\; \mathcal{Z}_{P_{w,q,q_1}} \; A_{P_{w,q,q_1}} \; (q^{-L_0} \bar{q}^{-\tilde{L}_0} e') \otimes (q_1^{-L_0} \bar{q}_1^{-\tilde{L}_0} e). \tag{7.17}$$

The combination with the powers of L_0 and $\tilde{L}_0$ assures the independence of the expression of q and q_1. The vectors e can be recovered from the operators $\Phi(e; w)$ by acting with them on the vacuum vector

$$\lim_{w \to 0} \Phi(e; w)\, \Omega \;=\; e. \tag{7.18}$$

Pictorially, this corresponds to filling up the central whole of P_{w,q,q_1} by gluing a disc to its boundary. The operators $\Phi(e; w)$ satisfy an important relation:

$$\Phi(e; z)\, \Phi(e'; w) \;=\; \Phi(\Phi(e; z - w)\, e'; w). \tag{7.19}$$

The above identity holds for $0 < |w| < |z|$ and $0 < |z - w| < 1$. It results from the two ways that one may obtain the disc with three holes by gluing two discs with two holes, see Fig. 20.

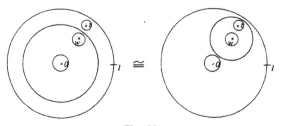

Fig. 20

---

[11] $\mathcal{Z}_{A_q}$ is a ratio of two partition function on the Riemann sphere and may be easily found from the relation (4.17) to be equal to $|q|^{-\frac{c}{12}}$

The relation (7.19) may be viewed as a global form of the operator product expansion. The local forms may be extracted from it by expanding the vector $\Phi(e; z - w)$ into terms homogeneous in $(z - w)$. In particular, for specially chosen vectors e and e' one obtains the operator versions of the relations (4.25) and (4.31), hence the name of the latter. The vector-operator correspondence together with the operator product expansion (7.19) are the cornerstones of the non-perturbative approach to CFT.

The amplitudes corresponding to the surfaces with boundary of Fig. 18(e) represent the action of the Virasoro algebra in the open string spaces $\mathcal{H}_{\lambda\lambda'}$. The surfaces (f) give rise, in turn, to the amplitudes which, applied to vectors $q_1^{-L_0}\bar{q}_1^{-\bar{L}_0}e \in \mathcal{H}$, define the action of the closed string sector fields $\Phi(e, w)$ in the open string spaces $\mathcal{H}_{\lambda\lambda'}$. Finally, the amplitudes of the disc (g) with three labeled and three free boundary intervals define 3-linear forms on the corresponding open string spaces. As before in the closed string sector, one may interpret them in terms of boundary operators labeled by vectors in, say, $\mathcal{H}_{\lambda''\lambda}$ and mapping from $\mathcal{H}_{\lambda\lambda'}$ to $\mathcal{H}_{\lambda'\lambda''}$.

The gluing properties give rise to non-trivial relations between various amplitudes. For example, gluing along the free sides a rectangle with a local BC imposed on the two other sides, see Fig. 21, one obtain a finite cylinder Z_L with the BC's imposed on the boundary components.

Fig. 21

Its amplitude $\mathcal{A}_{Z_L}$ (in the flat metric) may be computed in two ways. On one hand side, using the decomposition (7.13), we infer that

$$\mathcal{A}_{Z_L} = \text{tr}_{\mathcal{H}_{\lambda\lambda'}} q^{L_0 - \frac{c}{24}} = \sum_{\widehat{R}} \widehat{N}_{\widehat{R}_\lambda R_{\lambda'}}^{\widehat{R}} \, \text{tr}_{V_{\widehat{R}}} q^{L_0 - \frac{c}{24}} = \sum_{\widehat{R}} \widehat{N}_{\widehat{R}_\lambda R_{\lambda'}}^{\widehat{R}} \, \chi_{\widehat{R}}(\tau, 1) \qquad (7.20)$$

with $\tau = \frac{Li}{2\pi}$ and $q = e^{2\pi i \tau}$. On the other hand, we may express this amplitude as a matrix element of the close string amplitude between the boundary states $\widehat{\delta}_\lambda$ and $\widehat{\delta}_{\lambda'}$. With $q' = e^{2\pi i \tau'}$ and $\tau' = -\frac{1}{\tau} = \frac{2\pi i}{L}$, we obtain

$$\mathcal{A}_{Z_L} = \left(\widehat{\delta}_\lambda, \, (q')^{\frac{1}{2}(L_0 - \frac{c}{24})} (\overline{q'})^{\frac{1}{2}(\bar{L}_0 - \frac{c}{24})} \widehat{\delta}_{\lambda'} \right). \qquad (7.21)$$

Upon the substitution of Cardy's expression (7.7) for the boundary states $\widehat{\delta}_\lambda$, this becomes

$$\mathcal{A}_{Z_L} = \sum_{\widehat{R}, \widehat{R'}} (S_R^1)^{-\frac{1}{2}} S_R^{\widehat{R}_\lambda} (S_{R'}^1)^{-\frac{1}{2}} S_{R'}^{R_{\lambda'}} \left(e_{\widehat{R}}^{\widehat{i}} \otimes \overline{e_{\widehat{R}}^i}, \, (q')^{\frac{1}{2}(L_0 - \frac{c}{24})} (\overline{q'})^{\frac{1}{2}(\bar{L}_0 - \frac{c}{24})} e_{\widehat{R'}}^{\widehat{i'}} \otimes \overline{e_{\widehat{R'}}^{i'}} \right)$$

52

$$
= \sum_{\widehat{R'}} (S^1_{R'})^{-1} S^{\overline{R}_\lambda}_{R'} S^{R_{\lambda'}}_{R'} \left(e^{\widehat{i}}_{\widehat{R'}}, (q')^{\frac{1}{2}(L_0 - \frac{c}{24})} e^{\widehat{i'}}_{\widehat{R'}} \right) \left(e^{\widehat{i'}}_{\widehat{R'}}, (\overline{q'})^{\frac{1}{2}(L_0 - \frac{c}{24})} e^{\widehat{i}}_{\widehat{R'}} \right)
$$

$$
= \sum_{\widehat{R'}} (S^1_{R'})^{-1} S^{\overline{R}_\lambda}_{R'} S^{R_{\lambda'}}_{R'} \; \mathrm{tr}_{V_{\widehat{R'}}} (q')^{L_0 - \frac{c}{24}} = \sum_{\widehat{R'}} (S^1_{R'})^{-1} S^{\overline{R}_\lambda}_{R'} S^{R_{\lambda'}}_{R'} \; \chi_{\widehat{R'}}(\tau', 1) . \quad (7.22)
$$

With the use of the modular transformation property (5.7), we finally obtain:

$$
\mathcal{A}_{Z_L} = \sum_{\widehat{R}, \widehat{R'}} (S^1_{R'})^{-1} S^{\overline{R}_\lambda}_{R'} S^{R_{\lambda'}}_{R'} S^R_{R'} \; \chi_{\widehat{R}}(\tau, 1) .
$$

By virtue of the Verlinde formula (5.12), the last identity coincides with Eq. (7.20). We have, in fact, inverted here the logic of reference [26], where the consistency of the two ways of computing the amplitude $\mathcal{A}_{Z_L}$ was used to obtain the expression (7.7) for the boundary states $\widehat{\delta}_\lambda$.

The whole system of elementary amplitudes represents an intriguing algebraic structure which is common to all (rational) boundary CFT's. Already the case of boundary topological field theories, where the amplitudes depend only on the surface topology, leads to an interesting construction that remains to be fully understood. It entangles a commutative algebra structure on the closed string space of states and a non-commutative algebroid in the open string sector. An example of such a structure was inherent in the work of Kontsevich [37] on the deformation quantization of general Poisson manifolds, see [38]. Certainly, the two-dimensional CFT did not unveal yet all of its secrets.

References

[1] A. A. Belavin, A. M. Polyakov, A. B. Zamolodchikov, *Infinite conformal symmetry in two-dimensional quantum field theory*, Nucl. Phys. B **241**, 333-380 (1984)

[2] E. Witten, *Non-abelian bosonization in two dimensions*, Commun. Math. Phys. **92**, 455-472 (1984)

[3] V. Knizhnik, A. B. Zamolodchikov, *Current algebra and Wess-Zumino model in two dimensions*, Nucl. Phys. B **247**, 83-103 (1984)

[4] D. Gepner, E. Witten, *String theory on group manifolds*, Nucl. Phys. B **278**, 493-549 (1986)

[5] G. Segal, *Two-dimensional conformal field theories and modular functors*, in: IXth International Congress on Mathematical Physics (Swansee 1988), Hilger, Bristol 1989, pp. 22-37

[6] G. Felder, K. Gawędzki, A. Kupiainen, *Spectra of Wess-Zumino-Witten models with arbitrary simple groups*, Commun. Math. Phys. **117**, 127-158 (1988)

[7] E. Witten, *Quantum field theory and the Jones polynomial*, Commun. Math. Phys. **121**, 351-399 (1989)

[8] E. Verlinde, *Fusion rules and modular transformations in 2D conformal field theory*. Nucl. Phys. B **300** [FS **22**], 360-376 (1988)

[9] K. Gawędzki, *Quadrature of conformal field theories*, Nucl. Phys. B **328**, 733-752 (1989)

[10] E. Witten, *On holomorphic factorization of WZW and coset models*, Commun. Math. Phys. **144**, 189-212 (1992)

[11] D. Bernard, *On the Wess-Zumino-Witten models on the torus*, Nucl. Phys. B **303** 77-93 (1988)

[12] D. Bernard, *On the Wess-Zumino-Witten models on Riemann surfaces*, Nucl. Phys. B **309** 145-174 (1988)

[13] P. Christe, R. Flume, *The four-point correlations of all primary operators of the d=2 conformal invariant SU(2) σ-model with Wess-Zumino term*, Nucl. Phys. B **282** 466-494 (1987)

[14] Bernard, D., Felder, G., *Fock representations and BRST cohomology in SL(2) current algebra*, Commun. Math. Phys. **127**, 145-168 (1990)

[15] P. Goddard, A. Kent, D. Olive, *Virasoro algebras and coset space models*, Phys. Lett. B **152**, 88-92 (1985)

[16] P. Goddard, A. Kent, D. Olive, *Unitary representations of the Virasoro and super-Virasoro algebras*, Commun. Math. Phys. **103**, 105-119 (1986)

[17] K. Bardakçi, E. Rabinovici, B. Säring, *String models with c < 1 components*, Nucl. Phys. B **299**, 151-182 (1988)

[18] K. Gawędzki, A. Kupiainen, *G/H conformal field theory from gauged WZW model*, Phys. Lett. B **215**, 119-123 (1988)

[19] D. Karabali, Q. Park, H. J. Schnitzer, Z. Yang, *A GKO construction based on a path integral formulation of gauged Wess-Zumino-Witten actions*, Phys. Lett. B **216**, 307-312 (1989)

[20] K. Gawędzki, A. Kupiainen, *Coset construction from functional integrals*, Nucl. Phys. B **320**, 625-668 (1989)

[21] K. Gawędzki, P. Tran-Ngoc-Bich, *Hitchin systems at low genera*, IHES/P/98/21 preprint, hep-th/9803101

[22] I. B. Frenkel, N. Yu. Reshetikhin, *Quantum affine algebras and holomorphic difference equations*, Commun. Math. Phys. **146**, 1-60 (1992)

[23] M. Kontsevich, *Vassiliev's knot invariants*, Adv. Sov. Math. **16**, 137-150 (1993)

[24] G. Felder, C. Wieczerkowski, *Topological representations of the quantum group* $U_q(sl_2)$, Commun. Math. Phys. **138**, 583-605 (1991)

[25] V. G. Drinfel'd, *Quasi-Hopf algebras*, Leningr. Math. J. **1**, 1419-1457 (1990)

[26] J. L. Cardy, *Boundary conditions, fusion rules and the Verlinde formula*, Nucl. Phys. B **324**, 581-598 (1989)

[27] J. L. Cardy, D. C. Lewellen, *Bulk and boundary operators in conformal field theory*, Phys. Lett. B **259**, 274-278 (1991)

[28] I. Affleck, *Conformal field theory approach to the Kondo effect*, Acta Phys. Polon. B **26**, 1869-1932 (1995)

[29] M. B. Green, J. H. Schwarz, E. Witten, *Superstring Theory*, Cambridge University Press, Cambridge 1987

[30] J. Polchinski, *TASI lectures on D-branes*, hep-th/9611050

[31] A. Recknagel, V. Schomerus, *Moduli spaces of D-branes in CFT-backgrounds*, hep-th/9903139

[32] C. Klimčik, P. Ševera, *Open strings and D-branes in WZNW model*, Nucl. Phys. B **488**, 653-676 (1997)

[33] A. Alekseev, V. Schomerus, *D-branes in the WZW model*, hep-th/9812193

[34] M. Kato, T. Okada, *D-branes on group manifolds*, Nucl. Phys. B **499**, 583-595 (1997)

[35] N. Ishibashi, *The boundary and crosscap states in conformal field theories*, Mod. Phys. Lett. A **4**, 251-263 (1989)

[36] K. Gawędzki, *Topological actions in two-dimensional quantum field theory*, in: Non-perturbative quantum field theory, eds. G. 't Hooft, A. Jaffe, G. Mack, P. Mitter ,R. Stora, Plenum, New York 1988, pp. 101-142

[37] M. Kontsevich, *Deformation quantization of Poisson manifolds, I*, q-alg/9709040

[38] A. S. Cattaneo, G. Felder, *Path integral approach to the Kontsevich quantization formula*, math/9902090

Non-Perturbative Dynamics Of Four-Dimensional Supersymmetric Field Theories

Philip C. Argyres

Newman Laboratory, Cornell University, Ithaca NY 14853
argyres@mail.lns.cornell.edu

Abstract

An introduction to the construction and interpretation of supersymmetric low energy effective actions in four space-time dimensions is given. These effective actions are used to extract exact strong-coupling information about $N = 4$ and $N = 2$ supersymmetric gauge theories. The M-theory 5-brane construction which derives the effective action of certain $N = 2$ theories is described.

1   Introduction

The aim of these lectures is to introduce some of the arguments that have been used successfully in the last five years to obtain exact information about strongly coupled field theories. I will focus on four-dimensional field theories without gravity, although the techniques described here have been applied to theories in other dimensions and to string/M theory as well. I will also focus on theories with at least $N = 2$ supersymmetry (8 conserved supercharges) since these are physically rich theories with many open problems, but are still highly constrained by the symmetry.

The basic notion is that of a *low energy* (or *Wilsonian*) *effective action*. This is simply a local action describing a theory's degrees of freedom at energies below a given scale E. An example is the low energy effective action for QCD, chiral perturbation theory describing the interactions of pions at energies $E < \Lambda_{QCD}$. In such a theory particles heavier than Λ_{QCD} are included in the pion theory as classical sources. Other examples are the various ten and eleven-dimensional supergravity theories, which appear as effective actions for string/M theory at energies below their Planck scales. The effective action is obtained by averaging over (integrating out) the short distance fluctuations of the theory. If there is a sufficiently small ratio E/Λ between the cutoff energy scale E and the energy scale Λ characteristic of the dynamics of the degrees of freedom being averaged over, renormalization group arguments imply that the effective action can be systematically expanded as a power series in E/Λ—essentially an expansion in the number of derivatives of the fields.

We will use low energy effective actions to analyze four dimensional field theories by taking the limit as the cutoff energy scale E goes to zero, or equivalently, by just keeping the leading terms (up to two derivatives) in the low energy fields. I will call such $E \to 0$ low energy effective actions *infrared effective actions* (IREAs). The idea is to guess an IR effective field content for the microscopic (UV) theory in question and write down all possible IREAs built from these fields consistent with the global symmetries of the UV theory. For a "generic" UV theory this is no better than doing chiral perturbation theory for QCD, and would seem to give little advantage for obtaining exact results. However, if the theory has a continuous set of inequivalent vacua, it turns out that selection rules from global symmetries of the UV theory can sometimes constrain the IREA sufficiently to deduce exact results. There are a number of reviews deriving these exact results [1] assuming the constraints from supersymmetry. In particular, these lectures are a continuation of [2], where the construction of four-dimensional IREAs is explained in a relatively non-technical way.

1

We start in Sec. 2 with a brief review of IREAs with various amounts of supersymmetry. The constraints on the IREAs become progressively more restrictive as the number of supersymmetries is increased. In the $N = 2$ case they are strong enough to allow quite general and restrictive properties of the moduli space of vacua of gauge theories to be deduced. The remainder of these lectures is devoted to using these IREAs to extract exact strong-coupling information about supersymmetric gauge theories. In particular, Sec. 3 discusses the exact IREAs of $N = 4$ theories, while Sec. 4 discusses some general things that can be said about those of $N = 2$ theories (Seiberg-Witten theory [3]). Sec. 5 will elaborate on the mathematical formulation of $N = 2$ IREAs in preparation for Sec. 6 which presents an account of an M-theory 5-brane construction [4] which allows one to derive the IREA of certain $N = 2$ theories.

Important topics omitted include the properties of interacting IREAs—the representation theory of superconformal algebras [5] and their use in analyzing IREAs [6]; instead these lectures concentrate on IR free effective actions. Also missing are details of supersymmetry algebras and the construction of their representations—many good texts and review articles cover this material [7]—or the application of the ideas presented here to theories in other dimensions [8].

2   IREAs

Since an IREA describes physics only for arbitrarily low energies, it is, by definition, scale invariant: we simply take the cutoff scale E below any finite scale in the theory. Scale invariant theories and therefore IREAs fall into one of the following categories:

Trivial theories in which all fields are massive, so there are no propagating degrees of freedom in the far IR.

Free theories in which all massless fields are non-interacting in the far IR. (They can still couple to massive sources, but these sources should not be treated dynamically in the IREA.)

Interacting theories of massless degrees of freedom which are usually assumed to be conformal field theories [9].

We generally have no effective description of interacting conformal field theories in four dimensions [10] so we must limit ourselves to free or trivial theories in the IR. A large class of these is given by the Coleman-Gross theorem [11] which states that for small enough couplings any theory of scalars, spinors, and $U(1)$ vectors in four dimensions flows in the IR to a free theory. We thus take the field content of our IREA to be a collection of real scalars ϕ^i, Weyl spinors ψ^a_α, and $U(1)$ vector fields A^I_μ. Here α and μ are the space-time spinor and vector indices, while i, a, and I label the different field species.

Since this theory is free in the IR, no interesting dynamics involving the spinor fields can occur, so the vacuum structure of this theory is governed by the scalar potential. Dropping the other fields we write the general Lagrangian with up to two derivatives for a set of real scalars

$$\mathcal{L} = -V(\phi) + \tfrac{1}{2}g_{ij}(\phi)\partial_\mu\phi^i\partial^\mu\phi^j. \tag{1}$$

Here the potential V is an arbitrary real function of the ϕ^i which is bounded below (for stability), while the coefficient g_{ij} of the generalized kinetic term is a real, symmetric and positive definite tensor (for unitarity). We assume V attains its minimum value, which without loss of generality we take to be $V = 0$.

Minimizing the generalized kinetic energy term implies that in the vacuum the scalars should all be constant. Denoting these constant values by the same symbols as for the fields, the set of all possible vacua is then seen to naturally have the structure of a Riemannian manifold $\mathcal{M}_0 = \{\phi^i\}$ with metric g_{ij} since an arbitrary non-singular field redefinition $\phi^i \rightarrow \tilde{\phi}^i(\phi)$ transforms g_{ij} in the same way as a metric transforms under a change of coordinates.

If $V = 0$ identically, then $\mathcal{M}_0$ would describe a manifold of vacua of this theory. We call such a manifold of vacua the *moduli space* of the theory. Without any extra symmetries to constrain it, generically $V \neq 0$, so $\mathcal{M}_0$ is not the moduli space, but instead $\mathcal{M}_V = \mathcal{M}_0/\{V = 0\}$ is. At least locally $\mathcal{M}_V$ has the structure of a submanifold of $\mathcal{M}_0$.

Now let us incorporate the $U(1)$ gauge fields into our discussion of the moduli space. Some of the scalar fields may be charged under the $U(1)^n$ gauge group of the IREA. The infinitesimal $U(1)^n$ action of the gauge group on the scalars then generates a diffeomorphism of $\mathcal{M}_0$. For this to be a symmetry of the IREA it must both leave V invariant and be an isometry of the metric g_{ij}. In that case the IREA can be written (excluding the spinors) as

$$\mathcal{L} = -V(\phi) + \tfrac{1}{2}g_{ij}(\phi)D_\mu\phi^i D^\mu\phi^j - \frac{1}{16\pi}\mathrm{Im}\left[\tau_{IJ}(\phi)\mathcal{F}^I \wedge *\mathcal{F}^J\right], \tag{2}$$

where $D_\mu = \partial_\mu + A^I_\mu\xi_I$, treating the ξ^i_I as Killing vectors generating the isometry. The last term in Eq. 2 is a generalized Maxwell term for the $U(1)$ field strengths $F^I_{\mu\nu} = \partial_\mu A^I_\nu - \partial_\nu A^I_\mu$, where we have defined

$$\mathcal{F}^I = F^I - i*F^I \tag{3}$$

3

in terms of 2-form field strengths and the Hodge star operator $*F_{\mu\nu} = \frac{1}{2}\epsilon_{\mu\nu\rho\sigma}F^{\rho\sigma}$. τ_{IJ} is a complex (gauge invariant) function of the ϕ^i symmetric in I and J and whose imaginary part is positive definite (for unitarity). Defining the real and imaginary parts of the couplings by

$$\tau_{IJ} = \frac{\theta_{IJ}}{2\pi} + i\frac{4\pi}{(e^2)_{IJ}}, \tag{4}$$

the generalized Maxwell term can be expanded as

$$\mathcal{L}_{U(1)} = -\frac{1}{2(e^2)_{IJ}}F^I \wedge *F^J + \frac{\theta_{IJ}}{64\pi^2}F^I \wedge F^J, \tag{5}$$

showing that the imaginary part of τ_{IJ} is a matrix of couplings and the real part are theta angles.

The addition of the $U(1)$ gauge fields affects the moduli space because two points of $\mathcal{M}_0$ which are related by a gauge transformation must be identified. Thus $\mathcal{M}_0$ or $\mathcal{M}_V$ (since V is gauge invariant) is replaced by $\mathcal{M}$, formed by dividing by the action of the gauged isometry group $U(1)^n$: $\mathcal{M} = \mathcal{M}_V/U(1)^n$.

Note that the vacuum expectation values (vevs) of charged scalars can not parameterize the moduli space, because when a charged scalar gets a nonzero vev it Higgses the $U(1)$ it is charged under and thereby gets a mass. It is therefore not a flat direction—*i.e.* changing its vev takes us off the moduli space $\mathcal{M}$. Since we are interested only in the extreme IR limit, we only need to keep the *neutral* scalars which parameterize $\mathcal{M}$. In this case the IREA (2) simplifies since $V = 0$ on $\mathcal{M}$ by definition and $D_\mu = \partial_\mu$ on neutral scalars. Thus only the metric $g_{ij}(\phi)$ and couplings $\tau_{IJ}(\phi)$ need to be specified. (If we included the fermions, there would also be the coefficient functions of their kinetic terms as well.)

The IR free low energy $U(1)^n$ dynamics is form-invariant under *electric-magnetic duality* transformations. These are simply relabellings of the fields, interchanging electric and magnetic fields and charges, and, because of the Dirac quantization condition relating electric and magnetic charges [12], also inverting the couplings $\tau_{IJ} \to -\tau^{IJ}$, where τ^{IJ} is the matrix inverse of τ_{IJ}: $\tau^{IJ}\tau_{JK} = \delta^I_K$. This electric-magnetic duality transformation together with the invariance of the physics under 2π shifts of the theta angles (integer shifts of Reτ_{IJ}) $\tau_{IJ} \to \tau_{IJ} + \delta^K_I\delta^L_J + \delta^L_I\delta^K_J$, generate a discrete group of duality transformations:

$$\tau_{IJ} \to (A_I{}^L\tau_{LM} + B_{IM})(C^{JN}\tau_{NM} + D^J{}_M)^{-1}, \tag{6}$$

where

$$M \equiv \begin{pmatrix} A & B \\ C & D \end{pmatrix} \in Sp(2n, \mathbf{Z}). \tag{7}$$

The conditions on the $n \times n$ integer matrices A, B, C, and D for M to be in $Sp(2n, \mathbf{Z})$ are

$$
\begin{aligned}
AB^T = B^T A, \qquad & B^T D = D^T B, \\
A^T C = C^T A, \qquad & D^T C = C D^T, \\
A^T D - C^T B = \;& A D^T - B C^T = 1,
\end{aligned}
\tag{8}
$$

and imply that

$$
M^{-1} = \begin{pmatrix} D^T & -B^T \\ -C^T & A^T \end{pmatrix}.
\tag{9}
$$

The action of an electric-magnetic duality transformation on the $2n$-component row vector of magetic and electric charges $(n_m^I, n_{e,J})$ of massive states is

$$
(n_m \; n_e) \to (n_m \; n_e) \cdot M^{-1}.
\tag{10}
$$

Electric-magnetic duality transformations are not symmetry transformations since they acts on the couplings. Instead, electric-magnetic duality simply expresses the equivalence of free $U(1)$ field theories coupled to classical (massive) sources under $Sp(2n, \mathbf{Z})$ redefinitions of electric and magnetic charges. The importance of this redundancy in the Lagrangian description of IREAs becomes apparent when there is a moduli space $\mathcal{M}$ of inequivalent vacua. In that case, upon traversing a closed loop in $\mathcal{M}$ the physics must, by definition, be the same at the beginning and end of the loop, but the Lagrangian description need not—it may have suffered an electric-magnetic duality transformation. This possibility is often expressed by saying that the coupling matrix τ_{IJ}, in addition to being symmetric and having positive definite imaginary part, is also a section of a (flat) $Sp(2n, \mathbf{Z})$ bundle with action given by (6).

2.1 N=2 Supersymmetric IREAs

The basic (no central charges) $N = 2$ superalgebra is, in an indexless notation,

$$
\{Q_m, \overline{Q}_n\} = \delta_{mn} P, \qquad \{Q_m, Q_n\} = 0, \qquad m, n = 1, 2, \cdot
\tag{11}
$$

where Q_m are two Weyl spinor supercharges, and P is the energy-momentum vector. Note that the $N = 2$ algebra has an $SU(2)_R$ group of automorphisms under which Q_m transforms as a doublet. (Global symmetries under which the supercharges transform are called R *symmetries*.)

On shell irreducible representations of (11) are easy to construct. There are two solutions with no spins greater than one: the *hypermultiplet*, containing two propagating complex scalars, ϕ and $\tilde{\phi}$, as well as two Weyl fermions, ψ and $\tilde{\psi}$; and the *vector multiplet*, made from one complex scalar a, two Weyl spinors λ and $\tilde{\lambda}$, and a vector field A_μ. An important distinguishing feature of the hypermultiplet is that its scalars form a complex $SU(2)_R$ doublet. The bosonic degrees of freedom in a vector multiplet, by contrast, are a single complex scalar and a vector field, both transforming in the adjoint of the gauge group, and both singlets under $SU(2)_R$. In particular, in the case of $U(1)^n$ gauge group, which we are interested in for describing IREAs, the vector multiplet scalars are necessarily neutral.

An $N = 2$ IREA with Abelian gauge group and neutral hypermultiplets, *a priori* has an action of the form (2) where the ϕ^i fields run over all the bosons (in both the vector and hypermultiplets), and $\mathcal{F}^I$ run over the $U(1)$ gauge fields. Compatibility with the $N = 2$ global supersymmetry tightly constrains this action; see, for example, [2]. The result is that the general $N = 2$ IREA gauge group $U(1)^n$ (labelled by indices $I, J = 1, \ldots, n$) and n_f neutral hypermultiplets (labelled by indices $i, j = 1, \ldots, n_f$) has the form

$$\mathcal{L} = g_{i\bar{j}}(\phi, \tilde{\phi}) \left(\partial \phi^i \cdot \partial \overline{\phi}^{\bar{j}} + \partial \tilde{\phi}^i \cdot \partial \overline{\tilde{\phi}}^{\bar{j}} \right) + \mathrm{Im} \tau_{IJ}(a) \left(\partial a^I \cdot \partial \overline{a}^J + \mathcal{F}^I \cdot \mathcal{F}^J \right), \qquad (12)$$

with

$$\partial_{[I} \tau_{J]K} = 0, \qquad (13)$$

and τ_{IJ} a holomorphic function (really $Sp(2n, \mathbf{Z})$ section) of the vector multiplet scalars a^I.

This form of the IREA of $N = 2$ supersymmetric theories has many important consequences. The first is the absence of any potential terms for the scalars which implies that in $N = 2$ theories there will be a moduli space of vacua as long as $U(1)$ vector multiplets or neutral hypermultiplets can be shown to occur in the IREA.

The next $N = 2$ selection rule follows from the fact that there are no kinetic cross terms between the vector and hypermultiplets, implying that the moduli space has a natural (local) product structure $\mathcal{M} = \mathcal{M}_H \times \mathcal{M}_V$, where $\mathcal{M}_H$ is the subspace of $\mathcal{M}$ along which only the hypermultiplet vevs vary while the vector multiplet vevs remain fixed, and *vice versa* for $\mathcal{M}_V$. In cases where $\mathcal{M}_V$ is trivial (a point), $\mathcal{M} = \mathcal{M}_H$ is called a *Higgs branch* of the moduli space; when $\mathcal{M}_H$ is trivial $\mathcal{M}_V$ is called the *Coulomb branch* (since there are always the massless $U(1)$ vector bosons from the vector multiplets). Cases where both $\mathcal{M}_H$ and $\mathcal{M}_V$ are non-trivial are called *mixed branches*.

6

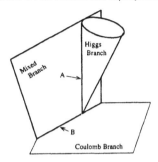

Figure 1: Cartoon of a classical $N = 2$ moduli space. The Higgs and mixed branches intersect along a Higgs submanifold A, while the mixed branch intersects the Coulomb branch along a Coulomb submanifold B.

In general the total moduli space of a given theory need not be a smooth manifold—it may have "jumps" where submanifolds of different dimensions meet. Classically this occurs as a result of the Higgs mechanism: a charged scalar vev Higgses some vector multiplets, typically lifting them (making them massive). But at the special point where the charged vev is zero, the vector multiplets become massless, leading to extra flat directions and a jump in the dimensionality of the moduli space. Hence, at least classically, the general picture of an $N = 2$ moduli space is a collection of intersecting manifolds, which can be Higgs, Coulomb, or mixed branches [16, 17], see Fig. 1.

This classical picture is, of course, modified quantum mechanically. A microscopic (UV) theory is characterized by some parameters (*e.g.* masses, strong coupling scales, theta angles, dimensionless couplings); we can always take ratios of these parameters to describe them by at most one scale Λ and a set of dimensionless parameters λ_k. The coefficient functions g_{ij} and τ_{IJ} of the IREA will, in general, depend on Λ and the λ_k. Determining this dependence of these IR quantities on UV parameters is the ultimate goal of the techniques reviewed in these lectures.

For asymptotically free gauge theories, the important UV parameter is the (complex) strong coupling scale of the theory, Λ (whose definition we'll recall in Section 4, below). The important property of asymptotically free theories is that they are nearly free at energy scales above Λ, so the classical theory is obtained in the limit $\Lambda \to 0$. Since Λ appears in τ_{IJ} (at, say, one loop), it appears in the Lagrangian in the same way a scalar vev a^I of an $N = 2$ vector multiplet would. Therefore, we can think of Λ as a background $U(1)$ vector superfield—in other words it is consistent to *assign* Λ

7

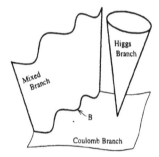

Figure 2: Cartoon of a quantum $N = 2$ moduli space. The Higgs branch and the Higgs (hypermultiplet vev) directions of the mixed branch remain unmodified from their classical geometries, though they may be deformed in the Coulomb (vector multiplet vev) directions. The Coulomb branch is generally different from its classical geometry.

supersymmetry transformation properties as if it were the lowest component of a chiral superfield. This implies that whatever strong dynamics takes place upon flowing to the IR, Λ will only enter the IREA in the way chiral multiplet scalars do. In particular, τ_{IJ} will be a holomorphic function of Λ [18]. Also, since the metric on the Higgs branch is independent of vector superfields, it is independent of Λ. Finally, we can use the fact that the classical theory is obtained in the limit $\Lambda \to 0$ to conclude that *the Higgs branch metric is given exactly by the classical answer* [16]. We thus learn that only the Coulomb branch can receive quantum corrections, and that any mixed branches will retain their classical product structure of a hypermultiplet manifold times the vector multiplet manifold corresponding to the subspace of the Coulomb branch along which the mixed and Coulomb branches intersect; see Fig. 2. Since the hypermultiplet manifolds can be determined classically in $N = 2$ supersymmetric gauge theories, we will not consider them further.

(It is worth examining more closely the logic of this argument. We are *assuming* that the IREA will be described by a nonlinear sigma model of some set of light chiral fields which are not necessarily simply a subset of those of the UV theory. We have no derivation of this hypothesis—we can only test it to see if it gives consistent answers. The couplings of the effective theory will be some functions of the couplings of the microscopic theory, which we would like to solve for. The next step of thinking of the couplings in the superpotential as background chiral superfields is just a trick—we are certainly allowed to do so if we like since the couplings enter in the microscopic theory in the same way a background chiral superfield would. The point of this trick is that

8

it makes the restrictions on possible quantum corrections allowed by supersymmetry apparent. These restrictions are just a supersymmetric version of the familiar "selection rules" of quantum mechanics.)

Finally, a key fact about the Coulomb branch is that though it can be corrected quantum mechanically, it is never wholly lifted in asymptotically free $N = 2$ gauge theories. We will see this when we describe $N = 2$ non-Abelian gauge theories in Section 4, below. This means that $N = 2$ supersymmetric theories *generically* have a moduli space of vacua. The challenge of solving for the vacuum structure of $N = 2$ gauge theories is thus that of determining the geometry on the Coulomb branch. This geometry is encoded in the IREA (12), the integrability condition (13), and the $Sp(2n, \mathbf{Z})$ transformation properties (6), and is known as *rigid special Kähler* geometry. We will develop the mathematics of these manifolds further in Section 5 below. But for the moment, let us move on to $N = 4$ supersymmetry.

2.2   N=4 IREAs

The $N = 4$ superalgebra is

$$\{Q_m, \overline{Q}_n\} = \delta_{mn} P, \qquad \{Q_m, Q_n\} = 0, \qquad m, n = 1, \ldots, 4. \tag{14}$$

This algebra has an $SU(4)_R$ group of automorphisms under which Q_m transforms as a **4**. There is one on-shell irreducible representation with no spins greater than one, which decomposes under an $N = 2$ subalgebra as a vector multiplet plus a hypermultiplet. Its field content can be organized into six real scalars a_i transforming as a **6** of $SU(4)_R$, four Weyl fermions ψ_n in the **4** of $SU(4)_R$, and a vector field A_μ. All these fields must transform in the adjoint of the gauge group since they are in the same multiplet as a vector boson; in the case of $U(1)^n$ gauge group the scalars are necessarily neutral.

The $N = 4$ IREA with Abelian gauge group has the same form as the $N = 2$ IREA (12). But now since the $N = 2$ hypermultiplet and vector multiplet scalars are related by the $SU(4)_R$ global symmetry, they must have the same metric, τ^{IJ}:

$$\mathcal{L} = \mathrm{Im}\tau_{IJ} \left(\partial a_i^I \cdot \partial \overline{a}_i^J + \mathcal{F}^I \cdot \mathcal{F}^J \right). \tag{15}$$

Furthermore, since by the $N = 2$ selection rules, the vector multiplet metric and the hypermultiplet metric cannot depend on the same fields, we must have

$$\tau_{IJ} = \mathrm{constant}. \tag{16}$$

9

This has the immediate consequence that the moduli space of the $N = 4$ theory must locally be flat:

$$\mathcal{M} = \mathbf{R}^{6n} = \{a_i^I\}. \tag{17}$$

3 N=4 Exact Results

We will now turn to our main task of using the supersymmetric IREAs found in the last section to deduce exact non-perturbative information about supersymmetric gauge theories. We start with the $N = 4$ case since it is the most constrained, and so gives the simplest illustration of the basic idea.

Consider an $N = 4$ super Yang-Mills theory. Its (UV) bosonic action is given by

$$\mathcal{L} = \text{Im} \left\{ \tau \text{tr} \left(D_\mu \Phi_i D^\mu \Phi_i + \mathcal{F} \cdot \mathcal{F} + \sum_{i>j} [\Phi_i, \Phi_j]^2 \right) \right\}, \tag{18}$$

where $i, j = 1, \ldots, 6$ for the six adjoint scalars in the $N = 4$ vector multiplet. For definiteness, let us take the gauge group to be $SU(n + 1)$.

Classically, the vacua of this theory occur for Φ_i vevs of the form (up to gauge transformations)

$$\langle \Phi_i \rangle = \begin{pmatrix} \alpha_i^1 & & \\ & \ddots & \\ & & \alpha_i^{n+1} \end{pmatrix}, \tag{19}$$

with

$$\sum_{K=1}^{n+1} \alpha_i^K = 0. \tag{20}$$

This tracelessness condition is required for an $(n + 1) \times (n + 1)$ matrix representation of the adjoint representation of $SU(n + 1)$. By the usual Higgs mechanism, a generic such vev spontaneously breaks $SU(n + 1) \rightarrow U(1)^n$. There are special vacua where two or more of the α_i^K are equal where $SU(n + 1)$ is not completely broken down to $U(1)$'s, but has some $SU(m)$ subgroups left unbroken.

The classical moduli space is thus the flat $6n$-dimensional manifold $\mathcal{M} = \{\alpha_i^I, I = 1, \ldots, n\}$. Actually, choosing the vevs of the form (19) does not completely fix the gauge invariance: the Weyl subgroup of $SU(n + 1)$ acts on the α_i^K by permutations on the K index. Thus the moduli space must be divided out by this S_{n+1} group of permutations, so

$$\mathcal{M} = \mathbf{R}^{6n} / S_{n+1}. \tag{21}$$

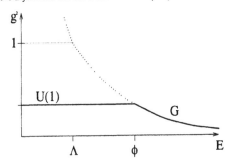

Figure 3: Running of the coupling of an asymptotically free gauge theory with gauge group G Higgsed to $U(1)$'s at a scale $\langle \Phi \rangle \gg \Lambda$. The $U(1)$ couplings do not run below $\langle \Phi \rangle$ because there are no charged fields lighter than ϕ.

The orbifold submanifolds at the fixed points of the S_{n+1} action occur at precisely the places where the low energy $U(1)^n$ gauge group is enhanced.

Now we turn to the quantum mechanical theory. The first question is whether the UV coupling τ (which is classically dimensionless) suffers some renormalization group running, thus generating some strong coupling scale Λ? Our IREA selection rules can immediately rule this out, however. For we have seen that the IR effective $U(1)^n$ coupling τ_{IJ} must be a constant, independent of the values of any of the vevs a_i^K. But if the UV coupling ran at high energies, we would detect this in a vev-dependence of the τ_{IJ}, for at weak enough UV coupling (large vevs in an asymptotically free theory) the classical Higgs mechanism picture of the classical picture can be made arbitrarily precise, implying that the τ_{IJ} will be equal to the value of the microscopic τ at the scale $\langle \Phi \rangle$; see Fig. 3. Since, in fact, the τ_{IJ} are independent of $\langle \Phi \rangle$, so τ must be exactly (even non-perturbatively) independent of scale.

We have thus learned that the $N = 4$ super Yang-Mills theory is a scale invariant, or conformal, field theory. Indeed, it is easy to check at one loop that the beta-function for the running of the gauge coupling vanishes, and can also be verified to all orders in perturbation theory. The form of the $N = 4$ IREA also shows it to be true non-perturbatively.

To make further progress on the quantum vacuum structure of these $N = 4$ theories, consider such theories at weak coupling,

$$\tau \to +i\infty. \tag{22}$$

Then the classical description of the $SU(n + 1) \to U(1)^n$ Higgs mechanism is good,

11

giving an IREA

$$\mathcal{L} = \text{Im}\tau_{IJ}\left(\partial a_i^I \cdot \partial \overline{a}_i^J + \mathcal{F}^I \cdot \mathcal{F}^J\right),$$ (23)

for $I, J = 1, \ldots, n$ where we have defined

$$a_i^I \equiv \sum_{K=1}^{I} \alpha_i^K,$$ (24)

and

$$\tau_{IJ} = \tau \begin{pmatrix} 2 & -1 & & \\ -1 & 2 & -1 & \\ & -1 & 2 & \ddots \\ & & \ddots & \ddots \end{pmatrix},$$ (25)

which is proportional to the Cartan matrix of $SU(n+1)$. Note that the change of basis (24) is integer-valued, $i.e.$ an element of $GL(n, \mathbf{Z})$; this was a necessary restriction in order to preserve the integrality of the magnetic and electric charges (n_m, n_e) of any massive states in the theory.

The moduli space of this theory is just the classical one (21), including dividing by the S_{n+1} Weyl group action on the a_i^I. For example, the simplest case is $SU(2)$ gauge group, where $n = 1$. Then the moduli space is $\mathcal{M} = \mathbf{R}^6/S_2$ where the $S_2 \simeq \mathbf{Z}_2$ acts on the six coordinates a_i as $S_2 : a_i \rightarrow -a_i$. This has a single fixed point at the origin. Thus the vacuum structure is simple: at the origin of moduli space there is a scale invariant vacuum with an unbroken $SU(2)$ gauge invariance, and there is a six dimensional space of flat directions leading away from it where the scale invariance is spontaneously broken by the non-zero a_i vevs and the low energy theory is the $N = 4$ $U(1)^n$ theory (23).

At points in the moduli space where new degrees of freedom (not included in the fields of the IREA) become massless, the IREA description of the physics breaks down. Generally this break down is signalled by a singularity in the metric on the moduli space. In the above example the origin was singular in this way: the $W^{\pm}$ bosons (and their superpartners) filling out the $SU(2)$ adjoint multiplet became massless there, and the metric was singular there (it is a $\mathbf{Z}_2$ orbifold point).

Finally, we can deduce what happens to the vacuum structure for couplings not near weak coupling. Since the low energy τ_{IJ} cannot depend on the vevs, it can only be a function of the UV coupling τ. Treating the IREA (23) as an $N = 2$ supersymmetric action, τ must enter only holomorphically in τ_{IJ}. Furthermore, by the angularity of the theta angle, $i.e.$ the invariance of the physics under $\tau \rightarrow \tau + 1$, τ_{IJ} can depend on

12

τ only as

$$\tau_{IJ} = \tau C_{IJ} + \sum_{\ell=0}^{\infty} C_{IJ}^{(\ell)} e^{2\pi i \ell \tau}, \tag{26}$$

where C_{IJ} is the Cartan matrix appearing in (25), and $C_{IJ}^{(\ell)}$ are arbitrary independent complex matrices. Note that the first term is not invariant under $\tau \to \tau + 1$, but shifts by the interger matrix C_{IJ}. This has no effect on the physics since it is just a low-energy electric-magnetic duality transformation in the $U(1)^n$ theory.

I do not know of a first principles argument to determine the $C_{IJ}^{(\ell)}$'s, but various indirect arguments from string and M theory (mentioned below) imply that they are all proportional to the Cartan matrix C_{IJ}. In that case we have $\tau_{IJ} = f(\tau)C_{IJ}$, with $f(\tau) \to \tau$ as $\tau \to +i\infty$. In particular, $f(\tau)$ differs from τ only by nonperturbative terms. Since we have no alternative non-perturbative definition of the UV coupling τ, *we are free to define $f(\tau)$ itself to be the UV coupling: $f(\tau) = \tau$*. So, finally, the IR $U(1)^n$ couplings are

$$\tau_{IJ} = \tau C_{IJ}. \tag{27}$$

The $Sp(2n, \mathbf{Z})$ electric-magnetic duality transformations (6) include transformations taking $\tau_{IJ} \to \tau'_{IJ}$ such that $\tau \to \tau + 1$ and $\tau \to -1/\tau$. These generate an $SL(2, \mathbf{Z})$ group of transformations on the microscopic coupling which leave the IR physics invariant. This is evidence for the existence of an *S-duality* of the $N = 4$ theories [19, 20] which is simply the statement that the theories with UV couplings τ related by $SL(2, \mathbf{Z})$ transformations are physically equivalent. S-dualities, also known as strong-weak coupling dualities, or Montonen-Olive dualities, and are conceptually distinct from electric-magnetic dualities. It is worth emphasizing that electric-magnetic dualities are equivalences of the free IR effective $U(1)^n$ theories, whereas S-dualities identify interacting theories with *a priori* distinct couplings.

Further evidence for $N = 4$ S-duality comes from the spectrum of BPS states in the theories. BPS states are states preserving some of the supersymmetries. The masses of states preserving half the supersymmetries are known exactly in terms of the *central charges* of the supersymmetry algebra [20]. For example, for $SU(2)$, the mass of a $\frac{1}{2}$-BPS state with magnetic and electric charges n_m and n_e is

$$M^2 = \frac{1}{\mathrm{Im}\tau} |n_e + \tau n_m|^2 (a_i^I C_{IJ} a_i^J). \tag{28}$$

It is easy to check that this formula is invariant under electric-magnetic duality transformations. Furthermore, under S-duality transformations it takes states with given

(n_m, n_e) into ones with different charges. In particular, given that the massive W-bosons of the $N = 4$ theory are $\frac{1}{2}$-BPS states with $(n_m, n_e) = (0, \pm 1)$, if S-duality is correct it follows that there must be also be a $\frac{1}{2}$-BPS state for all relatively prime choices of electric and magnetic charges. Such states have been constructed [21], adding to the evidence for the S-duality.

The best evidence for $N = 4$ S-duality comes from string and M theory. There are a set of similar dualities in these theories which fit together in an intricate way, and also imply the $N = 4$ S-duality. The self-consistency of this "web of dualities" thus lends strong support to the existence of $N = 4$ S-duality. Scale-invariant $N = 2$ theories also have S-dualities [22, 23, 24]. It is striking that, unlike their $N = 4$ counterparts, many of the $N = 2$ dualities can be proved using purely field theoretic arguments [25, 26].

4 Seiberg-Witten Theory

We now turn to deriving the vacuum properties of $N = 2$ supersymmetric gauge theories. Since there are now both hypermultiplets and vector multiplets at our disposal, we can construct a much richer set of $N = 2$ theories than $N = 4$ theories. For simplicity we will focus only on the $N = 2$ Yang-Mills theories, that is, those with only vector multiplets appearing in the microscopic action. The treatment of theories with hypermultiplets does not differ much from the pure Yang-Mills theories, especially as we are primarily interested in the Coulomb branch of the moduli space.

Taking $SU(n + 1)$ as our example again, denote the complex adjoint scalar field of the vector multiplet by Φ, an $(n + 1) \times (n + 1)$ complex traceless matrix. Then the $N = 2$ Yang-Mills action looks much like the $N = 4$ action (18),

$$\mathcal{L} = \text{Im}\left\{\tau \, \text{tr}\left(D_\mu \Phi D^\mu \Phi + [\Phi, \Phi]^2 + \mathcal{F} \cdot \mathcal{F}\right)\right\}. \tag{29}$$

Classically, the vacua of this theory occur for Φ vevs of the form (up to gauge transformations)

$$\langle \Phi \rangle = \begin{pmatrix} a^1 & & \\ & \ddots & \\ & & a^{n+1} \end{pmatrix}, \tag{30}$$

with

$$\sum_{K=1}^{n+1} a^K = 0, \tag{31}$$

and the a^K complex. Such a vev spontaneously breaks $SU(n + 1) \to U(1)^n$ except when two or more of the a^K are equal so that $SU(n + 1)$ is not completely broken down to $U(1)$'s, but has some $SU(m)$ subgroups left unbroken. Choosing the vevs of the form (30) does not completely fix the gauge invariance since the Weyl subgroup $S_{n+1} \subset SU(n+1)$ acts on the a^K by permutations on the K index. The classical moduli space is thus a flat n-complex-dimensional manifold with orbifold singularities

$$\mathcal{M} = \mathbf{C}^n/S_{n+1}. \qquad (32)$$

Gauge-invariant coordinates on this space can be taken to be the n independent complex symmetric polynomials in the a^K:

$$
\begin{aligned}
s_2 &= \sum_{J<K} a^J a^K, \\
s_3 &= \sum_{J<K<L} a^J a^K a^L, \\
&\vdots \\
s_{n+1} &= a^1 a^2 \cdots a^{n+1}. \qquad (33)
\end{aligned}
$$

(Note that there is no s_1 since the sum of the a^K's vanishes by the tracelessness condition.)

Now we turn to the quantum mechanical theory. Unlike the $N = 4$ super Yang-Mills theories, the $N = 2$ theories are asymptotically free and their UV coupling τ (which is classically dimensionless) runs with scale, generating a strong coupling scale Λ. Let us recall how this scale is defined. Consider an asymptotically free gauge theory with kinetic term $-(1/4g_0^2)\mathrm{tr}F^2$ in an effective action at a scale μ_0, with g_0 the coupling at that scale. For g_0 small enough we can calculate with arbitrary accuracy the renormalization group running of the coupling from the one loop result $8\pi^2 g^{-2}(\mu) \simeq -b_0 \log(|\Lambda|/\mu)$, where we have defined $|\Lambda| \equiv \mu_0 e^{-8\pi^2/b_0 g_0^2}$, the strong coupling scale of the gauge group. It is then convenient to introduce a complex "scale" $\Lambda \equiv |\Lambda| e^{i\theta/b_0}$ so that the complex coupling $\tau \equiv (\theta/2\pi) + i(4\pi/g^2) = (b_0/2\pi i) \log(\Lambda/\mu)$ at one loop.

(The coefficient of the one-loop beta function is given by

$$b_0 = \frac{11}{6}T(adj) - \frac{1}{3}\sum_a T(R_a) - \frac{1}{12}\sum_i T(R_i) \qquad (34)$$

where the indices a run over Weyl fermions in representations R_a of the gauge group, and i runs over real scalars in the representations R_i. $T(R)$ is the index of the representation R; for $SU(n+1)$, for example, the index of the fundamental representation is 1, and of the adjoint representation is $2(n+1)$. For the $N = 2$ Yang-Mills theory, all

15

fields are in the adjoint representation, and we have one complex scalar and two Weyl fermions, thus giving $b_0 = T(adj) = 2(n+1)$.)

Far out on the on the Coulomb branch, where $a^K \gg \Lambda$, the $SU(n+1)$ theory is Higgsed to the $U(1)^n$ gauge group at a scale where the microscopic theory is very weakly coupled; see Fig. 3. Thus, the low-energy effective $U(1)^n$ couplings τ_{IJ} will be proportional to the running microscopic coupling at the scale of the $a^K \sim \langle a \rangle$ vevs:

$$\tau_{IJ} \sim \frac{b_0 C_{IJ}}{2\pi i} \log\left(\frac{\Lambda}{\langle a \rangle}\right) \tag{35}$$

for C_{IJ} some constant matrix that can be computed in perturbation theory. We see here the parameter Λ enters the IREA along with the vector multiplet scalar vevs, so we can treat Λ as if it were such a vev. In particular, Λ can only enter τ_{IJ} holomorphically. Furthermore, due to the angular nature of the theta angle, as $\tau \to \tau + 1$, or $\Lambda^{bo} \to e^{2\pi i}\Lambda^{bo}$, the physics must remain invariant. Thus $\tau_{IJ} = \tau_{IJ}(s_k, \Lambda^{bo})$ is a holomorphic function of the s_k and Λ^{bo} which matches on to (35) as $\Lambda \to 0$.

We can now derive a key fact about the Coulomb branch: though it can be corrected quantum mechanically, it is never lifted in asymptotically free $N = 2$ gauge theories. This is because there is a Coulomb branch for large adjoint scalar vevs where the asymptotically free gauge theory is Higgsed to $U(1)^n$ at arbitrarily weak coupling. Quantum corrections in the resulting $N = 2$ IREA cannot lift these flat directions since the only way (at weak coupling) to give mass to the $U(1)$ photons in the vector multiplets is by the Higgs mechanism; but there are no charged scalars in the vector multiplet. Thus it is not lifted for large enough s_k, and so by analytic continuation it cannot be lifted even for $s_k \sim \Lambda$ where perturbation theory is no longer valid. In general, complex manifolds like the Coulomb branch of $N = 2$ theories can become singular only on complex submanifolds, that is to say submanifolds at least 2 real dimensions smaller than the moduli space. Thus these singularities cannot be barriers preventing analytic continuation into a region of strong coupling.

The simplest example is the Coulomb branch of the $SU(2)$ Yang-Mills theory. The microscopic potential terms imply the equation $[\bar\Phi, \Phi] = 0$ for the complex adjoint scalar field, implying that Φ can be diagonalized by color rotations:

$$\Phi = \begin{pmatrix} a & 0 \\ 0 & -a \end{pmatrix}, \tag{36}$$

and there is a discrete gauge identification $a \simeq -a$. The gauge-invariant variable is

$$s_2 = -a^2 \equiv -\frac{1}{2}U, \tag{37}$$

where we have introduced the traditional name "U" for this Coulomb branch coordinate. It is easy to see that (36) leaves the diagonal $U(1) \subset SU(2)$ unbroken, and the light field U is neutral under this $U(1)$. We can thus think of the light degrees of freedom appearing in the IREA as those of an $N = 2$ $U(1)$ vector multiplet with complex scalar field U and a vector boson A_μ, as well as two Weyl fermions. The IREA can thus be written as

$$\mathcal{L} = \operatorname{Im} \tau(U, \Lambda) \left(\partial a(U) \cdot \partial \overline{a(U)} + \mathcal{F} \cdot \mathcal{F} \right), \tag{38}$$

where, by the arguments of the preceeding paragraphs, $a(U)$ is some holomorphic function of U and Λ^4, and the effective $U(1)$ gauge coupling will have the form

$$\tau(U) = \frac{1}{2\pi i} \log\left(\frac{\Lambda^4}{U^2}\right) + \sum_{n=0}^{\infty} c_n \left(\frac{\Lambda^4}{U^2}\right)^n. \tag{39}$$

The fact that only U^2 enters this formula follows from matching dimension with Λ^4, whose power follows from the coefficient of the one-loop beta function; it reflects a global $\mathbf{Z}_2$ symmetry acting on the Coulomb branch under $U \to -U$.

Solving for the vacuum structure of the $SU(2)$ theory is thus reduced to determining this function $\tau(U)$. It is worth examining the formula (39) in some detail. The first, logarithm, term came from matching to the one-loop running of the microscopic coupling for $U \gg \Lambda^2$. Because under theta-angle rotations, corresponding to 2π phase rotations of Λ^4, the physics must remain invariant, the low energy $\tau(U)$ can at most suffer an $Sp(2, \mathbf{Z}) \simeq SL(2, \mathbf{Z})$ electric-magnetic duality transformation. The terms included in (39) imply that $\tau(U) \to \tau(U) + 1$ under such a rotation, which is indeed in $SL(2, \mathbf{Z})$. Any other terms containing multiple logarithms, or any non-constant coefficient of the single logarithm term are not allowed, since they would necessarily imply $\tau(U)$ transformations under theta-angle rotations which are U-dependent, and therefore not in $SL(2, \mathbf{Z})$ since $SL(2, \mathbf{Z})$ is a discrete group of transformations. The absence of these higer logarithm terms is equivalent to the absence of all higher-loop corrections to the running of the microscopic coupling.

The terms proportional to Λ^{4n} correspond to a non-perturbative n-instanton contribution. Since the model is Higgsed for large U, the instantons have an effective IR cutoff at the scale U, so these instanton effects are calculable; the first two coefficients have been calculated [27]. In principle one could compute $\tau(U)$ by calculating all the n-instanton contributions, and then analytically continuing (39) to the whole U-plane; in practice this is too hard. Instead, we follow N. Seiberg and E. Witten's more physical approach to determining $\tau(U)$ [3].

17

There are two puzzles which indicate that we are missing some basic physics:

(1.) The effective coupling $\tau(U)$ is holomorphic, implying that $\mathrm{Re}\tau$ and $\mathrm{Im}\tau$ are harmonic functions on the U-plane. Since they are not constant functions, they therefore must be unbounded both above and below. In particular this implies that $\mathrm{Im}\tau = \frac{1}{g^2}$ will be negative for some U, and the effective theory will be non-unitary!

(2.) If we were to add a tree-level mass m for the complex scalar Φ (and one of the Weyl fermions as well to preserve an $N = 1$ supersymmetry), then, for $m \gg \Lambda$, Φ can be integrated out leaving a low-energy pure $SU(2)$ $N = 1$ super-YM theory with scale $\hat\Lambda^6 = m^2\Lambda^4$. This theory has two vacua with mass gaps; in particular there are no massless photons. For nonzero $m \ll \Lambda$ by an $N = 1$ nonrenormalization argument one expects this qualitative behavior to persist. In that case our low-energy $N = 2$ theory on the U-plane should be approximately correct, and we should see some way to lift the degenerate vacua and create a mass gap. In particular we need to give the photon a mass, but there are no light charged degrees of freedom to Higgs the photon.

The next subsection will introduce the physical ingredient which resolves these puzzles and allows us to solve for $\tau(U)$.

4.1 Monopoles

The ingredient we need to be aware of is monopoles [28]. Monopoles can be constructed as finite-energy classical solutions of non-Abelian gauge theories spontaneously broken down to Abelian factors [29]. In particular they will occur in the $N = 2$ $SU(2)$ Yang-Mills theory. We illustrate this for simplicity in a (non-supersymmetric) $SU(2)$ theory broken down to $U(1)$ by a real adjoint Higgs:

$$\mathcal{L} = -\frac{1}{4g^2}F^a_{\mu\nu}F^{a\mu\nu} + \frac{1}{2}D^\mu\Phi^a D_\mu\Phi^a - V(\Phi) \tag{40}$$

where V has a minimum on the sphere in field space $\sum_a \Phi^a\Phi^a = v^2$. Different directions on this sphere are gauge-equivalent. In the vacuum $\langle\Phi^a\rangle$ lies on this sphere, Higgsing $SU(2) \rightarrow U(1)$ and giving a mass $m_W = gv$ to the $W^\pm$ gauge bosons. The unbroken $U(1)$ has coupling g, so satisfies Gauss's law $\vec{D}\cdot\vec{E} = g^2 j_e^0$, where j_e^μ is the electric current density. Thus the electric charge is computed as $Q_e = \frac{1}{g^2}\int_{S^2_\infty} \vec{E}\cdot d\vec{S}$. In the vacuum, the unbroken $U(1)$ is picked out by the direction of the Higgs vev, so $\vec{E} = \frac{1}{v}\Phi^a\vec{E}^a$. With this normalization of the electic charge, we find that the $W^{\pm}$ bosons have $Q_e = \pm 1$.

Static, finite-energy configurations must approach the vacuum at spatial infinity. Thus for a finite energy configuration the Higgs field Φ^a, evaluated as $r \to \infty$, provides a map from the S^2 at spatial infinity into the S^2 of the Higgs vacuum. Such maps are characterized by an integer, n_m, which measures the winding of one S^2 around the other. Mathematically, the second homotopy group of S^2 is the integers, $\pi_2(S^2) = \mathbf{Z}$. The winding, n_m, is the magnetic charge of the field configuration. To see this, the total energy from the Higgs field configuration:

$$\text{Energy} = \int d^3x \tfrac{1}{2}D_\mu\Phi^a D^\mu\Phi^a + V(\Phi) \geq \int d^3x \tfrac{1}{2}D_\mu\Phi^a D^\mu\Phi^a. \tag{41}$$

To have finite energy configurations we must therefore ensure that the covariant derivative of Φ^a falls off faster than $1/r$ at infinity. The general solution for the gauge field consistent with this behavior is

$$A_\mu^a \sim -\frac{1}{v^2}\epsilon^{abc}\Phi^b\partial_\mu\Phi^c + \frac{1}{v}\Phi^a A_\mu \tag{42}$$

with A_μ arbitrary. The leading-order behavior of the field strength is then

$$F^{a\mu\nu} = \frac{1}{v}\Phi^a F^{\mu\nu} \tag{43}$$

with

$$F^{\mu\nu} = -\frac{1}{v^3}\epsilon^{abc}\Phi^a\partial^\mu\Phi^b\partial^\nu\Phi^c + \partial^\mu A^\nu - \partial^\nu A^\mu \tag{44}$$

and the equations of motion imply $\partial_\mu F^{\mu\nu} = \partial_\mu * F^{\mu\nu} = 0$. Thus we learn that outside the core of the monopole the non-Abelian gauge field is purely in the direction of Φ^a, that is the direction of the unbroken $U(1)$. The magnetic charge of this field configuration is then computed to be

$$Q_m = \int_{S_\infty^2} \vec{B} \cdot d\vec{S} = \frac{1}{2v^3}\int_{S_\infty^2}\epsilon^{ijk}\epsilon^{abc}\Phi^a\partial^j\Phi^b\partial^k\Phi^c dS^i = 4\pi n_m \tag{45}$$

where n_m is the winding number of the Higgs field configuration, recovering the Dirac quantization condition.[1]

---

[1]This is actually the Dirac quantization condition only for even values of n_m since in this theory we could add fields in the fundamental **2** representation of $SU(2)$, which would carry electric charge $Q_e = \pm 1/2$.

Note that for such non-singular field configurations, the electric and magnetic charges can be rewritten as

$$Q_e = \frac{1}{g^2}\int_{S_\infty^2} \vec{E}\cdot d\vec{S} = \frac{1}{g^2 v}\int_{S_\infty^2} \Phi^a \vec{E}^a\cdot d\vec{S} = \frac{1}{g^2 v}\int d^3x \vec{E}^a\cdot(\vec{D}\Phi)^a$$

$$Q_m = \int_{S_\infty^2} \vec{B}\cdot d\vec{S} = \frac{1}{v}\int_{S_\infty^2}\Phi^a\vec{B}^a\cdot d\vec{S} = \frac{1}{v}\int d^3x \vec{B}^a\cdot(\vec{D}\Phi)^a \qquad (46)$$

using the vacuum equation of motion and the Bianchi identity $\vec{D}\cdot\vec{E}^a = \vec{D}\cdot\vec{B}^a = 0$ and integration by parts.

If we consider a static configuration with vanishing electric field the energy (mass) of the configuration is given by

$$m_M = \int d^3x \left(\frac{1}{2g^2}\vec{B}^a\cdot\vec{B}^a + \frac{1}{2}\vec{D}\Phi^a\cdot\vec{D}\Phi^a + V(\Phi)\right) \geq \int d^3x \left(\frac{1}{2g^2}\vec{B}^a\cdot\vec{B}^a + \frac{1}{2}\vec{D}\Phi^a\cdot\vec{D}\Phi^a\right)$$

$$= \frac{1}{2}\int d^3x \left(\frac{1}{g}\vec{B}^a - \vec{D}\Phi^a\right)^2 + \frac{vQ_m}{g}, \qquad (47)$$

giving the BPS bound

$$m_M \geq \left|\frac{vQ_m}{g}\right|. \qquad (48)$$

This semi-classical bound can be extended to *dyons* (solitonic states carrying both electric and magnetic charges):

$$m_D \geq gv\left|Q_e + i\frac{Q_m}{g^2}\right|. \qquad (49)$$

A theta angle has a non-trivial effect in the presence of magnetic monopoles: it shifts the allowed values of electric charge in the monopole sector of the theory [30]. To see this, consider gauge transformations, constant at infinity, which are rotations in the $U(1)$ subgroup of $SU(2)$ picked out by the Higgs vev, that is, rotations in $SU(2)$ about the axis $\hat{\Phi}^a = \Phi^a/|\Phi^a|$. The action of such an infinitesimal gauge transformation on the field is

$$\delta A_\mu^a = \frac{1}{v}(D_\mu\Phi)^a \qquad (50)$$

with Φ the background monopole Higgs field. Let $\mathcal{N}$ denote the generator of this gauge transformation. Then if we rotate by 2π about the $\hat{\Phi}$ axis we must get the identity

$$e^{2\pi i\mathcal{N}} = 1. \qquad (51)$$

Including the θ term, it is straightforward to compute $\mathcal{N}$ using the Noether method,

$$\mathcal{N} = \frac{\partial\mathcal{L}}{\partial\partial_0 A_\mu^a}\delta A_\mu^a = Q_e - \frac{\theta Q_m}{8\pi^2}, \qquad (52)$$

20

where we have used the definitions (46) of the electric and magnetic charge operators. This result implies

$$Q_e = n_e + n_m \frac{\theta}{2\pi} \tag{53}$$

where n_e is an arbitrary integer and $n_m = Q_m/4\pi$ determines the magnetic charge of the monopole. We will henceforth label dyons by the integers (n_e, n_m). Note that the BPS bound becomes

$$M_D \geq gv \left| \left(n_e + n_m \frac{\theta}{2\pi} \right) + i n_m \frac{4\pi}{g^2} \right| = gv|n_e + \tau n_m|. \tag{54}$$

This result is classical; quantum mechanically, the coupling τ runs, and gv and $g\tau$ will be replaced by functions of the strong coupling scale Λ and the vevs. In theories with extended supersymmetry the (quantum-corrected) BPS bound can be computed exactly, and states saturating the bound can be identified [20]. For example, in the $N = 2$ $SU(2)$ theory the BPS mass formula becomes [3]

$$M_D = |a(U)n_e + b(U)n_m|, \tag{55}$$

where a and b are holomorphic functions of U and Λ^4 satisfying

$$\frac{\partial b(U)}{\partial a(U)} = \tau(U), \tag{56}$$

with $a(U)$ the same function as appeared in the IREA (38).[2]

4.2   Solution to the SU(2) Theory

Returning to the $N = 2$ $SU(2)$ Yang-Mills theory, we have learned that this theory can have magnetic monopoles. Indeed, one can show that there are BPS solitons with charges $(n_e, n_m) = (0, \pm1)$ in this theory, and they turn out to lie in hypermultiplets of the supersymmetry algebra. Furthermore, from (39) we see that changing the phase of U shifts the effective theta angle. In particular under the global $\mathbf{Z}_2$: $U \rightarrow e^{i\pi}U$, $\tau \rightarrow \tau-1$. From the associated duality transformation on the charges of any massive states (53), we see that there will be $(\mp1, \pm1)$ dyons in the spectrum. Repeating this procedure, we find there must be a whole tower of semi-classically stable dyons of charges $(n, \pm1)$ for arbitrary integers n.

The existence of these dyon states suggests a possible resolution to one of our puzzles: perhaps at some strong coupling point on the moduli space, for example $U = U_0$ with

---

[2]$b(U)$ is often called $a_D(U)$ in the literature.

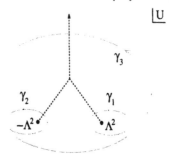

Figure 4: Cut U-plane with three loops. The cuts have been placed in an arbitrary manner connecting the two possible strong-coupling singularities, and a possible singularity at weak coupling ($U = \infty$).

$U_0 \sim \Lambda^2$, one of these dyons becomes massless, thereby providing the light charged scalar fields needed to Higgs the $U(1)$. Since we expect to recover the two gapped vacua of the $N = 1$ $SU(2)$ super-YM theory, and recalling the $\mathbf{Z}_2$ symmetry of the theory, it is natural to assume that there are two points on the U-plane where charged fields become massless, and they are at $U = \pm U_0$. Since Λ is the only scale in the theory, we take $U_0 = \Lambda^2$. (We can take this as the definition of our normalization of Λ, if we like.)

We can check this assumption by examining the behavior of τ as a function of U. Recall the other puzzle we had about the physics on the Coulomb branch: since $\tau(U)$ is holomorphic, $1/g^2 \sim \text{Im}\tau$ is harmonic and therefore unbounded from below, violating unitarity.

This puzzle is resolved by noting that τ is not, in fact, a holomorphic function of U. In particular, by electric-magnetic duality, as we traverse closed loops in the U-plane, τ need not come back to the same value, only one related to it by an $SL(2, \mathbf{Z})$ transformation. Mathematically, this is described by saying that τ is a section of a flat $SL(2, \mathbf{Z})$ bundle. This multi-valuedness of τ can be described by saying that τ is a holomorphic function on a cut U-plane, with cuts emanating from some singularities, and with the jump in τ across the cuts being an element of $SL(2, \mathbf{Z})$. The two points $U = \pm \Lambda^2$ at which we are assuming there are massless charged fields are the natural candidates for the branch points, see Fig. 4. The presence of these cuts allows us to avoid the conclusion that $\text{Im}\tau$ is unbounded.

Upon traversing the various loops γ_i in the above figure, τ will change by the action of an $SL(2, \mathbf{Z})$ element. These elements are called the *monodromies* of τ, and will be denoted $\mathcal{M}_i$.

We first calculate $\mathcal{M}_3$, the monodromy around the weak-coupling singularity at infinity. By taking γ_3 of large enough radius, τ will be accurately given by its one-loop value, the first term in (39). Taking $U \to e^{2\pi i} U$ in this formula gives $\tau \to \tau - 2$, giving for the monodromy at infinity[3]

$$\mathcal{M}_3 = \begin{pmatrix} -1 & 2 \\ 0 & -1 \end{pmatrix}. \tag{57}$$

In order to calculate the $\mathcal{M}_{1,2}$ monodromies, let us first calculate the monodromy we would expect if the field becoming massless at the associated singularity had charge (n_e, n_m). By a duality transformation we can change to a basis where this charge is purely electric: $(\tilde{n}_e, 0)$. In this basis the physics near the $U = U_0$ singularity is just that of QED with the electron becoming massless. This theory is IR free, so the behavior of the low-energy effective coupling will be dominated by its one-loop expression, at least sufficiently near U_0 where the mass of the charged field $\sim U - U_0$ is arbitrarily small:

$$\tilde{\tau} = \frac{\tilde{n}_e^2}{\pi i} \log(U - U_0) + \mathcal{O}(U - U_0)^0. \tag{58}$$

By traversing a small loop around U_0, $(U - U_0) \to e^{2\pi i}(U - U_0)$, we find the monodromy

$$\tilde{\tau} \to \tilde{\tau} + 2\tilde{n}_e^2 \quad \Longrightarrow \quad \widetilde{\mathcal{M}} = \begin{pmatrix} 1 & 2\tilde{n}_e^2 \\ 0 & 1 \end{pmatrix}. \tag{59}$$

Now let us duality-transform this answer back to the basis where the charges are (n_e, n_m). The required $SL(2, \mathbf{Z})$ element will be denoted $\mathcal{N} = \begin{pmatrix} a & b \\ c & d \end{pmatrix}$, and satisfies

$$\begin{pmatrix} a & b \\ c & d \end{pmatrix} \begin{pmatrix} n_e \\ n_m \end{pmatrix} = \begin{pmatrix} \tilde{n}_e \\ 0 \end{pmatrix}, \quad \text{and} \quad ad - bc = 1 \quad \text{with} \quad a, b, c, d \in \mathbf{Z}. \tag{60}$$

The transformed monodromy is then

$$\mathcal{M} = \mathcal{N} \widetilde{\mathcal{M}} \mathcal{N}^{-1} = \begin{pmatrix} 1 + 2n_e n_m & 2n_e^2 \\ -2n_m^2 & 1 - 2n_e n_m \end{pmatrix}. \tag{61}$$

Now, by deforming the γ_i contours in the U-plane, we find that the three monodromies must be related by

$$\mathcal{M}_3 = \mathcal{M}_1 \mathcal{M}_2. \tag{62}$$

---

[3]This actually only determines the monodromy up to an overall sign. The sign is determined by noting that $U \to e^{2\pi i} U$ has the effect of $\Phi \to -\Phi$ on the elementary Higgs field, so it reverses the sign of the low-energy electromagnetic field which in terms of $SU(2)$ variables is proportional to $\text{tr}(\Phi F)$. Thus it reverses the sign of electric and magnetic charges, giving an "extra" factor of $-1 \in SL(2, \mathbf{Z})$.

Assuming that a field with charges (n_{e1}, n_{m1}) becomes massless at $U = \Lambda^2$, while one with charges (n_{e2}, n_{m2}) does so at $U = -\Lambda^2$, and substituting into (62) using (57) and (61) gives as solutions

$$(n_{e1}, n_{m1}) = \pm(n, 1), \qquad (n_{e2}, n_{m2}) = \pm(n-1, 1), \qquad \text{for all} \quad n \in \mathbf{Z}. \qquad (63)$$

This set of charges actually represents a single physical solution. This is because taking $U \to e^{i\pi} U$ takes us to an equivalent theory by the $\mathbf{Z}_2$ symmetry; but this corresponds to shifting the low-energy theta-angle by 2π which in turn shifts all dyon electric charges by their magnetic charges. Repeated applications of this shift can take any of the above solutions to the solution

$$(n_{e1}, n_{m1}) = \pm(0, 1), \qquad (n_{e2}, n_{m2}) = \pm(-1, 1). \qquad (64)$$

The plus and minus sign solutions must both be there by anomaly cancellation in the low-energy $U(1)$. We thus learn that there is a consistent solution with a monopole becoming massless at $U = \Lambda^2$ and a charge $(-1, 1)$ dyon becoming massless at $U = -\Lambda^2$. Some progress has been made in weakening the initial assumption that there are just two strong-coupling singularities [31].

With the monodromies around the singularities in hand, we now turn to finding the low-energy coupling τ on the U-plane. The basic idea is that τ is determined by holomorphy and demanding that it match onto the behavior we have determined above at $U = \infty$ and $U = \pm\Lambda^2$. Seeing how to solve this "analytic continuation" problem analytically is not obvious, however. Seiberg and Witten did it by introducing an auxiliary mathematical object: a family of tori varying over the Coulomb branch.

This is a useful construction because the low-energy effective coupling τ has the same properties as the complex structure of a 2-torus. In particular, the complex structure of a torus can be described by its *modulus*, a complex number τ, with $\mathrm{Im}\tau > 0$. In this description, the torus can be thought of as a parallelogram in the complex plane with opposite sides identified, see Fig. 5. Furthermore, the modulus τ of such a torus gives equivalent complex structures modulo $SL(2, \mathbf{Z})$ transformations acting on τ. Therefore, if we associate to each point in the U-plane a holomorphically-varying torus, its modulus will automatically be a holomorphic section of an $SL(2, \mathbf{Z})$ bundle with positive imaginary part, which are just the properties we want for the effective coupling τ.

At $U = \pm\Lambda^2$, magnetically charged states become massless, implying that the effective coupling $\mathrm{Im}\tau \to 0$. (Recall that by $U(1)$ IR freedom, when an electrically charged

24

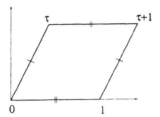

Figure 5: A complex torus as a parallelogram in the complex plane with opposite sides identified.

state becomes massless, the coupling $g \to 0$, implying $\tau \to +i\infty$. Doing the duality transform $\tau \to -1/\tau$ gives the above result for a magnetic charge becoming massless.) From the parallelogram, we see this implies that the torus is degenerating: one of its cycles is vanishing.

Now, a general torus can be described analytically as the Riemann surface which is the solution $y(x)$ to the complex cubic equation

$$y^2 = (x - e_1)(x - e_2)(x - e_3). \tag{65}$$

We can think of this as a double-sheeted cover of the x-plane, branched over the three points e_i and the point at infinity. We let this torus vary over the U-plane by letting the e_i vary: $e_i = e_i(U, \Lambda)$. By choosing the cuts to run between pairs of these branch points, and "gluing" the two sheets together along these cuts, one sees that the Riemann surface is indeed topologically a torus. Furthermore, the condition for a nontrivial cycle on this torus to vanish is that two of the branch points collide. Since we want this to happen at the two points $U = \pm \Lambda^2$, it is natural to choose $e_1 = \Lambda^2$, $e_2 = -\Lambda^2$, and $e_3 = U$:

$$y^2 = (x - \Lambda^2)(x + \Lambda^2)(x - U). \tag{66}$$

Note that this choice has a manifest $U \to -U$ symmetry, under which $x \to -x$ and $y \to \pm iy$.

Given this family of tori, one can compute their moduli as a ratio of line integrals:

$$\tau(U) = \frac{\oint_\beta \omega}{\oint_\alpha \omega}, \tag{67}$$

where ω is the (unique) holomorphic one-form on the Riemann surface,

$$\omega = \frac{dx}{y} = \frac{dx}{\sqrt{(x^2 - \Lambda^4)(x - U)}}, \tag{68}$$

25

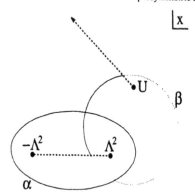

Figure 6: Cut x-plane with α and β cycles.

and α and β are any two non-trivial cycles on the torus which intersect once. For example, we might take α to be a cycle on the x-plane which loops around the branch points at $\pm\Lambda^2$, while β is the one which loops around the branch points at Λ^2 and U. If we chose the cuts on the x-plane to run between $\pm\Lambda^2$ and between U and ∞, then the α cycle would lie all on one sheet, while the β cycle would go onto the second sheet as it passes through the cut, see Fig. 6. Since the integrand in (68) is a closed one form ($d\omega = 0$), the value of τ does not depend on the exact locations of α and β, but only on how they loop around the branch points.

We can now check that our family of tori (66) indeed gives rise to the correct low-energy τ. By taking $U \to \infty$, it is not hard to explicitly evaluate (68) to find agreement with the first term in the weak-coupling expansion (39).[4] Also, without having to explicitly evaluate the integrals in (68), one can check that it reproduces the correct monodromies as U goes around the singularities at $\pm\Lambda^2$ by tracking how the α and β cycles are deformed as U varies. Finally, it turns out that the family of tori (66) is the unique one with these properties [3].

5 Geometry of N=2 Coulomb Branches

We would like to generalize the above arguments to other gauge groups and matter representations. To state the problem clearly:

---

[4]Though perhaps only up to an $SL(2,\mathbf{Z})$ transformation if I made the "wrong" choice for my α and β cycles.

Given: the field theory data specifying an $N = 2$ supersymmetric gauge theory, namely a gauge group G (not necessarily simple), a matter (hypermultiplet) representation R (not necessarily irreducible), bare masses m for the matter, and UV coupling constant(s) τ or strong coupling scales Λ for the vector multiplets,

Find: the $N = 2$ IREA on the Coulomb branch, namely the $U(1)^n$ couplings τ_{IJ} and the "special coordinates" a^I as functions of the microscopic field theory data and the gauge-invariant coordinates on the Coulomb branch.

We have not emphasized the *special coordinates* above, so let us define them here. Recall from Section 2 that the $N = 2$ IREA on the Coulomb branch has the form

$$\mathcal{L} = \text{Im}\tau_{IJ}\left(\partial a^I \cdot \partial \bar{a}^J + \mathcal{F}^I \cdot \mathcal{F}^J\right), \tag{69}$$

with τ_{IJ} satisfying the conditions

$$\partial_{[I}\tau_{J]K} = 0, \tag{70}$$

where $\partial_I = \partial/\partial a^I$, and τ_{IJ} a holomorphic $Sp(2n, \mathbf{Z})$ section of the vector multiplet scalars a^I. Clearly we could make a non-singular field redefinition on the scalars, effectively changing the coordinates we use to describe the Coulomb branch, and changing the form of the IREA (69). The choice of scalar fields such that the IREA has the above form where τ_{IJ} plays the role of both the Coulomb branch metric and the $U(1)^n$ effective couplings, are called special coordinates. In general the special coordinates can become singular, as they do at the monopole and dyon points in the $SU(2)$ example, so it is useful to choose well-behaved global coordinates on the Coulomb branch—the s_n. At weak coupling the special coordinates and the global coordinates are related by (33), but at strong coupling no such simple relation need exist. The special coordinates also appear in the BPS mass formula

$$M = |n_{e,I}a^I + n_m^J b_J|, \tag{71}$$

where the b_J are defined by

$$\partial_I b_J = \tau_{IJ}, \tag{72}$$

and exist by virtue of the integrability condition for this equation, (70).

This problem of determining the Coulomb branch IREA given the UV field theory data has not been solved, though many infinite series of solutions are known. Most of the known solutions were found essentially by (educated) guessing. In section 6 we will discuss one method which, although it is not known how to use it to solve the general problem, permits a *derivation* of the solutions when it works. In order to get to the point where we can discuss this method, we first need to reformulate the geometry of the vector multiplet manifolds; this is of interest also for the light it sheds on the general problem.

The Coulomb branch moduli space of the $N = 2$ IREA (69) satisfying condition (70) and the $Sp(2n, \mathbf{Z})$ properties of τ_{IJ} defines a *rigid special Kähler* (RSK) manifold [32]. Abstracting away from the IREA, we can thus define an RSK manifold as an n-complex-dimensional manifold $\mathcal{M}$ with certain properties. Choose some global complex coordinates s_K, $K = 1, \ldots, n$ on $\mathcal{M}$.[5] Then an RSK manifold has "special coordinates" $a^I(s_K)$, $I = 1, \ldots, n$, which are local holomorphic coordinates almost everywhere on $\mathcal{M}$, and a symmetric, holomorphic section τ_{IJ} of an $Sp(2n, \mathbf{Z})$ bundle on $\mathcal{M}$,[6] such that the metric in special coordinates is $g_{I\overline{J}} = \mathrm{Im}\tau_{IJ}$ and $\partial_{[I}\tau_{J]K} = 0$, where the derivative is with respect to the special coordinates. Note that $\mathrm{Im}\tau_{IJ}$ must be positive definite for the metric to be non-singular.

Several properties of RSK manifolds can immediately be deduced from this definition. The first is the existence of the "dual" special coordinates b_I, satisfying (72). Then, defining $\mathcal{K} = i(a^I\overline{b_I} - \overline{a^I}b_I)$, it is easy to check that $g_{I\overline{J}} = \partial_I\overline{\partial_J}\mathcal{K}$, which is the defining condition for a Kähler manifold. Defining the $2n$-component column vector $\mathbf{c}$ by

$$\mathbf{c} = \begin{pmatrix} b_I \\ a^I \end{pmatrix}, \tag{73}$$

the expression for the Kähler potential can be written compactly as $\mathcal{K} = \langle \mathbf{c}, \overline{\mathbf{c}} \rangle$, where the brackets denote the symplectic inner product

$$\langle \mathbf{c}, \mathbf{d} \rangle = \mathbf{c}^T \cdot J \cdot \mathbf{d} \quad \text{with} \quad J = \begin{pmatrix} 0 & 1 \\ -1 & 0 \end{pmatrix}. \tag{74}$$

Under transformations $M \in Sp(2n, \mathbf{Z})$ it is not hard to see that $\mathbf{c}$ transforms in the $2n$-dimensional representation:[7]

$$\mathbf{c} \to M \cdot \mathbf{c}, \tag{75}$$

and so the special coordinates are really part of a holomorphic $Sp(2n, \mathbf{Z})$ bundle $\mathcal{M}$ in the fundamental representation.

We will now describe three reformulations of RSK geometry. The first will be to show that RSK geometry is equivalent to having a family of algebraic varieties varying

---

[5] Or a patch of $\mathcal{M}$; this definition can be applied patch by patch to an atlas covering $\mathcal{M}$.

[6] With the usual action (6) on τ_{IJ}.

[7] Actually, the $\mathbf{c}$'s can in general also shift by constants under electric-magnetic duality transformations, which is important when there are hypermultiplet masses in the problem [22].

holomorphically with the s_K along with some extra structures; these are in turn equivalent to algebraically completely integrable Hamiltonian systems [33]. The second will be to show that a class of RSK manifolds are described by n-complex-dimensional families of Riemann surfaces of genus n with certain meromorphic one-forms [3]. The third will be to show that at least a subset of the RSK manifolds described in the second way can also be described by families of Riemann surfaces embedded in hyperKähler manifolds [4].

Though it is not known whether all RSK manifolds can be described in the second or third ways, in fact all RSK manifolds that have been found as Coulomb branches of $N = 2$ gauge theories do fall into the third category.

5.1   RSK and Families of Abelian Varieties

A straightforward generalization of the complex torus construction introduced in our discussion of the $SU(2)$ theory where the Coulomb branch was one-complex-dimensional to the case where the Coulomb branch is n-complex-dimensional, is to think of τ_{ij} as specifying the complex structure of an n-complex-dimensional torus [33]. Such a torus is specified by $2n$ linearly independent vectors forming the basis of a lattice Γ in $\mathbf{C}^n$. so the torus is $T^{2n} = \mathbf{C}^n/\Gamma$. Global linear complex changes of variables on $\mathbf{C}^n$ do not change the complex structure of T^{2n}, and can be used to set half of the basis vectors of Γ to real unit vectors. Thus the complex structure of T^{2n} is encoded in the $n \times n$ complex matrix, τ_{ij}, of coordinates of the remaining n basis vectors. It is easy to check that this τ_{ij} is really only defined up to $GL(2n, \mathbf{Z})$ fractional linear transformations reflecting the ability to choose a different set of n lattice vectors to set to the real unit vectors.

The τ_{ij}'s describing RSK geometry have four further constraints, however: they are symmetric, have positive definite imaginary part, are a section of an $Sp(2n, \mathbf{Z})$ bundle, and satisfy the integrability condition (70). The third constraint can be encoded in the geometry of T^{2n} by introducing an extra structure, a *polarization*, which is a non-degenerate $(1, 1)$-form t on T^{2n} with integral periods, and can be thought of as defining a symplectic inner product on the periods of 1-cycles on the torus as in (74). Complex tori obeying the first three conditions are known as *Abelian varieties*, which are essentially tori that can be described by algebraic equations involving generalized theta functions.

The fourth condition can be incorporated as the additional structure of a meromorphic 1-form, λ, on T^{2n} with the property that

$$\frac{\partial}{\partial s_K} \lambda = \omega^K, \tag{76}$$

up to total derivatives, where ω^K are a basis of n holomorphic one-forms on T^{2n}. This is related to τ_{ij} as follows. Choose a symplectic (or canonical) homology basis of one-cycles on T^{2n}. This is a basis of $2n$ one-cycles $\{\beta_I, \alpha^J\}$ such that

$$\int_{\alpha^I \wedge \alpha^J} t = \int_{\beta^I \wedge \beta^J} t = 0, \qquad \int_{\alpha^I \wedge \beta^J} t = \delta^I_J, \tag{77}$$

where the "wedge product" of one-cycles refers to the two-cycle spanned by them. Then the periods of λ (the integrals of λ over this basis of one-cycles) is the $2n$-component vector $\mathbf{c}$ introduced in (73).[8]

This reformulation of RSK geometry as complex manifolds with a family of Abelian varieties with meromorphic one form is quite general. Furthermore, the exterior derivative of the one-form on the total space of the RSK manifold plus its T^{2n} fibers is a symplectic two-form of a complex integrable system [33]. This equivalence of RSK geometry to integrable systems has led to the solution of many $N = 2$ IREAs [34, 33, 35]. However the procedure essentially involves matching an integrable system to the appropriate $N = 2$ field theory data, and no systematic way is known to do this matching.

5.2 RSK and Families of Riemann Surfaces

More systematic control over the construction of RSK geometries is obtained by specializing to classes of RSK manifolds whose geometry can be naturally encoded in simpler structures. One such specialization is to RSK manifolds whose associated Abelian variety T^{2n} can be realized as the Jacobian variety of a genus-n Riemann surface. For $n \geq 4$, these varieties form a subset of measure zero in the space of all Abelian varieties. Whether all families of Abelian varieties admitting the existence of an appropriate meromorphic one-form (to describe RSK geometry) are actually Jacobian varieties is an open question. In any case, to date all known constructions of RSK geometry are in terms of families of Jacobian varieties.

The connection between genus-n Riemann surfaces, Σ_n, and Jacobian varieties,

---

[8]For these periods to depend only on the homology class of the cycles, the one-form λ must have vanishing residues. Actually, λ's with non-vanishing residues are allowed, and are interpreted physically as bare hypermultiplet masses [22]. In what follows we will assume zero bare masses.

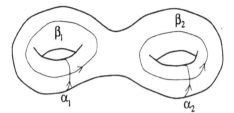

Figure 7: A genus 2 Riemann surface with a canonical homology basis of α and β cycles.

$Jac(\Sigma_n) \simeq T^{2n}$, is through the Jacobian map,

$$P \in \Sigma_n \longrightarrow \left\{ \int_{P_0}^{P} \omega^1, \dots, \int_{P_0}^{P} \omega^n \right\} \quad \text{mod periods}, \tag{78}$$

where P_0 is some argbitrary fixed base point and ω^K is a basis of the n holomorphic one-forms on Σ_n [36]. Under this map one-cycles on Σ_n are pushed forward to one-cycles on T^{2n}, the symplectic inner product (polarization) two-form t is pulled back to the intersection form on Σ_n, and the basis of holomorphic one forms ω^K and the meromorphic one-form λ on T^{2n} are pulled back to one-forms on Σ_n (which we call by the same names).

Thus τ_{IJ} is just the period matrix of the Riemann surface, and is given by

$$\tau_{IJ} = \left(\int_{\beta_I} \omega^K \right) \left(\int_{\alpha^J} \omega^K \right)^{-1} \tag{79}$$

where the second factor is to be interpreted as a matrix inverse on the JK indices and K is to be summed over. The symmetry and positive-definiteness conditions on τ_{IJ} follow from the Riemann bilinear relations, while the $Sp(2n, \mathbf{Z})$ structure follows from the intersection form on Riemann surfaces. In particular, one can always choose a canonical homology basis of $2n$ one-cycles $\{\alpha^I, \beta_J\}$ such that their intersections obey $\alpha^I \cdot \alpha^J = \beta_I \cdot \beta_J = 0$ and $\alpha^I \cdot \beta_J = \delta^I_J$; see Fig. 7.

To summarize, we have encoded the RSK geometry of an n complex-dimensional Coulomb branch in a family of genus-n Riemann surfaces varying holomorphically over the Coulomb branch and endowed with a meromorphic one-form λ satisfying (76). This formulation has been used to solve for many $N = 2$ IREAs essentially by guessing a form for the family of Riemann surfaces and matching to $N = 2$ field theory data [37, 23, 24]. Again, as with the integrable system formulation, this matching procedure has not been made systematic.

5.3 RSK and Riemann Surfaces in HyperKähler Manifolds

We now turn to one further reformulation of RSK geometry which, when combined with some string theory ideas, has allowed a more (though not completely) systematic approach.

The previous encoding of RSK geometry in a family of Riemann surfaces failed to "geometrize" the meromorphic one-form λ. This geometrization can be performed as follows [4, 38]. Suppose the family of Riemann surfaces Σ_n can be embedded in a fixed (independent of the s_K) hyperKähler 4-manifold Q. Now a hyperKähler manifold is a manifold Kähler with respect to three complex structures, I, J, and K, satisfying the quaternion algebra

$$I^2 = J^2 = K^2 = -1, \qquad IJ = -JI = K, \quad \text{and cyclic permutations.} \qquad (80)$$

Each complex structure can be thought of either as a rank-2 tensor acting on the tangent space to the manifold, *e.g.* $I = I_j^i$, or, using the metric to lower one of the indices, as an antisymmetric 2-tensor (a 2-form) on Q. Furthermore, the Kähler condition implies that these 2-forms are closed. Then, with respect to the complex structure I, $\omega \equiv J + iK$ is a closed holomorphic 2-form. Thus locally $\omega = d\lambda$ and λ pulls back to a meromorphic one-form on Σ_n. Because the family of Σ_n obtained as we vary the s_K are all embedded holomorphically in the fixed manifold Q, the RSK condition (76) on λ is automatically satisfied.

RSK manifolds which can be described in this way are clearly a subset of those that can be described just in terms of those described in terms of a family of Reimann surfaces and a one-form λ. But this restricted class has the great advantage that everything appears geometrically, requiring only a choice of a fixed "background" hyperKähler 4-manifold Q.

6 M-theory 5-Brane Construction of the Coulomb Branch

In this section we will outline the construction of solutions to $N = 2$ IREAs (Coulomb branch geometries) using the encoding of RSK geometry in a holomorphically varying family of Riemann surfaces Σ_n embedded in a hyperKähler four-manifold Q, following [4]. First we will outline an argument using the M-theory/IIA string theory equivalence to identify Q and some gross topological properties of the embedding corresponding to some $N = 2$ field theory data. Then we will show, in an example, how easy it is to solve for the specific family of embedded surfaces given this data, thus solving for the $N = 2$ IREA on the Coulomb branch of $SU(n + 1)$ Yang-Mills theory.

6.1 5-Branes in M-Theory

The basic idea [4] is to interpret the geometrical objects Q and Σ_n as physical objects in a supergravity theory such that at energies far below the Planck scale where the gravity decouples we are left with an $N = 2$ field theory.

Choose the supergravity theory to be the unique 11-dimensional supergravity theory, which is the low energy effective theory of M-theory [39]. This theory has 32 supercharges, corresponding to $N = 8$ supergravity in four dimensions. A consistent background to this theory is $\mathbf{R}^{6+1} \times Q$; if Q is hyperKähler this background breaks half the supersymmetries to $N = 4$ in four dimensions.

Now M theory has 5-branes, which are excitations of the theory extended in $5 + 1$ dimensions. On length scales much larger than the Planck scale of this theory, we can think of the 5-brane as a mathematical 6-manifold embedded in the 11-dimensional space-time. The 5-brane has field theory degrees of freedom which are constrained to propagate only on the brane. Furthermore, as long as the 5-brane is holomorphically embedded in the background 11-dimensional space, a solution to the supergravity equations of motion are obtained with only half of the remaining supersymmetries broken. Thus the 5-brane (six-dimensional) field theory has 8 conserved supercharges, corresponding to $N = 2$ supersymmetry in four dimensions. In the limit that curvature length scales of the brane are much greater than the Planck scale, the "bulk" 11-dimensional supergravity degrees of freedom decouple, leaving a unitary six-dimensional field theory on the brane.

The final step in this M-theory construction is to interpret the embedded Riemann surface Σ_n as part of the 5-brane world-volume. In particular, take the 5-brane world-volume to be the manifold $\mathbf{R}^{3+1} \times \Sigma_n$ with $\mathbf{R}^{3+1} \subset \mathbf{R}^{6+1}$ and $\Sigma_n \subset Q$. Then on distance scales large compared to the size of Σ_n the brane field theory is effectively an $N = 2$ four-dimensional field theory.

Thus we have incorporated all the mathematical ingredients needed to describe the RSK Coulomb branch geometry together with the associated physical degrees of freedom (the four-dimensional field theory) in a single supergravity configuration.

6.2   IIA/M-Theory Duality

To use this to solve a concrete $N = 2$ theory, we need to choose the background hyperKähler manifold Q. The simplest choice would be a flat four-manifold $Q = \mathbf{R}^4$, but this does not work. To see this, pick some complex structure on Q, say coordinates $v = x_1 + ix_2$ and $s = x_3 + ix_4$. Then any holomorphically embedded Riemann surface Σ_n will be described by some complex analytic equation in s and v: $F(s, v) = 0$. But, by the properties of analytic maps, the surface described by such an equation cannot be compact—it must extend to infinity in Q. Thus it would seem that we are really describing in this way a six-dimensional field theory on some curved background.

The key to connecting this construction to a four-dimensional interpretation is to use the equivalence of the (ten-dimensional) type IIA string theory to M theory compactified on a circle. In this equivalence, the string coupling g_s of the IIA theory is related to the radius of the compactified circle, R, by

$$g_s = (R/\ell_p)^{3/2}, \tag{81}$$

where ℓ_p is the 11-dimensional Planck length, and the defining string scale ℓ_s (related to the fundamental string tension) satisfies

$$g_s \ell_s = R. \tag{82}$$

Thus when the 11-dimensional supergravity description is good, that is when $\ell_p \ll R$, we have $g_s \gg 1$, so the string description is strongly coupled, and *vice versa*.

The connection to a four-dimensional description comes from taking the limit as the compactification radius $R \to 0$, so that the ten-dimensional string description becomes weakly coupled. In that limit the 5-brane reduces to either a 4-brane or a 5-brane in the ten-dimensional theory, depending on whether the M-theory 5-brane is or is not wrapped around the shrinking circle. The eventual brane configuration in ten dimensions will look like that shown in Fig. 8, with short 4-brane segments suspended between infinite 5-branes.

Now, at weak coupling IIA 5-branes are much heavier than 4-branes, and so can be considered as fixed objects, with any field theory degrees of freedom propagating on the 4-branes. Indeed, the typical length scales (inverse of the mass scales) of NS5-branes (ℓ_5) and D4-branes (ℓ_4) are

$$\ell_4 = g_s \ell_s \quad \text{and} \quad \ell_5 = g_s^2 \ell_s. \tag{83}$$

Furthermore, since the extent of the 4-branes is finite in one dimension, at long distances the 4-brane field theory will be effectively four-dimensional. Thus we recover

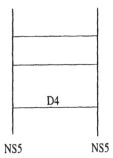

Figure 8: Three D4-branes suspended between two NS5-branes in Type IIA string theory. Only two of the ten dimensions are shown; the 4- and 5-branes are all parallel along an additional $3 + 1$ dimensions.

the four-dimensional $N = 2$ field theory. Finally, an important property of D-branes in string theory (of which the 4-branes are examples) is that the field theory degrees of freedom living on $n + 1$ parallel D-branes are described by an $SU(n + 1)$ theory Higgsed to $U(1)$'s, $i.e.$ a Yang-Mills theory on its Coulomb branch [40]. The size of the vevs Higgsing the gauge group ($i.e.$ the Coulomb branch coordinates) are proportional to the separations of the 4-branes. Thus we have learned that in order to describe the $SU(n + 1)$ $N = 2$ Yang-Mills Coulomb branch we should choose as our M-theory background $Q = \mathbf{R}^3 \times S^1$.

Before turning to the explicit construction of the $SU(n + 1)$ IREA, there is an important question to address in this construction, namely, why is an essentially classical 11-dimensional supergravity construction at all reliable to describe a field theory we only see in the $R \to 0$ limit, where the M-theory description should be strongly coupled? The answer lies in a supersymmetric selection rule. Denote by L a typical length scale of the brane configuration shown in Fig. 8, say the distance between some 4-branes or between the 5-branes. Now the typical length scale of the 4-brane dynamics is, from (83) and (82), $\ell_4 = g_s \ell_s = R$. So the relevant scales on the Coulomb branch are measured by the ratios L/R. In terms of the 11-dimensional picture, these ratios determine the shape (complex structure) of Σ_n but not its overall size—which is just as expected since only the complex structure of Σ_n encoded the RSK geometry. Furthermore, the overall size parameter enters as the vev of a hypermultiplet in the supergravity theory. By the $N = 2$ selection rule described in Section 2.1, hypermultiplet vevs do not affect the vector multiplet vevs (the Coulomb branch). Thus we learn that the size of R (or equivalently of the string coupling g_s) has no effect on the complex sructure of Σ_n,

35

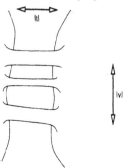

Figure 9: Two dimensions of an M-theory 5-brane embedded in the Q manifold. The three tubes wrap around the S^1 and extend along the $|t|$ direction. The two sheets extend to infinty in Q along the complex v direction far from the tubes.

which can therefore be computed in whatever limit is convenient. In physical terms, this argument shows that R is an irrelevant parameter in the Coulomb branch vacua of the 4-brane field theories.

6.3   The SU(n+1) Coulomb Branch

Let us choose complex coordinates on our hyperKähler 4-manifold $Q = \mathbf{R}^3 \times S^1$ to be $v = (x^1 + ix^2)/R$ and $s = (x^3 + iy)/R$ where y is a periodic coordinate along the S^1, $y \simeq y + 2\pi R$. Good global complex coordinates on Q can then be taken to be v and $t = e^s$. A holomorphically embedded Riemann surface Σ_n will be described by some complex analytic equation in t and v: $F(t,v) = 0$. Since upon shrinking the circle, the surface is supposed to reproduce the IIA brane configuration of Fig. 8, we expect that Σ_n will look globally something like two sheets connected by $n + 1$ tubes as in Fig. 9. Since the tubes are to collapse to D4-branes, they must be wrapped around the S^1, which is the phase of t, and extend along the modulus of t. The two sheets are to become NS5-branes so do not wrap the S^1; thus they should extend to infinity along the complex v direction.

Since this surface wraps $n+1$ times around the S^1 at intermediate $|t|$, by conservation of this winding number, it must also do so as $t \to 0, \infty$. The simplest way of satisfying this constraint is to demand that

$$t \sim v^{n+1} \qquad \text{as} \qquad t \to \infty, \tag{84}$$

36

and

$$t \sim v^{-n-1} \qquad \text{as} \qquad t \to 0. \tag{85}$$

(Other choices can also satisfy this constraint, but turn out to lead to $SU(n+1)$ $N=2$ theories with hypermultiplets, and correspond in the IIA picture to configurations with ssemi-infinite D4-branes extending to the left or right of the NS5-branes in Fig. 8.)

Now we can write determine the holomorphic equation $F(v,t) = 0$ for Σ_n. Since at fixed v there are two values of the t coordinate that lie on the surface in Fig. 9, F should be at most quadratic in t:

$$0 = F = A(v)t^2 + B(v)t + C(v). \tag{86}$$

Furthermore since at generic fixed t we found $n+1$ values of v on the surface, we see that A, B, and C can be at most $(n+1)$th order polynomials in v. Suppose the highest powers of v in A, B, and C are n_A, n_B, and n_C, respectively, with

$$0 \le n_A, n_B, n_C \le n+1. \tag{87}$$

Then the leading terms in (86) as $t \to \infty$ according to (84) give

$$v^{n_A+2n+2} + v^{n_B+n+1} + v^{n_C} = 0. \tag{88}$$

This has a solution as $v \to \infty$ with the $n_{A,B,C}$ in the range (87) only if

$$n_A = 0 \qquad \text{and} \qquad n_B = n+1. \tag{89}$$

A similar argument using (85) as $t \to 0$ gives

$$n_C = 0 \qquad \text{and} \qquad n_B = n+1. \tag{90}$$

Thus the equation for Σ_n must have the form $0 = \alpha t^2 + \beta(v^{n+1}+a_1 v^n + \cdots + a_n)t + \gamma$ with α, β, γ, and the a_i complex constants. Under holomorphic changes of variables which do not affect the asymptotic behavior of v and t, namely $t \to at$ and $v \to bv + c$, as well as an overall rescaling of F, we can finally put Σ_n in the form

$$t^2 + \frac{1}{\Lambda^{n+1}}\left(v^{n+1} + s_2 v^{n-1} + s_3 v^{n-2} + \cdots + s_n\right)t + 1 = 0. \tag{91}$$

We have identified the coefficients with the strong-coupling scale Λ of the Yang-Mills theory, and the gauge-invariant coordinates s_K on the $SU(n+1)$ Coulomb branch. This makes it clear that this curve indeed has precisely the right number of parameters to describe the Coulomb branch of the $SU(n+1)$ Yang-Mills theory. Furthermore, they can be assigned the right dimensions as well, by assigning v dimensions of energy. Many more detailed checks on this answer can be made by taking the s_K vevs large and comparing the resulting complex structure τ_{IJ} of this curve to that computed from loops in perturbation theory and instanton contributions semiclassically. Finally, it is easy to compute the meromorphic one-form from this data to be $\lambda = (v/t)dt$, thus allowing the computation of BPS masses.

To summarize, we have seen how interpreting geometrical structures in the RSK geometry of $N = 2$ Coulomb branches as physical objects in M-theory together with the type IIA/M-theory equivalence has allowed us to solve for the Coulomb branch IREA associated to particular field theory data in a simple algebraic way. This approach has been extended to solve for the IREAs of many infinite series of $N = 2$ field theory data. It is an open question whether all $N = 2$ field theory IREAs can be solved in this way.

Acknowledgments

It is a pleasure to thank the organizers and participants of the Feza Gürsey Institute Summer School for a pleasant and stimulating visit. This work was supported in part by NSF grant PHY-9513717 and an A.P. Sloan Foundation fellowship

References

References

[1] For an introduction to the $N = 1$ and 2 supersymmetric results, see *e.g.* K. Intriligator, N. Seiberg, *Nucl. Phys. Suppl.* BC **45**, 1 (1996), hep-th/9509066; M. Peskin, hep-th/9702094.

[2] P. Argyres, *Supersymmetric Effective Actions in Four Dimensions*, to appear in the proceedings of the Trieste Spring School, March 1998.

[3] N. Seiberg, E. Witten, *Nucl. Phys.* B **426**, 19 (1994), hep-th/9407087.

[4] E. Witten, *Nucl. Phys.* B **500**, 3 (1997), hep-th/9703166.

[5] See *e.g.* S. Minwalla, hep-th/9712074.

[6] See *e.g.* P. Argyres, R. Plesser, N. Seiberg, E. Witten, *Nucl. Phys.* B **461**, 71 (1996), hep-th/9511154.

[7] See *e.g.* J. Wess, J. Bagger, *Supersymmetry and Supergravity* (2nd edition, Princeton University Press, Princeton NJ, 1992); P. West, *Introduction to Supersymmetry and Supergravity* (2nd edition, World Scientific, Singapore, 1990); M. Sohnius, *Phys. Rep.* **128**, 39 (1985).

[8] See *e.g.* N. Seiberg in *Proceedings of the Trieste Spring School* (1997), hep-th/9705117.

[9] See, however, J. Polchinski, *Nucl. Phys.* B **303**, 226 (1988).

[10] See, however, J. Maldacena, *Adv. Theor. Math. Phys.* **2**, 231 (1998), hep-th/9711200; S. Gubser, I. Klebanov, A. Polyakov, *Phys. Lett.* B **428**, 105 (1998), hep-th/9802109; E. Witten, *Adv. Theor. Math. Phys.* **2**, 253 (1998), hep-th/9802150.

[11] S. Coleman, D. Gross, *Phys. Rev. Lett.* **31**, 851 (1973).

[12] P. Dirac, *Proc. R. Soc.* A **133**, 60 (1931); J. Schwinger, *Phys. Rev.* **144**, 1087 (1966), **173**, 1536 (1968); D. Zwanziger, *Phys. Rev.* **176**, 1480,1489 (1968).

[13] E. Witten, *Phys. Lett.* B **86**, 283 (1979).

[14] A. Shapere, F. Wilczek, *Nucl. Phys.* B **320**, 669 (1989).

[15] In the $N = 1$ context, K. Intriligator, N. Seiberg, *Nucl. Phys.* B **431**, 551 (1994), hep-th/9408155.

[16] P. Argyres, R. Plesser, N. Seiberg, *Nucl. Phys.* B **471**, 159 (1996), hep-th/9603042.

[17] P. Argyres, R. Plesser, A. Shapere, *Nucl. Phys.* B **483**, 172 (1997), hep-th/9608129.

[18] N. Seiberg, *Phys. Lett.* B **206**, 75 (1988); N. Seiberg, *Phys. Lett.* B **318**, 469 (1993), hep-ph/9309335.

[19] C. Montonen, D. Olive, *Phys. Lett.* B **72**, 117 (1977); H. Osborn, *Phys. Lett.* B **83**, 321 (1979).

[20] E. Witten, D. Olive, *Phys. Lett.* B **78**, 97 (1978);

[21] A. Sen, *Phys. Lett.* B **329**, 217 (1994), hep-th/9402032.

[22] N. Seiberg, E. Witten, *Nucl. Phys.* B **431**, 484 (1994), hep-th/9408099.

[23] P. Argyres, R. Plesser, A. Shapere, *Phys. Rev. Lett.* **75**, 1699 (1995), hep-th/9505100.

[24] P. Argyres, A. Shapere, *Nucl. Phys.* B **461**, 437 (1996), hep-th/9509175.

[25] P. Argyres, *Adv. Theor. Math. Phys.* **2**, 293 (1998), hep-th/9706095.

[26] P. Argyres and A. Buchel, *Phys. Lett.* B **431**, 317 (1998), hep-th/9804007.

[27] N. Dorey, V. Khoze, and M. Mattis, *Phys. Lett.* B **388**, 324 (1996), hep-th/9607066.

[28] J. Harvey, in *1995 Summer School in High-Energy Physics and Cosmology: Proceedings*, edited by E. Gava *et. al.* (World Scientific, 1997), hep-th/9603086.

[29] G. 't Hooft, *Nucl. Phys.* B **79**, 276 (1974); A. Polyakov, *J.E.T.P. Lett.* **20**, 194 (1974).

[30] E. Witten, *Phys. Lett.* B **86**, 283 (1979).

[31] R. Flume, M. Magro, L. O'Raifeartaigh, I. Sachs, and O. Schnetz, *Nucl. Phys.* B **494**, 331 (1997), hep-th/9611123.

[32] For discussions of the various definitions of rigid special Kähler geometry with references, see for example P. Fré, *Nucl. Phys. Proc. Suppl.* BC **45**, 59 (1996), hep-th/9512043; I. Antoniadis, B. Pioline, *Int. J. Mod. Phys.* A **12**, 4907 (1997), hep-th/9607058; B. Craps, F. Roose, W. Troost, A. Van Proeyen, *Nucl. Phys.* B **503**, 565 (1997), hep-th/9703082; D. Freed, hep-th/9712042.

[33] R. Donagi, E. Witten, *Nucl. Phys.* B **460**, 299 (1996), hep-th/9510101.

[34] For the first such papers, see A. Gorsky, I. Krichever, A. Marshakov, A. Mironov, and A. Morozov, *Phys. Lett.* B **355**, 466 (1995), hep-th/9505035; E. Martinec and N. Warner, *Phys. Lett.* B **549**, 97 (1996), hep-th/9509161; T. Nakatsu and K. Takasaki, *Mod.Phys.Lett.* A **11**, 157 (1996), hep-th/9509162; E. Martinec, *Phys. Lett.* B **367**, 91 (1996), hep-th/9510204.

[35] For a review of these constructions, see *e.g.* R. Donagi in *Surveys in Differential Geometry*, ed. S.T. Yau, alg-geom/9705010.

[36] For a review of Riemann surfaces, see for example, D. Mumford, *Tata Lectures on Theta I* (Birkhäuser, 1983).

[37] P. Argyres and A. Faraggi, *Phys. Rev. Lett.* **74**, 3931 (1995), hep-th/9411057; A. Klemm, W. Lerche, S. Theisen, and S. Yankielowicz, *Phys. Lett.* B **344**, 169 (1995), hep-th/9411048; U. Danielsson and B. Sundborg, *Phys. Lett.* B **358**, 273 (1995), hep-th/9504102; A. Hanany and Y. Oz, *Nucl. Phys.* B **452**, 283 (1995), hep-th/9505075; A. Brandhuber and K. Landsteiner, *Phys. Lett.* B **358**, 73 (1995), hep-th/9507008.

[38] A. Klemm, W. Lerche, P. Mayr, C. Vafa, N. Warner, *Nucl. Phys.* B **477**, 746 (1996), hep-th/9604034.

[39] For an introduction to the string/M theory properties used in this subsection, see for example Chapter 14 of J. Polchinksi, *String Theory*, vol. II (Cambridge, 1998).

[40] E. Witten, *Nucl. Phys.* B **460**, 335 (1996), hep-th/9510135.

DAMTP-1999-55

An introduction to meromorphic conformal field theory and its representations

Matthias R Gaberdiel[*] and Peter Goddard[†]

Department of Applied Mathematics and Theoretical Physics
University of Cambridge, Silver Street,
Cambridge CB3 9EW, U.K.

April 1999

Abstract

An introduction to meromorphic conformal field theory and its representation theory is given.

[*]E-mail address: M.R.Gaberdiel@amtp.cam.ac.uk
[†]E-mail address: pg@damtp.cam.ac.uk

1   Introduction

Conformal field theory has been at the centre of much attention during the last fifteen years since it plays an important role in at least three different areas of modern theoretical physics: conformal field theory provided the first examples of non-trivial exactly solvable quantum field theories, the critical points of models in statistical mechanics are described by them, and they play a central role in string theory as the theory that describes the string excitations from the point of view of the world-sheet. Conformal field theory has also had a major impact on various aspects of modern mathematics, in particular, the theory of vertex operator algebras and Borcherds algebras, finite groups and number theory.

Much of the activity in conformal field theory was initiated by the seminal work of Belavin, Polyakov and Zamolodchikov [1] who demonstrated that a certain class of models can be solved exactly. They also set up a general framework for studying chiral conformal field theory which was developed further, in the context of string theory, by Moore and Seiberg [2] in particular. Their work was mainly motivated by an attempt to prove a remarkable identity discovered by Verlinde [3]. On the mathematical side, in essence three different rigorous approaches to conformal field theory have been developed: a geometrical approach initiated by Segal [4]; an algebraic approach due to Borcherds [5, 6], Frenkel, Lepowsky and Meurman [7] and developed further by Frenkel, Huang and Lepowsky [8] Zhu [9] and Kac [10]; and an functional analytic approach in which techniques from algebraic quantum field theory are employed and which has been pioneered by Wassermann [11] and Gabbiani and Fröhlich [12]. Each of these three approaches produces a different perspective on conformal field and each facilitates the appreciation of its deep connections with other parts of mathematics, different in the three cases. In these lecture notes we present a rigorous approach that is closely related to the way conformal field theory arose at the birth of string theory [13, 14]. It is a development of earlier studies of meromorphic conformal field theory [15] and forms a slightly more pedagogical expansion of the material presented in [16].

All of these approaches concern what physicists sometimes call 'chiral conformal field theory', and that describes in essence half of a local conformal field theory. Every local conformal field theory possesses two meromorphic subtheories (that can be canonically defined), and the space of states of the whole theory can usually be described in terms of tensor products of representations of these two meromorphic subtheories; a chiral conformal field theory is then a meromorphic conformal field theory together with all of its representations. Unfortunately not much is known about general constructions that allow one to associate a consistent local conformal field theory to two chiral conformal field theories. However, the condition that the theory be well defined on higher genus Riemann surfaces imposes stringent conditions on the possible spectrum of a local conformal field theory, and allows, in typical cases, only for a finite number of different possibilities; an overview of recent developments in studying this constraint can be found in Terry's lectures [17].

The starting point of our discussion here is a family of amplitudes that define a meromorphic conformal field theory. These amplitudes are functions of n complex variables and describe the vacuum expectation values of n fields associated with certain basic states. The full space of states of the meromorphic theory is then obtained by factorising these amplitudes

2

in a certain sense.

The process of reconstructing the space of states from the vacuum expectation values of fields is familiar from axiomatic quantum field theory. In the usual Osterwalder-Schrader framework of Euclidean quantum field theory [18], the reflection positivity axiom guarantees that the resulting space of states has the structure of a Hilbert space. (In the context of conformal field theory this approach has been developed by Felder, Fröhlich and Keller [19].) In the present approach, the construction of the space of states depends only on the meromorphicity of the given family of amplitudes, $\mathcal{A}$, and positivity is not required for the basic development of the theory.

The spaces of states that are naturally defined are not Hilbert spaces but topological vector spaces, their topology being determined by requirements designed to ensure meromorphic amplitudes. (Recently Huang has also introduced topological vector spaces, which are related to ours, but from a different point of view [20].) They are also such that one can introduce fields ('vertex operators') for the basic states which are continuous operators. The locality property of these vertex operators is a direct consequence of the locality assumption about the family of amplitudes, $\mathcal{A}$, and this is then sufficient to prove the duality property (or Jacobi identity) of the vertex operators [15].

To develop the theory further, we need to assume that the basic amplitudes, $\mathcal{A}$, are Möbius invariant. This enables us to define vertex operators for more general states, modes of vertex operators and a Fock space which contains the essential algebraic content of the theory. This Fock space also enables us to define the concept of the equivalence of conformal field theories. The assumptions made so far are very general but if we assume that the amplitudes satisfy a cluster decomposition property we place much more severe restrictions on the theory, enabling us, in particular, to prove the uniqueness of the vacuum state.

Nothing assumed so far implies that the theory has a conformal structure, only one of Möbius symmetry. However, we show that it is always possible to extend the theory in such a way that it acquires a conformal structure. (For theories with a conformal structure this leaves the theory unchanged.) A conformal structure is necessary if we want to be able to define the theory on higher genus Riemann surfaces (although this is not discussed in these lectures). For this purpose, we also need to introduce the concept of a representations of a conformal (or rather a Möbius) field theory. Developing an idea of Montague, we show that any representation corresponds to a state in the space of states of the theory [21]. This naturally poses the question of what conditions a state has to satisfy in order to define a representation. For the case of highest weight representations, the conditions define an associative algebra which is that originally introduced by Zhu [9]. It is the main content of Zhu's Theorem that this algebra can be defined in terms of the algebraic Fock space.

These lectures are arranged as follows. In Section 2 we sketch the general structure of a local conformal field theory, and explain how the meromorphic subtheory and its representations naturally arise. We then start in Section 3 with an axiomatic introduction into the structure of a meromorphic conformal field theory. In particular, we introduce the basic assumptions about the family of amplitudes, $\mathcal{A}$, that define a meromorphic conformal field theory, and construct the topological vector space of states and the vertex operators for the basic states. In Section 4, Möbius invariance and its consequences are discussed. In Section 5, we define modes of vertex operators and use them to construct Fock spaces and thus to define the equivalence of theories. Examples of conformal field theories are provided in Section 6: the $U(1)$ theory; affine Lie algebra theory; the Virasoro theory; lattice theories; and an example which does not have a conformal structure. In Section 7, the assumption of cluster decomposition is introduced and in Section 8 we show how to extend a Möbius invariant theory to make it conformally invariant. All of the material thus far concerns meromorphic conformal field theory, but in Section 9 we introduce the concept of a representation and explain how every representation can be characterised by a state in the meromorphic theory. In Section 10, we define the idea of a Möbius covariant representation and the notion of equivalence for representations. An example of a representation is given in Section 11. In Section 12, we introduce Zhu's algebra and explain the significance of Zhu's Theorem in our context. Further developments and open problems are surveyed in Section 13. There are seven appendices in which some of the more technical details are described.

2 Local and chiral conformal field theory

Let us begin by describing somewhat sketchily what the general structure of a local conformal field theory is, and how meromorphic conformal field theory and its representations emerge in this context.

2.1 The space of states

In essence, a two-dimensional conformal field theory (like any other field theory) is determined by its space of states and the collection of its correlation functions. The space of states is a vector space $\mathcal{H}$ (that may or may not be a Hilbert space), and the correlation functions are defined for collections of vectors in some dense subspace $\mathcal{F}$ of $\mathcal{H}$. These correlation functions are defined on a two-dimensional space-time, which we shall always assume to be of Euclidean signature and to be compact (although the case of a surface with boundary is also interesting). These surfaces are classified (topologically) by their genus g which counts the number of 'handles'; the simplest such surface is the sphere with $g = 0$, the surface with $g = 1$ is the torus, *etc.* In a first step we shall therefore consider conformal field theories that are defined on the sphere; under certain conditions it is possible to associate to such a theory families of theories that are defined on surfaces of arbitrary genus. This is important in the context of string theory where the perturbative expansion consists of a sum over all such theories (and the genus of the surface plays the rôle of the loop order).

One of the special features of conformal field theory is the fact that the theory is naturally defined on a *complex* (or *Riemann*) surface, *i.e.* on a surface that possesses suitable complex coordinates. In the case of the sphere, the complex coordinates can be taken to be those of the complex plane that cover the sphere except for the point at infinity; complex coordinates for infinity are defined by means of the coordinate function $\gamma(z) = 1/z$ that maps a neighbourhood of infinity to a neighbourhood of 0. With this choice of complex coordinates, the sphere is usually referred to as the *Riemann sphere*, and this choice of complex coordinates is up to some suitable class of reparametrisations, unique. The correlation

4

functions of a conformal field theory that is defined on the sphere are thus of the form

$$\langle V(\psi_1; z_1, \bar{z}_1) \cdots V(\psi_n; z_n, \bar{z}_n) \rangle \,, \tag{2.1}$$

where $\psi_i \in \mathcal{F} \subset \mathcal{H}$, and z_i and $\bar{z}_i$ are complex numbers (or infinity). These correlation functions are assumed to be *local*, *i.e.* independent of the order in which the fields appear in (2.1).

One of the properties that makes two-dimensional conformal field theories exactly solvable is the fact that the theory contains a large (typically infinite-dimensional) symmetry algebra with respect to which the states in $\mathcal{H}$ fall into representations. This symmetry algebra is directly related (in a way we shall describe below) to a certain preferred subspace $\mathcal{F}$ of $\mathcal{F}$ that is characterised by the property that the correlation functions (2.1) of its states depend only on the complex parameter z, but not on its complex conjugate $\bar{z}$. More precisely, a state $\psi \in \mathcal{F}$ is in $\mathcal{F}$ if for any collection of $\psi_i \in \mathcal{F} \subset \mathcal{H}$, the correlation functions

$$\langle V(\psi; z, \bar{z}) V(\psi_1; z_1, \bar{z}_1) \cdots V(\psi_n; z_n, \bar{z}_n) \rangle \tag{2.2}$$

do not depend on $\bar{z}$. The correlation functions that involve only states in $\mathcal{F}$ are then analytic functions on the sphere. These correlation functions define the meromorphic (sub)theory [15] that will be the main focus of these lectures.

Similarly, we can consider the subspace of states $\overline{\mathcal{F}}$ that consists of those states for which the correlation functions of the form (2.2) do not depend on z. These states define an (anti)-meromorphic conformal field theory which can be analysed by essentially the same methods as a meromorphic conformal field theory. The two meromorphic conformal subtheories encode all the information about the symmetries of the theory, and for the most interesting class of theories, the so-called *finite* (or *rational*) theories, the whole theory can actually be reconstructed (up to some finite ambiguity) from them. In essence, this means that the whole theory is determined by symmetry considerations alone, and this is at the heart of the solvability of the theory.

The correlation functions of the theory determine the *operator product expansion* (OPE) of the conformal fields that expresses the operator product of two fields in terms of a single field. If ψ_1 and ψ_2 are two arbitrary states in $\mathcal{F}$ then the OPE of ψ_1 and ψ_2 is an expansion of the form

$$V(\psi_1; z_1, \bar{z}_1) V(\psi_2; z_2, \bar{z}_2)$$
$$= \sum_i (z_1 - z_2)^{\Delta_i} (\bar{z}_1 - \bar{z}_2)^{\bar{\Delta}_i} \sum_{r,s \geq 0} V(\phi^i_{r,s}; z_2, \bar{z}_2)(z_1 - z_2)^r (\bar{z}_1 - \bar{z}_2)^s \,, \tag{2.3}$$

where Δ_i and $\bar{\Delta}_i$ are real numbers and $\phi^i_{r,s} \in \mathcal{F}$. The actual form of this expansion can be read off from the correlation functions of the theory since the identity (2.3) has to hold in *all* correlation functions, *i.e.*

$$\left\langle V(\psi_1; z_1 \bar{z}_1) V(\psi_2; z_2, \bar{z}_2) V(\phi_1; w_1, \bar{w}_1) \cdots V(\phi_n; w_n, \bar{w}_n) \right\rangle$$
$$= \sum_i (z_1 - z_2)^{\Delta_i} (\bar{z}_1 - \bar{z}_2)^{\bar{\Delta}_i} \sum_{r,s \geq 0} (z_1 - z_2)^r (\bar{z}_1 - \bar{z}_2)^s \tag{2.4}$$

$$\langle V(\phi^\iota_{r,s}; z_2, \bar{z}_2) V(\phi_1; w_1, \bar{w}_1) \cdots V(\phi_n; w_n, \bar{w}_n) \rangle$$

for all $\phi_j \in \mathcal{F}$. If both states ψ_1 and ψ_2 belong to the meromorphic subtheory $\mathcal{F}$, (2.4) only depends on z_i, and $\phi^i_{r,s}$ belongs also to the meromorphic subtheory $\mathcal{F}$. The OPE therefore defines a certain product on the meromorphic fields. Since the product involves the complex parameters z_i in a non-trivial way, it does not directly define an algebra; the resulting structure is usually called a *vertex (operator) algebra* in the mathematical literature [5, 7].

By virtue of its definition in terms of (2.4), the operator product expansion is *associative*, *i.e.*

$$\Big(V(\psi_1; z_1, \bar{z}_1)V(\psi_2; z_2, \bar{z}_2)\Big)V(\psi_3; z_3, \bar{z}_3) = V(\psi_1; z_1, \bar{z}_1)\Big(V(\psi_2; z_2, \bar{z}_2)V(\psi_3; z_3, \bar{z}_3)\Big), \quad (2.5)$$

where the brackets indicate which OPE is evaluated first. If we consider the case where both ψ_1 and ψ_2 are meromorphic fields (*i.e.* in $\mathcal{F}$), then the associativity of the OPE implies that the states in $\mathcal{F}$ form a *representation* of the vertex operator algebra. The same also holds for the vertex operator algebra associated to the anti-meromorphic fields, and we can thus decompose the whole space $\mathcal{F}$ (or $\mathcal{H}$) as

$$\mathcal{H} = \bigoplus_{(j,\bar{j})} \mathcal{H}^{(j,\bar{j})}, \quad (2.6)$$

where each $\mathcal{H}^{(j,\bar{j})}$ is an (irreducible) representation of the two vertex operator algebras. Finite theories are characterised by the property that only finitely many representations of the vertex operator algebras occur in (2.6).

2.2 Modular invariance

The decomposition of the space of states in terms of representations of the two vertex operator algebras throws considerable light on the problem of whether the theory is well-defined on higher Riemann surfaces. One necessary constraint for this (which is believed also to be sufficient [2]) is that the vacuum correlator on the torus is independent of its parametrisation. Every two-dimensional torus can be described as the quotient space of $\mathbb{R}^2 \simeq \mathbb{C}$ by the relations $z \sim z + w_1$ and $z \sim z + w_2$, where w_1 and w_2 are not parallel. The complex structure of the torus is invariant under rotations and rescalings of $\mathbb{C}$, and therefore every torus is conformally equivalent (*i.e.* has the same complex structure) as a torus for which the relations are $z \sim z + 1$ and $z \sim z + \tau$, and τ is in the upper half plane of $\mathbb{C}$. It is also easy to see that τ, $T(\tau) = \tau + 1$ and $S(\tau) = -1/\tau$ describe conformally equivalent tori; the two maps T and S generate the group of SL$(2, \mathbb{Z})$ that consists of matrices of the form

$$A = \begin{pmatrix} a & b \\ c & d \end{pmatrix} \quad \text{where} \quad a, b, c, d \in \mathbb{Z} \quad \text{and} \quad ad - bc = 1, \quad (2.7)$$

and the action on τ is defined by

$$\tau \mapsto A\tau = \frac{a\tau + b}{c\tau + d}. \quad (2.8)$$

The parameter τ is sometimes called the modular parameter of the torus, and the group $SL(2, \mathbf{Z})$ is called the modular group (of the torus).

Given a conformal field theory that is defined on the Riemann sphere, the vacuum correlator on the torus can be determined as follows. We cut the torus along one of its non-trivial cycles; the resulting surface is a cylinder (or an annulus), and can therefore be described in terms of the conformal field theory on the sphere. In order to obtain the torus from the cylinder, the two ends have to be glued together, and in terms of the conformal field theory this means that one has to sum over a complete set of states; this leads therefore to a trace over the whole space of states, the *partition function* of the theory,

$$\sum_{(j,\bar{j})} \mathrm{Tr}_{\mathcal{H}^{(j,\bar{j})}} \left(\mathcal{O}(q, \bar{q}) \right) , \tag{2.9}$$

where $\mathcal{O}(q, \bar{q})$ is the operator that describes the propagation of the states along the annulus,

$$\mathcal{O}(q, \bar{q}) = q^{L_0 - \frac{c}{24}} \bar{q}^{\bar{L}_0 - \frac{\bar{c}}{24}} , \tag{2.10}$$

and L_0 and $\bar{L}_0$ are the scaling operators of the two vertex operator algebras (that will be discussed in more detail in the following sections). This propagator depends on the actual shape of the the annulus that can be described in terms of a complex parameter q. For a given torus that is described by τ, there is a natural choice for how to cut the torus into an annulus, and the complex parameter q that is associated to this annulus is $q = e^{2\pi i \tau}$. Since the torus that is described by τ and $A\tau$ (where $A \in SL(2, \mathbf{Z})$) are equivalent tori, the vacuum correlator is only well-defined provided that (2.9) is invariant under this transformation. This provides strong constraints on the spectrum of the theory.

For most conformal field theories (although not for all, see for example [22]) each of the spaces $\mathcal{H}^{(j,\bar{j})}$ is a tensor product of a representation $\mathcal{H}^j$ of the meromorphic vertex operator algebra and a representation $\bar{\mathcal{H}}^j$ of the anti-meromorphic vertex operator algebra. In this case, the vacuum correlator on the torus (2.9) takes the form

$$\sum_{(j,\bar{j})} \chi_j(q) \bar{\chi}_{\bar{j}}(\bar{q}) , \tag{2.11}$$

where χ_j is the *character* of the representation $\mathcal{H}^j$ of the meromorphic vertex operator algebra,

$$\chi_j(\tau) = \mathrm{Tr}_{\mathcal{H}^j}(q^{L_0}) \qquad \text{where} \quad q = e^{2\pi i \tau} , \tag{2.12}$$

and likewise for $\bar{\chi}_{\bar{j}}$. One of the remarkable facts about many vertex operator algebras (that has now been proved for a certain class of them [9]) is the property that the characters transform into one another under modular transformations,

$$\chi_j(-1/\tau) = \sum_k S_{jk} \chi_k(\tau) \qquad \text{and} \qquad \chi_j(\tau + 1) = \sum_k T_{jk} \chi_k(\tau) , \tag{2.13}$$

where S and T are *constant* matrices, *i.e.* independent of τ. In this case, writing

$$\mathcal{H} = \bigoplus_{i,\bar{j}} M_{i\bar{j}} \, \mathcal{H}^i \otimes \bar{\mathcal{H}}^{\bar{j}} , \tag{2.14}$$

where $M_{ij} \in \mathbb{N}$ denotes the multiplicity with which the tensor product $\mathcal{H}^i \otimes \bar{\mathcal{H}}^j$ appears in $\mathcal{H}$, the vacuum correlation function is well defined provided that

$$\sum_{i,j} S_{il} M_{ij} \bar{S}_{j\bar{k}} = \sum_{i,j} T_{il} M_{ij} \bar{T}_{j\bar{k}} = M_{l\bar{k}} , \qquad (2.15)$$

and $\bar{S}$ and $\bar{T}$ are the matrices defined as in (2.13) for the representations of the anti-meromorphic vertex operator algebra. This provides very powerful constraints for the multiplicity matrices M_{ij}. In particular, in the case of a finite theory (for which each of the two vertex operator algebras has only finitely many irreducible representations) these conditions typically only allow for a finite number of solutions that can be classified; this has been done for the case of the so-called minimal models and the affine theories with group $SU(2)$ by Capelli, Itzykson and Zuber [23, 24] (for a modern proof involving some Galois theory see [27]), and for the affine theories with group $SU(3)$ and the $N = 2$ superconformal models by Gannon [25, 26].

It may be worthwhile at this stage to stress that whilst the condition of modular invariance is very powerful indeed, it is not sufficient to select the consistent local conformal field theories. First of all, it is known that interesting consistent local conformal field theories exist that are only well defined on the sphere, but do not extend to the torus.[1] Secondly, it is not clear whether every modular invariant combination of characters of some vertex operator algebra can be obtained as the partition function of a consistent local conformal field theory. And finally, even if all modular invariant combination of characters arises as the partition function of a local conformal field theory, it is by no means clear whether this theory is uniquely determined in terms of the characters. For example, the heterotic $SO(32)$ and $E_8 \times E_8$ string theory have the same partition function but are clearly inequivalent conformal field theories. If one modifies the definition of the paritition function by including the zero modes of the Cartan generators, the characters of the two theories can be distinguished, but it is difficult to see how this prescription should be generalised and what it means physically. At any rate, it seems worthwhile to analyse the structure of a local (and in particular of a meromorphic) conformal field theory without imposing, from the ouset, the condition of modular invariance; this is what we shall do in the following.

3 Amplitudes, Spaces and Vertex Operators

The starting point for our approach to meromorphic conformal field theory is a collection of functions, which are eventually to be regarded as the vacuum expectation values of the fields associated with a certain basic set of states which generate the whole theory. We shall denote the space spanned by such states by V. In terms of the usual concepts of conformal field theory, V would be a subspace of the space of quasi-primary states. V can typically be taken to be finite-dimensional but this is not essential in what follows. (If it is infinite-dimensional, we shall at least assume that the algebraic dimension of V is countable, that is that the elements of V consist of finite linear combinations of a countable basis.)

---

[1] For example the subtheory of the Ising model where we only consider the fields associated to the vacuum sector and the energy operator is such an example.

We suppose that V can be regarded as the direct sum of a collection of subspaces, V_h, to each of which we can attach an integer, h, called the conformal weight of the states in that subspace, so that $V = \bigoplus_h V_h$. This is equivalent to saying that we have a diagonalisable operator $\delta : V \to V$, with eigenspaces $V_h = \{\psi \in V : \delta\psi = h\psi\}$.

We also suppose that for any positive integer n, and any finite collection of vectors $\psi_i \in V_{h_i}$, and $z_i \in \mathbb{P}$ (the Riemann Sphere), where $i = 1, \ldots, n$, we have a density

$$f(\psi_1, \ldots, \psi_n; z_1, \ldots, z_n) \equiv \langle V(\psi_1, z_1) V(\psi_2, z_2) \cdots V(\psi_n, z_n) \rangle \prod_{j=1}^{n} (dz_j)^{h_j} . \qquad (3.1)$$

Here $\langle V(\psi_1, z_1) V(\psi_2, z_2) \cdots V(\psi_n, z_n) \rangle$ is merely a suggestive notation for what will in the end acquire an interpretation as the vacuum expectation value of a product of fields. These "amplitudes" are assumed to be multilinear in ψ_i, invariant under the exchange of (ψ_i, z_i) with (ψ_j, z_j), and analytic in z_i, save only for possible singularities occurring at $z_i = z_j$ for $i \neq j$, which we shall assume to be poles of finite order (although one could consider generalisations in which the amplitudes are allowed to have essential singularities). Because of the independence of order of the (ψ_j, z_j), we can use the notation

$$f(\psi_1, \ldots, \psi_n; z_1, \ldots, z_n) = \left\langle \prod_{j=1}^{n} V(\psi_j, z_j) \right\rangle \prod_{j=1}^{n} (dz_j)^{h_j} . \qquad (3.2)$$

We denote the collection of these densities, and the theory we develop from them, by $\mathcal{A} = \{f\}$. We may assume that if all amplitudes in $\mathcal{A}$ involving a given $\psi \in V$ vanish then $\psi = 0$ (for, if this is not so, we may replace V by its quotient by the space of all vectors $\psi \in V$ which are such that all amplitudes involving ψ vanish).

We use these amplitudes to define spaces of states associated with certain subsets $\mathcal{C}$ of the Riemann Sphere $\mathbb{P}$. We can picture these spaces as consisting of states generated by fields acting at points of $\mathcal{C}$. First introduce the set, $\mathcal{B}_\mathcal{C}$, whose elements are labelled by finite collections of $\psi_i \in V_{h_i}$, $z_i \in \mathcal{C} \subset \mathbb{P}$, $i = 1, \ldots, n$, $n \in \mathbb{N}$ and $z_i \neq z_j$ if $i \neq j$; we denote a typical element $\boldsymbol{\psi} \in \mathcal{B}_\mathcal{C}$ by

$$\boldsymbol{\psi} = V(\psi_1, z_1) V(\psi_2, z_2) \cdots V(\psi_n, z_n)\Omega \equiv \prod_{i=1}^{n} V(\psi_i, z_i)\Omega . \qquad (3.3)$$

We shall immediately identify $\boldsymbol{\psi} \in \mathcal{B}_\mathcal{C}$ with the other elements of $\mathcal{B}_\mathcal{C}$ obtained by replacing each ψ_j in (3.3) by $\mu_j\psi_j$, $1 \leq j \leq n$, where $\mu_j \in \mathbb{C}$ and $\prod_{j=1}^{n} \mu_j = 1$.

Next we introduce the free (complex) vector space on $\mathcal{B}_\mathcal{C}$, i.e. the complex vector space with basis $\mathcal{B}_\mathcal{C}$ that is consisting of formal finite linear combinations $\Psi = \sum_j \lambda_j \boldsymbol{\psi}_j$, $\lambda_j \in \mathbb{C}$, $\boldsymbol{\psi}_j \in \mathcal{B}_\mathcal{C}$; we denote this space by $\mathcal{V}_\mathcal{C}$.

The vector space $\mathcal{V}_\mathcal{C}$ is enormous, and, intuitively, as we consider more and more complex combinations of the basis vectors, $\mathcal{B}_\mathcal{C}$, we generate vectors which are very close to one another. To measure this closeness, we need in essence to use suitably chosen amplitudes as test functions. To select a collection of linear functionals which we may use to construct from $\mathcal{V}_\mathcal{C}$ a space in which we have some suitable idea of topology, we select another subset

$\mathcal{O} \subset \mathbb{P}$ with $\mathcal{O} \cap \mathcal{C} = \varnothing$, where $\mathcal{O}$ is open, and we suppose further that the interior of $\mathcal{C}$, $\mathcal{C}^o$, is not empty. Let

$$\phi = V(\phi_1, \zeta_1) V(\phi_2, \zeta_2) \cdots V(\phi_m, \zeta_m) \Omega \in \mathcal{B}_\mathcal{O} , \tag{3.4}$$

where $\phi_j \in V_{k_j}$, $j = 1, \ldots, m$. Each $\phi \in \mathcal{B}_\mathcal{O}$ defines a map on $\psi \in \mathcal{B}_\mathcal{C}$ by

$$\eta_{\phi}(\psi) = (\phi, \psi) = \left\langle \prod_{i=1}^{m} V(\phi_i, \zeta_i) \prod_{j=1}^{n} V(\psi_j, z_j) \right\rangle , \tag{3.5}$$

which we can use as a contribution to our measure of nearness of vectors in $\mathcal{V}_\mathcal{C}$. [Strictly speaking, this map defines a density rather than a function, so that we should really be considering $\eta_{\phi}(\psi) \prod_{i=1}^{m} (d\zeta_i)^{k_i} \prod_{j=1}^{n} (dz_j)^{h_j}$.]

For each $\phi \in \mathcal{B}_\mathcal{O}$, η_{ϕ} extends by linearity to a map $\mathcal{V}_\mathcal{C} \to \mathbb{C}$, provided that $\mathcal{O} \cap \mathcal{C} = \varnothing$. We use these linear functionals to define our concept of closeness or, more precisely, the topology of our space. To make sure that we end up with a space which is complete, we need to consider sequences of elements of $\mathcal{V}_\mathcal{C}$ which are convergent in a suitable sense. Let $\tilde{\mathcal{V}}_\mathcal{C}$ be the space of sequences $\Psi = (\Psi_1, \Psi_2, \ldots)$, $\Psi_j \in \mathcal{V}_\mathcal{C}$. We consider the subset $\tilde{\mathcal{V}}_\mathcal{C}^\mathcal{O}$ of such sequences Ψ for which $\eta_{\phi}(\Psi_j)$ converges on subsets of ϕ of the form

$$\{\phi = V(\phi_1, \zeta_1) V(\phi_2, \zeta_2) \cdots V(\phi_m, \zeta_m) \Omega : \zeta_j \in K, |\zeta_i - \zeta_j| \geq \epsilon, i \neq j\}, \tag{3.6}$$

where for each collection of ϕ_j, $\epsilon > 0$ and a compact subset $K \subset \mathcal{O}$, the convergence is uniform in the (compact) set

$$\{(\zeta_1, \ldots, \zeta_m) : \zeta_j \in K, |\zeta_i - \zeta_j| \geq \epsilon, i \neq j\} . \tag{3.7}$$

If $\Psi \in \tilde{\mathcal{V}}_\mathcal{C}^\mathcal{O}$, the limit

$$\lim_{j \to \infty} \eta_{\phi}(\Psi_j) \tag{3.8}$$

is necessarily an analytic function of the ζ_j, for $\zeta_j \in \mathcal{O}$, with singularities only at $\zeta_i = \zeta_j$, $i \neq j$. (Again these could in principle be essential singularities, but the assumption of the cluster decompostion property, made in Section 7, will imply that these are only poles of finite order.) We denote this function by $\eta_{\phi}(\Psi)$. [A necessary and sufficient condition for uniform convergence on the compact set (3.7) is that the functions $\eta_{\phi}(\Psi_j)$ should be both convergent in the compect set and locally uniformly bounded, $i.e.$ each point of (3.7) has a neighbourhood in which $\eta_{\phi}(\Psi_j)$ is bounded independently of j; see Appendix A for further details.]

It is natural that we should regard two such sequences $\Psi^1 = (\Psi_i^1)$ and $\Psi^2 = (\Psi_i^2)$ as equivalent if

$$\lim_{j \to \infty} \eta_{\phi}(\Psi_j^1) = \lim_{j \to \infty} \eta_{\phi}(\Psi_j^2), \tag{3.9}$$

$i.e.$ $\eta_{\phi}(\Psi^1) = \eta_{\phi}(\Psi^2)$, for each $\phi \in \mathcal{B}_\mathcal{O}$. We identify such equivalent sequences, and denote the space of them by $\mathcal{V}_\mathcal{C}^\mathcal{O}$.

The space $\mathcal{V}_\mathcal{C}^\mathcal{O}$ has a natural topology: we define a sequence $\chi_j \in \mathcal{V}_\mathcal{C}^\mathcal{O}$, $j = 1, 2, \ldots$, to be convergent if, for each $\phi \in \mathcal{B}_\mathcal{O}$, $\eta_\phi(\chi_j)$ converges uniformly on each (compact) subset of the form (3.7). The limit

$$\lim_{j \to \infty} \eta_\phi(\chi_j) \tag{3.10}$$

is again necessarily a meromorphic function of the ζ_j, for $\zeta_j \in \mathcal{O}$, with poles only at $\zeta_i = \zeta_j$, $i \neq j$. Provided that the limits of such sequences are always in $\mathcal{V}_\mathcal{C}^\mathcal{O}$, i.e.

$$\lim_{j \to \infty} \eta_\phi(\chi_j) = \eta_\phi(\chi), \qquad \text{for some } \chi \in \mathcal{V}_\mathcal{C}^\mathcal{O} \tag{3.11}$$

we can define the topology by defining its closed subsets to be those for which the limit of each convergent sequence of elements in the subset is contained within it. In fact we do not have to incorporate the need for the limit to be in $\mathcal{V}_\mathcal{C}^\mathcal{O}$, because it is so necessarily; we show this in Appendix B. [As we note in this Appendix, this topology on $\mathcal{V}_\mathcal{C}^\mathcal{O}$ can be induced by a countable family of seminorms of the form $||\chi||_n = \max_{1 \leq i \leq n} \max_{\zeta_{i_j}} |\eta_{\phi_i}(\chi)|$, where the ϕ_{i_j} in ϕ_i are chosen from finite subsets of a countable basis and the ζ_{i_j} are in a compact set of the form (3.7).]

$\mathcal{B}_\mathcal{C}$ can be identified with a subset of $\tilde{\mathcal{V}}_\mathcal{C}^\mathcal{O}$ (using constant sequences), and this has an image in $\mathcal{V}_\mathcal{C}^\mathcal{O}$. It can be shown that this image is necessarily faithful provided that we assume the cluster property introduced in Section 7. In any case, we shall assume that this is the case in what follows and identify $\mathcal{B}_\mathcal{C}$ with its image in $\mathcal{V}_\mathcal{C}^\mathcal{O}$. There is a common vector $\Omega \in \mathcal{B}_\mathcal{C} \subset \mathcal{V}_\mathcal{C}^\mathcal{O}$ for all $\mathcal{C}, \mathcal{O}$ which is called the *vacuum vector*. The linear span of $\mathcal{B}_\mathcal{C}$ is dense in $\mathcal{V}_\mathcal{C}^\mathcal{O}$, (*i.e.* it is what is called a *total space*). With this identification, the image of $\mathcal{B}_\mathcal{C}$ in $\mathcal{V}_\mathcal{C}^\mathcal{O}$, ψ, defined as in (3.3), depends linearly on the vectors $\psi_j \in V$.

A key result in our approach is that, for suitable $\mathcal{O}$, $\mathcal{V}_\mathcal{C}^\mathcal{O}$ does not depend on $\mathcal{C}$. This is an analogue of the Reeh-Schlieder Theorem of Axiomatic Quantum Field Theory. In our context it is basically a consequence of the fact that any meromorphic function is determined by its values in an arbitrary open set. Precisely, we have the result:

Theorem 1: $\mathcal{V}_\mathcal{C}^\mathcal{O}$ is independent of $\mathcal{C}$ if the complement of $\mathcal{O}$ is path connected.

The proof is given in Appendix C. In the following we shall mainly consider the case where the complement of $\mathcal{O}$ is path-connected and, in this case, we denote $\mathcal{V}_\mathcal{C}^\mathcal{O}$ by $\mathcal{V}^\mathcal{O}$.

The definition of $\eta_\phi : \mathcal{V}_\mathcal{C} \to \mathbb{C}$, $\tilde{\mathcal{V}}_\mathcal{C}$ and, in particular, $\mathcal{V}_\mathcal{C}^\mathcal{O}$ all depend, at least superficially, on the particular coordinate chosen on $\mathbb{P}$, that is the particular identification of $\mathbb{P}$ with $\mathbb{C} \cup \{\infty\}$. However the coefficients with which elements of $\mathcal{B}_\mathcal{C}$ are combined to constitute elements of $\mathcal{V}_\mathcal{C}$ should be regarded as densities and then a change of coordinate on $\mathbb{P}$ induces an endomorphism of $\mathcal{V}_\mathcal{C}$ which relates the definitions of the space $\mathcal{V}_\mathcal{C}^\mathcal{O}$ which we would get with the different choices of coordinates, because η_ϕ only changes by an overall factor (albeit a function of the ζ_i). In this way $\mathcal{V}_\mathcal{C}^\mathcal{O}$, etc. can be regarded as coordinate independent.

Suppose that $\mathcal{O} \subset \mathcal{O}'$ and $\mathcal{C} \cap \mathcal{O}' = \varnothing$ with $\mathcal{C}^o \neq \varnothing$. Then if a sequence $\Psi = (\Psi_j) \in \tilde{\mathcal{V}}_\mathcal{C}$ is such that $\eta_\phi(\Psi_j)$ is convergent for all $\phi \in \mathcal{B}_{\mathcal{O}'}$ it follows that it is convergent for all $\phi \in \mathcal{B}_\mathcal{O} \subset \mathcal{B}_{\mathcal{O}'}$. In these circumstances, if $\eta_\phi(\Psi)$ vanishes for all $\phi \in \mathcal{B}_\mathcal{O}$, it follows that $\eta_{\phi'}(\Psi)$ will vanish for all $\phi' \in \mathcal{B}_{\mathcal{O}'}$, because each $\eta_\phi(\Psi)$ is the analytic continuation of

$\eta_{\boldsymbol{\phi}}(\boldsymbol{\Psi})$, for some $\boldsymbol{\phi} \in \mathcal{B}_{\mathcal{O}}$; the converse is also immediate because $\mathcal{B}_{\mathcal{O}} \subset \mathcal{B}_{\mathcal{O}'}$. Thus members of an equivalence class in $\mathcal{V}_{\mathcal{C}}^{\mathcal{O}'}$ are also in the same equivalence class in $\mathcal{V}_{\mathcal{C}}^{\mathcal{O}}$. We thus have an injection $\mathcal{V}^{\mathcal{O}'} \to \mathcal{V}^{\mathcal{O}}$, and we can regard $\mathcal{V}^{\mathcal{O}'} \subset \mathcal{V}^{\mathcal{O}}$. Since $\mathcal{B}_c$ is dense in $\mathcal{V}^{\mathcal{O}}$, it follows that $\mathcal{V}^{\mathcal{O}'}$ is also.

Given a subset $\mathcal{C} \subset \mathbb{P}$ with $\mathcal{C}^o \neq \varnothing$, $\mathcal{B}_{\mathcal{C}}$ is dense in a collection of spaces $\mathcal{V}^{\mathcal{O}}$, with $\mathcal{C} \cap \mathcal{O} = \varnothing$. Given open sets $\mathcal{O}_1$ and $\mathcal{O}_2$ such that the complement of $\mathcal{O}_1 \cup \mathcal{O}_2$ contains an open set, $\mathcal{B}_{\mathcal{C}}$ will be dense in both $\mathcal{V}^{\mathcal{O}_1}$ and $\mathcal{V}^{\mathcal{O}_2}$ if $\mathcal{C}$ is contained in the complement of $\mathcal{O}_1 \cup \mathcal{O}_2$ and $\mathcal{C}^o \neq \varnothing$. The collection of topological vector spaces $\mathcal{V}^{\mathcal{O}}$, where $\mathcal{O}$ is an open subset of the Riemann sphere whose complement is path-connected, forms in some sense the space of states of the meromorphic field theory we are considering.

A *vertex operator* can be defined for $\psi \in V$ as an operator $V(\psi, z) : \mathcal{V}^{\mathcal{O}} \to \mathcal{V}^{\mathcal{O}'}$, where $z \in \mathcal{O}$ but $z \notin \mathcal{O}' \subset \mathcal{O}$, by defining its action on the dense subset $\mathcal{B}_{\mathcal{C}}$, where $\mathcal{C} \cap \mathcal{O} = \varnothing$

$$V(\psi, z)\psi = V(\psi, z)V(\psi_1, z_1)V(\psi_2, z_2) \cdots V(\psi_n, z_n)\Omega \tag{3.12}$$

and $\psi \in \mathcal{B}_{\mathcal{C}}$. The image is in $\mathcal{V}_{\mathcal{C}'}$ for any $\mathcal{C}' \supset \mathcal{C}$ which contains z, and we can choose $\mathcal{C}'$ such that $\mathcal{C}' \cap \mathcal{O}' = \varnothing$. This then extends by linearity to a map $\mathcal{V}_{\mathcal{C}} \to \mathcal{V}_{\mathcal{C}'}$. To show that it induces a map $\mathcal{V}_{\mathcal{C}}^{\mathcal{O}} \to \mathcal{V}_{\mathcal{C}'}^{\mathcal{O}'}$ we need to show that if $\Psi^j, \in \mathcal{V}_{\mathcal{C}}^{\mathcal{O}}, \to 0$ as $j \to \infty$, then $V(\psi, z)\Psi^j \to 0$ as $j \to \infty$; *i.e.* if $\eta_{\boldsymbol{\phi}}(\Psi^j) \to 0$ for all $\boldsymbol{\phi} \in \mathcal{B}_{\mathcal{O}}$, then $\eta_{\boldsymbol{\phi}'}(V(\psi, z)\Psi^j) \to 0$ for all $\boldsymbol{\phi}' \in \mathcal{B}_{\mathcal{O}'}$. But $\eta_{\boldsymbol{\phi}'}(V(\psi, z)\Psi^j) = \eta_{\boldsymbol{\phi}}(\Psi^j)$ where $\boldsymbol{\phi} = V(\psi, z)\boldsymbol{\phi}' \in \mathcal{B}_{\mathcal{O}}$ and so tends to zero as required. It is straightforward to show that the vertex operator $V(\psi, z)$ is continuous. We shall refer to these vertex operators also as *meromorphic fields*.

It follows directly from the invariance of the amplitudes under permutations that

Proposition 2: If $z, \zeta \in \mathcal{O}$, $z \neq \zeta$, and $\phi, \psi \in V$, then

$$V(\phi, z)V(\psi, \zeta) = V(\psi, \zeta)V(\phi, z) \tag{3.13}$$

as an identity on $\mathcal{V}^{\mathcal{O}}$.

This result, that the vertex operators, $V(\psi, z)$, commute at different z in a (bosonic) meromorphic conformal field theory, is one which should hold morally, but normally one has to attach a meaning to it in some other sense, such as analytic continuation (compare for example [15]).

4 Möbius Invariance

In order to proceed much further, without being dependent in some essential way on how the Riemann sphere is identified with the complex plane (and infinity), we shall need to assume that the amplitudes $\mathcal{A}$ have some sort of Möbius invariance. We shall say that the densities in $\mathcal{A}$ are invariant under the Möbius transformation γ, where

$$\gamma(z) = \frac{az + b}{cz + d}, \tag{4.1}$$

(and we can take $ad - bc = 1$), provided that the densities in (3.2) satisfy

$$\left\langle \prod_{j=1}^{n} V(\psi_j, z_j) \right\rangle \prod_{j=1}^{n} (dz_j)^{h_j} = \left\langle \prod_{j=1}^{n} V(\psi_j, \zeta_j) \right\rangle \prod_{j=1}^{n} (d\zeta_j)^{h_j}, \quad \text{where } \zeta_j = \gamma(z_j), \qquad (4.2)$$

i.e.

$$\left\langle \prod_{j=1}^{n} V(\psi_j, z_j) \right\rangle = \left\langle \prod_{j=1}^{n} V(\psi_j, \gamma(z_j)) \right\rangle \prod_{j=1}^{n} (\gamma'(z_j))^{h_j} . \qquad (4.3)$$

Here $\psi_j \in V_{h_j}$. The Möbius transformations form the group $\mathcal{M} \cong \mathrm{SL}(2, \mathbb{C})/\mathbb{Z}_2$.
If $\mathcal{A}$ is invariant under the Möbius transformation γ, we can define an operator $U(\gamma)$:
$\mathcal{V}^{\mathcal{O}} \to \mathcal{V}^{\mathcal{O}_\gamma}$, where $\mathcal{O}_\gamma = \{\gamma(z) : z \in \mathcal{O}\}$, by defining it on the dense subset $\mathcal{B}_\mathcal{C}$ for some $\mathcal{C}$
with $\mathcal{C} \cap \mathcal{O} = \varnothing$, by

$$U(\gamma)\psi = \prod_{j=1}^{n} V(\psi_j, \gamma(z_j)) \prod_{j=1}^{n} (\gamma'(z_j))^{h_j} \Omega, \qquad (4.4)$$

where $\psi = V(\psi_1, z_1) \cdots V(\psi_n, z_n)\Omega \in \mathcal{B}_\mathcal{C}$. Again, this extends by linearity to a map defined
on $\mathcal{V}_\mathcal{C}$, and to show that it defines a map $\mathcal{V}_\mathcal{C}^{\mathcal{O}} \to \mathcal{V}_{\mathcal{C}_\gamma}^{\mathcal{O}_\gamma}$, where $\mathcal{C}_\gamma = \{\gamma(z) : z \in \mathcal{C}\}$, we again
need to show that if $\eta_{\boldsymbol{\phi}}(\Psi_j) \to 0$ for all $\boldsymbol{\phi} \in \mathcal{B}_\mathcal{O}$, then $\eta_{\boldsymbol{\phi}'}(U(\gamma)\Psi_j) \to 0$ for all $\boldsymbol{\phi}' \in \mathcal{B}_{\mathcal{O}_\gamma}$.
By the assumed invariance under γ, we have $\eta_{\boldsymbol{\phi}'}(U(\gamma)\Psi_j) = \eta_{\boldsymbol{\phi}}(\Psi_j)$, where $\boldsymbol{\phi} = U(\gamma^{-1})\boldsymbol{\phi}'$,
and the result follows.
It follows immediately from the definition of $U(\gamma)$ that $U(\gamma)\Omega = \Omega$ (where we have identified
$\Omega \in \mathcal{V}^{\mathcal{O}}$ with $\Omega \in \mathcal{V}^{\mathcal{O}_\gamma}$ as explained in Section 3). Furthermore,

$$U(\gamma)V(\psi, z)U(\gamma^{-1}) = V(\psi, \gamma(z))\gamma'(z)^h, \quad \text{for } \psi \in V_h. \qquad (4.5)$$

By choosing a point $z_0 \notin \mathcal{O}$, we can identify V with a subspace of $\mathcal{V}^{\mathcal{O}}$ by the map $\psi \mapsto$
$V(\psi, z_0)\Omega$; this map is an injection provided that $\mathcal{A}$ is invariant under an infinite subgroup
of $\mathcal{M}$ which maps z_0 to an infinite number of distinct image points. For, if

$$\langle \prod_{i=1}^{n} V(\psi_i, z_i)V(\psi, \zeta)\rangle \qquad (4.6)$$

vanishes for $\zeta = z_0$ for all ψ_i and z_i, then by the invariance property, the same holds for an
infinite number of ζ's. Regarded as a function of ζ, (4.6) defines a meromorphic function
with infinitely many zeros; it therefore vanishes identically, thus implying that $\psi = 0$.
In the following we shall use elements of $\mathrm{SL}(2, \mathbb{C})$ to denote the corresponding elements of
$\mathcal{M}$ where no confusion will result, so that

$$\text{if} \quad \gamma = \begin{pmatrix} a & b \\ c & d \end{pmatrix} \qquad \gamma(z) = \frac{az + b}{cz + d}. \qquad (4.7)$$

An element of $\mathcal{M}$ has either one or two fixed points or is the identity. The one-parameter
complex subgroups of $\mathcal{M}$ are either conjugate to the translation group $z \mapsto z + \lambda$ (one fixed
point) or the dilatation group $z \mapsto e^\lambda z$ (two fixed points).

Now, first, consider a theory which is invariant under the translation group $z \mapsto \tau_\lambda(z) = z + \lambda$. Then, if $\tau_\lambda = e^{\lambda L_{-1}}$, and we do not distinguish between $U(L_{-1})$ and L_{-1} in terms of notation, from (4.5) we have

$$e^{\lambda L_{-1}} V(\psi, z) e^{-\lambda L_{-1}} = V(\psi, z + \lambda). \tag{4.8}$$

[If, instead, we had a theory invariant under a subgroup of the Möbius group conjugate to the translation group, $\{\gamma_0^{-1} \tau_\lambda \gamma_0 : \lambda \in \mathbb{C}\}$ say, and if $\zeta = \gamma_0(z)$, $\zeta \mapsto \zeta' = \zeta + \lambda$ under $\gamma_0^{-1} \tau_\lambda \gamma_0$; then, if $\widehat{V}(\psi, \zeta) = V(\psi, z) \gamma_0'(z)^{-h}$ and $\hat{L}_{-1} = \gamma_0^{-1} L_{-1} \gamma_0$, then $e^{\lambda \hat{L}_{-1}} \widehat{V}(\psi, \zeta) e^{-\lambda \hat{L}_{-1}} = \widehat{V}(\psi, \zeta + \lambda)$.]

Consider now a theory which is invariant under the whole Möbius group. We can pick a group conjugate to the translation group, and we can change coordinates so that $z = \infty$ is the fixed point of the selected translation group. (In particular, this defines an identification of $\mathbb{P}$ with $\mathbb{C} \cup \{\infty\}$ up to a Euclidean or scaling transformation of $\mathbb{C}$.) If we select a point z_0 to define the injection $V \to V^\mathcal{O}$, $z_0 \notin \mathcal{O}$, we have effectively selected two fixed points. Without loss of generality, we can choose $z_0 = 0$. Then

$$\psi = V(\psi, 0)\Omega \in V^\mathcal{O}. \tag{4.9}$$

We can then introduce naturally two other one-parameter groups, one generated by L_0 which fixes both 0 and ∞ (the group of dilatations or scaling transformations), and another which fixes only 0, generated by L_1 (the group of special conformal transformations). Then

$$e^{\lambda L_{-1}}(z) = z + \lambda, \qquad e^{\lambda L_0}(z) = e^\lambda z, \qquad e^{\lambda L_1}(z) = \frac{z}{1 - \lambda z}, \tag{4.10}$$

$$e^{\lambda L_{-1}} = \begin{pmatrix} 1 & \lambda \\ 0 & 1 \end{pmatrix}, \qquad e^{\lambda L_0} = \begin{pmatrix} e^{\frac{1}{2}\lambda} & 0 \\ 0 & e^{-\frac{1}{2}\lambda} \end{pmatrix}, \qquad e^{\lambda L_1} = \begin{pmatrix} 1 & 0 \\ -\lambda & 1 \end{pmatrix}, \tag{4.11}$$

and thus

$$L_{-1} = \begin{pmatrix} 0 & 1 \\ 0 & 0 \end{pmatrix}, \qquad L_0 = \begin{pmatrix} \frac{1}{2} & 0 \\ 0 & -\frac{1}{2} \end{pmatrix}, \qquad L_1 = \begin{pmatrix} 0 & 0 \\ -1 & 0 \end{pmatrix}. \tag{4.12}$$

In particular, it then follows that

$$[L_m, L_n] = (m - n) L_{m+n}, \qquad m, n = 0, \pm 1. \tag{4.13}$$

We also have that $L_n \Omega = 0$, $n = 0, \pm 1$. With this parametrisation, the operator corresponding to the Möbius transformation γ, defined in (4.7), is given as (see [15])

$$U(\gamma) = \exp\left(\frac{b}{d} L_{-1}\right) \left(\frac{\sqrt{ad - bc}}{d}\right)^{L_0} \exp\left(-\frac{c}{d} L_1\right). \tag{4.14}$$

For $\psi \in V_h$, by (4.5), $U(\gamma) V(\psi, z) U(\gamma^{-1}) = V(\psi, \gamma(z)) \gamma'(z)^h$, and so, by (4.9), $U(\gamma)\psi = \lim_{z \to 0} V(\psi, \gamma(z)) \Omega \gamma'(z)^h$. From this it follows that,

$$L_0 \psi = h\psi, \qquad L_1 \psi = 0, \qquad L_{-1} \psi = V'(\psi, 0)\Omega. \tag{4.15}$$

14

Thus $L_0 = \delta$ acting on V.

Henceforth we shall assume that our theory defined by $\mathcal{A}$ is Möbius invariant.

Having chosen an identification of $\mathbb{P}$ with $\mathbb{C} \cup \{\infty\}$ and of V with a subspace of $\mathcal{V}^{\mathcal{O}}$, we can now also define vertex operators for $\boldsymbol{\psi} = \prod_{j=1}^{n} V(\psi_j, z_j)\Omega \in \mathcal{B}_C$ by

$$V(\boldsymbol{\psi}, z) = \prod_{j=1}^{n} V(\psi_j, z_j + z). \tag{4.16}$$

Then $V(\boldsymbol{\phi}, z)$ is a continuous operator $\mathcal{V}^{\mathcal{O}_1} \to \mathcal{V}^{\mathcal{O}_2}$, provided that $z_j + z \notin \mathcal{O}_2 \subset \mathcal{O}_1$ but $z_j + z \in \mathcal{O}_1$, $1 \leq j \leq n$.

We can further extend the definition of $V(\boldsymbol{\psi}, z)$ by linearity from $\boldsymbol{\psi} \in \mathcal{B}_C$ to vectors $\Psi \in \mathcal{V}_C^{\mathcal{O}}$, the image of $\mathcal{V}_C$ in $\mathcal{V}_C^{\mathcal{O}}$, to obtain a continuous linear operator $V(\Psi, z) : \mathcal{V}^{\mathcal{O}_1} \to \mathcal{V}^{\mathcal{O}_2}$, where $C_z \cap \mathcal{O}_2 = \varnothing$, $\mathcal{O}_2 \subset \mathcal{O}_1$ and $C_z \subset \mathcal{O}_1$ for $C_z = \{\zeta + z : \zeta \in C\}$. One might be tempted to try to extend the definition of the vertex operator even further to states in $\mathcal{V}_C^{\mathcal{O}} \cong \mathcal{V}^{\mathcal{O}}$, but the corresponding operator will then only be well-defined on a suitable dense subspace of $\mathcal{V}^{\mathcal{O}_1}$. For the vertex operator associated to $\Psi \in \mathcal{V}_C^{\mathcal{O}}$, we again have

$$e^{\lambda L_{-1}} V(\Psi, z) e^{-\lambda L_{-1}} = V(\Psi, z + \lambda), \qquad V(\Psi, 0)\Omega = \Psi. \tag{4.17}$$

Furthermore,

$$V(\Psi, z)V(\phi, \zeta) = V(\phi, \zeta)V(\Psi, z), \tag{4.18}$$
$$V(\Psi, z)\Omega = e^{zL_{-1}}\Psi \tag{4.19}$$

for any $\phi \in V$, $\zeta \notin C_z$. [In (4.18), the lefthand and righthand sides are to be interpreted as maps $\mathcal{V}^{\mathcal{O}_1} \to \mathcal{V}^{\mathcal{O}_2}$, with $V(\phi, \zeta) : \mathcal{V}^{\mathcal{O}_1} \to \mathcal{V}^{\mathcal{O}_L}$ and $V(\Psi, z) : \mathcal{V}^{\mathcal{O}_L} \to \mathcal{V}^{\mathcal{O}_2}$ on the lefthand side and $V(\Psi, z) : \mathcal{V}^{\mathcal{O}_1} \to \mathcal{V}^{\mathcal{O}_R}$ and $V(\phi, \zeta) : \mathcal{V}^{\mathcal{O}_R} \to \mathcal{V}^{\mathcal{O}_2}$ on the righthand side, where $\mathcal{O}_2 \subset \mathcal{O}_L \subset \mathcal{O}_1$, $\mathcal{O}_2 \subset \mathcal{O}_R \subset \mathcal{O}_1$, $\zeta \in \mathcal{O}_R \cap \mathcal{O}_L^c \cap \mathcal{O}_2^c$ and $C_z \subset \mathcal{O}_L \cap \mathcal{O}_R^c \cap \mathcal{O}_2^c$ (where $\mathcal{O}_2^c$ denotes the complement of $\mathcal{O}_2$, etc.). Equation (4.19) holds in $\mathcal{V}^{\mathcal{O}}$ with $C_z \cap \mathcal{O} = \varnothing$.]

Actually, these two conditions characterise the vertex operator already uniquely:

Theorem 3 [Uniqueness]: For $\Psi \in \mathcal{V}_C^{\mathcal{O}}$, the operator $V(\Psi, z)$ is uniquely characterised by the conditions (4.18) and (4.19).

The proof is essentially that contained in [15]: If $W(z)V(\phi, \zeta) = V(\phi, \zeta)W(z)$ for $\phi \in V$, $\zeta \notin C_z$, and $W(z)\Omega = e^{zL_{-1}}\Psi$, it follows that, for $\Phi \in \mathcal{V}_{C'}^{\mathcal{O}'}$, $W(z)V(\Phi, \zeta) = V(\Phi, \zeta)W(z)$ provided that $C_\zeta' \cap C_z = \varnothing$ and so

$$
\begin{aligned}
W(z)e^{\zeta L_{-1}}\Phi &= W(z)V(\Phi, \zeta)\Omega = V(\Phi, \zeta)W(z)\Omega \\
&= V(\Phi, \zeta)e^{zL_{-1}}\Psi \\
&= V(\Phi, \zeta)V(\Psi, z)\Omega \\
&= V(\Psi, z)V(\Phi, \zeta)\Omega \\
&= V(\Psi, z)e^{\zeta L_{-1}}\Phi
\end{aligned}
$$

for all $\Phi \in \mathcal{V}_{C'}^{\mathcal{O}'}$, which is dense in $\mathcal{V}^{\mathcal{O}'}$, showing that $W(z) = V(\Psi, z)$.

15

From this uniqueness result and (4.5) we can deduce the commutators of vertex operators $V(\psi, z)$, $\psi \in V_h$, with L_{-1}, L_0, L_1:

$$[L_{-1}, V(\psi, z)] = \frac{d}{dz} V(\psi, z) \tag{4.20}$$

$$[L_0, V(\psi, z)] = z \frac{d}{dz} V(\psi, z) + h V(\psi, z) \tag{4.21}$$

$$[L_1, V(\psi, z)] = z^2 \frac{d}{dz} V(\psi, z) + 2hz V(\psi, z) . \tag{4.22}$$

We recall from (4.15) that $L_1 \psi = 0$ and $L_0 \psi = h\psi$ if $\psi \in V_h$; if $L_1 \psi = 0$, ψ is said to be *quasi-primary*.

The definition (4.16) immediately implies that, for states $\psi, \phi \in V$, $V(\psi, z)V(\phi, \zeta) = V(V(\psi, z - \zeta)\phi, \zeta)$. This statement generalises to the key duality result of Theorem 4, which can be seen to follow from the uniqueness theorem:

Theorem 4 [Duality]: If $\Psi \in V_{\mathcal{C}}^{\mathcal{O}}$ and $\Phi \in V_{\mathcal{C}'}^{\mathcal{O}'}$, where $\mathcal{C}_z \cap \mathcal{C}'_\zeta = \varnothing$, then

$$V(\Psi, z)V(\Phi, \zeta) = V(V(\Psi, z - \zeta)\Phi, \zeta) . \tag{4.23}$$

[In (4.23), the lefthand and righthand sides are to be interpreted as maps $\mathcal{V}^{\mathcal{O}_1} \rightarrow \mathcal{V}^{\mathcal{O}_2}$, with $V(\Phi, \zeta) : \mathcal{V}^{\mathcal{O}_1} \rightarrow \mathcal{V}^{\mathcal{O}_L}$ and $V(\Psi, z) : \mathcal{V}^{\mathcal{O}_L} \rightarrow \mathcal{V}^{\mathcal{O}_2}$ on the lefthand side and $V(\Psi, z - \zeta)\Phi \in V_{\mathcal{C}_z - \zeta \cup \mathcal{C}'}$ where $\mathcal{O}_2 \subset \mathcal{O}_L \subset \mathcal{O}_1$, $\mathcal{C}_z \subset \mathcal{O}_L \cap \mathcal{O}_2^c$ and $\mathcal{C}'_\zeta \subset \mathcal{O}_1 \cap \mathcal{O}_L^c$.]

The result follows from the uniqueness theorem on noting that

$$V(\Phi, z)V(\Psi, \zeta)\Omega = V(\Phi, z)e^{\zeta L_{-1}}\Psi = e^{\zeta L_{-1}}V(\Phi, z - \zeta)\Psi = V(V(\Phi, z - \zeta)\Psi), \zeta)\Omega . \tag{4.24}$$

5 Modes, Fock Spaces and the Equivalence of Theories

The concept of equivalence between two meromorphic field theories in our definition could be formulated in terms of the whole collection of spaces $\mathcal{V}^{\mathcal{O}}$, where $\mathcal{O}$ ranges over the open subsets of $\mathbb{P}$ with path-connected complement, but this would be very unwieldy. In fact, each meromorphic field theory has a Fock space at its heart and we can focus on this in order to define (and, in practice, test for) the equivalence of theories. To approach this we first need to introduce the concept of the modes of a vertex operator.

It is straightforward to see that we can construct contour integrals of vectors in $\mathcal{V}^{\mathcal{O}}$, *e.g.* of the form

$$\int_{C_1} dz_1 \int_{C_2} dz_2 \ldots \int_{C_r} dz_r \mu(z_1, z_2, \ldots, z_r) \prod_{i=1}^{n} V(\psi_i, z_i)\Omega , \tag{5.1}$$

where $r \leq n$ and the weight function μ is analytic in some neighbourhood of $C_1 \times C_2 \times \cdots \times C_r$ and the distances $|z_i - z_j|$, $i \neq j$, are bounded away from 0 on this set. In this way we can define the modes

$$V_n(\psi) = \oint_C z^{h+n-1} V(\psi, z) dz , \qquad \text{for } \psi \in V_h , \tag{5.2}$$

as linear operators on $\mathcal{V}_\mathcal{C}^\mathcal{O}$, where C encircles $\mathcal{C}$ and $C \subset \mathcal{O}$ with $\infty \in \mathcal{O}$ and $0 \in C$, and we absorb a factor of $1/2\pi i$ into the definition of the symbol $\oint$. The meromorphicity of the amplitudes allows us to establish

$$V(\psi, z) = \sum_{n=-\infty}^{\infty} V_n(\psi) z^{-n-h} \tag{5.3}$$

with convergence with respect to the topology of $\mathcal{V}^{\mathcal{O}'}$ for an appropriate $\mathcal{O}'$.

The definition of $V_n(\psi)$ is independent of C if it is taken to be a simple contour encircling the origin once positively. Further, if $\mathcal{O}_2 \subset \mathcal{O}_1$, $\mathcal{V}^{\mathcal{O}_1} \subset \mathcal{V}^{\mathcal{O}_2}$ and if $\infty \in \mathcal{O}_2, 0 \notin \mathcal{O}_1$, the definition of $V_n(\psi)$ on $\mathcal{V}^{\mathcal{O}_1}$, $\mathcal{V}^{\mathcal{O}_2}$, agrees on $\mathcal{V}^{\mathcal{O}_1}$, which is dense in $\mathcal{V}^{\mathcal{O}_2}$, so that we may regard the definition as independent of $\mathcal{O}$ also. $V_n(\psi)$ depends on our choice of 0 and ∞ but different choices can be related by Möbius transformations.

We define the Fock space $\mathcal{H}^\mathcal{O} \subset \mathcal{V}^\mathcal{O}$ to be the space spanned by finite linear combinations of vectors of the form

$$\Psi = V_{n_1}(\psi_1) V_{n_2}(\psi_2) \cdots V_{n_N}(\psi_N)\Omega, \tag{5.4}$$

where $\psi_j \in V$ and $n_j \in \mathbb{Z}$, $1 \le j \le N$. Then, by construction, $\mathcal{H}^\mathcal{O}$ has a countable basis. It is easy to see that $\mathcal{H}^\mathcal{O}$ is dense in $\mathcal{V}^\mathcal{O}$. Further it is clear that $\mathcal{H}^\mathcal{O}$ is independent of $\mathcal{O}$, and, where there is no ambiguity, we shall denote it simply by $\mathcal{H}$. It does however depend on the choice of 0 and ∞, but different choices will be related by the action of the Möbius group again.

It follows from (4.19) that

$$V(\psi, 0)\Omega = \psi \tag{5.5}$$

which implies that

$$V_n(\psi)\Omega = 0 \qquad \text{if } n > -h \tag{5.6}$$

and

$$V_{-h}(\psi)\Omega = \psi . \tag{5.7}$$

Thus $V \subset \mathcal{H}$.

Since ∞ and 0 play a special role, it is not surprising that L_0, the generator of the subgroup of $\mathcal{M}$ preserving them, does as well. From (4.21) it follows that

$$[L_0, V_n(\psi)] = -nV_n(\psi), \tag{5.8}$$

so that for Ψ defined by (5.4),

$$L_0\Psi = h\Psi, \qquad \text{where } h = -\sum_{j=1}^{N} n_j . \tag{5.9}$$

Thus

$$\mathcal{H} = \bigoplus_{h \in \mathbb{Z}} \mathcal{H}_h, \qquad \text{where } V_h \subset \mathcal{H}_h , \tag{5.10}$$

where $\mathcal{H}_h$ is the subspace spanned by vectors of the form (5.4) for which $h = \sum_j n_j$.

Thus L_0 has a spectral decomposition and the $\mathcal{H}_h$, $h \in \mathbf{Z}$, are the eigenspaces of L_0. They have countable dimensions but here we shall only consider theories for which their dimensions are finite. (This is not guaranteed by the finite-dimensionality of V; in fact, in practice, it is not easy to determine whether these spaces are finite-dimensional or not, although it is rather obvious in many examples.)

We can define vertex operators for the vectors (5.4) by

$$V(\Psi, z) = \oint_{\mathcal{C}_1} z_1^{h_1+n_1-1} V(\psi_1, z+z_1) dz_1 \cdots \oint_{\mathcal{C}_N} z_N^{h_N+n_N-1} V(\psi_N, z+z_N) dz_N, \qquad (5.11)$$

where the $\mathcal{C}_j$ are contours about 0 with $|z_i| > |z_j|$ if $i < j$. We can then replace the densities (3.1) by the larger class $\mathcal{A}'$ of densities

$$\langle V(\Psi_1, z_1) V(\Psi_2, z_2) \cdots V(\Psi_n, z_n) \rangle \prod_{j=1}^{n} (dz_j)^{h_j}, \qquad (5.12)$$

where $\Psi_j \in \mathcal{H}_{h_j}$, $1 \le j \le n$. It is not difficult to see that replacing $\mathcal{A}$ with $\mathcal{A}'$, *i.e.* replacing V with $\mathcal{H}$, does not change the definition of the spaces $\mathcal{V}^O$. Theorem 3 (Uniqueness) and Theorem 4 (Duality) will still hold if we replace $\mathcal{V}_C^O$ with H_C^O, the space we would obtain if we started with $\mathcal{H}$ rather than V, *etc.* These theorems enable the Möbius transformation properties of vertex operators to be determined (see Appendix D).

However, if we use the whole of $\mathcal{H}$ as a starting point, the Möbius properties of the densities $\mathcal{A}'$ can not be as simple as in Section 4 because not all $\psi \in \mathcal{H}$ have the *quasi-primary* property $L_1 \psi = 0$. But we can introduce the subspaces of quasi-primary vectors within $\mathcal{H}$ and $\mathcal{H}_h$,

$$\mathcal{H}^Q = \{\Psi \in \mathcal{H} : L_1 \Psi = 0\}, \qquad \mathcal{H}_h^Q = \{\Psi \in \mathcal{H}_h : L_1 \Psi = 0\}, \qquad \mathcal{H}^Q = \bigoplus_h \mathcal{H}_h^Q. \qquad (5.13)$$

$V \subset \mathcal{H}^Q$ and $\mathcal{H}^Q$ is the maximal V which will generate the theory with the same spaces $\mathcal{V}^O$ and with agreement of the densities. [Under the cluster decomposition assumption of Section 7, $\mathcal{H}$ is generated from $\mathcal{H}^Q$ by the action of the Möbius group or, more particularly, L_{-1}. See Appendix D.]

We are now in a position to define the equivalence of two theories. A theory specified by a space V and amplitudes $\mathcal{A} = \{f\}$, leading to a quasi-primary space $\mathcal{H}^Q$, is said to be equivalent to the theory specified by a space $\hat{V}$ and amplitudes $\hat{\mathcal{A}} = \{\hat{f}\}$, leading to a quasi-primary space $\hat{\mathcal{H}}^Q$, if there are graded injections $\iota : V \to \hat{\mathcal{H}}^Q$ (*i.e.* $\iota(V_h) \subset \hat{\mathcal{H}}_h^Q$) and $\hat{\iota} : \hat{V} \to \mathcal{H}^Q$ which map amplitudes to amplitudes.

Many calculations in conformal field theory are most easily performed in terms of modes of vertex operators which capture in essence the algebraic structure of the theory. In particular, the modes of the vertex operators define what is usually called a W-algebra; this can be seen as follows.

The duality property of the vertex operators can be rewritten in terms of modes as

$$\begin{aligned}
V(\Phi, z) V(\Psi, \zeta) &= V(V(\Phi, z-\zeta)\Psi, \zeta) \\
&= \sum_n V(V_n(\Phi)\Psi, \zeta)(z-\zeta)^{-n-h_\phi}, \qquad (5.14)
\end{aligned}$$

where $L_0 \Psi = h_\Psi \Psi$ and $L_0 \Phi = h_\Phi \Phi$, and $\Psi, \Phi \in \mathcal{H}$. We can then use the usual contour techniques of conformal field theory to derive from this formula commutation relations for the respective modes. Indeed, the commutator of two modes $V_m(\Phi)$ and $V_n(\Psi)$ acting on $\mathcal{B}_\mathcal{C}$ is defined by

$$
[V_m(\Phi), V_n(\Psi)] = \oint_{|z|>|\zeta|} dz \oint d\zeta \; z^{m+h_\Phi-1} \zeta^{n+h_\Psi-1} V(\Phi, z) V(\Psi, \zeta)
$$
$$
- \oint_{|\zeta|>|z|} dz \oint d\zeta \; z^{m+h_\Phi-1} \zeta^{n+h_\Psi-1} V(\Phi, z) V(\Psi, \zeta) \qquad (5.15)
$$

where the contours on the right-hand side encircle $\mathcal{C}$ anti-clockwise. We can then deform the two contours so as to rewrite (5.15) as

$$
[V_m(\Phi), V_n(\Psi)] = \oint_0 \zeta^{n+h_\Psi-1} d\zeta \oint_\zeta z^{m+h_\Phi-1} dz \sum_l V(V_l(\Phi)\Psi, \zeta)(z-\zeta)^{-l-h_\Phi}, \qquad (5.16)
$$

where the z contour is a small positive circle about ζ and the ζ contour is a positive circle about $\mathcal{C}$. Only terms with $l \geq 1 - h_\Phi$ contribute, and the integral becomes

$$
[V_m(\Phi), V_n(\Psi)] = \sum_{N=-h_\Phi+1}^{\infty} \binom{m+h_\Phi-1}{m-N} V_{m+n}(V_N(\Phi)\Psi). \qquad (5.17)
$$

In particular, if $m \geq -h_\Phi + 1$, $n \geq -h_\Psi + 1$, then $m - N \geq 0$ in the sum, and $m + n \geq N + n \geq N - h_\Psi + 1$. This implies that the modes $\{V_m(\Psi) : m \geq -h_\Psi + 1\}$ close as a Lie algebra; the same also holds for $\{V_m(\Psi) : 0 \geq m \geq -h_\Psi + 1\}$.

As we shall discuss below in Section 7, in conformal field theory it is usually assumed that the amplitudes satisfy another property which guarantees that the spectrum of L_0 is bounded below by 0. If this is the case then the sum in (5.17) is also bounded above by h_Ψ.

6 Some Examples

Before proceeding further, we shall give a number of examples of theories that satisfy the axioms that we have specified so far.

6.1 The $U(1)$ theory

The simplest example is the case where V is a one-dimensional vector space, spanned by a vector J of weight 1, in which case we write $J(z) \equiv V(J, z)$. The amplitude of an odd number of J-fields is defined to vanish, and in the case of an even number it is given by

$$
\langle J(z_1) \cdots J(z_{2n}) \rangle = \frac{k^n}{2^n n!} \sum_{\pi \in S_{2n}} \prod_{j=1}^{n} \frac{1}{(z_{\pi(j)} - z_{\pi(j+n)})^2}, \qquad (6.1)
$$

$$
= k^n \sum_{\pi \in S'_{2n}} \prod_{j=1}^{n} \frac{1}{(z_{\pi(j)} - z_{\pi(j+n)})^2}, \qquad (6.2)
$$

where k is an arbitrary (real) constant and, in (6.1), S_{2n} is the permutation group on $2n$ object, whilst, in (6.2), the sum is restricted to the subset S'_{2n} of permutations $\pi \in S_{2n}$ such that $\pi(i) < \pi(i+n)$ and $\pi(i) < \pi(j)$ if $1 \le i < j \le n$. (This defines the amplitudes on a basis of V and we extend the definition by multilinearity.) It is clear that the amplitudes are meromorphic in z_j, and that they satisfy the locality condition. It is also easy to check that they are Möbius covariant, with the weight of J being 1.

From the amplitudes we can directly read off the operator product expansion of the field J with itself as

$$J(z)J(w) \sim \frac{k}{(z-w)^2} + O(1).$$ (6.3)

Comparing this with (5.14), and using (5.17) we then obtain

$$[J_n, J_m] = nk\delta_{n,-m}.$$ (6.4)

This defines a representation of the affine algebra $\hat{u}(1)$. It is clear from (6.4) that J_0 is a central element of this Lie algebra, $i.e.$ J_0 commutes with every mode J_n. Because of Schur's lemma, J_0 therefore takes a definite value in each irreducible representation. Indeed, since we have defined J_n as a mode of the vertex operator $J(z)$ acting on $\mathcal{V}^\mathcal{O}$, we have that $J_0 \equiv 0$; the modes of J acting in $\mathcal{V}^\mathcal{O}$ therefore define a specific representation of the algebra (6.4) (namely the representation where $J_0 = 0$). As we shall mention below, the meromorphic conformal field theory has other representations for which the mode J_0 acts non-trivially. It is therefore sometimes useful to consider the whole algebra (6.4) rather than the specific representation of it that is obtained from modes acting on $\mathcal{V}^\mathcal{O}$. The definition of this algebra requires however knowledge about the representation theory of the corresponding meromorphic conformal field theory.

6.2 Affine Lie algebra theory

Following Frenkel and Zhu [28], we can generalise this example to the case of an arbitrary finite-dimensional Lie algebra g. Suppose that the matrices t^a, $1 \le a \le \dim g$, provide a finite-dimensional representation of g so that $[t^a, t^b] = f^{ab}{}_c t^c$, where $f^{ab}{}_c$ are the structure constants of g. In this case, the space V is of dimension $\dim g$ and has a basis consisting of weight one states J^a, $1 \le a \le \dim g$. Again, we write $J^a(z) = V(J^a, z)$.

If K is any matrix which commutes with all the t^a, define

$$\kappa^{a_1 a_2 \cdots a_m} = \text{tr}(Kt^{a_1}t^{a_2}\cdots t^{a_m}).$$ (6.5)

The $\kappa^{a_1 a_2 \cdots a_m}$ have the properties that

$$\kappa^{a_1 a_2 a_3 \cdots a_{m-1} a_m} = \kappa^{a_2 a_3 \cdots a_{m-1} a_m a_1}$$ (6.6)

and

$$\kappa^{a_1 a_2 a_3 \cdots a_{m-1} a_m} - \kappa^{a_2 a_1 a_3 \cdots a_{m-1} a_m} = f^{a_1 a_2}{}_b \kappa^{b a_3 \cdots a_{m-1} a_m}.$$ (6.7)

With a cycle $\sigma = (i_1, i_2, \ldots, i_m) \equiv (i_2, \ldots, i_m, i_1)$ we associate the function

$$f_\sigma^{a_{i_1} a_{i_2} \cdots a_{i_m}}(z_{i_1}, z_{i_2}, \ldots, z_{i_m}) = \frac{\kappa^{a_{i_1} a_{i_2} \cdots a_{i_m}}}{(z_{i_1} - z_{i_2})(z_{i_2} - z_{i_3}) \cdots (z_{i_{m-1}} - z_{i_m})(z_{i_m} - z_{i_1})}.$$ (6.8)

If the permutation $\rho \in S_n$ has no fixed points, it can be written as the product of cycles of length at least 2, $\rho = \sigma_1\sigma_2\ldots\sigma_M$. We associate to ρ the product f_ρ of functions $f_{\sigma_1}f_{\sigma_2}\ldots f_{\sigma_M}$ and define $\langle J^{a_1}(z_1)J^{a_2}(z_2)\ldots J^{a_n}(z_n)\rangle$ to be the sum of such functions f_ρ over permutations $\rho \in S_n$ with no fixed point. Graphically, we can construct these amplitudes by summing over all graphs with n vertices where the vertices carry labels a_j, $1 \leq j \leq n$, and each vertex is connected by two directed lines (propagators) to other vertices, one of the lines at each vertex pointing towards it and one away. Thus, in a given graph, the vertices are divided into directed loops or cycles, each loop containing at least two vertices. To each loop, we associate a function as in (6.8) and to each graph we associate the product of functions associated to the loops of which it is composed.

Again, this defines the amplitudes on a basis of V and we extend the definition by multilinearity. The amplitudes are evidently local and meromorphic, and one can verify that they satisfy the Möbius covariance property with the weight of J^a being 1.

The amplitudes determine the operator product expansion to be of the form

$$J^a(z)J^b(w) \sim \frac{\kappa^{ab}}{(z-w)^2} + \frac{f^{ab}{}_c J^c(w)}{(z-w)} + O(1)\,, \tag{6.9}$$

and the algebra therefore becomes

$$[J^a_m, J^b_n] = f^{ab}{}_c J^c_{m+n} + m\kappa^{ab}\delta_{m,-n}\,. \tag{6.10}$$

This is (a representation of) the affine algebra $\hat{g}$ which plays a prominent rôle in Mark's and Terry's lectures [17, 29]. In the particular case where g is simple, $\kappa^{ab} = \mathrm{tr}(Kt^at^b) = k\delta^{ab}$, for some k (which is usually called the level), if we choose a suitable basis.

6.3 The Virasoro Theory

Again following Frenkel and Zhu [28], we can construct the Virasoro theory in a similar way. In this case, the space V is one-dimensional, spanned by a vector L of weight 2 and we write $L(z) = V(L, z)$. We can again construct the amplitudes graphically by summing over all graphs with n vertices, where the vertices are labelled by the integers $1 \leq j \leq n$, and each vertex is connected by two lines (propagators) to other vertices. In a given graph, the vertices are now divided into loops, each loop containing of at least two vertices. To each loop $\ell = (i_1, i_2, \ldots, i_m)$, we associate a function

$$f_\ell(z_{i_1}, z_{i_2}, \ldots, z_{i_m}) = \frac{c/2}{(z_{i_1} - z_{i_2})^2 (z_{i_2} - z_{i_3})^2 \cdots (z_{i_{m-1}} - z_{i_m})^2 (z_{i_m} - z_{i_1})^2}\,, \tag{6.11}$$

where c is a real number, and, to a graph, the product of the functions associated to its loops. [Since it corresponds a factor of the form $(z_i - z_j)^{-2}$ rather than $(z_i - z_j)^{-1}$, each line or propagator might appropriately be represented by a double line.] The amplitudes $\langle L(z_1)L(z_2)\ldots L(z_n)\rangle$ are then obtained by summing the functions associated with the various graphs with n vertices. [Note graphs related by reversing the direction of any loop contribute equally to this sum.]

21

These amplitudes determine the operator product expansion to be

$$L(z)L(\zeta) \sim \frac{c/2}{(z-\zeta)^4} + \frac{2L(\zeta)}{(z-\zeta)^2} + \frac{L'(\zeta)}{z-\zeta} + \qquad (6.12)$$

which leads to the Virasoro algebra

$$[L_m, L_n] = (m-n)L_{m+n} + \frac{c}{12}m(m^2-1)\delta_{m,-n}. \qquad (6.13)$$

6.4 Lattice Theories

Suppose that Λ is an even n-dimensional Euclidean lattice, so that, if $k \in \Lambda$, k^2 is an even integer. We introduce a basis $e_1, e_2, \ldots, e_n$ for Λ, so that any element k of Λ is an integral linear combination of these basis elements. We can introduce an algebra consisting of matrices γ_j, $1 \le j \le n$, such that $\gamma_j^2 = 1$ and $\gamma_i\gamma_j = (-1)^{e_i \cdot e_j}\gamma_j\gamma_i$. If we define $\gamma_k = \gamma_1^{m_1}\gamma_2^{m_2}\ldots\gamma_n^{m_n}$ for $k = m_1e_1 + m_2e_2 + \ldots + m_ne_n$, we can define quantities $\epsilon(k_1, k_2, \ldots, k_N)$, taking the values ± 1, by

$$\gamma_{k_1}\gamma_{k_2}\ldots\gamma_{k_N} = \epsilon(k_1, k_2, \ldots, k_N)\gamma_{k_1+k_2+\ldots+k_N}. \qquad (6.14)$$

We define the theory associated to the lattice Λ by taking V to have a basis $\{\psi_k : k \in \Lambda\}$, where the weight of ψ_k is $\frac{1}{2}k^2$, and, writing $V(\psi_k, z) = V(k, z)$, the amplitudes to be

$$\langle V(k_1, z_1)V(k_2, z_2)\cdots V(k_N, z_n)\rangle = \epsilon(k_1, k_2, \ldots, k_N)\prod_{1 \le i < j \le N}(z_i - z_j)^{k_i \cdot k_j} \qquad (6.15)$$

if $k_1 + k_2 + \ldots + k_N = 0$ and zero otherwise. The $\epsilon(k_1, k_2, \ldots, k_N)$ obey the conditions

$$\epsilon(k_1, k_2, \ldots, k_{j-1}, k_j, \quad k_{j+1}, k_{j+2}, \ldots, k_N)$$
$$= (-1)^{k_j \cdot k_{j+1}}\epsilon(k_1, k_2, \ldots, k_{j-1}, k_{j+1}, k_j, k_{j+2}, \ldots, k_N), \qquad (6.16)$$

which guarantees locality, and

$$\epsilon(k_1, k_2, \ldots \quad, k_j)\epsilon(k_{j+1}, \ldots, k_N)$$
$$= \epsilon(k_1 + k_2 + \ldots + k_j, k_{j+1} + \ldots + k_N)\epsilon(k_1, k_2, \ldots, k_j, k_{j+1}, \ldots, k_N), \qquad (6.17)$$

which implies the cluster decomposition property of Section 7.

The theory associated to the Leech lattice Λ_{24} plays an important role in the orbifold construction of the moonshine module $V^\natural$ (see [17], Section 2.6).

6.5 A non-conformal example

The above examples actually define meromorphic *conformal* field theories, but since we have not yet defined what we mean by a theory to be conformal, it is instructive to consider also an example that satisfies the above axioms but is not conformal. The simplest such case is a slight modification of the $U(1)$ example described in 6.1: again we take V to be a one-dimensional vector space, spanned by a vector K, but now the grade of K is taken to be 2. Writing $K(z) \equiv V(K, z)$, the amplitudes of an odd number of K-fields vanishes, and in the case of an even number we have

$$\langle K(z_1) \cdots K(z_{2n}) \rangle = \frac{k^n}{2^n n!} \sum_{\pi \in S_{2n}} \prod_{j=1}^{n} \frac{1}{(z_{\pi(j)} - z_{\pi(j+n)})^4} \,. \tag{6.18}$$

It is not difficult to check that these amplitudes satisfy all the axioms we have considered so far. In this case the operator product expansion is of the form

$$K(z)K(w) \sim \frac{k}{(z-w)^4} + O(1) \,, \tag{6.19}$$

and the algebra of modes is given by

$$[K_n, K_m] = \frac{k}{6} n(n^2 - 1) \delta_{n,-m} \,. \tag{6.20}$$

7 Cluster Decomposition

So far the axioms we have formulated do not impose any restrictions on the relative normalisation of amplitudes involving for example a different number of vectors in V, and the class of theories we are considering is therefore rather flexible. This is mirrored by the fact that it does not yet follow from our considerations that the spectrum of the operator L_0 is bounded from below, and since L_0 is in essence the energy of the corresponding physical theory, we may want to impose this constraint. In fact, we would like to impose the slightly stronger condition that the spectrum of L_0 is bounded by 0, and that there is precisely one state with eigenvalue equal to zero. This will follow (as we shall show momentarily) from the *cluster decomposition property*, which states that if we separate the variables of an amplitude into two sets and scale one set towards a fixed point (*e.g.* 0 or ∞) the behaviour of the amplitude is dominated by the product of two amplitudes, corresponding to the two sets of variables, multiplied by an appropriate power of the separation, specifically

$$\left\langle \prod_i V(\phi_i, \zeta_i) \prod_j V(\psi_j, \lambda z_j) \right\rangle \sim \left\langle \prod_i V(\phi_i, \zeta_i) \right\rangle \left\langle \prod_j V(\psi_j, z_j) \right\rangle \lambda^{-\Sigma h_j} \qquad \text{as } \lambda \to 0 \,, \tag{7.1}$$

where $\phi_i \in V_{h'_i}, \psi_j \in V_{h_j}$. It follows from Möbius invariance, that this is equivalent to

$$\left\langle \prod_i V(\phi_i, \lambda \zeta_i) \prod_j V(\psi_j, z_j) \right\rangle \sim \left\langle \prod_i V(\phi_i, \zeta_i) \right\rangle \left\langle \prod_j V(\psi_j, z_j) \right\rangle \lambda^{-\Sigma h'_i} \qquad \text{as } \lambda \to \infty \,. \tag{7.2}$$

The cluster decomposition property extends also to vectors $\Phi_i, \Psi_j \in \mathcal{H}$. It is not difficult to check that the examples of the previous section satisfy this condition.

We can use the cluster decomposition property to show that the spectrum of L_0 is non-negative and that the vacuum is, in a sense, unique. To this end let us introduce the projection operators defined by

$$P_N = \oint_0 u^{L_0 - N - 1} du, \qquad \text{for } N \in \mathbb{Z}. \tag{7.3}$$

In particular, we have

$$P_N \prod_j V(\psi_j, z_j)\Omega = \oint u^{h-N-1} V(\psi_j, uz_j)\Omega \, du \,, \tag{7.4}$$

where $h = \sum_j h_j$. It then follows that the P_N are projection operators

$$P_N P_M = 0, \text{ if } N \neq M, \qquad P_N^2 = P_N, \qquad \sum_N P_N = 1 \tag{7.5}$$

onto the eigenspaces of L_0,

$$L_0 P_N = N P_N \,. \tag{7.6}$$

For $N \leq 0$, we then have

$$
\left\langle \prod_i V(\phi_i, \zeta_i) P_N \prod_j V(\psi_j, z_j) \right\rangle = \oint_0 u^{\Sigma h_j - N - 1} \left\langle \prod_i V(\phi_i, \zeta_i) \prod_j V(\psi_j, uz_j) \right\rangle du
$$
$$
\sim \left\langle \prod_i V(\phi_i, \zeta_i) \right\rangle \left\langle \prod_j V(\psi_j, z_j) \right\rangle \oint_{|u|=\rho} u^{-N-1} d(\overline{u}.7)
$$

which, by taking $\rho \to 0$, is seen to vanish for $N < 0$ and, for $N = 0$, to give

$$P_0 \prod_j V(\psi_j, z_j)\Omega = \Omega \left\langle \prod_j V(\psi_j, z_j) \right\rangle \,, \tag{7.8}$$

and so $P_0 \Psi = \Omega \langle \Psi \rangle$. Thus the cluster decomposition property implies that $P_N = 0$ for $N < 0$, i.e. the spectrum of L_0 is non-negative, and that $\mathcal{H}_0$ is spanned by the vacuum Ω, which is thus the unique state with $L_0 = 0$.

As we have mentioned before the absence of negative eigenvalues of L_0 gives an upper bound on the order of the pole in the operator product expansion of two vertex operators, and thus to an upper bound in the sum in (5.17): if $\Phi, \Psi \in \mathcal{H}$ are of grade $L_0 \Phi = h_\Phi \Phi$, $L_0 \Psi = h_\Psi \Psi$, we have that $V_n(\Phi)\Psi = 0$ for $n > h_\Psi$ because otherwise $V_n(\Phi)\Psi$ would have a negative eigenvalue, $h_\Psi - n$, with respect to L_0. In particular, this shows that the leading singularity in $V(\Phi, z)V(\Psi, \zeta)$ is at most $(z - \zeta)^{-h_\Psi - h_\Phi}$.

The cluster property also implies that the space of states of the meromorphic field theory does not have any proper invariant subspaces in a suitable sense. To make this statement precise we must first give a meaning to a subspace of the space of states of a conformal field theory. The space of states of the theory is really the collection of topological spaces $\mathcal{V}^\mathcal{O}$, where $\mathcal{O}$ is an open subset of $\mathbb{P}$ whose complement is path-connected. Recall that $\mathcal{V}^\mathcal{O} \subset \mathcal{V}^{\mathcal{O}'}$ if $\mathcal{O} \supset \mathcal{O}'$. By a subspace of the conformal field theory we shall mean subspaces $\mathcal{U}^\mathcal{O} \subset \mathcal{V}^\mathcal{O}$ specified for each open subset $\mathcal{O} \subset \mathbb{P}$ with path-connected complement, such that $\mathcal{U}^\mathcal{O} = \mathcal{U}^{\mathcal{O}'} \cap \mathcal{V}^\mathcal{O}$ if $\mathcal{O} \supset \mathcal{O}'$.

Proposition 5: Suppose $\{\mathcal{U}^\mathcal{O}\}$ is an invariant closed subspace of $\{\mathcal{V}^\mathcal{O}\}$, i.e. $\mathcal{U}^\mathcal{O}$ is closed; $\mathcal{U}^\mathcal{O} = \mathcal{U}^{\mathcal{O}'} \cap \mathcal{V}^\mathcal{O}$ if $\mathcal{O} \supset \mathcal{O}'$; and $V(\psi, z)\mathcal{U}^\mathcal{O} \subset \mathcal{U}^{\mathcal{O}'}$ for all $\psi \in V$ where $z \in \mathcal{O}$, $z \notin \mathcal{O}' \subset \mathcal{O}$. Then $\{\mathcal{U}^\mathcal{O}\}$ is not a proper subspace, i.e. either $\mathcal{U}^\mathcal{O} = \mathcal{V}^\mathcal{O}$ for all $\mathcal{O}$, or $\mathcal{U}^\mathcal{O} = \{0\}$.

Proof: Suppose that $\phi \in \mathcal{U}^{\mathcal{O}}$, $\psi_j \in V$, $z_j \in \mathcal{O}$, $z_j \notin \mathcal{O}' \subset \mathcal{O}$ and consider

$$\prod_{j=1}^{n} V(\psi_j, z_j)\phi \in \mathcal{U}^{\mathcal{O}'} . \tag{7.9}$$

Now, taking a suitable integral of the left hand side,

$$P_0 \prod_{j=1}^{n} V(\psi_j, z_j)\phi = \lambda\Omega = \left\langle \prod_{j=1}^{n} V(\psi_j, z_j)\phi \right\rangle \Omega . \tag{7.10}$$

Thus either all the amplitudes involving ϕ vanish for all $\phi \in \mathcal{U}$, in which case $\mathcal{U} = \{0\}$, or $\Omega \in \mathcal{U}^{\mathcal{O}'}$ for some $\mathcal{O}'$, in which case it is easy to see that $\Omega \in \mathcal{U}^{\mathcal{O}}$ for all $\mathcal{O}$ and it follows that $\mathcal{U}^{\mathcal{O}} = \mathcal{V}^{\mathcal{O}}$ for all $\mathcal{O}$.

The cluster property also implies that the image of $\mathcal{B}_C$ in $\mathcal{V}_C^{\mathcal{O}}$ is faithful. To show that the images of the elements $\psi, \psi' \in \mathcal{B}_C$ are distinct we note that otherwise $\eta_\phi(\psi) = \eta_\phi(\psi')$ for all $\phi \in \mathcal{V}_{\mathcal{O}}$ with ϕ as in (3.4). By taking m in (3.4) to be sufficiently large, dividing the ζ_i, $1 \le i \le m$, into n groups which we allow to approach the z_j, $1 \le j \le n$, successively. The cluster property then shows that these must be the same points as the z'_j, $1 \le j \le n'$ in ψ' and that $\psi_i = \mu_j \psi'_j$ for some $\mu_j \in \mathbb{C}$ with $\prod_{j=1}^{n} \mu_j = 1$, establishing that $\psi = \psi'$ as elements of $\mathcal{V}_C^{\mathcal{O}}$.

8   Conformal Symmetry

So far our axioms do not require that our amplitudes correspond to a conformal field theory, only that the theory have a Möbius invariance, and indeed, as we shall see, the example in 6.3 is not conformally invariant. Further, what we shall discuss in the sections which follow the present one will not depend on a conformal structure, except where we explicitly mention it; in this sense, the present section is somewhat of an interlude. On the other hand, the conformal symmetry is crucial for more sophisticated considerations, in particular the theory on higher genus Riemann surfaces, and therefore forms a very important part of the general framework.

Let us first describe a construction be means of which a potentially new theory can be associated to a given theory, and explain then in terms of this construction what it means for a theory to be conformal.

Suppose we are given a theory that is specified by a space V and amplitudes $\mathcal{A} = \{f\}$. Let us denote by $\hat{V}$ the vector space that is obtained from V by appending a vector L of grade two, and let us write $V(L, z) = L(z)$. The amplitudes involving only fields in V are given as before, and the amplitude

$$\langle \prod_{j=1}^{m} L(w_j) \prod_{i=1}^{n} V(\psi_i, z_i) \rangle , \tag{8.1}$$

where $\psi_i \in V_{h_i}$ is defined as follows: we associate to each of the $n + m$ fields a point, and then consider the (ordered) graphs consisting of loops where each loop contains at most one of the points associated to the ψ_i, and each point associated to an L is a vertex of precisely one loop. (The points associated to ψ_i may be vertices of an arbitrary number of loops.) To each loop whose vertices only consist of points corresponding to L we associate the same function as before in Section 6.3, and to the loop $(z_i, w_{\pi(1)}, \ldots, w_{\pi(l)})$ we associate the expression

$$\prod_{j=1}^{l-1} \frac{1}{(w_{\pi(j)} - w_{\pi(j+1)})^2} \left(\frac{h_i}{(w_{\pi(1)} - z_i)(w_{\pi(l)} - z_i)} + \frac{1}{2} \left[\frac{1}{(w_{\pi(1)} - z_i)} \frac{d}{dz_i} + \frac{1}{(w_{\pi(l)} - z_i)} \frac{d}{dz_i} \right] \right) . \tag{8.2}$$

We then associate to each graph the product of the expressions associated to the different loops acting on the amplitude which is obtained from (8.1) upon removing $L(w_1) \cdots L(w_m)$, and the total amplitude is the sum of the functions associated to all such (ordered) graphs. (The product of the expressions of the form (8.2) is taken to be 'normal ordered' in the sense that all derivatives with respect to z_i only act on the amplitude that is obtained from (8.1) upon excising the Ls; in this way, the product is independent of the order in which the expressions of the from (8.2) are applied.)

We extend this definition by multilinearity to amplitudes defined for arbitrary states in $\hat{V}$. It follows immediately that the resulting amplitudes are local and meromorphic; in Appendix E we shall give a more explicit formula for the extended amplitudes, and use it to prove that the amplitudes also satisfy the Möbius covariance and the cluster property. In terms of conventional conformal field theory, the construction treats all quasiprimary states in V as primary with respect to the Virasoro algebra of the extended theory; this is apparent from the formula given in Appendix E.

We can generalise this definition further by considering in addition graphs which contain 'double loops' of the form (z_i, w_j) for those points z_i which correspond to states in V of grade two, where in this case neither z_i nor w_j can be a vertex of any other loop. We associate the function

$$\frac{c_\psi / 2}{(z_i - w_j)^4} \tag{8.3}$$

to each such loop (where c_ψ is an arbitrary linear functional on the states of weight two in V), and the product of the different expressions corresponding to the different loops in the graph act in this case on the amplitude (8.1), where in addition to all L-fields also the fields corresponding to $V(\psi_i, z_i)$ (for each i which appears in a double loop) have been removed. It is easy to see that this generalisation also satisfies all axioms.

This construction typically modifies the structure of the meromorphic field theory in the sense that it changes the operator product expansion (and thus the commutators of the corresponding modes) of vectors in V; this is for example the case for the 'non-conformal' model described in Section 6.5. If we introduce the field L as described above, we find the commutation relations

$$[L_m, K_n] = (m - n) K_{m+n} + \frac{c_K}{12} m (m^2 - 1) \delta_{m,-n} . \tag{8.4}$$

However, this is incompatible with the original commutator in (6.20): the Jacobi identity requires that

$$0 = [L_m, [K_n, K_l]] + [K_n, [K_l, L_m]] + [K_l, [L_m, K_n]]$$

$$
\begin{aligned}
&= (l - m)\, [K_n, K_{l+m}] + (m - n)\, [K_l, K_{m+n}] \\
&= \frac{k}{6} \delta_{l+m+n,0} \left[-(l - m)\, (l + m)\, \left((l + m)^2 - 1 \right) + (2m + l)\, l\, (l^2 - 1) \right] \\
&= \frac{k}{6} \delta_{l+m+n,0}\, m\, (m^2 - 1)\, (2l + m)
\end{aligned}
\tag{8.5}
$$

and this is not satisfied unless $k = 0$ (in which case the original theory is trivial). In fact, the introduction of L modifies (6.20) as

$$
\begin{aligned}
{} [K_m, K_n] &= \frac{k}{6} m(m^2 - 1)\delta_{m,-n} + \frac{k}{a}(m - n)Z_{m+n} \\
{} [L_m, K_n] &= \frac{c_K}{12} m(m^2 - 1)\delta_{m,-n} + (m - n)K_{m+n} \\
{} [Z_m, K_n] &= 0 \\
{} [L_m, Z_n] &= \frac{a}{12} m(m^2 - 1)\delta_{m,-n} + (m - n)Z_{m+n} \\
{} [L_m, L_n] &= \frac{c}{12} m(m^2 - 1)\delta_{m,-n} + (m - n)L_{m+n} \\
{} [Z_m, Z_n] &= 0,
\end{aligned}
\tag{8.6}
$$

where a is non-zero and can be set to equal k by rescaling Z, and the Z_n are the modes of a field of grade two. This set of commutators then satisfies the Jacobi identities. It also follows from the fact that the commutators of Z with K and Z vanish, that amplitudes that involve only K-fields and at least one Z-field vanish; in this way we recover the original amplitudes and commutators.

The construction actually depends on the choice of V (as well as the values of c_ψ and c), and therefore does not only depend on the equivalence class of meromorphic field theories. However, we can ask whether a given equivalence class of meromorphic field theories contains a representative $(V, \mathcal{A})$ (*i.e.* a choice of V that gives an equivalent description of the theory) for which $(\hat{V}, \hat{\mathcal{A}})$ is equivalent to $(V, \mathcal{A})$; if this is the case, we call the meromorphic field theory *conformal*. It follows directly from the definition of equivalence that a meromorphic field theory is conformal if and only if there exists a representative $(V, \mathcal{A})$ and a vector $L^0 \in V$ (of grade two) so that

$$
\left\langle \left(L(w) - L^0(w) \right) \prod_{j=1}^{n} V(\psi_j, z_j) \right\rangle = 0
\tag{8.7}
$$

for all $\psi_j \in V$, where L is defined as above. In this case, the linear functional c_ψ is defined by

$$
c_\psi = 2(w - z)^4 \langle L^0(w) V(\psi, z) \rangle .
\tag{8.8}
$$

In the case of the non-conformal example of Section 6.5, it is clear that (8.7) cannot be satisfied as the Fock space only contains one vector of grade two, $L^0 = \alpha K_{-2}\Omega$, and L^0 does not satisfy (8.7) for any value of α. On the other hand, for the example of Section 6.1, we can choose

$$
L^0 = \frac{1}{2k}\, J_{-1} J_{-1}\Omega ,
\tag{8.9}
$$

27

and this then satisfies (8.7). Similarly, in the case of the example of Section 6.2, we can choose

$$L^0 = \frac{1}{2(k+Q)} \sum_a J^a_{-1} J^a_{-1} \Omega \,, \tag{8.10}$$

where Q is the dual Coxeter number of g (*i.e.* the value of the quadratic Casimir in the adjoint representation), and again (8.7) is satisfied for this choice of L^0 (and the above choice of V). This construction is known as the 'Sugawara construction'.

For completeness it should be mentioned that the modes of the field L (that is contained in the theory in the conformal case) satisfy the Virasoro algebra

$$[L_m, L_n] = (m - n)L_{m+n} + \frac{c}{12} m(m^2 - 1)\delta_{m,-n} \,, \tag{8.11}$$

where c is the number that appears in the above definition of L. Furthermore, the modes L_m with $m = 0, \pm 1$ agree with the Möbius generators of the theory.

9 Representations

In order to introduce the concept of a representation of a meromorphic conformal field theory or conformal algebra, we consider a collection of densities more general than those used in Section 3 to define the meromorphic conformal field theory itself. The densities we now consider are typically defined on a cover of the Riemann sphere, $\mathbb{P}$, rather than $\mathbb{P}$ itself. We consider densities which are functions of variables u_i, $1 \leq i \leq N$, and z_j, $1 \leq j \leq n$, which are analytic if no two of these $N + n$ variables are equal, may have poles at $z_i = z_j$, $i \neq j$, or $z_i = u_j$, and may be branched about $u_i = u_j$, $i \neq j$. To define a representation, we need the case where $N = 2$, in which the densities are meromorphic in all but two of the variables.

Starting again with $V = \oplus_h V_h$, together with two finite-dimensional spaces W_α and W_β (which may be one-dimensional), we suppose that, for each integer $n \geq 0$, and $z_i \in \mathbb{P}$ and u_1, u_2 on some branched cover of $\mathbb{P}$, and for any collection of vectors $\psi_i \in V_{h_i}$ and $\chi_1 \in W_\alpha, \chi_2 \in W_\beta$, we have a density

$$\begin{aligned} g(\psi_1 &\,, \ldots, \psi_n; z_1, \ldots, z_n; \chi_1, \chi_2; u_1, u_2) \\ &\equiv \langle V(\psi_1, z_1)V(\psi_2, z_2)\cdots V(\psi_n, z_n)W_\alpha(\chi_1, u_1)W_\beta(\chi_2, u_2)\rangle \prod_{j=1}^n (dz_j)^{h_j} (du_1)^{r_1}(du_2)^{r_2} \,, \end{aligned} \tag{9.1}$$

where r_1, r_2 are real numbers, which we call the *conformal weights* of χ_1 and χ_2, respectively. The amplitudes

$$\langle V(\psi_1, z_1)V(\psi_2, z_2)\cdots V(\psi_n, z_n)W_\alpha(\chi_1, u_1)W_\beta(\chi_2, u_2)\rangle \tag{9.2}$$

are taken to be multilinear in the ψ_j and χ_1, χ_2, and invariant under the exchange of (ψ_i, z_i) with (ψ_j, z_j), and meromorphic as a function of the z_j, analytic except for possible poles at $z_i = z_j$, $i \neq j$, and $z_i = u_1$ or $z_i = u_2$. As functions of u_1, u_2, the amplitudes are analytic except for the possible poles at $u_1 = z_i$ or $u_2 = z_i$ and a possible branch cut at $u_1 = u_2$. We denote a collection of such densities by $\mathcal{R} = \{g\}$.

Just as before, given an open set $C \subset \mathbb{P}$ we introduced spaces $\mathcal{B}_C$, whose elements are of the form (3.3), so we can now introduce sets, $\mathcal{B}_{C\alpha\beta}$, labelled by finite collections of $\psi_i \in V_{h_i}$, $z_i \in C \subset \mathbb{P}$, $i = 1, \ldots, n$, $n \in \mathbb{N}$ and $z_i \neq z_j$ if $i \neq j$, together with $\chi_1 \in W_\alpha, \chi_2 \in W_\beta$ and $u_1, u_2 \in C$, $u_1 \neq u_2$ and $z_i \neq u_j$, denoted by

$$
\begin{aligned}
\chi &= V(\psi_1, z_1)V(\psi_2, z_2) \cdots V(\psi_n, z_n)W_\alpha(\chi_1, u_1)W_\beta(\chi_2, u_2)\Omega \\
&\equiv \prod_{i=1}^{n} V(\psi_i, z_i)W_\alpha(\chi_1, u_1)W_\beta(\chi_2, u_2)\Omega \, .
\end{aligned}
\tag{9.3}
$$

We again immediately identify different $\chi \in \mathcal{B}_{C\alpha\beta}$ with the other elements of $\mathcal{B}_{C\alpha\beta}$ obtained by replacing each ψ_j in (9.3) by $\mu_j\psi_j$, $1 \leq j \leq n$, χ_i by $\lambda_i\chi_i$, $i = 1, 2$, where $\lambda_1, \lambda_2, \mu_j \in \mathbb{C}$ and $\lambda_1\lambda_2 \prod_{j=1}^{n} \mu_j = 1$.

Proceeding as before, we introduce the vector space $\mathcal{V}_{C\alpha\beta}$ with basis $\mathcal{B}_{C\alpha\beta}$ and we cut it down to size *exactly* as before, *i.e.* we note that if we introduce another open set $\mathcal{O} \subset \mathbb{P}$, with $\mathcal{O} \cap C = \varnothing$, and, as in (3.4) write

$$
\phi = V(\phi_1, \zeta_1)V(\phi_2, \zeta_2) \cdots V(\phi_m, \zeta_m)\Omega \in \mathcal{B}_\mathcal{O} \, ,
\tag{9.4}
$$

where $\phi_j \in V_{k_j}$, $j = 1, \ldots m$, each $\phi \in \mathcal{B}_\mathcal{O}$ defines a map on $\mathcal{B}_{C\alpha\beta}$ by

$$
\eta_\phi(\chi) = (\phi, \chi) = \left\langle \prod_{i=1}^{m} V(\phi_i, \zeta_i) \prod_{i=1}^{n} V(\psi_i, z_i)W_\alpha(\chi_1, u_1)W_\beta(\chi_2, u_2) \right\rangle .
\tag{9.5}
$$

Again η_ϕ extends by linearity to a map $\mathcal{V}_{C\alpha\beta} \to \mathbb{C}$ and we consider the space, $\tilde{\mathcal{V}}_{C\alpha\beta}$, of sequences $\mathbf{X} = (X_1, X_2, \ldots)$, $X_j \in \mathcal{V}_{C\alpha\beta}$, for which $\eta_\phi(X_j)$ converges uniformly on each of the family of compact sets of the form (3.7). We write $\eta_\phi(\mathbf{X}) = \lim_{j\to\infty} \eta_\phi(X_j)$ and define the space $\mathcal{V}^\mathcal{O}_{C\alpha\beta}$ as being composed of the equivalence classes of such sequences, identifying two sequences $\mathbf{X}_1, \mathbf{X}_2$, if $\eta_\phi(\mathbf{X}_1) = \eta_\phi(\mathbf{X}_2)$ for all $\phi \in \mathcal{B}_\mathcal{O}$. Using the same arguments as in the proof of Theorem 1 (see Appendix C), it can be shown that the space $\mathcal{V}^\mathcal{O}_{C\alpha\beta}$ is independent of C, provided that the complement of $\mathcal{O}$ is path-connected; in this case we write $\mathcal{V}^\mathcal{O}_{\alpha\beta} \equiv \mathcal{V}^\mathcal{O}_{C\alpha\beta}$. We can define a family of seminorms for $\mathcal{V}^\mathcal{O}_{\alpha\beta}$ by $||\mathbf{X}||_\phi = |\eta_\phi(\mathbf{X})|$, where ϕ is an arbitrary element of $\mathcal{B}_\mathcal{O}$, and the natural topology on $\mathcal{V}^\mathcal{O}_{\alpha\beta}$ is the topology that is induced by this family of seminorms. (This is to say, that a sequence of states in $\mathbf{X}_j \in \mathcal{V}^\mathcal{O}_{\alpha\beta}$ converges if and only if $\eta_\phi(\mathbf{X}_j)$ converges for every $\phi \in \mathcal{B}_\mathcal{O}$.)

So far we have not specified a relationship between the spaces $\mathcal{V}^\mathcal{O}$, which define the conformal field theory, and the new spaces $\mathcal{V}^\mathcal{O}_{\alpha\beta}$, which we have now introduced to define a representation of it. Such a relation is an essential part of the definition of a representation; it has to express the idea that the two spaces define the same relations between combinations of vectors in the sets $\mathcal{B}_C$. To do this consider the space of all continuous linear functionals on $\mathcal{V}^\mathcal{O}$, the *dual* space of $\mathcal{V}^\mathcal{O}$, which we will denote $(\mathcal{V}^\mathcal{O})'$, and also the dual, $(\mathcal{V}^\mathcal{O}_{\alpha\beta})'$, of $\mathcal{V}^\mathcal{O}_{\alpha\beta}$. It is natural to consider these dual spaces as topological vector spaces with the weak topology: for each $f \in (\mathcal{V}^\mathcal{O})'$, we can consider the (uncountable) family of seminorms defined by $||f||_\Psi \equiv |f(\Psi)|$, where Ψ is an arbitrary element of $\mathcal{V}^\mathcal{O}$ (and similarly for

$(\mathcal{V}_{\alpha\beta}^{\mathcal{O}})'$). The weak topology is then the topology that is induced by this family of seminorms (so that $f_n \to f$ if and only if $f_n(\Psi) \to f(\Psi)$ for each $\Psi \in \mathcal{V}^{\mathcal{O}}$).

Every element of $\phi \in \mathcal{B}_{\mathcal{O}}$ defines a continuous linear functional both on $\mathcal{V}^{\mathcal{O}}$ and on $\mathcal{V}_{\alpha\beta}^{\mathcal{O}}$, each of which we shall denote by η_{ϕ}, and the linear span of the set of all linear functionals that arise in this way is dense in both $(\mathcal{V}^{\mathcal{O}})'$ and $(\mathcal{V}_{\alpha\beta}^{\mathcal{O}})'$. We therefore have a map from a dense subspace of $(\mathcal{V}^{\mathcal{O}})'$ to a dense subspace of $(\mathcal{V}_{\alpha\beta}^{\mathcal{O}})'$, and the condition for the amplitudes (9.1) to define a *representation* of the meromorphic (conformal) field theory whose spaces of states are given by $\mathcal{V}^{\mathcal{O}}$ is that this map extends to a *continuous* map between the dual spaces, *i.e.* that there exists a continuous map

$$\iota : (\mathcal{V}^{\mathcal{O}})' \to (\mathcal{V}_{\alpha\beta}^{\mathcal{O}})' \quad \text{such that} \quad \iota(\eta_{\phi}) = \eta_{\phi}. \tag{9.6}$$

This in essence says that $\mathcal{V}_{\alpha\beta}^{\mathcal{O}}$ will not distinguish limits of linear combinations of $\mathcal{B}_{\mathcal{O}}$ not distinguished by $\mathcal{V}^{\mathcal{O}}$.

Given a collection of densities $\mathcal{R}$ we can construct (in a similar way as before for the collection of amplitudes $\mathcal{A}$) spaces of states $\mathcal{V}_{\alpha}^{\mathcal{O}}$ and $\mathcal{V}_{\beta}^{\mathcal{O}}$, on which the vertex operators of the meromorphic theory are well-defined operators. By the by now familiar scheme, let us introduce the set $\mathcal{B}_{\mathcal{C}\alpha}$ that is labelled by finite collections of $\psi_i \in V_{h_i}$, $z_i \in \mathcal{C} \subset \mathbb{P}$, $i = 1, \ldots, n$, $n \in \mathbb{N}$ and $z_i \neq z_j$ if $i \neq j$, together with $\chi \in W_{\alpha}$ and $u \in \mathcal{C}$, $z_i \neq u$, denoted by

$$\begin{aligned} \chi &= V(\psi_1, z_1)V(\psi_2, z_2) \cdots V(\psi_n, z_n)W_{\alpha}(\chi, u)\Omega \\ &\equiv \prod_{i=1}^{n} V(\psi_i, z_i)W_{\alpha}(\chi, u)\Omega. \end{aligned} \tag{9.7}$$

We again immediately identify different $\chi \in \mathcal{B}_{\mathcal{C}\alpha}$ with the other elements of $\mathcal{B}_{\mathcal{C}\alpha}$ obtained by replacing each ψ_j in (9.7) by $\mu_j\psi_j$, $1 \leq j \leq n$, χ by $\lambda\chi$, where $\lambda, \mu_j \in \mathbb{C}$ and $\lambda \prod_{j=1}^{n} \mu_j = 1$. We also define $\mathcal{B}_{\mathcal{C}\beta}$ analogously (by replacing $\chi \in W_{\alpha}$ by $\chi \in W_{\beta}$).

We then introduce the vector space $\mathcal{V}_{\mathcal{C}\alpha}$ with basis $\mathcal{B}_{\mathcal{C}\alpha}$, and we cut it down to size exactly as before by considering the map analogous to (9.5), where now $\phi \in \mathcal{B}_{\mathcal{O}\beta}$. The resulting space is denoted by $\mathcal{V}_{\mathcal{C}\alpha}^{\mathcal{O}\beta}$, and is again independent of $\mathcal{C}$ provided that the complement of $\mathcal{O}$ is path-connected; in this case we write $\mathcal{V}_{\alpha}^{\mathcal{O}\beta} \equiv \mathcal{V}_{\mathcal{C}\alpha}^{\mathcal{O}\beta}$. It also has a natural topology induced by the seminorms $|\eta_{\phi}(\mathbf{X})|$, where now $\phi \in \mathcal{B}_{\mathcal{O}\beta}$. With respect to this topology, the span of $\mathcal{B}_{\mathcal{C}\alpha}$ is dense in $\mathcal{V}_{\alpha}^{\mathcal{O}\beta}$. We can similarly consider the spaces $\mathcal{V}_{\beta}^{\mathcal{O}\alpha}$ by exchanging the rôles of W_{α} and W_{β}.

For $\psi \in V$, a vertex operator $V(\psi, z)$ can be defined as an operator $V(\psi, z) : \mathcal{V}_{\alpha}^{\mathcal{O}\beta} \to \mathcal{V}_{\alpha}^{\mathcal{O}'\beta}$, where $z \in \mathcal{O}$ but $z \notin \mathcal{O}' \subset \mathcal{O}$, by defining its action on the total subset $\mathcal{B}_{\mathcal{C}\alpha}$, where $\mathcal{C} \cap \mathcal{O} = \varnothing$

$$V(\psi, z)\chi = V(\psi, z)V(\psi_1, z_1) \cdots V(\psi_n, z_n)W_{\alpha}(\chi, u)\Omega, \tag{9.8}$$

and $\chi \in \mathcal{B}_{\mathcal{C}\alpha}$ is as in (9.7). The image is in $\mathcal{B}_{\mathcal{C}'\alpha}$ for any $\mathcal{C}' \supset \mathcal{C}$ which contains z, and we can choose $\mathcal{C}'$ such that $\mathcal{C}' \cap \mathcal{O}' = \varnothing$. This then extends by linearity to a map $\mathcal{V}_{\mathcal{C}\alpha} \to \mathcal{V}_{\mathcal{C}'\alpha}$, and we can show, by analogous arguments as before, that it induces a map $\mathcal{V}_{\mathcal{C}\alpha}^{\mathcal{O}\beta} \to \mathcal{V}_{\mathcal{C}'\alpha}^{\mathcal{O}'\beta}$.

By the same arguments as before in Section 5, this definition can be extended to vectors Ψ of the form (5.4) that span the Fock space of the meromorphic theory. The actual Fock space $\mathcal{H}^{\mathcal{O}'}$ (that is typically a quotient space of the free vector space spanned by the vectors of the from (5.4)) is a subspace of $(\mathcal{V}^{\mathcal{O}})'$ provided that $\mathcal{O}' \cup \mathcal{O} = \mathbb{P}$, and if the amplitudes define a representation, $\iota(\mathcal{H}^{\mathcal{O}'}) \subset (\mathcal{V}^{\mathcal{O}}_{\alpha\beta})'$ because of (9.6). In this case it is then possible to define vertex operators $V(\Psi, z) : \mathcal{V}^{\mathcal{O}\beta}_{\alpha} \to \mathcal{V}^{\mathcal{O}'\beta}_{\alpha}$ for arbitrary elements of the Fock space $\mathcal{H}$, and this is what is usually thought to be the defining property of a representation. By the same argument the vertex operators are also well-defined for elements in $V_{\mathcal{C}}$ for suitable $\mathcal{C}$. There exists an alternative criterion for a set of densities to define a representation, which is in essence due to Montague [21], and which throws considerable light on the nature of conformal field theories and their representations. (Indeed, we shall use it to construct an example of a representation for the $u(1)$-theory below.)

Theorem 6: The densities (9.1) define a representation provided that, for each open set $\mathcal{O} \subset \mathbb{P}$ with path-connected complement and $u_1, u_2 \notin \mathcal{O}$, there is a state

$$\Sigma_{\alpha\beta}(u_1, u_2; \chi_1, \chi_2) \in \mathcal{V}^{\mathcal{O}} \tag{9.9}$$

that is equivalent to $W_{\alpha}(u_1, \chi_1)W_{\beta}(u_2, \chi_2)$ in the sense that the amplitudes of the representation are given by $\eta_{\phi}(\Sigma_{\alpha\beta})$:

$$\left\langle \prod_{i=1}^{m} V(\phi_i, \zeta_i) W_{\alpha}(u_1, \chi_1)W_{\beta}(u_2, \chi_2) \right\rangle = \left\langle \prod_{i=1}^{m} V(\phi_i, \zeta_i)\Sigma_{\alpha\beta}(u_1, u_2; \chi_1, \chi_2) \right\rangle, \tag{9.10}$$

where $\zeta_i \in \mathcal{O}$.

The proof of this theorem depends on the following:

Lemma: There exist sequences $e_i \in \mathcal{V}^{\mathcal{O}}_{\mathcal{C}}, f_i \in (\mathcal{V}^{\mathcal{O}}_{\mathcal{C}})'$, dense in the appropriate topologies, such that $f_j(e_i) = \delta_{ij}$ and such that $\sum_{i=1}^{\infty} e_i f_i(\Psi)$ converges to Ψ for all $\Psi \in \mathcal{V}^{\mathcal{O}}_{\mathcal{C}}$.

To prove the Lemma, take the $\{e_i\}$ to be formed from the union of the bases of the eigenspaces $\mathcal{H}_N$ of L_0, which we have taken to be finite-dimensional, taken in order, $N = 0, 1, 2, \ldots$. Using the projection operators P_N defined by (7.3), we have that $\sum_{N=0}^{\infty} P_N \Psi = \Psi$ and $P_n \Psi$ can be written as a sum of the e_i which are basis elements of $\mathcal{H}_N$, with coefficients $f_i(\Psi)$ which depend continuously and linearly on Ψ. It is then clear that $\sum_{i=1}^{\infty} e_i f_i(\Psi) = \Psi$ and, if $\eta \in (\mathcal{V}^{\mathcal{O}})'$, $\eta = \sum_{i=1}^{\infty} f_i \eta(e_i)$, showing that $\{e_i\}$ is dense in $\mathcal{V}^{\mathcal{O}}$ and $\{f_i\}$ is dense in $(\mathcal{V}^{\mathcal{O}})'$.

Proof of Theorem 6: Assuming we have a continuous map $\iota : (\mathcal{V}^{\mathcal{O}})' \to (\mathcal{V}^{\mathcal{O}}_{\alpha\beta})'$, let us define $\Sigma_{\alpha\beta}(u_1, u_2; \chi_1, \chi_2)$ by

$$\Sigma_{\alpha\beta}(u_1, u_2; \chi_1, \chi_2) = \sum_i e_i \iota(f_i)\Big(W_{\alpha}(u_1, \chi_1)W_{\beta}(u_2, \chi_2)\Big). \tag{9.11}$$

Then, if $\eta_{\phi} = \sum_j \lambda_j f_j$,

$$\left\langle \prod_{i=1}^{m} V(\phi_i, \zeta_i) W_{\alpha}(u_1, \chi_1)W_{\beta}(u_2, \chi_2) \right\rangle = \eta_{\phi}\Big(\Sigma_{\alpha\beta}(u_1, u_2; \chi_1, \chi_2)\Big)$$

31

$$\begin{aligned}
&= \sum_{ji} \lambda_j f_j(e_i)\, \iota(f_i) \Big(W_\alpha(u_1, \chi_1) W_\beta(u_2, \chi_2) \Big) \\
&= \iota(\eta_{\boldsymbol{\phi}}) \Big(W_\alpha(u_1, \chi_1) W_\beta(u_2, \chi_2) \Big) \\
&= \Big\langle \prod_{i=1}^{m} V(\phi_i, \zeta_i) \Sigma_{\alpha\beta}(u_1, u_2; \chi_1, \chi_2) \Big\rangle ,
\end{aligned} \qquad (9.12)$$

and the convergence of (9.11) can be deduced from this.
Conversely, suppose that (9.10) holds; then

$$\prod_{i=1}^{n} V(\psi_i, \zeta_i) W_\alpha(u_1, \chi_1) W_\beta(u_2, \chi_2) \to \prod_{i=1}^{n} V(\psi_i, \zeta_i) \Sigma_{\alpha\beta}(u_1, u_2; \chi_1, \chi_2) \qquad (9.13)$$

defines a continuous map $\mathcal{V}_{\alpha\beta}^{\mathcal{O}} \to \mathcal{V}^{\mathcal{O}}$ (where $\zeta_i, u_1, u_2 \notin \mathcal{O}$), and this induces a dual map $\iota : (\mathcal{V}^{\mathcal{O}})' \to (\mathcal{V}_{\alpha\beta}^{\mathcal{O}})'$, continuous in the weak topology, satisfying $\iota(\eta_{\boldsymbol{\phi}}) = \eta_{\boldsymbol{\phi}}$, $i.e.$ (9.6) holds. The map (9.13) defines an isomorphism of $\mathcal{V}_{\alpha\beta}^{\mathcal{O}}$ onto $\mathcal{V}^{\mathcal{O}}$: it is onto for otherwise its image would define an invariant subspace of $\mathcal{V}^{\mathcal{O}}$ and the argument of Proposition 5 shows that this must be the whole space; and it is an injection because if it maps a vector $\mathbf{X}$ to zero, $\eta_{\boldsymbol{\phi}}(\mathbf{X})$ must vanish for all $\boldsymbol{\phi} \in \mathcal{B}_{\mathcal{O}}$, implying $\mathbf{X} = 0$.

10 Möbius covariance, Fock spaces and the Equivalence of Representations

We shall now assume that each density in the collection $\mathcal{R}$ is invariant under the action of the Möbius transformations, $i.e.$ that the amplitudes satisfy

$$\Big\langle \prod_{i=1}^{n} V(\psi_i, z_i) W_\alpha(\chi_1, u_1) W_\beta(\chi_2, u_2) \Big\rangle = \Big\langle \prod_{i=1}^{n} V(\psi_i, \gamma(z_i)) W_\alpha(\chi_1, \gamma(u_1)) W_\beta(\chi_2, \gamma(u_2)) \Big\rangle$$

$$\prod_{l=1}^{2} (\gamma'(u_l))^{r_l} \prod_{i=1}^{n} (\gamma'(z_i))^{h_i} , \qquad (10.1)$$

where r_l are the real numbers which appear in the definition of the densities, and h_i is the grade of ψ_i.
In this case, we can define operators $U(\gamma)$, mapping $\mathcal{V}_\alpha^{\mathcal{O}\beta}$ to $\mathcal{V}_\alpha^{\mathcal{O},\beta}$; on the total subset $\mathcal{B}_{\mathcal{C}\alpha}$, where $\mathcal{C} \cap \mathcal{O} = \varnothing$, these operators are defined by

$$U(\gamma)\chi = V(\psi_1, \gamma(z_1)) \cdots V(\psi_n, \gamma(z_n)) W_\alpha(\chi, \gamma(u))\Omega \prod_{i=1}^{n} (\gamma'(z_i))^{h_i} \, \gamma'(u)^{r_1} , \qquad (10.2)$$

where χ is defined as in (9.7), and h_i is the grade of ψ_i, $i = 1, \ldots, n$. This definition extends by linearity to operators being defined on $\mathcal{V}_{\mathcal{C}\alpha}$, and by analogous arguments to those in Section 4, this extends to a well-defined map $\mathcal{V}_\alpha^{\mathcal{O}\beta} \to \mathcal{V}_\alpha^{\mathcal{O},\beta}$. If we choose two points z_∞ and z_0 as before, we can introduce the Möbius generators $L_0^M, L_{\pm 1}^M$ which are well-defined on these spaces.

We define the Fock space $\mathcal{H}_\alpha^\mathcal{O} \subset \mathcal{V}_\alpha^\mathcal{O}$ to be the space spanned by finite linear combinations of vectors of the form

$$\Phi = V_{n_1}(\psi_1) \cdots V_{n_N}(\psi_N) W_\alpha(\chi, 0)\Omega, \tag{10.3}$$

where $\psi_j \in V$, $\chi \in W_\alpha$ and $n_j \in \mathbb{Z}$, $1 \leq j \leq N$. Here the modes $V_n(\psi)$ are defined as before in (5.2) where the contour encircles the point $0 \in \mathcal{C}$, and this still makes sense since the amplitudes $\mathcal{R}$ are not branched about $u_i = z_j$. It is clear that $\mathcal{H}_\alpha^\mathcal{O}$ is a dense subspace of $\mathcal{V}_\alpha^\mathcal{O}$, and that it is independent of $\mathcal{O}$; where no ambiguity arises we shall therefore denote it by $\mathcal{H}_\alpha$. By construction, $W_\alpha \subset \mathcal{H}_\alpha$. We can also define $W_\beta \subset \mathcal{H}_\beta$ in the same way.

As before it is then possible to extend the amplitudes $\mathcal{R}$ to amplitudes being defined for $\chi_1 \in \mathcal{H}_\alpha$ and $\chi_2 \in \mathcal{H}_\beta$ (rather than $\chi_1 \in W_\alpha$ and $\chi_2 \in W_\beta$), and for the subset of quasiprimary states in $\mathcal{H}_\alpha$ and $\mathcal{H}_\beta$ (i.e. for the states that are annihilated by L_1^M defined above), the Möbius properties are analogous to those in (10.1).

As in the case of the meromorphic theory we can then define the equivalence of two representations. Let us suppose that for a given meromorphic field theory specified by V and $\mathcal{A} = \{f\}$, we have two collections of densities, one specified by W_α, W_β with the amplitudes given by $\mathcal{R} = \{g\}$, and one specified by $\hat{W}_\alpha, \hat{W}_\beta$ and $\hat{\mathcal{R}}$. We denote the corresponding Fock spaces by $\mathcal{H}_\alpha, \mathcal{H}_\beta$ in the case of the former densities, and by $\hat{\mathcal{H}}_\alpha$ and $\hat{\mathcal{H}}_\beta$ in the case of the latter. We say that the two densities define *equivalent* representations if there exist graded injections

$$\iota_\alpha : W_\alpha \to \hat{\mathcal{H}}_\alpha \qquad \iota_\beta : W_\beta \to \hat{\mathcal{H}}_\beta, \tag{10.4}$$

and

$$\hat{\iota}_\alpha : \hat{W}_\alpha \to \mathcal{H}_\alpha \qquad \hat{\iota}_\beta : \hat{W}_\beta \to \mathcal{H}_\beta, \tag{10.5}$$

that map amplitudes to amplitudes. We similarly define two representations to be *conjugate* to one another if instead of (10.4) and (10.5) the amplitudes are mapped to each other under

$$\iota_\alpha : W_\alpha \to \hat{\mathcal{H}}_\beta \qquad \iota_\beta : W_\beta \to \hat{\mathcal{H}}_\alpha, \tag{10.6}$$

and

$$\hat{\iota}_\alpha : \hat{W}_\alpha \to \mathcal{H}_\beta \qquad \hat{\iota}_\beta : \hat{W}_\beta \to \mathcal{H}_\alpha. \tag{10.7}$$

A representation is called *highest weight*, if the equivalence class of collections of densities contains a representative which has the *highest weight property*: for each density g and each choice of $\chi_1 \in W_\alpha, \chi_2 \in W_\beta$ and $\psi_i \in V_{h_i}$, the pole in $(z_i - u_l)$ is bounded by h_i. This definition is slightly more general than the definition which is often used, in that it is not assumed that the highest weight vectors transform in any way under the zero modes of the meromorphic fields.

In Section 7, we showed, using the cluster property, that the meromorphic conformal field theory does not have any proper ideals. This implies now

Proposition 7: Every non-trivial representation is faithful.

Proof: Suppose that $V(\Phi, z)$, where $\Phi \in \mathcal{V}_\mathcal{C}^{\mathcal{O}'}$, $\mathcal{C} \cap \mathcal{O}' = \varnothing$ and $\mathcal{C}_z \subset \mathcal{O}$, acts trivially on the representation $\mathcal{V}_\alpha^\mathcal{O}$, i.e. that

$$V(\Phi, z)\Psi = 0 \qquad \text{for every} \quad \Psi \in \mathcal{V}_\alpha^\mathcal{O}. \tag{10.8}$$

Then, for any $\psi \in V$ and $\zeta \in \mathcal{O}'$ for which $\zeta + z \in \mathcal{O}$ we have

$$V\left(V(\psi,\zeta)\Phi, z\right)\Psi = V(\psi, \zeta + z)V(\Phi, z)\Psi = 0\,, \tag{10.9}$$

and thus $V(\psi,\zeta)\Phi$ also acts trivially on $\mathcal{V}_\alpha^\mathcal{O}$. This implies that the subspace of states in $\mathcal{V}_\mathcal{C}^{\mathcal{O}'}$ that act trivially on $\mathcal{V}_\alpha^\mathcal{O}$ is an ideal. Since there are no non-trivial ideals in $\mathcal{V}_\mathcal{C}^\mathcal{O}$, this implies that the representation is faithful.

11 An Example of a Representation

Let us now consider the example of the $U(1)$ theory which was first introduced in Section 6.1. In this section we want to construct a family of representations for this meromorphic conformal field theory.

Let us first define the state

$$\Psi_n = \int_a^b dw_1 \cdots \int_a^b dw_n : J(w_1)\cdots J(w_n):\,, \tag{11.1}$$

where $a,b \in \mathcal{C} \subset \mathbb{C}$, the integrals are chosen to lie in $\mathcal{C}$, and the normal ordering prescription $:\,\cdot\,:$ means that all poles in $w_i - w_j$ for $i \neq j$ are subtracted. We can deduce from the definition of the amplitudes $(6.1, 6.2)$ and (11.1) that the amplitudes involving Ψ_n are of the form

$$\left\langle \Psi_n \prod_{j=1}^N J(\zeta_j) \right\rangle = k^n \sum_{\substack{i_1,\ldots,i_n \in \{1,\ldots,N\} \\ i_j \neq i_l}} \prod_{l=1}^n \frac{(b-a)}{(a-\zeta_{i_l})(b-\zeta_{i_l})} \left\langle \prod_{j \notin \{i_1,\ldots i_n\}} J(\zeta_j) \right\rangle, \tag{11.2}$$

where $\zeta_j \in \mathcal{O} \subset \mathbb{C}$ and $\mathcal{C} \cap \mathcal{O} = \varnothing$. By analytic continuation of (11.2) we can then calculate the contour integral $\oint_{C_a} J(z)dz\Psi_n$, where C_a is a contour in $\mathcal{C}$ encircling a but not b, and we find that

$$\oint_{C_a} J(z)dz\Psi_n = -nk\Psi_{n-1}\,, \tag{11.3}$$

and

$$\oint_{C_a} (z-a)^n J(z)dz\Psi_n = 0 \qquad \text{for} \quad n \geq 1\,, \tag{11.4}$$

where the equality holds in $\mathcal{V}^\mathcal{O}$. Similar statements also hold for the contour integral around b,

$$\oint_{C_b} J(z)dz\Psi_n = nk\Psi_{n-1}\,, \tag{11.5}$$

and

$$\oint_{C_b} (z-b)^n J(z)dz\Psi_n = 0 \qquad \text{for} \quad n \geq 1\,. \tag{11.6}$$

Next we define

$$\Psi_\alpha = \sum_{n=0}^\infty \frac{\alpha^n}{n!\,k^n}\Psi_n =: \exp\left(\frac{\alpha}{k}\int_a^b J(w)dw\right):\,, \tag{11.7}$$

34

where α is any (real) number. This series converges in $\mathcal{V}^{\mathcal{O}}$, since for any amplitude of the form

$$\langle \Psi_\alpha J(\zeta_1) \cdots J(\zeta_N) \rangle \tag{11.8}$$

only the terms in (11.7) with $n \leq N$ contribute, as follows from (11.2).

We can use Ψ to define amplitudes as in (9.10), and in order to show that these form a representation, it suffices (because of Theorem 6) to demonstrate that the functions so defined have the appropriate analyticity properties. The only possible obstruction arises from the singularity for $\zeta_i \to a$ and $\zeta_i \to b$, but it follows from (11.3-11.6) that

$$J(\zeta)\Psi \sim \frac{-\alpha}{(\zeta - a)} + O(1) \quad \text{as} \quad \zeta \to a, \tag{11.9}$$

and

$$J(\zeta)\Psi \sim \frac{\alpha}{(\zeta - b)} + O(1) \quad \text{as} \quad \zeta \to b, \tag{11.10}$$

and thus that the singularities are only simple poles. This proves that the amplitudes defined by (9.10) give rise to a representation of the $U(1)$ theory. From the point of view of conventional conformal field theory, this representation (and its conjugate) is the highest weight representation that is generated from a state of $U(1)$-charge $\pm\alpha$.

It may be worthwhile to point out that we can rescale all amplitudes of a representation of a meromorphic field theory by

$$g \mapsto C(u_1 - u_2)^{2\delta} g, \tag{11.11}$$

where C and δ are fixed constants (that are the same for all g), without actually violating any of the conditions we have considered so far. (The only effect is that r_1 and r_2 are replaced by $\hat{r}_l = r_l - \delta$, $l = 1, 2$.) For the representation of a meromorphic *conformal* field theory, the ambiguity in δ can however be canonically fixed: since the meromorphic fields contain the stress-energy field L (whose modes satisfy the Virasoro algebra L_n), we can require that

$$L_n = L_n^M \quad \text{for} \quad n = 0, \pm 1, \tag{11.12}$$

when acting on $\mathcal{H}_\alpha$. The action of L_0 on $\mathcal{H}_\alpha$ is not modified by (11.11), but since $\hat{r}_l = r_l - \delta$, the action of L_0^M is, and (11.12) therefore fixes the choice of δ in (11.11).

In the above example, in order to obtain a representation of the meromorphic conformal field theory (with L^0 being given by (8.9)), we have to modify the amplitudes as in (11.11) with $\delta = -\alpha^2/2k$. This can be easily checked using (11.3-11.6).

12 Zhu's Algebra

The description of representations in terms of collections of densities has a large redundancy in that many different collections of densities define the same representation. Typically we are only interested in highest weight representations, and for these we may restrict our attention to the representatives for which the highest weight property holds. In this section we want to analyse the conditions that characterise the corresponding states $\Sigma_{\alpha\beta}$; this approach is in essence due to Zhu [9].

Suppose we are given a highest weight representation, *i.e.* a collection of amplitudes that are described in terms of the states $\Sigma_{\alpha\beta}(u_1, u_2; \chi_1, \chi_2) \in \mathcal{V}^{\mathcal{O}}$, where $u_l \notin \mathcal{O}$. Each such state defines a linear functional on the Fock space $\mathcal{H}^{\mathcal{O}'}$, where $\mathcal{O} \cup \mathcal{O}' = \mathbb{P}$. But, for given u_1, u_2, the states $\Sigma_{\alpha\beta}(u_1, u_2; \chi_1, \chi_2)$, associated with the various possible representations, satisfy certain linear conditions: they vanish on a certain subspace $O_{u_1,u_2}(\mathcal{H}^{\mathcal{O}'})$. Thus they define, and are characterised by, linear functionals on the quotient space $\mathcal{H}^{\mathcal{O}'}/O_{u_1,u_2}(\mathcal{H}^{\mathcal{O}'})$. This is a crucial realisation, because it turns out that, in cases of interest, this quotient is finite-dimensional. Further the quotient has the structure of an algebra, first identified by Zhu [9], in terms of which the equivalence of representations, defined by these linear functionals, can be characterised.

Let us consider the case where $u_1 = \infty$ and $u_2 = -1$, for which we can choose $\mathcal{O}$ and $\mathcal{O}'$ so that $0 \in \mathcal{O}$ and $0 \notin \mathcal{O}'$. We want to characterise the subspace of $\mathcal{H} = \mathcal{H}^{\mathcal{O}'}$ on which the linear functional defined by $\Sigma_{\alpha\beta}(\infty, -1; \chi_1, \chi_2)$ vanishes identically. Given ψ and χ in $\mathcal{H}$, we define the state $V^{(N)}(\psi)\chi$ in $\mathcal{H}$ by

$$V^{(N)}(\psi)\chi = \oint_0 \frac{dw}{w^{N+1}} V\left[(w+1)^{L_0}\psi, w\right]\chi, \qquad (12.1)$$

where N is an arbitrary integer, and the contour is a small circle (with radius less than one) around $w = 0$. If $\Sigma_{\alpha\beta}$ has the highest weight property then

$$\langle \Sigma_{\alpha\beta}(\infty, -1; \chi_1, \chi_2) V^{(N)}(\psi)\phi \rangle = 0 \qquad \text{for } N > 0. \qquad (12.2)$$

This follows directly from the observation that the integrand in (12.2) does not have any poles at $w = -1$ or $w = \infty$.

Let us denote by $O(\mathcal{H})$ the subspace of $\mathcal{H}$ that is generated by states of the form (12.1) with $N > 0$, and define the quotient space $A(\mathcal{H}) = \mathcal{H}/O(\mathcal{H})$. Then it follows that every highest weight representation defines a linear functional on $A(\mathcal{H})$. If two representations induce the same linear functional on $A(\mathcal{H})$, then they are actually equivalent representations, and thus the number of inequivalent representations is always bounded by the dimension of $A(\mathcal{H})$. In fact, as we shall show below, the vector space $A(\mathcal{H})$ has the structure of an associative algebra, where the product is defined by (12.1) with $N = 0$. In terms of the states $\Sigma_{\alpha\beta}$ this product corresponds to

$$\langle \Sigma_{\alpha\beta}(\infty, -1; \chi_1, \chi_2) V^{(0)}(\psi)\phi \rangle = \langle \Sigma_{\alpha\beta}(\infty, -1; V_0(\psi)\chi_1, \chi_2)\phi \rangle. \qquad (12.3)$$

One may therefore expect that the different highest weight representations of the meromorphic conformal field theory are in one-to-one correspondence with the different representations of the algebra $A(\mathcal{H})$, and this is indeed true [9]. Most conformal field theories of interest have the property that $A(\mathcal{H})$ is a finite-dimensional algebra, and there exist therefore only finitely many inequivalent highest weight representations of the corresponding meromorphic conformal field theory; we shall call a meromorphic conformal field theory for which this is true *finite*.

In the above discussion the two points, $u_1 = \infty$ and $u_2 = -1$ were singled out, but the definition of the quotient space (and the algebra) is in fact independent of this choice. Let us consider the Möbius transformation γ which maps $\infty \mapsto u_1$, $-1 \mapsto u_2$ and $0 \mapsto 0$ (where $u_l \neq 0$); it is explicitly given as

$$\gamma(\zeta) = \frac{u_1 u_2 \zeta}{u_2(\zeta + 1) - u_1} \qquad \leftrightarrow \qquad \begin{pmatrix} u_1 u_2 & 0 \\ u_2 & u_2 - u_1 \end{pmatrix}, \tag{12.4}$$

with inverse

$$\gamma^{-1}(z) = \frac{u_1 - u_2}{u_2} \frac{z}{(z - u_1)} \qquad \leftrightarrow \qquad \begin{pmatrix} u_1 - u_2 & 0 \\ u_2 & -u_1 u_2 \end{pmatrix}. \tag{12.5}$$

Writing $\psi' = U(\gamma)\psi$ and $\chi' = U(\gamma)\chi$ we then find (see Appendix F)

$$V_{u_1,u_2}^{(N)}(\psi')\chi' = U(\gamma)V^{(N)}(\psi)\chi, \tag{12.6}$$

where $V_{u_1,u_2}^{(N)}(\psi)$ is defined by

$$V_{u_1,u_2}^{(N)}(\psi)\chi = \oint_0 \frac{dw}{w} \frac{u_1}{(u_1 - w)} \left(\frac{u_2}{(u_2 - u_1)} \frac{(u_1 - w)}{w} \right)^N$$
$$V\left[\left(\frac{(u_1 - w)(u_2 - w)}{u_1 u_2} \right)^{L_0} e^{\frac{w}{u_1 u_2}L_1} \psi, w \right] \chi, \tag{12.7}$$

and the contour encloses $w = 0$ but not $w = u_l$. We can then also define $O_{u_1,u_2}(\mathcal{H})$ to be the space that is generated by states of the form (12.7) with $N > 0$, and $A_{u_1,u_2}(\mathcal{H}) = \mathcal{H}/O_{u_1,u_2}(\mathcal{H})$.
As $z = 0$ is a fixed point of γ, $U(\gamma) : \mathcal{H} \to \mathcal{H}$, and because of (12.6), $U(\gamma) : O(\mathcal{H}) \to O_{u_1,u_2}(\mathcal{H})$. It also follows from (12.6) with $N = 0$ that the product is covariant, and this implies that the different algebras $A_{u_1,u_2}(\mathcal{H})$ for different choices of u_1 and u_2 are isomorphic. To establish that the algebra action is well-defined and associative, it is therefore sufficient to consider the case corresponding to $u_1 = \infty$ and $u_2 = -1$. In this case we write $V^{(0)}(\psi)\chi$ also as $\psi * \chi$.
Let us first show that $O_{u_1,u_2}(\mathcal{H}) = O_{u_2,u_1}(\mathcal{H})$. Because of the Möbius covariance it is sufficient to prove this for the special case where $u_1 = \infty$ and $u_2 = -1$. For this case we have $V_{\infty,-1}^{(N)}(\psi) = V^{(N)}(\psi)$ as before, and

$$V_{-1,\infty}^{(N)}(\psi) \equiv V_c^{(N)}(\psi) = (-1)^N \oint_c \frac{dw}{w} \frac{1}{(w + 1)} \left(\frac{w + 1}{w} \right)^N V\left((w + 1)^{L_0}\psi, w \right). \tag{12.8}$$

In particular, $V^{(1)}(\psi) = -V_c^{(1)}(\psi)$, and we denote by $N(\psi)$ this operator, i.e. $N(\psi) = V^{(1)}(\psi) = -V_c^{(1)}(\psi)$. If ψ is of definite conformal weight we have

$$\begin{aligned}
V^{(N)}(L_{-1}\psi) &= \oint_0 \frac{dw}{w^{N+1}} (w + 1)^{h_\psi + 1} \frac{dV(\psi, w)}{dw} \\
&= -\oint_0 dw \frac{d}{dw} \left(\frac{(w + 1)^{h_\psi + 1}}{w^{N+1}} \right) V(\psi, w) \\
&= (N - h_\psi)V^{(N)}(\psi) + (N + 1)V^{(N+1)}(\psi),
\end{aligned} \tag{12.9}$$

and this implies, for $N \neq -1$,

$$V^{(N+1)}(\psi) = \frac{1}{N+1} V^{(N)}(L_{-1}\psi) - \frac{N-h_\psi}{N+1} V^{(N)}(\psi). \tag{12.10}$$

Similarly, we have

$$V_c^{(N+1)}(\psi) = -\frac{1}{N+1} V_c^{(N)}(L_{-1}\psi) - \frac{N+h_\psi}{N+1} V_c^{(N)}(\psi). \tag{12.11}$$

Because of (12.10) $O_{\infty,-1}(\mathcal{H})$ is spanned by the states of the form $N(\psi)\chi$, and the same is also the case for $O_{-1,\infty}(\mathcal{H})$ because of (12.11); thus, in particular, $O_{\infty,-1}(\mathcal{H}) = O_{-1,\infty}(\mathcal{H})$. It then follows that (12.2) also holds if $V^{(N)}(\psi)$ is replaced by $V_c^{(N)}(\psi)$, and because of the definition of $V_c^{(N)}(\psi)$, the analogue of (12.3) is now

$$\langle \Sigma_{\alpha\beta}(\infty, -1; \chi_1, \chi_2) V_c^{(0)}(\psi)\phi \rangle = \langle \Sigma_{\alpha\beta}(\infty, -1; \chi_1, V_0(\psi)\chi_2)\phi \rangle. \tag{12.12}$$

One should therefore expect that the action of $V_L(\psi) \equiv V^{(0)}(\psi)$ and $V_R(\psi) \equiv V_c^{(0)}(\chi)$ commute up to elements in $N(\mathcal{H})$. To prove this, it is again sufficient to consider the case, where ψ and χ are eigenvectors of L_0 with eigenvalues h_ψ and h_χ, respectively. Then we have

$$
\begin{aligned}
[V_L(\psi), V_R(\chi)] &= \oint\oint_{|\zeta|>|w|} \frac{d\zeta}{\zeta}(\zeta+1)^{h_\psi} \frac{dw}{w}(w+1)^{h_\chi-1} V(\psi,\zeta) V(\chi,w) \\
&\quad - \oint\oint_{|z|>|\zeta|} \frac{dw}{w}(w+1)^{h_\chi-1} \frac{d\zeta}{\zeta}(\zeta+1)^{h_\psi} V(\chi,w) V(\psi,\zeta) \\
&= \oint_0 \left\{ \oint_w \frac{d\zeta}{\zeta}(\zeta+1)^{h_\psi} V(\psi,\zeta) V(\chi,w) \right\} \frac{dw}{w}(w+1)^{h_\chi-1} \\
&= \sum_n \oint_0 \left\{ \oint_w \frac{d\zeta}{\zeta}(\zeta+1)^{h_\psi} V(V_n(\psi)\chi,w)(\zeta-w)^{-n-h_\psi} \right\} \frac{dw}{w}(w+1)^{h_\chi-1} \\
&= \sum_{h_\chi \geq n \geq 0} \sum_{l=0}^{n+h_\psi-1} (-1)^l \binom{h_\psi}{l+1-n} \\
&\qquad \oint_0 \frac{dw}{w(w+1)} \left(\frac{w+1}{w}\right)^{l+1} (w+1)^{h_\chi-n} V(V_n(\psi)\chi,w) \\
&\in N(\mathcal{H}). \tag{12.13}
\end{aligned}
$$

Because of (12.10), every element in $N(\mathcal{H})$ can be written as $V_R(\phi)$ for a suitable ϕ, and (12.13) thus implies that $[V_L(\psi), V^{(N)}(\chi)] \in N(\mathcal{H})$ for $N > 0$; hence $V_L(\psi)$ defines an endomorphism of $A(\mathcal{H})$.

For two endomorphisms, Φ_1, Φ_2, of $\mathcal{H}$, which leave $O(\mathcal{H})$ invariant (so that they induce endomorphisms of $\mathcal{A}(\mathcal{H})$), we shall write $\Phi_1 \approx \Phi_2$ if they agree as endomorphisms of $\mathcal{A}(\mathcal{H})$, i.e. $(\Phi_1 - \Phi_2)\mathcal{H} \subset O(\mathcal{H})$. Similarly we write $\phi \approx 0$ if $\phi \in O(\mathcal{H})$.

In the same way in which the action of $V(\psi, z)$ is uniquely characterised by (4.18) and (4.19), we can now prove the following

Theorem 8 [Uniqueness Theorem for Zhu modes]: If Φ is an endomorphism of $\mathcal{H}$ leaving $O(\mathcal{H})$ invariant and if Φ satisfies the two conditions: (a) $\Phi\Omega = \psi$; and (b) $[\Phi, V_R(\chi)] \in N(\mathcal{H})$ for all $\chi \in \mathcal{H}$, then $\Phi \approx V_L(\psi)$.

Proof: This follows from

$$\Phi\chi = \Phi V_R(\chi)\Omega \approx V_R(\chi)\,\Phi\,\Omega = V_R(\chi)\psi = V_L(\psi)\chi \,. \tag{12.14}$$

It is then an immediate consequence that

$$V_L(V_L(\psi)\chi) \approx V_L(\psi)V_L(\chi) \,, \tag{12.15}$$

and a particular case of this is that $V_L(N(\psi)\chi) \approx N(\psi)V_L(\chi)$.

It follows from this last result that if we define a product $\phi *_L \psi$ on $\mathcal{H}$ by $\phi *_L \psi = V_L(\phi)\psi$, this induces a product on $\mathcal{A}(\mathcal{H})$ and (12.15) shows that this product is associative, *i.e.* that

$$(\phi * \psi) * \chi = \phi * (\psi * \chi) \,, \tag{12.16}$$

giving $\mathcal{A}(\mathcal{H})$ the structure of an algebra.

We can also define a product by $\phi *_R \psi = V_R(\phi)\psi$. Since

$$\phi *_L \psi = V_L(\phi)\psi = V_L(\phi)V_R(\psi)\Omega \approx V_R(\psi)V_L(\phi)\Omega = V_R(\psi)\phi = \psi *_R \phi \tag{12.17}$$

this defines the reverse ring structure. In terms of the algebraic structures on $A_{u_1,u_2}(\mathcal{H})$ and $A_{u_2,u_1}(\mathcal{H})$ we therefore have

$$A_{u_1,u_2}(\mathcal{H}) = \left(A_{u_2,u_1}(\mathcal{H}) \right)^o \,, \tag{12.18}$$

where A^o is the *reverse algebra*.

By a similar calculation to the above, we can also deduce that for $h_\phi > 0$

$$
\begin{aligned}
[V^{(0)}(\phi), V^{(0)}(\psi)] &\approx \sum_{m=0}^{h_\phi+h_\psi-1} V^{(0)}(V_{m+1-h_\phi}(\phi)\psi) \sum_{s=0}^{\min(h_\phi,m)} (-1)^{m+s} \binom{h_\phi}{s} \\
&= \sum_{m=0}^{h_\phi-1} V^{(0)}(V_{m+1-h_\phi}(\phi)\psi) \sum_{s=0}^{m} (-1)^{m+s} \binom{h_\phi}{s} \\
&= \sum_{m=0}^{h_\phi-1} \binom{h_\phi-1}{m} V^{(0)}(V_{m+1-h_\phi}(\phi)\psi) \\
&= \oint V^{(0)}(V(\phi,\zeta)\psi)(\zeta+1)^{h_\phi-1}d\zeta \,.
\end{aligned}
\tag{12.19}
$$

and

$$
\begin{aligned}
V^{(1)}(\psi)\Omega &= \oint V(\psi,\zeta)(\zeta+1)^{h_\psi}\frac{d\zeta}{\zeta^2}\Omega \\
&= \sum_{n=0}^{h_\psi} \binom{h_\psi}{n} V_{n-h_\psi-1}(\psi)\Omega \\
&= V_{-h_\psi-1}(\psi)\Omega + h_\psi V_{-h_\psi}(\psi)\Omega \\
&= (L_{-1} + L_0)V_{-h_\psi}(\psi)\Omega \,.
\end{aligned}
\tag{12.20}
$$

In particular, this implies that $(L_{-1} + L_0)\psi \approx 0$ for every $\psi \in \mathcal{H}$.
For the Virasoro field $L(z) = V(L, z)$ (12.19) becomes

$$
\begin{aligned}
[V^{(0)}(L), V^{(0)}(\psi)] &= \sum_{m=0}^{1} \begin{pmatrix} 1 \\ m \end{pmatrix} V^{(0)}(V_{m-1}(L)\psi) \\
&= (L_{-1} + L_0)\psi \approx 0,
\end{aligned} \tag{12.21}
$$

which thus implies that L is central in Zhu's algebra.

So far our considerations have been in essence algebraic, in that we have considered the conditions $\Sigma_{\alpha\beta}$ has to satisfy in terms of the linear functional it defines on the Fock space $\mathcal{H}$. If, however, we wish to reverse this process, and proceed from a linear functional on $A(\mathcal{H})$ to a representation of the conformal field theory, we need to be concerned about the analytic properties of the resulting amplitudes. To this end, we note that we can perform an analytic version of the construction as follows.

In fact, since $\Sigma_{\alpha\beta}$ is indeed an element of $\mathcal{V}^{\mathcal{O}}$ for a suitable $\mathcal{O}$, it actually defines a linear functional on the whole dual space $\overline{\mathcal{V}_{\mathcal{O}}} \equiv (\mathcal{V}^{\mathcal{O}})'$ (of which the Fock space is only a dense subspace). Let us denote by $\overline{O(\mathcal{V}_{\mathcal{O}})}$ the completion (in $\overline{\mathcal{V}_{\mathcal{O}}}$) of the space that is generated by states of the form (12.7) with $N > 0$ where now $\psi \in \mathcal{H}$ and $\chi \in \overline{\mathcal{V}_{\mathcal{O}}}$. By the same arguments as before, the linear functional associated to $\Sigma_{\alpha\beta}$ vanishes then on $\overline{O(\mathcal{V}_{\mathcal{O}})}$, and thus defines a linear functional on the quotient space

$$
A\left(\overline{\mathcal{V}_{\mathcal{O}}}\right) = \overline{\mathcal{V}_{\mathcal{O}}} / \overline{O(\mathcal{V}_{\mathcal{O}})}. \tag{12.22}
$$

It is not difficult to show (see [31] for further details) that a priori $A\left(\overline{\mathcal{V}_{\mathcal{O}}}\right)$ is a quotient space of $A(\mathcal{H})$; the main content of Zhu's Theorem [9] is equivalent to:

Theorem 9 [Żhu's Theorem]: The two quotient spaces are isomorphic vector spaces

$$
A\left(\overline{\mathcal{V}_{\mathcal{O}}}\right) \simeq A(\mathcal{H}). \tag{12.23}
$$

Proof. It follows from the proof in [9] that every non-trivial linear functional on $A(\mathcal{H})$ (that is defined by $\rho(a) = \langle w^*, aw \rangle$ where w is an element of a representation of $A(\mathcal{H})$, and w^* is an element of the corresponding dual space) defines a non-trivial state $\Sigma_{\alpha\beta} \in \mathcal{V}^{\mathcal{O}}$, and therefore a non-trivial element in the dual space of $A\left(\overline{\mathcal{V}_{\mathcal{O}}}\right)$.

The main importance of this result is that it relates the analytic properties of correlation functions (which are in essence encoded in the definition of the space $\mathcal{V}^{\mathcal{O}}$, etc.) to the purely algebraic Fock space $\mathcal{H}$.

Every linear functional on $A(\overline{\mathcal{V}_{\mathcal{O}}})$ defines a highest weight representation of the meromorphic conformal field theory, and two such functionals define equivalent representations if they are related by the action of Zhu's algebra as in (12.3). Because of Zhu's Theorem there is therefore a one-to-one correspondence between highest weight representations of the meromorphic conformal field theory whose Fock space is $\mathcal{H}$, and representations of the algebra $A(\mathcal{H})$; this (or something closely related to it) is the form in which Zhu's Theorem is usually stated.

Much of the structure of the meromorphic conformal field theory (and its representations) can be read off from properties of $A(\mathcal{H})$. For example, it was shown in [30] (see also Appendix G) that if $A(\mathcal{H})$ is semisimple, then it is necessarily finite-dimensional, and therefore there exist only finitely many irreducible representations of the meromorphic field theory.

13 Further Developments

In these lectures we have introduced a rigorous approach to conformal field theory taking the amplitudes of meromorphic fields as a starting point. We have shown how the paradigm examples of conformal field theories, *i.e.* lattice theories, affine Lie algebra theories and the Virasoro theory, all fit within this approach. We have shown how to introduce the concept of a representation of such a meromorphic conformal field theory by a using a collection of amplitudes which involve two non-meromorphic fields, so that the amplitudes may be branched at the corresponding points. We showed how this led naturally to the introduction of Zhu's algebra and why the condition that this algebra be finite-dimensional is a critical one in distinguishing interesting and tractable theories from those that appear to be less so.

To complete a treatment of the fundamental aspects of conformal field theory we should discuss subtheories, coset theories and orbifolds, all of which can be expressed naturally within the present approach [31]. It is also clear how to modify the axioms for theories involving fermions.

Zhu's algebra plays a central rôle in characterising the structure of a conformal field theory. In particular, one should expect that the interesting and tracktable theories are those for which Zhu's algebra is finite dimensional. (We have called such theories finite above; they include in particular what physicists normally call 'rational conformal field theories'.) Unfortunately, the determination of Zhu's algebra is usually rather difficult since the modes $N(\psi)$ that generate the space $O(\mathcal{H})$ are not homogeneous with respect to L_0. It would therefore be interesting to find an equivalent condition for the finiteness of a conformal field theory that is easier to analyse in practice.

One such condition that may be equivalent to the finiteness of a meromorphic conformal field theory is the so-called C_2 condition of Zhus [9]; this is the condition that the quotient space

$$A_{(1)}(\mathcal{H}) = \mathcal{H}/O_{(1)}(\mathcal{H}) \tag{13.1}$$

is finite dimensional[2] Here $O_{(1)}(\mathcal{H})$, is spanned by the states of the form $V_{(N)}(\psi)\chi$ with $N \geq 1$ where

$$V_{(N)}(\psi) \equiv V_{-h-n}(\psi). \tag{13.2}$$

It is not difficult to show that the dimension of Zhu's algebra is bounded by that of the above quotient space [9], *i.e.* $dim(A(\mathcal{H})) \leq dim(A_{(1)}(\mathcal{H}))$. For many examples (Virasoro minimal models, affine algebra with $g = su(2)$), the two dimensions are actually the same,

---

[2]Incidentally, this space has also the structure of an abelian algebra; the significance of this algebra structure is however not clear to us.

but this is not true in general. The simplest counterexample is the theory associated to the affine algebra for $g = e_8$ at level $k = 1$. As is well known, this meromorphic conformal field theory is 'self-dual', *i.e.* its only representation is the vacuum representation itself; thus Zhu's algebra is one-dimensional. On the other hand, it is clear that the dimension of $A_{(1)}(\mathcal{H})$ is at least 249 since the vacuum state and the 248 vectors of the form $J^a_{-1}\Omega$ (where a runs over a basis of the 248-dimensional adjoint representation of e_8) are linearly independent in $A_{(1)}(\mathcal{H})$.

Although the dimenions of Zhu's algebra and $A_{(1)}(\mathcal{H})$ differ in general, it may still be the case that one is finite dimensional if and only if the other one is; it would be interesting if this could be proven.

It is relatively straightforward to generalise the discussion of representations to correlation functions involving $N > 2$ non-meromorphic fields. The only difference is that in this case, there are more than two points u_l, $l = 1, \ldots, N$, at which the amplitudes are allowed to have branch cuts. The condition that a collection of such amplitudes defines an N-point correlation function of the meromorphic conformal field theory can then be described analogously to the case of $N = 2$: we consider the vector space $\mathcal{V}_{\mathcal{C}\alpha}$ (where $\alpha = (\alpha_1, \ldots, \alpha_N)$ denotes the indices of the N vector spaces W_{α_i} that are associated to the N points $u_1, \ldots, u_N$), whose elements are finite linear combinations of vectors of the form

$$V(\psi_1, z_1) \cdots V(\psi_n, z_n)\, W_{\alpha_1}(\chi_1, u_1) \cdots W_{\alpha_N}(\chi_N, u_N)\Omega\,. \tag{13.3}$$

We complete this space (and cut it down to size) using the standard construction with respect to $\mathcal{B}_{\mathcal{O}}$ and the above set of amplitudes, and we denote the resulting space by $\mathcal{V}^{\mathcal{O}}_{\mathcal{C}\alpha}$. The relevant condition is then that $\mathcal{B}_{\mathcal{O}}$ induces a continuous map

$$(\mathcal{V}^{\mathcal{O}})' \to (\mathcal{V}^{\mathcal{O}}_\alpha)'\,. \tag{13.4}$$

There also exists a formulation of this condition analogous to (9.10): a collection of amplitudes defines an N-point correlation function if there exists a family of states

$$\Sigma_\alpha(u_1, \ldots, u_N; \chi_1, \ldots, \chi_N) \in \mathcal{V}^{\mathcal{O}}_{\mathcal{C}} \tag{13.5}$$

for each $\mathcal{O}, \mathcal{C}$ with $\mathcal{O} \cap \mathcal{C} = \varnothing$ that is equivalent to $W_{\alpha_1}(\chi_1, u_1) \cdots W_{\alpha_N}(\chi_N, u_N)$ in the sense that

$$\begin{aligned}
\Big\langle \textstyle\prod_{i=1}^n V(\psi_i, z_i)\ &W_{\alpha_1}(\chi_1, u_1) \cdots W_{\alpha_N}(\chi_N, u_N) \Big\rangle \\
&= \langle \textstyle\prod_{i=1}^n V(\psi_i, z_i)\Sigma_\alpha(u_1, \ldots, u_N; \chi_1, \ldots, \chi_N)\rangle\,,
\end{aligned} \tag{13.6}$$

where $z_i \in \mathcal{O}$. An argument analogous to Theorem 6 then implies that (13.4) is equivalent to (13.6).

In the case of the two-point correlation functions (or representations) we introduced a quotient space (12.22) of the vector space $\overline{\mathcal{V}_{\mathcal{O}}} = (\mathcal{V}^{\mathcal{O}})'$ that classified the different highest weight representations. We can now perform an analogous construction. Let us consider the situation where the N highest weight states are at $u_1 = \infty, u_2, \ldots, u_N$, and define

$$V_N^{(M)}(\psi)\chi = \oint_0 \frac{d\zeta}{\zeta^{1+M}} \left(\frac{\prod_{j=2}^N (\zeta - u_j)}{\zeta^{N-2}} \right)^{h_\psi} V(\psi, \zeta)\chi\,, \tag{13.7}$$

where M is an integer, $\psi \in \mathcal{H}$, $\chi \in \overline{\mathcal{V}_\mathcal{O}}$, and the contour encircles 0, but does not encircle $u_1, \ldots, u_N$. We denote by $\overline{O_N(\mathcal{V}_\mathcal{O})}$ the completion of the space (in $\overline{\mathcal{V}_\mathcal{O}}$) that is spanned by states of the form (13.7) with $M > 0$, and we denote by $A_N(\overline{\mathcal{V}_\mathcal{O}})$ the quotient space $\overline{\mathcal{V}_\mathcal{O}}/\overline{O_N(\mathcal{V}_\mathcal{O})}$. (In this terminology, Zhu's algebra is the space $A_2(\overline{\mathcal{V}_\mathcal{O}})$.) By the same arguments as in (12.2) is is easy to see that every state Σ_α that corresponds to an N-point function of N highest weights (where the highest weight property is defined as before) defines a functional on $\overline{\mathcal{V}_\mathcal{O}}$ that vanishes on $\overline{O_N(\mathcal{V}_\mathcal{O})}$, and thus defines a linear functional on $A_N(\overline{\mathcal{V}_\mathcal{O}})$.

One can show (see [31] for more details) that the space $A_N(\overline{\mathcal{V}_\mathcal{O}})$ carries N commuting actions of Zhu's algebra $A(\mathcal{H})$ which are naturally associated to the N non-meromorphic points $u_1, \ldots, u_N$. For example the action corresponding to u_1 is given by (13.7) with $M = 0$ and $\psi \in \mathcal{H}$, $\psi \circ \phi = V_N^{(0)}(\psi)\phi$. This action is actually well-defined for $\psi \in A(\mathcal{H})$ since we have for $L \geq 0$

$$V_N^{(0)}\left(V^{(L)}(\psi_1)\,\psi_2\right)\,\phi \approx V_N^{(L)}(\psi_1)\cdot V_N^{(0)}(\psi_2)\,\phi\,, \tag{13.8}$$

where we denote by $\approx$ equality in $\overline{\mathcal{V}_\mathcal{O}}$ up to states in $\overline{O_N(\mathcal{V}_\mathcal{O})}$. Applying (13.8) for $L = 0$ implies that the algebra relations of $A(\mathcal{H})$ are respected, i.e. that

$$(\psi_1 * \psi_2) \circ \phi = \psi_1 \circ (\psi_2 \circ \phi)\,, \tag{13.9}$$

where $*$ denotes the multiplication of $A(\mathcal{H})$. Every N-point correlation function determines therefore N representations of Zhu's algebra, and because of Zhu's Theorem, we can associate N representations of the meromorphic conformal field theory to it. Conversely, every linear functional on $A_N(\overline{\mathcal{V}_\mathcal{O}})$ defines an N-point correlation function, and two functionals define equivalent such functions if they are related by the actions of Zhu's algebra; in this way the different N-point correlation functions of the meromorphic conformal field theory are classified by $A_N(\overline{\mathcal{V}_\mathcal{O}})$.

There exists also an 'algebraic' version of this quotient space, $A_N(\mathcal{H}) = \mathcal{H}/O_N(\mathcal{H})$, where $O_N(\mathcal{H})$ is generated by the states of the form (13.7) where now ψ and ϕ are in $\mathcal{H}$. This space is more amenable for study, and one may therefore hope that in analogy to Zhu's Theorem, the two quotient spaces are isomorphic vector spaces,

$$A_N(\overline{\mathcal{V}_\mathcal{O}}) \simeq A_N(\mathcal{H})\,; \tag{13.10}$$

it would be interesting if this could be established.

Finally, there exists a homogeneous version of this quotient space which again is much more easy to analyse. Its dimension bounds the dimension of $A_N(\mathcal{H})$ in the same way the dimension of (13.1) bounds that of Zhu's algebra. Again, it would be interesting to prove that the finite dimensionality of this homogeneous quotient space is equivalent to that of $A_N(\mathcal{H})$.

It is also rather straightforward to apply the above techniques to an analysis of correlation functions on higher genus Riemann surfaces. Again, it is easy to see that the correlation functions on a genus g surface can be described in terms of a state of the meromorphic conformal field theory on the sphere, in very much the same way in which N-point correlation functions can be defined by (9.10). The corresponding state induces a linear functional on

$\overline{\mathcal{V_O}}$ (or $\mathcal{H}$), and since it vanishes on a certain subspace thereof, defines a linear functional on a suitable quotient space. For the case of the genus $g = 1$ surface, the torus, the corresponding quotient space is very closely related to Zhu's algebra, and one may expect that similar relations hold more generally.

Acknowledgements

We would like to thank Ben Garling, Terry Gannon, Graeme Segal and Anthony Wassermann for useful conversations. M.R.G. is grateful to the organizers of the Turkish Summer School on Theoretical Physics, Istanbul, August 1998 for the opportunity to present these results.

M.R.G. is grateful to Jesus College, Cambridge, for a Research Fellowship, to Harvard University for hospitality during the tenure of a NATO Fellowship in 1996/97, and to Fitzwilliam College, Cambridge for a College Lectureship. P.G. is grateful to the Mathematische Forschungsinstitut Oberwolfach and the Aspen Center for Physics for hospitality in January 1995 and August 1996, respectively. The visit to Aspen was partially funded by EPSRC grant GR/K30667.

A   Sequences of holomorphic functions

A sequence of functions $\{f_n\} = f_1, f_2, f_3, \ldots$, each defined on a domain $D \subset \mathbb{C}$, is said to be *uniformly bounded* on D if there exists a real number M such that $|f_n(z)| < M$ for all n and $z \in D$. The sequence is said to be *locally uniformally bounded* on D if, given $z_0 \in D$, $\{f_n\}$ is uniformly bounded on $N_\delta(z_0, D) = \{z \in D : |z - z_0| < \delta\}$ for some $\delta \equiv \delta(z_0) > 0$. A sequence of functions $\{f_n\}$ defined on D is said to be *uniformly convergent* to $f : D \to \mathbb{C}$ if, given $\epsilon > 0$, $\exists\, N$ such that $|f_n(z) - f(z)| < \epsilon$ for all $z \in D$ and $n > N$. We write $f_n \to f$ uniformly in D. The sequence is said to be *locally uniformly convergent* to $f : D \to \mathbb{C}$ if, given $z_0 \in D$, $f_n \to f$ uniformally in $N_\delta(z_0, D)$ for some $\delta \equiv \delta(z_0) > 0$.

Clearly, by the Heine-Borel Theorem, a sequence is locally uniformly bounded on D if and only if it is uniformly bounded on every compact subset of D, and locally uniformly convergent on D if and only if it is uniformly convergent on every compact subset of D. Local uniformity of the convergence of a sequence of continuous functions guarantees continuity of the limit and, similarly, analyticity of the limit of a sequence of analytic functions is guaranteed by local uniformity of the convergence [32].

The following result [33] is of importance in the approach to conformal field theory developed in these lectures:

Theorem. If D is an open domain and $f_n : D \to \mathbb{C}$ is analytic for each n, $f_n(z) \to f(z)$ at each $z \in D$, and the sequence $\{f_n\}$ is locally uniformly bounded in D, then $f_n \to f$ locally uniformly in D and f is analytic in D.

[Proof: Again, given $z_0 \in D$, $\overline{N_\delta(z_0)} = \{z \in \mathbb{C} : |z - z_0| \leq \delta\} \subset D$ for some $\delta > 0$ because D is open. Because $\overline{N_\delta(z_0)}$ is compact, $\exists M$, such that $|f_n(z)| < M$ for all n and all $z \in \overline{N_\delta(z_0)}$.

First we show that the sequence $f_n'(z)$ is uniformally bounded on $N_\rho(z_0)$ for all $\rho < \delta$. For

$$|f_n'(z)| = \left| \frac{1}{2\pi i} \oint_{C_\delta(z_0)} \frac{f_n(\zeta) d\zeta}{(\zeta - z)^2} \right| \leq \frac{M\delta}{(\delta - \rho)^2} = M_1(\rho), \quad \text{say.} \tag{A.1}$$

Thus for fixed ρ, given $\epsilon > 0$, $\exists \, \delta_1 > 0$ such that $|f_n(z) - f_n(z')| < \frac{1}{3}\epsilon$ for all values of n provided that $z, z' \in N_\rho(z_0)$ are such that $|z - z'| < \delta_1(\epsilon)$. Then also $|f(z) - f(z')| < \frac{1}{3}\epsilon$ for $|z - z'| < \delta_1(\epsilon)$. Now we can find a finite number K of points $z_j \in N_\rho(z_0)$, $1 \leq j \leq K$, such that given any point in $z \in N_\rho(z_0)$, $|z - z_j| < \delta_1$ for some j. Now each $f_n(z_j) \to f(z_j)$ and so we can find integers L_j such that $|f(z_j) - f_n(z_j)| < \frac{1}{3}\epsilon$ for $n > L = \max_{1 \leq j \leq K} \{L_j\}$ and $z \in N_\rho(z_0)$, $|f_n(z) - f(z)| \leq |f_n(z) - f_n(z_j)| + |f_n(z_j) - f(z_j)| + |f(z_j) - f(z)| < \epsilon$, establishing uniform convergence on $N_\rho(z_0)$ and so local uniform convergence. This is sufficient to deduce that f is analytic.

B   Completeness of $\mathcal{V}_C^\mathcal{O}$

We prove that, if a sequence $\chi_j \in \mathcal{V}_C^\mathcal{O}$, $j = 1, 2, \ldots$, $\eta_\phi(\chi_j)$ converges on each subset of ϕ of the form (3.6) (where the convergence is uniform on (3.7)), the limit $\lim_{j \to \infty} \eta_\phi(\chi_j)$ necessarily equals $\eta_\phi(\chi)$ for some $\chi \in \mathcal{V}_C^\mathcal{O}$.

To see this, note that uniform convergence in this sense is implied by the (uniform) convergence on a countable collection of such sets, taken by considering $\epsilon = 1/N$, N a positive integer, K one of a collection of compact subsets of $\mathcal{O}$ and the ϕ_j to be elements of some countable basis. Taken together we obtain in this way a countable number of conditions for the uniform convergence. Defining $\|\Psi\|_n = \max_\phi \max_\zeta |\eta_\phi(\Psi)|$, where ϕ ranges over the first n of these countable conditions for each $\mathcal{O}$ and the maximum is taken over ζ_j within (3.7), we have a sequence of semi-norms, $\|\Psi\|_n$, on $\mathcal{V}_C$, with $\|\Psi\|_n \leq \|\Psi\|_{n+1}$. Given such a sequence of semi-norms, we can define a Cauchy sequence (Ψ_j), $\Psi_j \in \mathcal{V}_C$, by the requirement that $\|\Psi_i - \Psi_j\|_n \to 0$ as $i, j \to \infty$ for each fixed n. This requirement is equivalent to uniform convergence on each set of the form (3.6). Moreover, the space $\mathcal{V}_C^\mathcal{O}$ is obtained by adding in the limits of these Cauchy sequences (identifying points zero distance apart with respect to all of the semi-norms). This space is necessarily complete because if $\chi_j \in \mathcal{V}_C^\mathcal{O}$ is Cauchy, i.e. $\|\chi_i - \chi_j\|_n \to 0$ as $i, j \to \infty$ for each fixed n, and $\Psi_i^m \to \chi_i$ as $m \to \infty$, $\Psi_i^m \in \mathcal{V}_C$, then selecting I_N so that $\|\chi_i - \chi_j\|_N < 1/3N$ for $i, j \geq I_N$, and $I_{N+1} \geq I_N$, we can find an integer m_N such that $\|\psi_{I_N}^{m_N} - \chi_{I_N}\|_N < 1/3N$, and if $\psi_N = \psi_{I_N}^{m_N}$,

$$\begin{aligned}
\|\psi_M - \psi_N\|_p &\leq \|\psi_M - \chi_{I_M}\|_p + \|\chi_{I_M} - \chi_{I_N}\|_p + \|\chi_{I_N} - \psi_N\|_p \\
&\leq \|\psi_M - \chi_{I_M}\|_M + \|\chi_{I_M} - \chi_{I_N}\|_N + \|\chi_{I_N} - \psi_N\|_N \\
&\leq 1/N
\end{aligned} \tag{B.1}$$

provided that $M \geq N \geq p$, implying that ψ_M is Cauchy. It is easy to see that its limit is the limit of χ_j, showing that $\mathcal{V}_C^\mathcal{O}$ is complete. The completeness of this space is equivalent to the condition (3.11).

C   Proof that $\tilde{\mathcal{V}}_\mathcal{C}^\mathcal{O}$ is independent of $\mathcal{C}$ (Theorem 1)

Proof: To prove that $\tilde{\mathcal{V}}_\mathcal{C}^\mathcal{O}$ is independent of $\mathcal{C}$, first note that we may identify a vector $\chi \in \tilde{\mathcal{V}}_\mathcal{C}^\mathcal{O}$ with $\psi = \prod_{i=1}^n V(\psi_i, z_i)\Omega$ (where $z_i \notin \mathcal{O}$ for $1 \le i \le n$ but it is not necessarily the case that $z_i \in \mathcal{C}$ for each i) if $\eta_\phi(\chi) = \eta_\phi(\psi)$, for all $\phi \in \mathcal{B}_\mathcal{O}$, i.e. the value of $\eta_\phi(\chi)$ is given by (3.5) for all $\phi \in \mathcal{B}_\mathcal{O}$. Consider then the set $\mathcal{Q}$ of values of $\mathbf{z} = (z_1, z_2, \ldots, z_n)$ for which $\psi(\mathbf{z}) = \prod_{i=1}^n V(\psi_i, z_i)\,\Omega$ is a member of $\tilde{\mathcal{V}}_\mathcal{C}^\mathcal{O}$. Then $\mathcal{D}' \subset \mathcal{Q} \subset \mathcal{D}$, where $\mathcal{D}' = \{\mathbf{z} : z_i, z_j \in \mathcal{C}, z_i \ne z_j, 1 \le i < j \le n\}$ and $\mathcal{D} = \{\mathbf{z} : z_i, z_j \in \mathcal{O}^c, z_i \ne z_j, 1 \le i < j \le n\}$, where $\mathcal{O}^c$ is the complement of $\mathcal{O}$ in $\mathbb{P}$. We shall show that $\mathcal{Q} = \mathcal{D}$.

If $\mathbf{z}_b$ is in $\mathcal{D}^o$, the interior of $\mathcal{D}$, but not in $\mathcal{Q}$, choose a point $\mathbf{z}_a \in \mathcal{D}' \subset \mathcal{Q}$ and join it to $\mathbf{z}_b$ by a path C inside $\mathcal{D}^o$, $\{\mathbf{z}(t) : 0 \le t \le 1\}$ with $\mathbf{z}(0) = \mathbf{z}^a$ and $\mathbf{z}(1) = \mathbf{z}^b$. (There is such a point $\mathbf{z}^a$ because $\mathcal{C}^o \ne \varnothing$; the path C exists because the interior of $\mathcal{O}^c$ is connected, from which it follows that $\mathcal{D}^o$ is.) Let t_c be the supremum of the values of t_0 for which $\{\mathbf{z}(t) : 0 \le t \le t_0\} \subset \mathcal{Q}^o$, the interior of $\mathcal{Q}$, and let $\mathbf{z}^c = \mathbf{z}(t_c)$. Then $\mathbf{z}^c = (z_1^c, z_2^c, \ldots, z_n^c)$ is inside the open set $\mathcal{D}^o$ and so we can find a neighbourhood of the form $N_1 = \{\mathbf{z} : |\mathbf{z} - \mathbf{z}^c| < 4\delta\}$ which is contained inside $\mathcal{D}^o$. Let $\mathbf{z}^d$, $\mathbf{z}^e$ be points each distant less than δ from $\mathbf{z}^c$, with $\mathbf{z}^d$ outside $\mathcal{Q}$ and $\mathbf{z}^e$ inside $\mathcal{Q}^o$. (There must be a point $\mathbf{z}^d$ outside $\mathcal{Q}$ in every neighbourhood of $\mathbf{z}^c$.) Then the set $N_2 = \{\mathbf{z}^e + (\mathbf{z}^d - \mathbf{z}^e)\omega : |\omega| < 1\}$ is inside N_1 but contains points outside $\mathcal{Q}$. We shall show that $N_2 \subset \mathcal{Q}$ establishing a contradiction to the assumption that there is a point $\mathbf{z}_b$ in $\mathcal{D}^o$ but not in $\mathcal{Q}$, so that we must have $\mathcal{D}^o \subset \mathcal{Q} \subset \mathcal{D}$.

The circle $\{\mathbf{z}^e + (\mathbf{z}^d - \mathbf{z}^e)\omega : |\omega| < \epsilon\}$ is inside $\mathcal{Q}$ for some ϵ in the range $0 < \epsilon < 1$. Now, we can form the integral

$$\chi = \int_S \psi(\mathbf{z})\mu(\mathbf{z})d^r\mathbf{z} \tag{C.1}$$

of $\psi(\mathbf{z})$ over any compact r-dimensional sub-manifold $S \subset \mathcal{Q}$, with continuous weight function $\mu(\mathbf{z})$, to obtain an element $\chi \in \tilde{\mathcal{V}}_\mathcal{C}^\mathcal{O}$, because the approximating sums to the integral will have the necessary uniform convergence property. So the Taylor coefficients

$$\psi_N = \int_{|\omega|=\epsilon} \psi(\mathbf{z}^e + (\mathbf{z}^d - \mathbf{z}^e)\omega)\omega^{-N-1}d\omega \in \tilde{\mathcal{V}}_\mathcal{C}^\mathcal{O} \tag{C.2}$$

and, since the $\sum_{N=0}^\infty \eta_\phi(\psi_N)\omega^N$ converges to $\eta_\phi(\psi(\mathbf{z}^e + (\mathbf{z}^d - \mathbf{z}^e)\omega))$ for $|\omega| < 1$ and all $\phi \in \mathcal{B}_\mathcal{O}$, we deduce that $N_2 \subset \mathcal{Q}$, hence proving that $\mathcal{D}^o \subset \mathcal{Q} \subset \mathcal{D}$. Finally, if $\mathbf{z}_j$ is a sequence of points in $\mathcal{Q}$ convergent to $\mathbf{z}_0 \in \mathcal{D}$, it is straightforward to see that $\psi(\mathbf{z}_i)$ will converge to $\psi(\mathbf{z}_0)$, so that $\mathcal{Q}$ is closed in $\mathcal{D}$, and so must equal $\mathcal{D}$.

D   Möbius Transformation of Vertices

By virtue of the Uniqueness Theorem we can establish the transformation properties

$$e^{\lambda L_{-1}}V(\Psi, z)e^{-\lambda L_{-1}} = V(\Psi, z + \lambda) \tag{D.1}$$

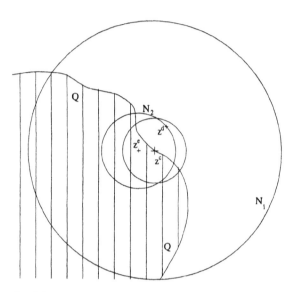

Figure 1: Sketch of the geometrical arrangement of $\mathbf{z}^c$, $\mathbf{z}^d$, $\mathbf{z}^e$ in relation to $\mathcal{Q}$, N_1 and N_2.

$$e^{\lambda L_0} V(\Psi, z) e^{-\lambda L_0} = e^{\lambda h} V(\Psi, e^{\lambda} z) \qquad \text{(D.2)}$$

$$e^{\lambda L_1} V(\Psi, z) e^{-\lambda L_1} = (1 - \lambda z)^{-2h} V\left(\exp(\lambda(1 - \lambda z) L_1)\Psi, z/(1 - \lambda z)\right), \qquad \text{(D.3)}$$

where we have used the relation

$$\begin{pmatrix} 1 & 0 \\ -\lambda & 1 \end{pmatrix} \begin{pmatrix} 1 & z \\ 0 & 1 \end{pmatrix} = \begin{pmatrix} 1 & \frac{z}{1-\lambda z} \\ 0 & 1 \end{pmatrix} \begin{pmatrix} 1 & 0 \\ -\lambda(1 - \lambda z) & 1 \end{pmatrix} \begin{pmatrix} \frac{1}{1-\lambda z} & 0 \\ 0 & 1 - \lambda z \end{pmatrix}. \qquad \text{(D.4)}$$

From these if follows that

$$\langle V(\Phi, z) \rangle = \qquad \text{(D.5)}$$

$$\langle V(\Phi, z) V(\Psi, \zeta) \rangle = \frac{\varphi(\Phi, \Psi)}{(z - \zeta)^{h_\Phi + h_\Psi}}, \qquad \text{(D.6)}$$

where $L_0 \Phi = h_\Phi \Phi$, $L_0 \Psi = h_\Psi \Psi$ and $h_\Phi \neq 0$. The bilinear form $\varphi(\Phi, \Psi)$, defined as being the constant of proportionality in (D.6), has the symmetry property

$$\varphi(\Phi, \Psi) = (-1)^{h_\Phi + h_\Psi} \varphi(\Psi, \Phi). \qquad \text{(D.7)}$$

If, in addition, $L_1 \Phi = L_1 \Psi = 0$, it follows from (D.3) applied to (D.6) that $\varphi(\Psi, \Phi) = 0$ unless $h_\Phi = h_\Psi$.

It follows from (D.3) that, if $\Psi \in \mathcal{H}_1$,

$$\langle V(\Psi, z) \rangle = (1 - \lambda z)^{-2} \langle V(\Psi, z/(1 - \lambda z)) \rangle + \frac{\lambda}{(1 - \lambda z)} \langle V(L_1 \Psi, z/(1 - \lambda z)) \rangle, \qquad \text{(D.8)}$$

and from (D.5) that both the left hand side and the first term on the right hand side of (D.8) vanish, implying that the second term on the right hand side also vanishes. But $L_1\Psi \in \mathcal{H}_0$ and, if we assume cluster decomposition, so that the vacuum is unique, $L_1\Psi = \kappa\Omega$ for some $\kappa \in \mathbb{C}$. We deduce that $\kappa = \langle V(L_1\Psi, \zeta)\rangle = 0$, so that Ψ is quasi-primary, $i.e.$ $\mathcal{H}_1 = \mathcal{H}_1^Q$. We can show inductively that $\mathcal{H}_h$ is the direct sum of spaces $L_{-1}^n\mathcal{H}_{h-n}^Q$ where $0 \leq n < h$, $i.e.$ $\mathcal{H}$ is composed of quasi-primary states and their descendants under the action of L_{-1}. Given $\Psi \in \mathcal{H}_h$, we can find $\Phi \in L_{-1}H_{h-1}$ such that $L_1(\Psi - \Phi) = 0$; then Ψ is the sum of the quasi-primary state $\Psi - \Phi$ and Φ which, by an inductive hypothesis, is the sum of descendants of quasi-primary states. To find Φ, note

$$L_1(\Psi + \sum_{n=1}^{h} a_n L_{-1}^n L_1^n \Psi) = \sum_{n=1}^{h}(a_{n-1} + a_n(2nh + n(n+1)))L_{-1}^{n-1}L_1^n\Psi, \qquad (D.9)$$

where $a_0 = 1$. So choosing $a_n = -a_{n-1}/(2nh+n(n+1))$, $1 \leq n \leq h$, we have $L_1(\Psi - \Phi) = 0$ for $\Phi = -\sum_{n=1}^{h} a_n L_{-1}^n L_1^n \Psi \in L_{-1}H_{h-1}$ establishing the result.

E Proof that the extended amplitudes $\hat{A}$ satisfy the axioms

An alternative description of the extended amplitudes can be given as follows: we define the amplitudes involving vectors in $\hat{V}$ recursively (the recursion being on the number of times L appears in an amplitude) by

$$\langle L(w)\rangle = 0 \qquad (E.1)$$

$$
\begin{aligned}
\langle L(w) \prod_{i=1}^{n} V(\psi_i, z_i)\rangle = {} & \sum_{l=1}^{n} \frac{c_{\psi_l}}{(w-z_l)^4} \langle \prod_{i\neq l} V(\psi_i, z_i)\rangle \\
& + \sum_{l=1}^{n} \frac{h_l}{(w-z_l)^2} \langle \prod_{i=1}^{n} V(\psi_i, z_i)\rangle \\
& + \sum_{l=1}^{n} \frac{1}{(w-z_l)} \frac{d}{dz_l} \langle V(\psi_1, z_1)\cdots V(\psi_l, z_l)\cdots V(\psi_n, z_n)\rangle
\end{aligned}
$$

$$(E.2)$$

and

$$\langle L(w) \prod_{j=1}^{m} L(w_j) \prod_{i=1}^{n} V(\psi_i, z_i) \rangle = \sum_{k=1}^{m} \frac{2}{(w - w_k)^2} \langle \prod_{j=1}^{m} L(w_j) \prod_{i=1}^{n} V(\psi_i, z_i) \rangle$$
$$+ \sum_{l=1}^{n} \frac{h_l}{(w - z_l)^2} \langle \prod_{j=1}^{m} L(w_j) \prod_{i=1}^{n} V(\psi_i, z_i) \rangle$$
$$+ \sum_{k=1}^{m} \frac{1}{(w - w_k)} \frac{d}{dw_k} \langle \prod_{j=1}^{m} L(w_j) \prod_{i=1}^{n} V(\psi_i, z_i) \rangle$$
$$+ \sum_{l=1}^{n} \frac{1}{(w - z_l)} \frac{d}{dz_l} \langle \prod_{j=1}^{m} L(w_j) \prod_{i=1}^{n} V(\psi_i, z_i) \rangle \qquad (E.3)$$
$$+ \sum_{k=1}^{m} \frac{c/2}{(w - w_k)^4} \langle \prod_{j \neq k}^{m} L(w_j) \prod_{i=1}^{n} V(\psi_i, z_i) \rangle$$
$$+ \sum_{l=1}^{n} \frac{c_{\psi_l}/2}{(w - z_l)^4} \langle \prod_{j=1}^{m} L(w_j) \prod_{i \neq l}^{m} V(\psi_i, z_i) \rangle .$$

Here h_i is the grade of the vector $\psi_i \in V$, c is an arbitrary (real) number, and c_ψ is zero unless ψ is of grade two. It is not difficult to see that the functions defined by (E.1)–(E.3) agree with those defined in the main part of the text: for a given set of fields, the difference between the two amplitudes does not have any poles in w_j, and therefore is constant as a function of w_j; this constant is easily determined to be zero.

The diagrammatical description of the amplitudes immediately implies that the amplitudes are local. We shall now use the formulae (E.1)–(E.3) to prove that they are also Möbius covariant. The Möbius group is generated by translations, scalings and the inversion $z \mapsto 1/z$. It is immediate from the above formulae that the amplitudes (with the grade of L being 2) are covariant under translations and scalings, and we therefore only have to check the covariance under the inversion $z \mapsto 1/z$. First, we calculate (setting for the moment $c_\psi = 0$ for all ψ of grade two)

$$\langle L(1/w) \prod_{i=1}^{n} V(\psi_i, 1/z_i) \rangle = \sum_{l=1}^{n} \frac{h_l}{(1/w - 1/z_l)^2} \langle \prod_{i=1}^{n} V(\psi_i, 1/z_i) \rangle$$
$$+ \sum_{l=1}^{n} \frac{1}{(1/w - 1/z_l)} \frac{d}{d\tilde{z}_l} \langle V(\psi_1, 1/z_1) \cdots V(\psi_l, \tilde{z}_l) \cdots V(\psi_n, 1/z_n) \rangle \Big|_{\tilde{z}_l = 1/z_l} . \qquad (E.4)$$

Using the Möbius covariance of the original amplitudes, we find

$$
\begin{aligned}
\frac{d}{d\tilde{z}_l}\langle V(\psi_1, 1/z_1) \cdots V(\psi_l, \tilde{z}_l) \cdots V(\psi_n, 1/z_n)\rangle\Big|_{\tilde{z}_l=1/z_l} &= -z_l^2 \frac{d}{dz_l}\langle \prod_{i=1}^{n} V(\psi_i, 1/z_i)\rangle \\
&= -z_l^2 \prod_{i\neq l} \left(\frac{-1}{z_i^2}\right)^{-h_i} \frac{d}{dz_l}\left[\left(\frac{-1}{z_l^2}\right)^{-h_l} \langle \prod_{i=1}^{n} V(\psi_i, z_i)\rangle\right] \\
&= -z_l^2 \prod_{i=1}^{n} \left(\frac{-1}{z_i^2}\right)^{-h_i} \frac{d}{dz_l}\langle V(\psi_1, z_1) \cdots V(\psi_l, z_l) \cdots V(\psi_n, z_n)\rangle \\
&\quad -z_l^2 \prod_{i=1}^{n} \left(\frac{-1}{z_i^2}\right)^{-h_i}(-h_l)\frac{2}{z_l^3}\left(-\frac{1}{z_l^2}\right)^{-1}\langle \prod_{i=1}^{n} V(\psi_i, z_i)\rangle \\
&= \prod_{i=1}^{n} \left(\frac{-1}{z_i^2}\right)^{-h_i}\left[-z_l^2 \frac{d}{dz_l}\langle V(\psi_1, z_1) \cdots V(\psi_l, z_l) \cdots V(\psi_n, z_n)\rangle\right. \\
&\quad \left. -2\,h_l\,z_l\,\langle \prod_{i=1}^{n} V(\psi_i, z_i)\rangle\right].
\end{aligned}
\tag{E.5}
$$

Inserting this formula in the above expression, we get

$$
\begin{aligned}
\langle\; L(1/w)\prod_{i=1}^{n} V(\psi_i, 1/z_i)\rangle \\
= \left(\frac{-1}{w^2}\right)^{-2}\prod_{i=1}^{n}\left(\frac{-1}{z_i^2}\right)^{-h_i}\left\{\sum_{l=1}^{n} h_l\left[\frac{z_l^2}{w^2(w-z_l)^2}+\frac{2z_l^2}{w^3(w-z_l)}\right]\langle\prod_{i=1}^{n} V(\psi_i, z_i)\rangle\right. \\
\left. +\sum_{l=1}^{n}\frac{z_l^3}{w^3(w-z_l)}\frac{d}{dz_l}\langle V(\psi_1, z_1)\cdots V(\psi_l, z_l)\cdots V(\psi_n, z_n)\rangle\right\} \\
= \left(\frac{-1}{w^2}\right)^{-2}\prod_{i=1}^{n}\left(\frac{-1}{z_i^2}\right)^{-h_i}\left\{\sum_{l=1}^{n} h_l\frac{3\,z_l^2\,w-2\,z_l^3}{w^3(w-z_l)^2}\langle\prod_{i=1}^{n} V(\psi_i, z_i)\rangle\right. \\
\left. +\sum_{l=1}^{n}\frac{z_l^3}{w^3(w-z_l)}\frac{d}{dz_l}\langle V(\psi_1, z_1)\cdots V(\psi_l, z_l)\cdots V(\psi_n, z_n)\rangle\right\}.
\end{aligned}
\tag{E.6}
$$

It remains to show that the expression in brackets actually agrees with (E.4). To prove this, we observe, that because of Möbius invariance of the amplitudes we have

$$
\sum_{l=1}^{n}\frac{d}{dz_l}\langle V(\psi_1, z_1)\cdots V(\psi_l, z_l)\cdots V(\psi_n, z_n)\rangle = 0,
\tag{E.7}
$$

$$
\sum_{l=1}^{n} h_l\,\langle\prod_{i=1}^{n} V(\psi_i, z_i)\rangle + \sum_{l=1}^{n} z_l\frac{d}{dz_l}\langle V(\psi_1, z_1)\cdots V(\psi_l, z_l)\cdots V(\psi_n, z_n)\rangle = 0
\tag{E.8}
$$

and

$$
\sum_{l=1}^{n} 2\,h_l\,z_l\,\langle\prod_{i=1}^{n} V(\psi_i, z_i)\rangle + \sum_{l=1}^{n} z_l^2\frac{d}{dz_l}\langle V(\psi_1, z_1)\cdots V(\psi_l, z_l)\cdots V(\psi_n, z_n)\rangle = 0.
\tag{E.9}
$$

The claim then follows from the observation that

$$
\sum_{l=1}^{n} h_l \left[\frac{1}{(w - z_l)^2} - \frac{3\, z_l^2\, w - 2\, z_l^3}{w^3\, (w - z_l)^2} \right] \langle \prod_{i=1}^{n} V(\psi_i, z_i) \rangle
$$

$$
+ \sum_{l=1}^{n} \left[\frac{1}{(w - z_l)} - \frac{z_l^3}{w^3\, (w - z_l)} \right] \frac{d}{dz_l} \langle V(\psi_1, z_1) \cdots V(\psi_l, z_l) \cdots V(\psi_n, z_n) \rangle
$$

$$
= \frac{1}{w^3} \left\{ w^2 \sum_{l=1}^{n} \frac{d}{dz_l} \langle \prod_i V(\psi_i, z_i) \rangle \right.
$$

$$
+ w \left[\sum_{l=1}^{n} h_l \langle \prod_{i=1}^{n} V(\psi_i, z_i) \rangle + \sum_{l=1}^{n} z_l \frac{d}{dz_l} \langle \prod_i V(\psi_i, z_i) \rangle \right]
$$

$$
\left. + \sum_{l=1}^{n} 2\, h_l\, z_l \langle \prod_{i=1}^{n} V(\psi_i, z_i) \rangle + \sum_{l=1}^{n} z_l^2 \frac{d}{dz_l} \langle \prod_{i=1}^{n} V(\psi_i, z_i) \rangle \right\}
$$

$$
= 0 .
$$

(E.10)

We have thus shown that the functions of the form (E.2) (for $c_\psi = 0$) have the correct transformation property under Möbius transformations. This implies, as L is quasiprimary, that the functions of the form (E.3) have the right transformation property for $c = 0$. However, the sum involving the c-terms has also (on its own) the right transformation property, and thus the above functions have. This completes the proof.

Finally, we want to show that the amplitudes $\hat{A}$ have the cluster property provided the amplitudes A do. We want to prove the cluster property by induction on the number N_L of L-fields in the extended amplitudes. If $N_L = 0$, then the result follows from the assumption about the original amplitudes. Let us therefore assume that the result has been proven for $N_L = N$, and consider the amplitudes with $N_L = N + 1$. For a given amplitude, we subdivide the fields into two groups, and we consider the limit, where the parameters z_i of one group are scaled to zero, whereas the parameters ζ_j of the other group are kept fixed. Because of the Möbius covariance, we may assume that the group whose parameters z_i are scaled to zero contain at least one L-field, $L(z_1)$, say, and we can use (E.2) (or (E.3)) to rewrite the amplitudes involving $L(z_1)$ in terms of amplitudes which do not involve $L(z_1)$ and which have $N_L \leq N$. It then follows from (E.2) (or (E.3)) together with the induction hypothesis that the terms involving $(z_1 - \zeta_j)^{-l}$ (where $l = 1, 2$ or $l = 4$) are not of leading order in the limit where the z_i are scaled to zero, whereas all terms with $(z_1 - z_i)^{-l}$ are. This implies, again by the induction hypothesis, that the amplitudes satisfy the cluster property for $N_L = N + 1$, and the result follows by induction.

F   Möbius transformation of Zhu's modes

We want to prove formula (12.7) in this Appendix. We have to show that

$$
V^{(N)}_{u_1, u_2}(\psi) = U(\gamma) V^{(N)}(U(\gamma)^{-1}\psi) U(\gamma)^{-1}
$$

$$
= U(\gamma) \oint_0 V\left[(\zeta + 1)^{L_0} U(\gamma)^{-1}\psi, \zeta \right] \frac{d\zeta}{\zeta^{N+1}} U(\gamma)^{-1}
$$

$$= \oint_0 U(\gamma)V\left[(\zeta+1)^{L_0}U(\gamma)^{-1}\psi,\zeta\right]U(\gamma)^{-1}\frac{d\zeta}{\zeta^{N+1}}. \tag{F.1}$$

We therefore have to find an expression for the transformed vertex operator. By the uniqueness theorem, it is sufficient to evaluate the expression on the vacuum; then we find

$$U(\gamma)V\left[(\zeta+1)^{L_0}U(\gamma)^{-1}\psi,\zeta\right]U(\gamma)^{-1}\Omega = U(\gamma)e^{\zeta L_{-1}}(\zeta+1)^{L_0}U(\gamma)^{-1}\psi. \tag{F.2}$$

To calculate the product of the Möbius transformations, we write them in terms of 2×2 matrices, determine their product and rewrite the resulting matrix in terms of the generators $L_0, L_{\pm 1}$. After a slightly lengthy calculation we then find

$$U(\gamma)\ e^{\zeta L_{-1}}(\zeta+1)^{L_0}U(\gamma)^{-1}\psi =$$
$$V\left[\left(\frac{\zeta+1}{(1-\frac{u_2\zeta}{(u_1-u_2)})^2}\right)^{L_0}\exp\left(\frac{\zeta}{u_2\zeta+(u_2-u_1)}L_1\right)\psi,\frac{u_1u_2\zeta}{u_2\zeta+(u_2-u_1)}\right]\Omega. \tag{F.3}$$

In the integral for $V^{(N)}_{u_1,u_2}$ we then change variables to

$$w = \frac{u_1u_2\zeta}{u_2\zeta+(u_2-u_1)} = \gamma(\zeta); \tag{F.4}$$

in terms of w the relevant expressions become

$$1+\zeta = \frac{u_1(w-u_2)}{u_2(w-u_1)} \qquad 1-\frac{u_2\zeta}{(u_1-u_2)} = \frac{u_1}{(u_1-w)} \qquad d\zeta = \frac{u_1(u_2-u_1)}{u_2}\frac{dw}{(w-u_1)^2}. \tag{F.5}$$

Putting everything together, we then obtain formula (12.7).

G   Rings and Algebras

In this Appendix we review various concepts in algebra; the treatment follows closely the book [34].

We restrict attention to rings, R which have a unit element, $1 \in R$. An algebra, A, over a field F, is a ring which is also vector space over F in such a way that the structures are compatible [$\lambda(xy) = (\lambda x)y$, $\lambda \in F$, $x,y \in R$]. The dimension of A is its dimension as a vector space. We shall in general consider complex algebras, *i.e.* algebras over $\mathbb{C}$. Since $1 \in A$ we have $F \subset A$.

A (left) module for a ring R is an additive group M with a map $R \times M \to M$, compatible with the structure of R [*i.e.* $(rs)m = r(sm)$, $(r + s)m = rm + sm$, $r, s \in R, m \in M$]. A module M for an algebra A, viewed as a ring, is necessarily a vector space over F (because $F \subset A$) and provides a representation of A as an algebra in terms of endomorphisms of the vector space M.

R provides a module for itself, the adjoint module. A submodule N of a module M for R is an additive subgroup of M such that $rN \subset N$ for all $r \in R$. A simple or irreducible module is one which has no proper submodules. A (left) ideal J of R is a submodule of the adjoint module, $i.e.$ an additive subgroup $J \subset R$ such that $rj \in I$ for all $r \in R, j \in J$. The direct sum $M_1 \oplus M_2$ of the R modules M_1, M_2 is the additive group $M_1 \oplus M_2$ with $r(m_1, m_2) = (rm_1, rm_2), r \in R, m_1 \in M_1, m_2 \in M_2$. The direct sum of a (possibly infinite) set $M_i, i \in I$, of R modules consisted of elements $(m_i, i \in I)$, with all but finitely many $m_i = 0$. The module M is decomposable if it can be written as the direct sum of two non-zero modules and completely reducible if it can be written as the direct sum of a (possibly infinite) sum of irreducible modules.

A representation of an algebra A is irreducible if it is irreducible as a module of the ring A. An ideal J is maximal in R if $K \supset J$ is another ideal in R then $K = R$. If M is an irreducible module for the ring R, then $M \cong R/J$ for some maximal ideal $J \subset R$. [Take $m \in M, m \neq 0$ and consider $Rm \subset M$. This is a submodule, so $Rm = M$. The kernel of the map $r \mapsto rm$ is an ideal, $J \subset R$. So $M \cong R/J$. If $J \subset K \subset R$ and K is an ideal then K/J defines a submodule of M, so that $K/J = M$ and $K = R$, $i.e.$ J is maximal.] Thus an irreducible representation of a finite-dimensional algebra A is necessarily finite-dimensional. The coadjoint representation A' of an algebra A is defined on the dual vector space to A consisting of linear maps $\rho : A \to F$ with $(r\rho)(s) = \rho(sr)$. If M is an n-dimensional irreducible representation of A and $d_{ij}(r)$ the corresponding representation matrices, the n elements $d_{ij}(r), 1 \leq j \leq n, i$ fixed, define an n-dimensional invariant subspace A' corresponding to a representation equivalent to M. So the sum of the dimensions of the inequivalent representations of A does not exceed $\dim A$. This shows that each irreducible representation of a finite-dimensional algebra is finite-dimensional and there are only finitely many equivalence classes of such representations.

This is not such a strong statement as it seems because A may have indecomposable representations. In fact A may have an infinite number of inequivalent representations of a given dimension even if $\dim A < \infty$. [$E.g$ consider the three dimensional complex algebra, consisting of $\lambda^{\cdot} + \mu x + \nu y$. $\lambda, \mu, \nu \in \mathbb{C}$, subject to $x^2 = y^2 = xy = yx = 0$, which has the faithful three dimensional representation

$$\begin{pmatrix} \lambda & 0 & \mu \\ 0 & \lambda & \nu \\ 0 & 0 & \lambda \end{pmatrix} \tag{G.1}$$

and the inequivalent two-dimensional representations

$$\begin{pmatrix} \lambda & \mu + \xi\nu \\ 0 & \lambda \end{pmatrix}, \tag{G.2}$$

for each $\xi \in \mathbb{C}$.] The situation is more under control if the algebra is semi-simple. The ring R is semi-simple if the adjoint representation is completely reducible. If R is semi-simple, 1 is the sum of a finite number of elements of R, one in each of a number of the summands in the expression of R as a sum of irreducible modules, $1 = \sum_{i=1}^{n} e_i$ and, since any $r = \sum_{i=1}^{n} re_i$, it follows that there is a finite number, n, of summands $R_i = Re_i$ and $R = \bigoplus_{i=1}^{n} R_i$. If R is semi-simple, every R module is completely reducible (though not necessarily into a finite number of irreducible summands). [Any module M is the quotient of the free module $\mathcal{R} = \bigoplus_{m \in M} R_m$, where R_m is a copy of the adjoint module R, by the ideal consisting of those $(r_m)_{m \in M}$ such that $\sum_{m \in M} r_m m = 0$. The result follows since $\mathcal{R}$

is completely reducible if R is and the quotient of a completely reducible module is itself completely reducible.]

An R module M is finitely-generated if $M = \{\sum_{i=1}^{n} r_i m_i : r_i \in R\}$ for a finite number, n, of fixed elements $m_i \in M$, $1 \le i \le n$. If R is semi-simple, any finitely generated R module is completely reducible into a finite number of summands. [This follows because M is the quotient of $\bigoplus_{i=1}^{n} R_i$, where $R_i \cong R$, by the ideal $\{(r_i) : \sum_{i=1}^{n} r_i m_i = 0\}$.]

If M and N are R modules, an R-homomorphism $f : M \to N$ is a map satisfying $rf = fr$. If M and N are simple modules, Schur's Lemma implies that the set of R-homomorphisms $\text{Hom}_R(M, N) = 0$ if M and N are not equivalent. If $M = N$, $\text{Hom}_R(M, M) \equiv \text{End}_R(M)$ is a division ring, that is every a ring in which every non-zero element has an inverse. In the case of an algebra, if $\dim M < \infty$, $\text{End}_A(M) = F$, the underlying field.

If the R-module M is completely decomposable into a finite number of irreducible submodules, we can write $M = \sum_{i=1}^{N} M_i^{n_i}$, where each M_i is irreducible and M_i and M_j are inequivalent if $i \ne j$. Since $\text{Hom}_R(M_i, M_j) = 0$ if $i \ne j$,

$$\text{End}_R(M) = \prod_{i=1}^{N} \text{End}_R(M_i^{n_i}) = \prod_{i=1}^{n} \mathcal{M}_{n_i}(D_i)\,, \tag{G.3}$$

where the division algebra $D_i = \text{End}_R(M_i)$ and $\mathcal{M}_n(D)$ is the ring of $n \times n$ matrices with entries in the division algebra D.

If R is a semi-simple ring, we can write $R = \bigoplus_{i=1}^{N} R_i^{n_i}$, where the R_i are irreducible as R modules and inequivalent for $i \ne j$. So $\text{End}_R R = \prod_{i=1}^{N} \mathcal{M}_{n_i}(D_i)$, where $D_i = \text{End}_R(R_i)$. But $\text{End}_R R = R^o$, the reverse ring to R defined on the set R by taking the product of r and s to be sr rather than rs. Since, evidently, $(R^o)^o = R$,

$$R = \prod_{i=1}^{N} \mathcal{M}_{n_i}(D_i^o)\,, \tag{G.4}$$

i.e. every semi-simple ring is isomorphic to the direct product of a finite number of finite-dimensional matrix rings over division algebras [Wedderburn's Structure Theorem].

In the case of a semi-simple algebra, each $D_i = F$, the underlying field, so

$$A = \prod_{i=1}^{N} \mathcal{M}_{n_i}(F)\,, \tag{G.5}$$

where $\mathcal{M}_n(F)$ is the algebra of $n \times n$ matrices with entries in the field F. In particular, any semi-simple algebra is finite-dimensional.

References

[1] A.A. Belavin, A.M. Polyakov and A.B. Zamolodchikov, *Infinite conformal symmetry in two-dimensional quantum field theory*, Nucl. Phys. **B241**, 333 (1984).

[2] G. Moore and N. Seiberg, *Classical and Quantum Conformal Field Theory*, Comm. Math. Phys. **123**, 177 (1989).

[3] E. Verlinde, *Fusion rules and modular transformations in 2D conformal field theory*, Nucl. Phys. **B300 [FS22]**, 360 (1988).

[4] G. Segal, *Notes on Conformal Field Theory*, unpublished manuscript.

[5] R.E. Borcherds, *Vertex algebras, Kac-Moody algebras and the monster*, Proc. Nat. Acad. Sci. U.S.A. **83**, 3068 (1986).

[6] R.E. Borcherds, *Monstrous moonshine and monstrous Lie algebras*, Invent. Math. **109**, 405 (1992).

[7] I. Frenkel, J. Lepowsky and A. Meurman, *Vertex Operator Algebras and the Monster* (Academic Press, 1988).

[8] I. Frenkel, Y.-Z. Huang and J. Lepowsky, *On axiomatic approaches to vertex operator algebras and modules*, Mem. Amer. Math. Soc. **104**, 1 (1993).

[9] Y. Zhu, *Vertex Operator Algebras, Elliptic Functions and Modular Forms*, Caltech preprint (1990), J. Amer. Math. Soc. **9**, 237 (1996).

[10] V. Kac, *Vertex algebras for beginners* (Amer. Math. Soc., 1997).

[11] A.J. Wassermann, *Operator algebras and conformal field theory* in *Proceedings of the I.C.M. Zürich 1994)* (Birkhäuser, 1995) 966.

[12] F. Gabbiani and J. Fröhlich, *Operator algebras and conformal field theory*, Commun. Math. Phys. **155**, 569 (1993).

[13] G. Veneziano, *Construction of a crossing-symmetric, Regge-behaved amplitude for linearly rising trajectories*, Nuovo Cim. **57A**, 190 (1968).

[14] Z. Koba and H.B. Nielsen, *Manifestly crossing-invariant parametrization of n-meson amplitude*, Nucl. Phys. **B12**, 517 (1969).

[15] P. Goddard, *Meromorphic conformal field theory* in *Infinite dimensional Lie algebras and Lie groups: Proceedings of the CIRM Luminy Conference, 1988* (World Scientific, Singapore, 1989) 556.

[16] M.R. Gaberdiel and P. Goddard, *Axiomatic Conformal Field Theory*, DAMTP-1998-135, hep-th/9810019.

[17] T. Gannon, *Monstrous moonshine and the classification of CFT*, lecture notes in this volume.

[18] K. Osterwalder and R. Schrader, *Axioms for euclidean Green's functions*, Commun. Math. Phys. **31**, 83 (1973).

[19] G. Felder, J. Fröhlich and G. Keller, *On the structure of unitary conformal field theory. I. Existence of conformal blocks*, Commun. Math. Phys. **124**, 417 (1989).

[20] Y.-Z. Huang, *A functional-analytic theory of vertex (operator) algebras*, math.QA/9808022.

[21] P. Montague, *On Representations of Conformal Field Theories and the Construction of Orbifolds*, Lett. Math. Phys. **38**, 1 (1996), hep-th/9507083.

[22] M.R. Gaberdiel and H.G. Kausch, *A Local Logarithmic Conformal Field Theory*, Nucl. Phys. **B538**, 631 (1999), hep-th/9807091.

[23] A. Cappelli, C. Itzykson and J.-B. Zuber, *Modular invariant partition functions in two dimensions*, Nucl. Phys. **B280 [FS18]**, 445 (1987).

[24] A. Cappelli, C. Itzykson and J.-B. Zuber, *The A-D-E classification of minimal and $A_1^{(1)}$ conformal invariant theories*, Commun. Math. Phys. **113**, 1 (1987).

[25] T. Gannon, *The Classification of SU(3) modular invariants revisited*, Ann. Poincare **65**, 15 (1996).

[26] T. Gannon, *$U(1)^m$ modular invariants, $N = 2$ minimal models, and the quantum Hall effect*, Nucl. Phys. **B491**, 659 (1997), hep-th/9608063.

[27] T. Gannon, *The Capelli-Itzyskon-Zuber A-D-E classification*, math.QA/9902064.

[28] I.B. Frenkel and Y. Zhu, *Vertex operator algebras associated to representations of affine and Virasoro algebras*, Duke Math. J. **66**, 123 (1992).

[29] M.A. Walton, lecture notes in this volume.

[30] C. Dong, H. Li, G. Mason, *Twisted Representations of Vertex Operator Algebras*, q-alg/9509005.

[31] M.R. Gaberdiel and P. Goddard, in preparation.

[32] E. Hille, *Analytic Function Theory I* (Blaisdell Publishing Company, 1959).

[33] E. Hille, *Analytic Function Theory II* (Blaisdell Publishing Company, 1962).

[34] B. Farb, R.K. Dennis, *Noncommutative Algebra*, (Springer, 1993).

Derivation of the Hadronic Structure Functions From Quantum Chromodynamics

Lectures at the the Feza Gürsey Institute, Istanbul
and the Mittag-Leffler Institute, Stockholm in 1998.

S. G. Rajeev[1]

*Department of Physics and Astronomy, University of Rochester, Rochester,
New York 14627*

Abstract

We solve a long-standing problem in particle physics: that of deriving the Deep Inelastic structure functions of the proton from the fundamental theory of strong interactions, Quantum ChromoDynamics (QCD). In the Bjorken limit, the momenta of the constituents of the proton (the partons) can be assumed to be in a two-dimensional plane in Minkowski space: a dimensional reduction of QCD to two space-time dimensions. Two dimensional QCD is then shown to be equivalent for all energies and values of number of colors N to a new theory of hadrons, Quantum HadronDynamics (QHD). The phase space of QHD is the Grassmannian (set of subspcaes) of the complex Hilbert space $L^2(R)$. The natural symplectic form along with a hamiltonian define a classical dynamical system, which is equivalent to the large N limit of QCD. 't Hooft's planar limit is the linear approximation to our theory: we recover his integral equation for the meson spectrum but also all the interactions of the mesons. The Grassmannian is a union of connected components labelled by an integer (the renormalized dimension of the subspace) which has the physical meaning of baryon number. The proton is the topological soliton: the minimum of the energy in the sector with baryon number one gives the structure functions of the proton. We solve the resulting integral equations numerically; the agreement with experimental data is quite good for values of the Bjorken variable $x_B > 0.2$.

---

[1]rajeev@pas.rochester.edu

i

Contents

List of Figures

Chapter 1

Introduction

The first indication that the atom contains a point–like nucleus came from experiments of Rutherford. He scattered alpha particles (which are positively charged) from a gold foil. Since the negative charges (electrons) were known to be much lighter than the alpha particles,they would not scatter the alpha particles very much. If the positive charges inside the atom were more or less uniformly distributed, the alpha particles would not be scattered through wide angles. Rutherford found to the contrary that the alpha particles were scattered by wide angles. In fact the probability of scattering through an angle θ is proportional to $\sin^{-4}\frac{\theta}{2}$, exactly what would happen if all the positive charge of the atom were concentrated at a point: he had discovered the atomic nucleus. Soon after, it was realized that the nucleus is of finite size, although small compared to the atom. It is composed of protons and neutrons bound together by the strong interaction.

Many years later, another celebrated series of experiments studied the structure of the proton itself. It was found that it was not an elementary particle either, in fact that it was made of point-like constituents. In such a 'Deep Inelastic Scattering' experiment an electron (or neutrino) beam is scattered by a proton (or a nucleus containing both protons and neutrons). The idea is to learn about the still mysterious strong interactions using the electroweak interactions as a probe.

A particle that can take part in the strong interactions is called a 'hadron'. There are two kinds of hadrons: those of half-integer angular momentum are called 'baryons' and those of integer angular momentum are the 'mesons'. The proton is the lightest baryon and the π -meson (pion) the lightest meson. There are an infinite number of baryons and mesons but the more

1

energetic ones are unstable against decay into the lighter hadrons. (We consider an idealized world in which the electromagnetic, weak and gravitational interactions of the hadrons with each other are ignored. Thus a particle is considered stable if its decays are purely electromagnetic or weak. This is the sense in which the pion is stable.)

The hadrons are composed of more elementary constituents: the quarks, anti-quarks and gluons. These constituents are collectively called partons[1, 2]. However, it has been found experimentaly that it is impossible to create the partons in isolation: they only exist inside hadrons, and thus, are not true particles. This is the phenomenon of 'confinement'. More precisely, only combinations of quarks and gluons invariant under the action of the color group $SU(N)$ (see below) exist as isolation. Thus hadrons can be defined to be states that are invariant under color. Some authors use the word hadron to include the quarks and gluons. We will use the term 'hadron' to refer to a bound state of quarks and gluons which can exist as an isolated particle. Of course such a particle may be unstable against decay into other hadrons.

Around the same time of these developments an entirely different picture of a baryon was proposed by Skyrme: that it is a topological soliton made of an infinite number of mesons. Unfortunately, this idea did not fit with the dominant view of the time and was largely ignored. In the mid-eighties Skyrme's idea was finally revived by a group at Syracuse University (including the author) [3, 4, 5] and integrated into the modern theory of strong interactions, i.e., Quantum Chromodynamics. Witten [6] clarified why the baryons are fermions (when N is odd) even though the underlying theory has only bosonic fields. (Witten had already arrived at the idea that baryons are solitons in the large N limit of 't Hooft [7] by independent arguments [8].)

But then the problem remains how to reconcile it with the picture of a baryon as a bound state of point-like constituents. We will show in these lectures how the parton model can be derived from the soliton model. It fact it had never been possible to derive the distribution functions of the partons inside the hadron from fundamental principles. Our picture will solve this problem, providing for the first time a quantitative theory of the structure of a proton.

The basic idea of our approach is to find a new description of strong interactions directly in terms of hadrons rather than in terms of quarks and gluons. It should be equivalent to the color singlet sector of QCD at all ener-

2

gies and all values of the number of colors. We will call this theory quantum hadrondynamics. In four dimensions this is still just an idea; we havent been able to show yet that such a new paradigm for strong interactions exists. However in the two dimensional case, I constructed such an equivalent alternative formalism some years ago. Part of the motivation for studying the two dimensional case was that it would provide a solution to the problem of deriving the structure functions of four dimensional QCD. (It has been known since the early work of Feynman and Bjorken that Deep Inelastic Scattering can be explained by a two dimensional theory of strong interactions.) The first report of the idea of such an equivalent theory was in Ref. [9]. The solutions of the integral equations for the baryon wavefunction was first studied in Ref. [10]. A more detailed description of the theory appeared in Ref. [11]. In another direction, these ideas were applied to spherically symmetric situations in Ref. [12]. The methods of geometrical quantization were applied to the problem in Ref. [13]. There a version with bosonic quarks was also studied. We returned to the study of the baryon wavefunctions in Ref. [14] where the derivation of the parton model, already mentioned in [11] was given in more detail. In papers which will soon appear [15] we will extend these results to include the anti-quark and sea quark distributions. Some of the ideas that go into this work are outlined in the last chapter. The basic mathematical ideas necessary to derive the gluon structure functions have also been developed [16]. We hope to solve that problem as well in the near future.

The ideas we introduce range from infinite dimensional geometry to the phenemenology of particle physics. Indeed a whole new set of tools had to be developed in order to implement the new paradigm we propose. Some of the methods are currently not associated with particle physics: they are more closely related to classical mechanics and are perhaps more familiar to mathematicians. But the new ideas of one generation become the standard lore of the next. I hope these lecture notes will prepare a new generation of theoretical physicists to pursue these ideas. Prior knowledge of quantum field theory is not essential, but will provide perspective. In general, a higher degree of mathematical maturity than specialized knowledge of particle physics is assumed.

In the next few sections of this introduction we give a quick summary of the physics background necessary. The reader who finds this boring should skip to the next chapter, after a glance at sections 1.4 and 1.5 which are essential to the argument.

1.1 Deep Inelastic Scattering

Let us consider in more detail the scattering of an electron by a proton. The electron emits a virtual photon of momentum $q = k_f - k_i$ where k_i and k_f are the initial and final electron momenta. This photon then interacts with the hadron producing some state $|X>$ which could contain many particles: several mesons and some excited baryons and anti-baryons. This is an *inelastic* scattering process. If we sum over all the states $|X>$ so produced, we will get the 'inclusive' cross-section for the scattering of the electron. Upto well–known electromagnetic effects, the cross-section for this scattering is given by the matrix element[17]

$$
\begin{aligned}
W_{\mu\nu}(P,q) &= \frac{1}{8\pi}\sum_\sigma\sum_X < P\sigma|J_\mu(0)|X >< X|J_\nu(0)|P\sigma > (2\pi)^4\delta^4(P + q - P_X \\
&= \frac{1}{8\pi}\sum_\sigma\int dx e^{-iq\cdot x} < P\sigma|J_\mu(x)J_\nu(0)|P\sigma > .
\end{aligned}
$$

Here, $J_\mu(x)$ is the electromagnetic current operator which is the one appropriate for electron scattering: the weak interactions of the electron are much smaller in comparison. For neutrino scattering it would instead be the charged weak current (correponding to $W^\pm$) if the neutrino is converted into an electron; and the neutral current (corresponding to Z^0) otherwise. Also, P is the momentum of the hadron and we assume that we have averaged over the spin σ of the hadron. (The target is usually not polarized, so that we must average over the values of the spin variables. More detailed measurements with polarized targets have been made more recently; we will not consider these for now.)

The tensor $W_{\mu\nu}$ is symmetric and transverse: conservation of the electric current gives

$$q^\mu W_{\mu\nu}(P,q) = 0 \tag{1.1}$$

Lorentz invariance and parity (which is a symmetry of the electromagnetic interactions) imply that the tensor $W_{\mu\nu}(p,q)$ has the form

$$
\begin{aligned}
W_{\mu\nu}(P,q) &= \left[\eta_{\mu\nu} - \frac{q_\mu q_\nu}{q^2}\right]W_1(x_B, Q^2) + \\
&\quad \frac{1}{m^2}\left(P_\mu - q_\mu\frac{P\cdot q}{q^2}\right)\left(P_\nu - q_\nu\frac{P\cdot q}{q^2}\right)W_2(x_B, Q^2).
\end{aligned}
\tag{1.2}
$$

4

Being Lorentz scalars, the 'structure functions' $W_{1,2}$ can depend only on the Lorentz scalars q^2 and $P \cdot q$ (P^2 is fixed to have the value m^2, where m is the mass of the proton). When an electron is scattered against a target, the momentum q of the photon it emits is space-like.It is conventional to use the positive number $Q^2 = -q^2$ to describe the energy of the photon. The higher the value of Q^2 the better the resolution of our measurement of the structure of the hadron. As the other independent variable we can take the dimensionless ratio ('Bjorken variable')

$$x_B = \frac{Q^2}{2P \cdot q}. \qquad (1.3)$$

If the hadron we are studying is stable against strong decays (which is always the case for experimental reasons) it will be the lightest particle with its quantum numbers. Then, the mass of the intermediate state will be greater than or equal to the mass of the target:

$$(P + q)^2 \geq m^2; \quad \Rightarrow [\frac{1}{x_B} - 1] \geq 0. \qquad (1.4)$$

This shows that the Bjorken variable takes values in the range

$$1 \geq x_B \geq 0. \qquad (1.5)$$

It is often more convenient to use another equivalent pair of structure functions

$$F_1(x_B, Q^2) = W_1(x, Q^2), \quad F_2(x_B, Q^2) = \frac{P \cdot q}{m^2} W_2(x_B, Q^2). \qquad (1.6)$$

In its rest frame the hadron has a certain 'size' a : it is of the order of the the charge radius of the proton, $a^{-1} \sim 100\text{MeV}$. If we take the limit $Q^2 >> a^{-2}$ keeping x_B fixed we are looking deep inside the hadron: this is the region of Deep Inelastic Scattering. The basic idea is much like that of a microscope: to see inside an object of size a we need light of a wavelength that is small compared to a ; or equivalently an energy for the photon that is large compared to a^{-1} .

When the states are normalized by the usual convention

$$< P|P' >= 2E_P(2\pi)^3\delta^3(P - P'), \qquad (1.7)$$

the functions $F_{1,2}(x_B, Q^2)$ are dimensionless. It is found experimentally that these functions are approximately independent of Q^2 : they depend essentially only on the dimensionless ratio x_B . The simplest explanation of this phenomenon is that the proton is made of pointlike massless constituents: the structure function would then depend only on the dimensionless ratio x_B

If we multiply $P \cdot q$ and Q^2 by the same number, the structure functions are approximately invariant: this is the approximate symmetry of 'scale invariance'. In the first versions of this model, the partons were concieved of as free particles. Of course they must interact to bind into a hadron, but the idea was that at high enough Q^2 the interaction would for some reason be small. This was the germ of the idea of 'asymptotic freedom': that the constituents of the hadrons behaved like free particles at short distances, or high energies.

This 'parton' model was subsequently derived [18, 19, 20], [21] from a much deeper fundamental theory of strong interactions, Quantum Chromodynamics (QCD). This is a non–abelian gauge theory which has the unusual property that at high Q^2 the coupling constant (which measures the strength of the interaction) vanishes like $\dfrac{1}{\log \frac{Q^2}{\Lambda^2}}$. Here Λ is a parameter with the dimensions of momentum which is a fundamental parameter of QCD. Hence at high energies (compared to Λ) QCD tends to a free theory, yielding asymptotic freedom and the parton model. In the limit of large Q^2 , the structure functions are predicted to be independent of Q^2 . At finite values of Q^2 , the structure functions have a slow dependence on Q^2 ; as a polynomial in the coupling constant, or equivalently, $\dfrac{1}{\log \frac{Q^2}{\Lambda^2}}$. Moreover, we can calculate this dependence on $\dfrac{1}{\log \frac{Q^2}{\Lambda^2}}$ using perturbation theory: scale invariance is broken in a way that is calculable. Perturbation theory allows us to calculate the Q^2 dependence of the structure functions: given the value of $F_{1,2}(x_B, Q^2)$ at one (large) value of Q^2 , we can calculate it for any other large value of Q^2 . (We mean that Q should be large compared to Λ , which is of the order of $100 \ MeV$. If $\frac{Q}{\Lambda}$ is not large, perturbatiion theory can no longer be used.)

It is one of the triumphs of modern particle physics that perturbative QCD accurately describes these scale violations.

However, the x_B dependence of the structure functions are not as well-understood. Perturbation theory is not sufficient to calculate it: $F_2(x_B, Q^2)$

is in a certain sense the probability distribution (in momentum space) of a parton inside a hadron. Thus it describes how partons bind together to form a hadron. Perturbation theory around a free field theory can never describe such bound states. Due to the lack of any fundamental understanding of the x_B dependence of the structure functions, physicists have been forced to extract them directly from experiment. A whole generation of experimentalists and phenemenologists have worked to produce a quite reliable extraction of the x_B dependence of structure functions from data. It will be our goal to explain this from the fundamental theory of strong interactions, QCD.

In the next section we will summarize the definition of QCD as a quantum field theory.

1.2   Quantum Chromodynamics

Quantum Chromodynamics is the fundamental theory of strong interactions. It is a non-abelian gauge theory with gauge group $SU(N)$ with matter fields (quarks) which are spin half fermions transforming under (N_f copies of) the fundamental representation of $SU(N)$. The natural number N is called the 'number of colors'. It has the value 3 in nature; but it will be convenient to leave it unspecified until the very end when we make comparisons with data. The 'number of flavors' N_f is 2 for most purposes although it can be as high as 6 in principle: the heavier flavors of quarks can be ignored for most purposes. Again, it is best to leave N_f as an arbitrary parameter for now.

The action principle that defines QCD is

$$S = \frac{N}{4\alpha} \int \text{tr} F_{\mu\nu} F^{\mu\nu} d^4x + \sum_{a=1}^{N_f} \int \bar{q}^a [-i\gamma \cdot \nabla + m_a] q_a d^4x \tag{1.8}$$

The Yang–Mills field strength is

$$F_{\mu\nu} = \partial_\mu A_\nu - \partial_\nu A_\mu + [A_\mu, A_\nu] \tag{1.9}$$

where A_μ are a set of four $N \times N$ anti-hermitean matrices: $A_\mu dx^\mu$ is a one-form taking values in the Lie algebra of the unitary group $U(N)$

In perturbation theory, this 'gauge field ' describes 'gluons': massless spin one particles analogous to the photon. q_a is a fermionic field (describing quarks) which is a Dirac spinor transforming in the fundamental representation of the 'color' group $U(N)$, for each a . The index takes values

7

$a = 1 \cdots N_f$, where N_f is the number of copies of such quarks: the number of flavors.

The parameters of the theory are (in addition to N and N_f), the quark masses m_a for $a = 1, \cdots N_f$ and the dimensionless 'coupling constant' α . In fact the quantum theory is defined by an additional set of rules for 'renormalization' which replace the constant α by a new parameter Λ which has the dimensions of mass as well. These issues are well-known and have been reviewed in several articles [21], so we will avoid them here. T he basic point is that the the coupling constant acquires a dependence on the energy scale $\alpha \sim \frac{1}{\log \frac{Q^2}{\Lambda^2}}$.

These $N_f + 1$ parameters should in principle determine the masses and decay rates as well as the structure functions of all the hadrons. It has turned out to be quite difficult to predict hadronic properties from this fundamental theory. We need to develop approximation methods which make the problem tractable. For the most part this problem is still not solved. In these talks we will describe how we solved a part of this puzzle: that of calculating the hadronic structure functions from QCD.

1.3   Structure Functions from QCD

We noted earlier that since the electromagnetic interactions are well-understood, it is sufficient to concentrate on the 'unknown' part of the problem, which is encoded in the structure functions $W_{1,2}(x_B, Q^2)$. These were defined in terms of the expectation values of the product of current operators. Now, part of QCD is tractable by perturbation theory; this allows us to reduce the 'unknown' part of the problem further to the expectation values of simpler operators: the structure functions can be expressed in terms of 'parton distribution functions'. The problem we will solve is that of calculating these distribution functions.

The basic result that makes this reduction possible is called the 'factorization theorem': the structure function can be written as a convolution of two factors: one whose x_B and Q^2 dependence can be calculated in perturbation theory (the 'hard' factor); and another (the 'soft' factor) which has an unknown x_B dependence but which is independent of Q^2 . This is a result of quantum field theory that can be proved to all orders of perturbation theory; so it is comparable in depth to the proof of renormalizability of gauge theories. We will just give an intuitive argument, pointing the reader

8

to the literature on perturbative QCD for details [17, 22].

The photon is scattered by a quark of charge e_a inside the hadron. To leading order, the probability of the photon being scattered by the proton can be written as the product of two pieces: the probablity that a quark of some momentum k will scatter the photon and the probability that there is such a quark inside the proton. (Then we of course sum over the intermediate momentum k.) This is the simplest version of the factorization theorem. In the next order of perturbation theory, we will have to include the possibility that quark might emit a gluon before absorbing the photon and then re-emit it. We can imagine the scattering of the photon by the quark as a subprocess with its own structure function, except that it can be calculated within perturbation theory. The probability that a hadron contains a quark of a given momentum is of course not calculable in perturbation theory: that is the distribution function which we will attempt to understand.

In other words, the structure function of the proton can be written as the convolution of two pieces: the structure function of a parton (which is computable perturbatively) and the distribution function of the parton inside the hadron (which is non-perturbative).

In more detail,

$$F_2(x_B, Q^2) = \sum_{i=a,\bar{a},G} \int_0^1 \frac{d\xi}{\xi} C_2^i(\frac{x}{\xi}, Q^2) \phi_i(\xi). \qquad (1.10)$$

Here $\phi_a(x_B)$ is the probability of finding a quark of flavor and momentum fraction x_B inside a proton; $\phi_{\bar{a}}$ is the probability for an anti-quark of flavor a, and ϕ_G for gluons. Note that all the Q^2 dependence is in the first factor; it can be calculated as a power series in $\frac{1}{\log \frac{Q^2}{\Lambda^2}}$ using the standard rules of perturbation theory. To leading order in perturbation theory, even the first factor is independent of Q^2:

$$C_2^a(x_B, Q^2) = C_2^{\bar{a}}(x_B, Q^2) = e_a^2 \delta(x_B - 1), \quad C_2^G(x_B, Q^2) = 0. \qquad (1.11)$$

If $b_i(k), b_i^\dagger(k)$ are the creation-annihilation operators for the parton, we have

$$\phi_i(x_B) = \sum_\sigma \int \frac{d^2 \mathbf{k}_T}{(2\pi)^2} < P\sigma|b_i^\dagger(x_B P, \mathbf{k}_T) b_i(x_B P, \mathbf{k}_T)|P\sigma >. \qquad (1.12)$$

Here, we are averaging over all possible values of the part of the momentum orthogonal to the plane spanned by P and q, (called the 'transverse momentum $\mathbf{k}_T$).

9

In terms of the field operators, we get the quark distribution functions

$$\phi_a(x_B) = \frac{1}{2} \sum_\sigma \int dy \, e^{-ix_B Py} < P\sigma |\bar{q}^a(y, \mathbf{0}_T) \gamma_- q_a(0)|P\sigma > \qquad (1.13)$$

and the anti-quark distribution functions

$$\phi_{\bar{a}}(x_B) = -\frac{1}{2} \sum_\sigma \int dy \, e^{-ix_B Py} < P\sigma |q_\alpha^a(y, \mathbf{0}_T) [\gamma_-]_{\beta\alpha} \bar{q}_\beta^a(0)|P\sigma > . \qquad (1.14)$$

The position arguments of the field operators are separated by a null line in the plane spanned by P and q. The averaging over transverse momenta implies that in position space, these field operators have the same transverse co-ordinates.

Since the gluons do not carry electric charge, to the leading order of perturbative QCD, the gluon distribution functions are not necessary in order to understand the structure of the proton. We will ignore them for now.

These formulae are not gauge-invariant and are to be understood in the null gauge $A_- = 0$. In a general gauge we should insert a parallel transport operator $P e^{\int A_-(y, \mathbf{0}_T) dy}$ along the null line connecting two field operators.

1.4 Reduction of QCD to Two Dimensions

We are probing the hadron (whose momentum is a time-like vector P) by a photon (or W, Z -boson) with a space-like momentum q. These two vectors define a two-dimensional subspace of Minkowski space. There is also a space-like vector a which characterizes the 'size' of the hadron: the length of a is the charge radius of the hadron. In the Deep Inelastic limit, $Q^2 = -q^2 >> |a|^{-1}$ keeping $x_B = \frac{Q^2}{2P \cdot q}$ fixed. This means that the hadron is of very large size in the directions transverse to the two dimensional subspace spanned by P and q. By the uncertainty principle, the transverse momentum of its constituents is of order $|a|^{-1}$, which is thus small compared to Q. A reasonable first approximation would be to let the hadron have infinite extent in the transverse directions; and to require the momenta of the constituents to lie entirely in the plane spanned by P, q. In other words the approximation is that the fields are independent of the transverse spatial co-ordinates.

This is analogous to the procedure of dimensional reduction popular in unified field theories of gravity. The main difference is that the dimensions

10

that are ignored do not form a subspace of small volume: instead they are infinite in extent. The point is that the momenta in these directions are small which is the same as requiring that the fields are independent of those spatial directions: a dimensional reduction.

Thus Deep Inelastic Scattering is described by the dimensional reduction of QCD to two dimensions. That is what makes this phenomenon accessible: two dimensional gauge theories have been understood by the large N method. With some further (less drastic) approximations we will be able to determine the spectrum and structure of hadrons in two dimensions. These should then be verifiable experimentally.

The dimensional reduction of QCD to two dimensions is given by the action principle

$$
\begin{aligned}
S \;=\; & \frac{N}{4\alpha_1} \int \mathrm{tr} F_{\mu\nu} F^{\mu\nu} d^2 x + \sum_{a=1}^{N_f} \int \bar{q}^{a\alpha}[-i\gamma \cdot \nabla + m_a] q_{a\alpha} d^4 x \\
& + \frac{N}{2\alpha_1} \int \mathrm{tr}(\nabla_\mu \phi_A)^2 d^2 x + \frac{N}{2\alpha_1} \int \mathrm{tr}[\phi_3, \phi_4]^2 d^2 x + \int \bar{q}^{a\alpha}(-i) \Gamma_\alpha^{A\beta} \phi_A q_{a\beta} d^2 x.
\end{aligned}
$$

Here we allow the indices μ, ν to take only the values $0, 1$. The gauge field now splits into a 1-form in the two dimensional space with components A_μ and a pair of scalar fields $\phi_3 = A_3, \phi_4 = A_4$ corresponding to the transverse polarization states of the gluon.

The four dimensional Dirac spinor q splits into a pair of two–dimensional spinors q^α, corresponding the two eigenvalues of $\gamma_3 \gamma_4$. Moreover α_1 is the coupling constant of the two-dimensional theory which has dimensions of (mass)^2. It is a combination of α and the size of the hadron: $\alpha_1 \sim \alpha a^{-2}$. Thus the reduction of QCD to two dimensions gives a gauge theory with scalar fields in the adjoint representation, and twice as many flavors of quarks as the original theory.

We will study for the most part two dimensional QCD, defined by the action

$$
S \;=\; \frac{N}{4\alpha_1} \int \mathrm{tr} F_{\mu\nu} F^{\mu\nu} d^2 x + \sum_{a=1}^{N_f} \int \bar{q}^{a\alpha}[-i\gamma \cdot \nabla + m_a] q_{a\alpha} d^4 x.
$$

This is not quite the same as the dimensional reduction of *four* dimensional QCD to two dimensions: there are no scalar fields. This truncated theory will be sufficient to to determine the quark and anti-quark structure functions to

11

the accuracy we need. The essential techniques required to solve the theory including the scalar fields have been developed as well. We will return to them in a later publication.

1.5   Two Dimensional QCD in Null Gauge

The essential simplification of two dimensional QCD is that the gauge fields can be removed completely from the problem, leaving just the quark fields q and $\bar{q}$ and the scalar gluons ϕ_A . This should not be too surprising: in D dimensional space-time, a gauge field has $D-2$ polarization states.(Of course, we still have the scalar fields ϕ which are the remnants in two dimensions of the two polarization states of the gluon in four dimensional gauge theory. But we will ignore them, as they are mostly relevant to the determination of the gluon structure functions: a problem we will postpone to a later publication.)

We will change notation slightly from the last section: the indices a, b etc. will denote the pairs $a\alpha$, $b\beta$, so they will range over $1, 2, \cdots 2N_f$. This simply reflects the fact that the two dimensional theory has twice as many flavors as the four dimensional theory: the transverse polarization label looks just just like a flavor index to the two-dimensional theory.

In the null gauge, the action of the theory becomes (see Appendix A)

$$L = \chi^{\dagger a}(-i\partial_t)\chi_a - \chi^{\dagger a}\left\{\frac{1}{2}[\hat{p} + \frac{m^2}{\hat{p}}] - iA_t\right\}\chi_a + \frac{N}{2\alpha_1}\,\mathrm{tr}[\partial_x A_t]^2. \qquad (1.15)$$

The field A_t carries no dynamical degrees of freedom: it can be eliminated in terms of χ by solving its equations of motion:

$$A^i_{tj}(x) = \frac{i}{N}\alpha_1 \int \frac{1}{2}|x - y| : \chi^{\dagger ai}(y)\chi_{aj}(y) : dy. \qquad (1.16)$$

(The Coulomb potential in one-dimensional space is given by $\frac{\partial^2}{\partial x^2}\frac{1}{2}|x - y| = \delta(x - y)$.) The hamiltonian of the resulting theory is

$$\begin{aligned} H &= \int dx \chi^{\dagger ai}\frac{1}{2}[\hat{p} + \frac{m^2}{\hat{p}}]\chi_{ai} \\ &\quad - \frac{1}{2N}\alpha_1 \int \frac{1}{2}|x - y| : \chi^{\dagger ai}(x)\chi_{aj}(x) :: \chi^{\dagger bj}(y)\chi_{bi}(y) : dxdy \end{aligned}$$

Now define the operator

$$\hat{M}_b^a(x, y) = -\frac{2}{N} : \chi^{\dagger ai}(x)\chi_{bi}(y) : \tag{1.17}$$

which is gauge invariant. It describes the creation of a quark at x and an anti-quark at y, but in a color invariant combination: in other words it describes a meson. The hamiltonian above can be expressed entirely in terms of this operator after some reordering of factors of χ.

We can rearrange the quartic operator in the χ's as a quadratic operator in M. First,

$$: \chi^{\dagger ai}(x)\chi_{aj}(x) :: \chi^{\dagger bj}(y)\chi_{bi}(y) : \; = \; : \chi^{\dagger ai}(x)\chi_{aj}(x)\chi^{\dagger bj}(y)\chi_{bi}(y) :$$
$$+ : \chi^{\dagger ai}(x)\chi_{bi}(y) :< 0|\chi_{aj}(x)\chi^{\dagger bj}(y)|0 >$$
$$+ < 0|\chi^{\dagger ai}(x)\chi_{bi}(y)|0 >: \chi_{aj}(x)\chi^{\dagger bj}(y) :$$

Now,(see Appendix A)

$$< 0|\chi^{\dagger ai}(x)\chi_{bi}(y)|0 >= N\delta_b^a \int_{-\infty}^{0} \frac{dp}{2\pi} e^{-ipx+ipy} = N\delta_b^a \frac{1}{2}[\delta(x - y) + \epsilon(x - y)]$$

where,

$$\epsilon(x - y) = \mathcal{P}\int \text{sgn} \, (p)e^{ip(x-y)}\frac{dp}{2\pi}.$$

Thus,

$$: \chi^{\dagger ai}(x)\chi_{aj}(x) :: \chi^{\dagger bj}(y)\chi_{bi}(y) : \; = \; : \chi^{\dagger ai}(x)\chi_{aj}(x)\chi^{\dagger bj}(y)\chi_{bi}(y) :$$
$$+ : \chi^{\dagger ai}(x)\chi_{ai}(y) :: \frac{N}{2}[\delta(x - y) + \epsilon(x - y)]$$
$$+ \frac{N}{2}[\delta(x - y) + \epsilon(x - y)] : \chi_{ai}(x)\chi^{\dagger ai}(y) :$$

On the other hand,

$$\hat{M}_b^a(x, y)\hat{M}_a^b(y, x) \; = \; \left(\frac{2}{N}\right)^2 : \chi^{\dagger ai}(x)\chi_{bi}(y) :: \chi^{\dagger bj}(y)\chi_{aj}(x) :$$
$$= \; \left(\frac{2}{N}\right)^2 : \chi^{\dagger ai}(x)\chi_{bi}(y)\chi^{\dagger bj}(y)\chi_{aj}(x) :$$

The terms corresponding to other orderings of χ will involve $\epsilon(y, y)$ which should be interpreted as zero. So,

$$: \chi^{\dagger ai}(x)\chi_{aj}(x) :: \chi^{\dagger bj}(y)\chi_{bi}(y) : \; = \; -\left(\frac{N}{2}\right)^2 \hat{M}_b^a(x, y)\hat{M}_a^b(y, x)$$

$$+\left(-\frac{N}{2}\right)\hat{M}_a^a(x,y)\frac{N}{2}[\delta(x-y)+\epsilon(x-y)]$$
$$+\frac{N}{2}[\delta(x-y)+\epsilon(x-y)]\left(-\frac{N}{2}\right)\hat{M}_a^a(y,x)$$

Thus,

$$H = \left(-\frac{N}{2}\right)\int\frac{1}{2}[p+\frac{\tilde{\mu}^2}{p}]\hat{M}_a^{\,a}(p,p)\frac{dp}{2\pi}+$$
$$-\frac{1}{2N}\alpha_1\int\frac{1}{2}|x-y|\left\{-\left(\frac{N}{2}\right)^2\hat{M}_b^a(x,y)\hat{M}_a^b(y,x)\right.$$
$$\left.+\left(-\frac{N^2}{2}\right)\hat{M}_a^a(x,y)\epsilon(x-y)\right\}dxdy$$

We have dropped terms involving $\delta(x-y)|x-y|$ since this product is just zero; moreover the two terms involving $\epsilon(x-y)$ are equal and have been combined.

Now we use the identity

$$\int\frac{1}{2}|x-y|\epsilon(x-y)f(x-y)dxdy = -\frac{1}{\pi}\mathcal{P}\int\frac{1}{p}\tilde{f}(p,p)\frac{dp}{2\pi}. \qquad (1.18)$$

Thus

$$\frac{H}{N} = -\frac{1}{2}\int\frac{1}{2}[p+\frac{\tilde{\mu}^2}{p}]\hat{M}_a^{\,a}(p,p)\frac{dp}{2\pi}+$$
$$\frac{1}{8}\alpha_1\int\frac{1}{2}|x-y|\hat{M}_b^a(x,y)\hat{M}_a^b(y,x)dxdy$$

where

$$\tilde{\mu}^2 = m^2 - \frac{\alpha_1}{\pi}. \qquad (1.19)$$

The commutation relations can also be expressed entirely in terms of the variables $\hat{M}$:

$$\{\tilde{M}_b^{\,a}(p,q), \tilde{M}_d^{\,c}(r,s)\} = \frac{1}{N}\left(\delta_b^c 2\pi\delta(q-r)[\delta_d^a \text{ sgn }(p-s)+\tilde{M}_d^{\,a}(p,s)]\right.$$
$$\left.-\delta_d^a 2\pi\delta(s-p)[\delta_b^c \text{ sgn }(r-q)+\tilde{M}_d^{\,c}(r,q)]\right).$$

We notice that in the limit of large N , these commutators become small: it is a sort of classical limit.

1.6 Constraint on the Variable $\hat{M}$

We have all the essential ingredients of a reformulation of two dimensional QCD in terms of the color singlet variable $\hat{M}$, eliminating quarks and gluons from the picture. The commutation relations and the hamiltonian together imply the time evolution equations for $\hat{M}$. However there is one more ingredient which we can miss at first: the set of allowed values of $\hat{M}$. Being bilinear in the fermionic variables, $\hat{M}$ behaves much like bosonic variables: they satisfy commutation relations (rather than anti-commutation relations). Also, they have a classical limit; in our case this is the large N limit when there commutators become small. But their origin as fermion bilinears have a residual effect: they satisfy a quadratic constraint. This constraint is ultimately an expression of the Pauli exclusion principle; even in the large N limit the fact the underlying degrees of freedom are fermions cannot be ignored.

Regard the classical variable $\rho(x, y)$ obtained by taking the large N limit of $\frac{1}{N}\chi^{i\dagger}(x)\chi_i(y)$ as a matrix in the variables x and y. It is in fact the density matrix of quarks. Its eigenvalues have to be between zero and one: zero when the state is completely empty and one when it is filled. For the color singlet states (hadrons), each state is either completely filled with quarks or completely empty. Hence the eigenvalue is either 0 or one, and this density matrix is a projection operator: $\rho^2 = \rho$ or,

$$\int \rho(x, y)\rho(y, z)dz = \rho(x, z). \tag{1.20}$$

This is a quadratic constraint on the classical variable.

Another (oversimplified) way to understand this constraint is that $\rho(x, y)$ corresponds to a meson state where a quark is created at x and an anti-quark at y. If we now create two mesons, with the position of the anti-quark of the first coinciding with the quark of the second, they will annihilate eachother leaving us with one meson!.

We will usually use the normal ordered variable $M(x, y) = \frac{1}{N} : \chi^{i\dagger}(x)\chi_j(y) :$, which has the advanatage that $M = 0$ on the vacuum. In terms of it the constraint becomes, in matrix notation

$$[\epsilon, M]_+ + M^2 = 0. \tag{1.21}$$

M has to satisfy some technical conditions as well, but we will talk about them later.

It is possible to prove these constraints by a straightforward calculation on the operators, even when N is held finite.] (See the appendix of Ref. [12].) It is important to note that they only hold in the color singlet subspace of the fermionic operators.

This means that it is possible to understand two dimensional QCD without ever mentioning quarks, antiquarks or gluons. The operator $\hat{M}$ is gauge invariant and can be the field variable of a theory that *directly* describes hadrons. From this point of of view quarks and gluons are just mathematical artifacts. This gives us a whole new paradigm for the theory of strong interactions. Instead of a 'Quantum ChromoDynamics" in terms of unobservable quarks and gluons but a new "Quantum HadronDynamics" in terms of the directly observable particles, the hadrons.

In the next chapter we will develop just such a point of view. We will first present a classical theory which is equivalent to the large N limit of the above theory. Upon quantization we will recover two-dimensional QCD.

Chapter 2

Two-Dimensional Quantum Hadron Dynamics

In this chapter we will present first the classical hadron theory and then its quantization. We will see that the quantum hadron theory is equivalent to QCD in two dimensions.

2.1 Grassmannians

Let $\mathcal{H} = L^2(R) \otimes C^{N_{2f}}$ be the complex Hilbert space of complex-valued functions on the real line[1]. Define the subspace of functions whose Fourier transforms vanish for negative momenta:

$$\mathcal{H}_+ = \left\{ \psi | \psi(x) = \int_0^\infty \tilde{\psi}(p) e^{ipx} \frac{dp}{2\pi} \right\} \tag{2.1}$$

and its orthogonal complement

$$\mathcal{H}_- = \left\{ \psi | \psi(x) = \int_{-\infty}^0 \tilde{\psi}(p) e^{ipx} \frac{dp}{2\pi} \right\}. \tag{2.2}$$

The functions in $\mathcal{H}_+$ are boundary values of analytic functions on the upper half of the complex plane in the variable x. (Of course $\mathcal{H}_-$ is related to the lower half of the complex plane in the same way.) Define the operator

---

[1] The number N_{2f} is the number of flavors of the two dimensional theory; it is twice the number of flavors of the four-dimensional theory.

ϵ ('the sign of the momentum') to be -1 on $\mathcal{H}_-$ and $+1$ on $\mathcal{H}_+$. Clearly

$$\epsilon^\dagger = \epsilon, \quad \epsilon^2 = 1. \tag{2.3}$$

The Grassmannian of $\mathcal{H}$ is the set of all its subspaces. To each subspace W, there is an operator Φ which is -1 on W and $+1$ on $W^\perp$. Again,

$$\Phi^\dagger = \Phi, \quad \Phi^2 = 1. \tag{2.4}$$

Conversely any such operator corresponds to an orthogonal splitting of $\mathcal{H}$: it will have eigenvalues ± 1 and W can be identified as the subspace with negative eigenvalue. Thus the set of such Φ's is the same as the set of subspaces of $\mathcal{H}$. The set of subspaces of $\mathcal{H}$ is the Grassmannian, which we can thus also regard as the set of operators Φ satisfying the above condition.

But we dont want to allow all such subspaces: since $\mathcal{H}$ is infinite dimensional, that would give a Grassmannian that is too big. For example, its tangent space will not be a Hilbert space, and it wouldnt admit a Riemann metric. We should only allow subspaces W that are 'at a finite distance' from $\mathcal{H}_-$. It is convenient to introduce the variable

$$M = \Phi - \epsilon \tag{2.5}$$

which measures the deviation from the standard point ϵ, which corresponds to the subspace $\mathcal{H}_-$. We will require that M be Hilbert-Schmidt. That is, the sum of the absolute magnitudes squared of all its matrix elements is finite. (See Appendix B for a rapid summary of some functional analysis we will need.) This sum is a measure of the distance of W from $\mathcal{H}_-$.

Thus we define the restricted Grassmannian to be,

$$\mathrm{Gr}(\mathcal{H}, \epsilon) = \{M^\dagger = M, \quad [\epsilon, M]_+ + M^2 = 0, \ \mathrm{tr}M^2 < \infty\}. \tag{2.6}$$

The set of all such M's forms an infinite dimensional Hilbert manifold. This manifold is defined by a quadratic equation in the (real) Hilbert space of self-adjoint Hilbert-Schmidt operators.

If we split the operator into 2×2 blocks (as in $\epsilon = \begin{pmatrix} -1 & 0 \\ 0 & 1 \end{pmatrix}$),

$$M = \begin{pmatrix} a & b \\ b^\dagger & d \end{pmatrix} \tag{2.7}$$

we get (using the constraint on M),

$$a = \frac{1}{2}(bb^\dagger + a^2), \quad d = -\frac{1}{2}(b^\dagger b + d^2), \quad ab + bd = 0. \tag{2.8}$$

Since M as a whole is Hilbert-Schmidt, $b : \mathcal{H}_+ \to \mathcal{H}_-$ is Hilbert-Schmidt as well. The above constraints then imply that a and d are trace-class. This means roughly speaking that the sum of their eigenvalues is absolutely convergent (see the appendix B for precise definition).

Note that a is a positive operator and d a negative operator. In particular, $\operatorname{tr} a \geq 0$ and $\operatorname{tr} d \leq 0$.

The tangent space at the origin is given by the special case where a, b, c, d are all infinitesimally small. The constraints then show that a and d are second order infinitesimals while b are first order. In the first order the only constraint on b is that it has to be a Hilbert-Schmidt operator. Thus, the tangent space at the origin is a complex Hilbert space, consisting of operators of the form $M = \begin{pmatrix} 0 & b \\ b^\dagger & 0 \end{pmatrix}$. In other words we have the identification,

$$T_0 \operatorname{Gr}(\mathcal{H}, \epsilon) = \mathcal{I}^2(\mathcal{H}_+ \to \mathcal{H}_-). \tag{2.9}$$

2.2 The Dirac Theory of Fermions

We have introduced the Grassmannian from a purely geometric point of view above. But it has a natural physical interpretation in terms of the Dirac theory of fermions.

In Dirac's theory, the complex Hilbert space $\mathcal{H}$ represents the set of states of a single fermion. There is a self-adjoint operator h representing energy. In the null co-ordinates we are using, energy is

$$p_0 = \frac{1}{2}[p + \frac{m^2}{p}]. \tag{2.10}$$

Thus it has the same sign as the null component of momentum p. The main physical obstacle to this interpretation is that the energy is not positive: all states are thus unstable with respect to decay into the negative energy states of lower and lower energy.

Dirac's main idea was that in any physical state all except a finite number of the negative energy states are occupied. Since fermions obey the Pauli exclusion principle, only one particle can occupy a given state: decay into states of very low energy is forbidden because those states are occupied. In particular the vacuum is not the state containing no particles. Instead the vacuum is the state where all the negative energy states are occupied and the positive energy states are empty. The filled negative energy states in the vacuum is called the 'Dirac Sea'. All physical quantities such as energy or charge are to be measured in terms of their departure from the vacuum value: even though the vacuum contains an infinite number of particles, it is still assiged zero energy, charge etc.

An excitation from the vacuum could be a state where a finite number of positive energy states are occupied and a finite number of negative energy states are empty. Such a state will have positive energy compared to the vacuum. The empty negative energy states (also called 'holes') have positive energy. The holes themselves behave just like the particles except that they have the opposite value for some observables such as electric charge. They are the 'anti-particles'. The number of particles minus the anti-particles is a conserved quantity which can take any integer value. An arbitrary physical state is a linear combination of such states containing a finite number of particles and anti-particles.

A mathematical interpretation of this situation can be given in terms of a modified exterior product of $\mathcal{H}$. In familiar non-relativistic theories, the space of states of a multi-particle system of fermions would be the exterior power of order r, where is the number of fermions: $\sum_{r=0}^{\infty} \Lambda^r \mathcal{H}$. Instead in Dirac's theory we split the one particle Hilbert space into two subspaces $\mathcal{H}_- \oplus \mathcal{H}_+$ of negative and positive energy. The space of states of the multi-particle system is

$$\mathcal{F} = \sum_{r,s=0}^{\infty} \Lambda^s \mathcal{H}'_- \otimes \Lambda^r \mathcal{H}_+. \tag{2.11}$$

This space of multi-particle states is called the Fock space.

There is a subset of states in this Fock space cosnisting of wedge products of single particle states. (A general state is a linear combination of such wedge products.) Suppose, W is a subspace which doesnt differ 'too much' from $\mathcal{H}_-$. That is, the intersection of W with $\mathcal{H}_+$ and the intersection of $W^\perp$ with $\mathcal{H}_-$ are both finite dimensional. Then we can form a state in $\mathcal{F}$ in which all states of W are occupied and all those in $W^\perp$ are empty. Suppose e_i is a basis in $W \cap \mathcal{H}_+$ and and f_j a basis in $W^\perp \cap \mathcal{H}_-$.

Then the state in $\mathcal{F}$ corresponding to occupying W will contain particles in $W \cap \mathcal{H}_+$ and holes in $W^\perp \cap \mathcal{H}_-$:

$$f^{1'} \wedge f^{2'} \wedge \cdots \wedge e_1 \wedge e_2 \wedge e_3 \wedge \cdots. \tag{2.12}$$

(Here, $f^{j'}$ is the dual basis in $[W^\perp \cup \mathcal{H}_-]'$). Moreover this is independent (upto multiplication by a complex number) of the choice of basis. Thus to each such subspace $W \subset \mathcal{H}$ there is a state (upto scalar multiple) in the Fock space $\mathcal{F}$. In other words we have an embedding of the Grassmannian into the Projective space $\mathcal{PF}$) . This is the infinite dimensionl version of the Plücker embedding familiar from algebraic geometry. [23, 24].

The Grassmannian describes this subset of states in Dirac's theory. The condition that the operator $M = \Psi - \epsilon$ corresponding to a subspace W be Hilbert-Schmidt is precisely what is required for the above construction to go through. (It is not necessary to require that the projection operator $\pi_+ : W \to \mathcal{H}_+$ be finite dimensional as we did above: the precise condtion for the construction to work is that it be Hilbert-Schmidt.) What is special about this family of states is that they are coherent states: they minimize the uncertainty in the physical observables. Hence they are the states that have a sensible classical limit. The set of such coherent states is the classical phase space of the theory. That is why it is sensible to choose the Grassmannian as the classical phase space of our theory of hadrons.

Dirac's theory was originally meant to describe electrons. We should think of it now as applied to quarks. Since the quarks are also fermions, this is reasonable. But there is one impotant twist: each quark comes in N colors: thus each state in $\mathcal{H}$ can be occupied by N quarks, not just one. Moreover, only states that are invariant under the action of the group $SU(N)$ are allowed: these are the states that describe the hadrons. The classical limit referred to above is the large N limit.

The operator Φ has eigenvalue -1 on states that are completely filled and $+1$ on states that are empty. The variable M measures the deviation from the vacuum state, in which all the negative energy states are completely filled. A baryon is essentially a color invariant bound state of N quarks. Thus the Dirac Sea of quarks can also be thought as a sea of baryons: each state is filled either by N quarks or by a baryon. An infinitesimal disturbance from this can be thought of either as the creation of a meson (a quark–anti-quark pair) or as the promotion of a quark from the negative energy sea to a positive energy state. This is an infinitesimal change if the number of colors N is large. The operator M thus has to be determined by a map $b^\dagger : \mathcal{H}_- \to \mathcal{H}_+$. The block-diagonal elements of M are zero in this infinitesimal limit.

In addition to such small deviations from the vacuum, the theory also allows for topological solitons which are not connected to the vacuum by any continuos path. These are very important, as they describe the baryons. To see how they arise we need to understand the topological properties of the Grassmannian. This is the subject of the next section.

2.3   Renormalized Dimension of a Subspace

Recall that $\Phi = \epsilon + M$ has eigenvalues ± 1 . Hence in some formal sense the trace of Φ is an even integer: the difference between the number of positive eigenvalues and the number of negative eigenvalues. But of course this trace is not convergent. Even if we subtract the contribution of the vacuum, it is divergent: the $\operatorname{tr} M$ can diverge, only the trace of its square needs to be convergent in general, since we only required it to be Hilbert-Schmidt. However the trace of M is conditionally convergent: we can define it to be the trace with respect to the above splitting into submatrices:

$$\operatorname{tr}_\epsilon M = \operatorname{tr} a + \operatorname{tr} d. \tag{2.13}$$

This trace exists since a and d are trace-class matrices. It can be shown to be an even integer, which is invariant under the continuous deformations of the operator M .

The interpretation of Φ in terms of subspaces will help us understand the meaning of this integer. Imagine that we take a state from $\mathcal{H}_+$ and add it to $\mathcal{H}_-$ to get a subspace W : then W has dimension one more than $\mathcal{H}_-$. This should change the trace of Φ by -2 : one of the eigenvalues of Φ has changed from $+1$ to -1 . Since $\operatorname{tr} M$ is essentially the difference between the traces of Φ and ϵ , we have $\operatorname{tr}_\epsilon M = -2$. Thus

$$B = -\frac{1}{2} \operatorname{tr} M \tag{2.14}$$

is the 'renormalized dimension' of the subspace it describes: the difference between its dimension and the dimension of the standard subspace $\mathcal{H}_-$.

In fact this is the only topological invariant of M : any two operators with the same conditional trace can be connected to each other by a continuous path. Thus the restricted Grassmannian is a union of connected components labelled by an integer, which is called the 'virtual rank' or 'renormalized dimension' of that component. Since it is invariant under all continuous deformations, in particular it will be invariant under time evolution. This is true for any reasonable definition of time evolution, independent of the choice of hamiltonian. Thus the renormalized dimension is a 'topologically conserved quantity'.

Each connected component of the Grassmannian by itself is an infinite dimensional manifold. Although the different components are the same (diffeomorphic) as manifolds, the component of rank zero has a special role as it contains the 'vacuum' $M = 0$. An example with renormalized dimension one is a 'factorizable' operator,

$$M = -2\psi \otimes \psi^\dagger. \tag{2.15}$$

The constraints on M are satisfied if ψ is a positive energy state of length one:

$$\epsilon\psi = \psi, \quad ||\psi||^2 = 1. \tag{2.16}$$

In the interpretation in terms of the Dirac theory, we have filled a positive energy state ψ in addition to all the negative energy states. Hence this state contains an additional fermion occupying the state ψ.

More generally, let ψ_a be an orthonormal system of states satisfying

$$\epsilon\psi = \epsilon_a\psi, \quad <\psi_a, \psi_b> = \delta_b^a. \tag{2.17}$$

Then

$$M = -2\sum_a \epsilon_a\psi_a \otimes \psi_a^\dagger \tag{2.18}$$

is a solution of the constraints with renormalized dimension $\sum_a \epsilon_a$. In this configuration, we have filled a certain number if positive energy states (ψ_a with positive μ_a) while creating holes in some others, ψ_a , with negative μ_a . The renormalized dimension is just the difference between the number of occupied positive energy states and the number of holes in the negative energy states.

Thus we see that the renormalized dimension is just the fermion number if we interpret the Grassmannian in terms of the Dirac Sea. We will see that the separable configuration is related to the valence quark model of the baryon.

As noted earlier, $a \geq 0$ and $d \leq 0$; thus a contributes negatively to the baryon number and d positively. The only configurations that have $b = 0$ are the separable ones above. To see this note that, $b = 0$ implies that a and d are proportional to projection operators. In order to be trace class, these have to be finite rank projections: corresponding to a certain number of baryons and anti-baryons.

The valence parton approximation of the parton model corresponds to the separable ansatz in our soliton model. In that approximation, $\phi_{\tilde{a}}(p)$ is zero: there are no antiquarks in the proton within the valence approximation.

2.4   Some Submanifolds of the Grassmannian

In principle a baryon can be in any of the configurations of renormalized dimension one: the ground state (proton) will be the one of least energy. Once we have determined the energy function (see below), the proton structure functions are determined by minimizing this energy over all possible configurations. Later on, we will describe a method to do just that numerically: the steepest descent method. However this is a computationally intensive and slow method. A much faster method will be to minimize the energy over some submanifold of the Grassmannian, which is chosen so that the minimum in this submanifold is close to the true minimum, yet is easier to find.

Physical intuition plays an important role in the choice of the ansatz of configurations explored this way. Such restricted phase spaces play an important role in our conceptual understanding as well: we will see that the theory restricted to rank one configurations of the type

$$M = -2\psi \otimes \psi^{\dagger}, \quad \epsilon\psi = \psi, \quad ||\psi||^2 = 1 \qquad (2.19)$$

is just the valence parton model: we will be able to derive the parton model from QHD this way.

Note that a change of phase $\psi \to e^{i\theta}\psi$ does not affect M. Indeed the set of rank one configurations is a submanifold of the Grassmannian diffeomorphic to $\mathcal{P}(\mathcal{H}_+)$. If we restrict the dynamics of our theory to this submanifold, we get an approximate theory which we will show is equivalent to the valence parton model.

24

Allowing for a slightly larger set of configurations will give us the parton model with Sea quarks and anti-quarks. We will describe here the submanifold of the Grassmannian that descibes this approximation to our theory. We generalize the above rank one ansatz to a finite rank ansatz

$$M = \mu_{\alpha\beta}\psi_\alpha\psi_\beta^\dagger. \tag{2.20}$$

Here, the ψ_α are a finite number r of eigenstates of ϵ which are orthonormal:

$$\epsilon\psi_\alpha = \epsilon_\alpha\psi_\alpha, \quad \psi_\alpha^\dagger\psi_\beta = \delta_\beta^\alpha. \tag{2.21}$$

The constraints

$$M = M^\dagger, \quad (\epsilon + M)^2 = 1, \quad -\frac{1}{2}\,\mathrm{tr}_\epsilon M = 1, \quad \mathrm{tr}M^2 < \infty \tag{2.22}$$

become the constraints on the $r \times r$ matrix μ :

$$\mu = \mu^\dagger, \quad [\tilde{\epsilon} + \mu]^2 = 1, \quad -\frac{1}{2}\,\mathrm{tr}\mu = 1. \tag{2.23}$$

Here, $\tilde{\epsilon} = \begin{pmatrix} \epsilon_1 & 0 & \cdots \\ 0 & \epsilon_2 & \cdots \\ \cdot & \cdot & \\ 0 & \cdots & \epsilon_r \end{pmatrix}$ is a diagonal $r \times r$ matrix, the restriction of ϵ to the finite dimensional subspace spanned by the ψ_α . The condition $\mathrm{tr}M^2 < \infty$ is of course automatic since M is now an operator of finite rank.

Thus, given a set of vectors ψ_α satisfying the above conditions, we get a point in the Grassmannian of renormalized dimension one if μ itself belongs to a *finite dimensional* Grassmannian. If there are r_+ vectors ψ_α with positive momentum and r_- with negative momentum, we have

$$\mathrm{tr}[\tilde{\epsilon} + \mu] = r_+ - r_- - 2. \tag{2.24}$$

The simplest solution is, $r_+ = 1, r_- = 0$, so that μ is just a number: it then has to be -2 . This is the rank one solution desribed earlier, which leads to the valence parton model.

If $r = r_+ + r_-$ is two, there is no solution to the above requirements (more precisely all solutions reduce to the rank one solution). The next simplest possibility is of rank three, with $r_+ = 2, r_- = 1$. Then each solution to the constraints on μ determines a one dimensional subspace of C^3 ; i.e., a point in CP^2 . In other words, for each such μ , there is a unit vector $\zeta = \begin{pmatrix} \zeta_- \\ \zeta_0 \\ \zeta_+ \end{pmatrix} \in C^3$ such that

$$\tilde{\epsilon} + \mu = -1 + 2\zeta \otimes \zeta^\dagger, \quad ||\zeta||^2 = 1. \tag{2.25}$$

Now, there is a $U(1) \times U(2)$ "gauge symmetry" in the problem: we can rotate the two positive energy vectors ψ_0 and ψ_+ into each other and change the phase of the negative energy vector ψ_- without changing M , provided we make the corresponding changes in μ as well. This $U(1) \times U(2)$ action changes the phase of ζ_- and rotates the other two components among each other. This freedom can be used to choose

$$\zeta = \begin{pmatrix} \zeta_- \\ 0 \\ \sqrt{[1 - \zeta_-^2]} \end{pmatrix}, \quad \zeta_- > 0. \tag{2.26}$$

In summary the submanifold of rank three configurations of renormalized dimension one is given by

$$\begin{aligned} M &= -2\psi_0 \otimes \psi_0^\dagger + 2\zeta_- \{\zeta_-[\psi_- \otimes \psi_-^\dagger - \psi_+ \otimes \psi_+^\dagger] \\ &\quad + \sqrt{[1 - \zeta_-^2]}[\psi_- \otimes \psi_+^\dagger + \psi_+ \otimes \psi_-^\dagger]\} \end{aligned}$$

where ψ_-, ψ_0, ψ_+ are three vectors in $\mathcal{H}$ satisfying

$$\epsilon\psi_- = -\psi_-, \quad \epsilon\psi_0 = \psi_0, \quad \epsilon\psi_+ = \psi_+, \tag{2.27}$$

$$||\psi_-||^2 = ||\psi_0||^2 = ||\psi_+||^2 = 1, \quad \psi_0^\dagger\psi_+ = 0. \tag{2.28}$$

Also,

$$0 \leq \zeta_- \leq 1. \tag{2.29}$$

The special case $\zeta_- = 0$ reduces to the rank one ansatz.

The physical meaning is clear if we consider the negative momentum components of M . The anti-quark distribution function is

$$\phi_{\bar{a}}(p) = \tilde{M}_{aa}(-p, -p) = \zeta_-^2 |\psi_{-a}(p)|^2. \tag{2.30}$$

Thus ζ_-^2 is the probability of finding an anti-quark inside the baryon. Also, $\zeta_-^2|\psi_{-a}(-p)|^2$ is the distribution function of the anti-quark of flavor a . In the same way $|\tilde{\psi}_{0a}(p)|^2 + \zeta_-^2|\tilde{\psi}_{+a}(p)|^2$ is the distribution function for quarks. In a loose sense $|\tilde{\psi}_{0a}(p)|^2$ is the valence quark wavefunction and $|\tilde{\psi}_{+a}(p)|^2$ is the Sea quark wavefunction. But the splitting of the quark distribution into a valence and a Sea distribution is rather arbitrary and has no real physical meaning.

By using the ansatz above with a larger and larger rank we can get better and better approximations to the Grassmannians. But the expressions get quite complicated beyond the rank three ansatz. We will find that this rank three ansatz holds the key to understanding the anti-quark and sea quark disributions in the quark model. A variational ansatz based on these configurations will provide a derivation of the parton model with sea quarks and anti-quarks from quantum hadrondynamics.

2.5   Integral Kernels

We can express the operator M in terms of its integral kernel in position space[2]:

$$M\psi(x) = \int M(x,y)\psi(y)dy. \qquad (2.31)$$

Alternately we can think in terms of its action on the momentum space wavefunctions:

$$\tilde{\psi}(p) = \int \psi(x)e^{-ipx}dx, \quad \psi(x) = \int \tilde{\psi}(p)e^{ipx}\frac{dp}{2\pi} \qquad (2.32)$$

and

$$[\widetilde{M\psi}](p) = \int \tilde{M}(p,q)\tilde{\psi}(q)\frac{dq}{2\pi}. \qquad (2.33)$$

The two points of view are of course related by Fourier transformation:

$$M(x,y) = \int \tilde{M}(p,q)e^{ipx-iqy}\frac{dpdq}{(2\pi)^2}. \qquad (2.34)$$

---

[2]We will often suppress the indices a,b etc. labelling the basis in C^{N_2}

The operator ϵ is diagonal in momentum space:

$$\widetilde{\epsilon\psi}(p) = \text{sgn }(p)\tilde{\psi}(p). \tag{2.35}$$

It can also be described in position space as a distribution:

$$\epsilon\psi(x) = \int \epsilon(x-y)\psi(y)dy = \frac{i}{\pi}\mathcal{P}\int \frac{1}{x-y}\psi(y)dy \tag{2.36}$$

In fact this is a well-known object in complex function theory: the Hilbert transform (except for a factor of i). It relates the real and imaginary parts of analytic functions.

The constraints on M become then,

$$\tilde{M}(p,q) = M^*(q,p)$$
$$[\text{ sgn }(p) + \text{ sgn }(q)]\tilde{M}(p,q) + \int \tilde{M}(p,r)\tilde{M}(r,q)\frac{dr}{2\pi} = 0$$
$$\int |M(p,q)|^2\frac{dpdq}{2\pi} < \infty$$

or equivalently, position space,

$$M(x,y) = M^*(y,x)$$
$$\int [\epsilon(x,y)M(y,z) + M(x,y)\epsilon(y,z) + M(x,y)M(y,z)]dy = 0$$
$$\int |M(x,y)|^2 dxdy < \infty.$$

It is useful to see what the tangent space at the origin is like. Since M is infinitesimally small, the constraint can be linearized:

$$[\text{ sgn }(p) + \text{ sgn }(q)]\tilde{M}(p,q) = 0. \tag{2.37}$$

Thus $\tilde{M}(p,q)$ is only non-zero if p and q are of opposite signs. In fact, the case where $p > 0, q < 0$ determines the opposite one because of the hermiticity condition on M. Thus the tangent space is just the space of square integrable functions of one positive variable and one negative variable $\tilde{M}(p,q)$.

2.6 The Infinite Dimensional Unitary group

Given any self-adjoint operator Φ, $g\Phi g^\dagger$ is also self-adjoint. If the transformation g is unitary, $gg^\dagger = g^\dagger g = 1$, it preserves the condition

$\Phi^2 = 1$. Indeed the action of the unitary group on such operators is transitive: any operator satisfying $\Phi^\dagger = \Phi$ and $\Phi^2 = 1$ can be taken to any other by a unitary transformation. In terms of the variable M , the unitary transformation is

$$M \mapsto gMg^\dagger + g[\epsilon, g^\dagger].$$ (2.38)

In the infinite dimensional case, we have required that M be Hilbert-Schmidt (H-S). This means that there is a corresponding condition on the family of allowed unitary transformations: we define the restricted unitary group,

$$U(\mathcal{H}, \epsilon) = \{g | gg^\dagger = 1, [\epsilon, g] \in \mathcal{I}^2\}.$$ (2.39)

It is straightforward to verify that the H-S condition on the commutator is preserved under the multiplication and inverse operations, since $\mathcal{I}^2$ is an ideal in the algebra of bounded operators.

Indeed the restricted Grassmannian is a homogenous space of this restricted unitary group:

$$Gr(\mathcal{H}, \epsilon) = U(\mathcal{H}, \epsilon)/U(\mathcal{H}_+) \times U(\mathcal{H}_-).$$ (2.40)

The Lie algebra of the restricted Unitary group is

$$\underline{U}(\mathcal{H}, \epsilon) = \{u | u = -u^\dagger, [\epsilon, u] \in \mathcal{I}^2\}.$$ (2.41)

The infinitesimal action on the variable M is:

$$M \mapsto [u, \epsilon + M].$$ (2.42)

2.7   Poisson Structure

We will regard the Grassmannian as the phase space of our dynamical system. The matrix elements of the operator M are then a complete set of observables. We will seek a set of Poisson brackets among these variables. Fortunately there is a natural choice: there is a unique choice that is invariant under the action of the restricted unitary group. In fact the Grassmannian can be viewed as the co-adjoint orbit of the (central extension of the) unitary group. We can regard the Poisson structure as induced by the Kirillov symplectic form on this orbit. However, the physical arguments use the Poisson brackets rather than the symplectic form so in thsi paper we will not say much aboutthe sympletic form. (See Refs. [11, 13] for an elaboration of this more geometric point of view.)

If the Poisson brackets are invariant under the action of the group, the infinitesimal action,

$$M \rightarrow [u, \epsilon + M] \tag{2.43}$$

would be a canonical transformation for any u satisfying

$$u = -u^\dagger, [\epsilon, u] \in \mathcal{I}^2. \tag{2.44}$$

The natural choice of a function that generates this canonical transformation is of the form

$$f_u = k \, \mathrm{tr} \, uM \tag{2.45}$$

for some constant k. There is a technical problem: the trace may not exist. Now remember that under the splitting $\mathcal{H} = \mathcal{H}_- \oplus \mathcal{H}_+$,

$$u = \begin{pmatrix} \alpha & \beta \\ -\beta^\dagger & \delta \end{pmatrix} \in \begin{pmatrix} B & \mathcal{I}^2 \\ \mathcal{I}^2 & B \end{pmatrix}, \quad M = \begin{pmatrix} a & b \\ b^\dagger & d \end{pmatrix} \in \begin{pmatrix} \mathcal{I}^1 & \mathcal{I}^2 \\ \mathcal{I}^2 & \mathcal{I}^1 \end{pmatrix} \tag{2.46}$$

so that the conditional trace

$$\mathrm{tr}_\epsilon uM = \mathrm{tr}\alpha a + \mathrm{tr}\beta b - \mathrm{tr}\beta^\dagger b^\dagger + \mathrm{tr}\delta d \tag{2.47}$$

exists. Moreover its value is some imaginary number, so that the constant k should also be purely imaginary in order that

$$f_u = k \, \mathrm{tr}_\epsilon uM \tag{2.48}$$

be a real-valued function.

Thus we postulate the Poisson brackets

$$\{f_u, M\} = [u, \epsilon + M]. \tag{2.49}$$

These imply, of course that

$$\{f_u, f_v\} = k \, \mathrm{tr}_\epsilon v[u, \epsilon + M] = -f_{[u,v]} + k \, \mathrm{tr}_\epsilon v[u, \epsilon]. \tag{2.50}$$

The commutation relations of the restricted unitary group are satisfied only upto a constant term: it is the central extension of the unitary group that acts on the Grassmannian, not the group itself. This has been studied at great length in the book by Pressley and Segal [25] so we wont go too far in that direction.

We can write the Poisson brackets also in terms of the integral kernel of the operator M in position space:

$$k\{M_b^a(x,y), M_d^c(z,u)\} = \delta_b^c \delta(y-z)[\epsilon_d^a(x,u) + M_d^a(x,u)] \\ -\delta_d^a \delta(x-u)[\epsilon_b^c(z,y) + M_d^c(z,y)],$$

or momentum space:

$$k\{\tilde{M}_b^a(p,q), \tilde{M}_d^c(r,s)\} = \delta_b^c 2\pi\delta(q-r)[\delta_d^a \text{ sgn } (p-s) + \tilde{M}_d^a(p,s)] \\ -\delta_d^a 2\pi\delta(s-p)[\delta_b^c \text{ sgn } (r-q) + \tilde{M}_d^c(r,q)].$$

The principle of invariance under the unitary group cannot determine the constant k in the Poisson brackets. In a sense it doesnt matter what value we choose for it, as long as the values of all the observables are also multiplied by k. However it will be convenient to choose the value $k = \frac{i}{2}$ as we will see in the next subsection.

Soon we will have to calculate the Poisson bracket of functions $f(M)$ which is not linear in M. The derivative of such a function can be thought of as an operator valued function $f'(M)$ of M:

$$df = \text{tr} f'(M) dM = \int \tilde{f}_a^{\prime b}(M) d\tilde{M}_b^a(p,q) \frac{dpdq}{(2\pi)^2} \tag{2.51}$$

Then we can use the above Poisson brackets to get

$$k\{f, M\} = [f'(M), \epsilon + M] \tag{2.52}$$

where the l.h.s. involves the commutator, as operators, of $f'(M)$ and $\epsilon + M$. We leave the proof as an exercise to the reader.

2.8   Momentum

The above Poisson brackets are invariant under translations. There must be a canonical transformation that implements this symmetry. Formally, this is given by chosing u to be the derivative operator with respect to position; or, i times the multiplication by momentum. (Strictly speaking this is not in the Lie algebra of the unitary group since it is not bounded, but let us ignore this technicality for the moment.) Thus, momentum is

$$P = ik \int p\tilde{M}(p,p)\frac{dp}{2\pi}.$$
(2.53)

Now, imagine calculating this quantity for the separable configuration of renormalized dimension one:

$$M = -2\psi \otimes \psi^\dagger, \quad ||\psi||^2 = 1, \quad \epsilon\psi = \psi.$$
(2.54)

With the choice $k = \frac{i}{2}$ we would have,

$$P = \int_0^\infty p|\tilde{\psi}(p)|^2\frac{dp}{2\pi}.$$
(2.55)

This has a simple physical interpretation: P is just the expectation value of momentum is a state with wavefunction $\tilde{\psi}(p)$.

Note that P is always a positive function: the quadratic constraint becomes, for $p = q$,

$$2\,\text{sgn}\,(p)\tilde{M}(p,p) = -\int \bar{|}M(p,r)|^2\frac{dr}{2\pi}$$
(2.56)

so that momentum can be written as

$$P = \frac{1}{4}\int |p|\,|\tilde{M}(p,r)|^2\frac{dp}{2\pi}\frac{dr}{2\pi} \geq 0.$$
(2.57)

This makes physical sense if we regard this as the null component of the momentum vector.

2.9 Kinetic Energy

For a free particle of mass μ , the time-like component of momentum (energy), p_0 is related to the spatial component p_1 by the condition

$$p_0^2 - p_1^2 = \mu^2.$$
(2.58)

If we introduce the variables

$$p = p_0 + p_1, \quad p_- = p_0 - p_1$$
(2.59)

this becomes

$$(2p_0 - p)p = \mu^2, \quad pp_- = \mu^2. \tag{2.60}$$

Thus the kinetic energy is related to the null component of momentum through the dispersion relation:

$$p_0 = \frac{1}{2}(p + \frac{\mu^2}{p}), \quad p_- = \frac{\mu^2}{p}. \tag{2.61}$$

Notice that in this point of view the sign of energy p_0 and of momentum p are the same: this will prove to be convenient when studying the structure of the ground state of relativistic fermion theories. (See the Appendix A for more details.)

Thus we will postulate the kinetic energy of our dynamical system on the Grassmannian to be

$$K = -\frac{1}{2} \int \frac{1}{2}[p + \frac{\mu^2}{p}]\tilde{M}(p,p)\frac{dp}{2\pi}. \tag{2.62}$$

By the same argument as for momentum we can see that

$$K = \frac{1}{8} \int [|p| + \frac{\mu^2}{|p|}] \, |\tilde{M}(p,r)|^2 \frac{dp}{2\pi} \frac{dr}{2\pi} \geq 0 \tag{2.63}$$

If we add to this an appropriate potential energy U we will get the hamiltonian of the system. This potential energy must transform like p_-, in order that $(p, p_- + U)$ transform like the null components of momentum.

Under Lorentz tranformations,

$$p \to \lambda p, \quad p_- \to \lambda^{-1} p_-. \tag{2.64}$$

The Poisson bracket be invariant under Lorentz transformations,

$$\tilde{M}(p,q) \mapsto \tilde{M}_\lambda(p,q) = \lambda \tilde{M}(\lambda p, \lambda q). \tag{2.65}$$

In position space, $x \to \lambda^{-1}x$ so that

$$M(x,y) \mapsto M_\lambda(x,y) = \lambda^{-1} M(\lambda^{-1}x, \lambda^{-1}y). \tag{2.66}$$

Lorentz invariance will constrain the form of the potential energy as well.

2.10 Hamiltonian

The total hamiltonian will be a sum of kinetic energy K and a potential energy U. While the kinetic energy is best understood in momentum space, potential energy is best written in position space. The simplest choice of U will be a quadratic function of $M(x, y)$. (Anything simpler will lead to linear equations of motion.) Thus we postulate

$$U[M] = \frac{1}{4} \int M_b^a(x, y) M_a^b(y, x) v(x, y) dx dy \qquad (2.67)$$
$$+ \frac{1}{4} \int M_a^a(x, x) M_b^b(y, y) v_1(x, y) dx dy.$$

This will have the requisite invariance under the internal symmetry $U(F)$.

Lorentz invariance requires that U transform like p_- or, equivalently, like $\frac{1}{p}$ or x. Thus,

$$v(\lambda^{-1}x, \lambda^{-1}y) = \lambda^{-1}v(x, y). \qquad (2.68)$$

Moreover, they can only depend on the difference $x - y$ due to translation invariance. Thus

$$v_1(x, y) = \frac{1}{2}\alpha_1|x - y|, \quad v_2(x, y) = \frac{1}{2}\alpha_2|x - y| \qquad (2.69)$$

for.some pair of constants α_1, α_2.

Thus the hamiltonian of our theory is

$$E[M] = -\frac{1}{4} \int [p + \frac{\mu^2}{p}]\tilde{M}(p, p)\frac{dp}{2\pi}$$
$$+ \frac{\alpha_1}{8} \int M_b^a(x, y) M_a^b(y, x)|x - y| dx dy$$
$$+ \frac{\alpha_2^2}{8} \int M_a^a(x, x) M_b^b(y, y)|x - y| dx dy.$$

The constants m, α_1, α_2 have to fixed later based on experimental data. They are the coupling constants of our theory.

We have already shown that the kinetic energy is positive. The potential energy is manifestly positive when $\alpha_1, \alpha_2 \geq 0$ the way we have written it in position space. Thus the minimum value of E is zero and it is attained at the point $M = 0$. (This is in fact why we choose to parametrize our system by the variable M and not Φ.) This of course lies in the connected component with renormalized dimension zero. In the other components, energy will have a minimum again, but it is not zero. In fact we will spend much time estimating this ground state energy.

34

2.11 Singular Integrals

It will be convenient for later purposes to express the hamiltonian in momentum space variables. This will require the use of some singular integrals. The basic singular integral is the Cauchy Principal value [26]

$$\mathcal{P} \int_a^b \frac{f(p)}{p-q} \frac{dp}{2\pi} = \lim_{\epsilon \to 0^+} \left[\int_a^{q-\epsilon} + \int_{q+\epsilon}^b \right] \frac{f(p)}{p-q} \frac{dp}{2\pi} \qquad (2.70)$$

This exists whenever f is Hölder continuous [3] of order greater than zero at the point $p=q$. (Of course we are assuming that $a < q < b$). The basic idea of the Principal value integral is to cut-off the integral by removing a small interval of width 2ϵ located *symmetrically* about the singular point and then take the limit as $\epsilon \to 0$. (If the function vanishes at the point $p=q$ the limit might exist even if not taken symmetrically: then we have an 'improper integral' rather than a 'singular integral'.)

If we had cut-off the integral asymmetrically, the contribution of the region close to the singularity would have been

$$f(q) \left[\int_a^{q-\epsilon_1} + \int_{q+\epsilon_1}^b \right] \frac{1}{p-q} \frac{dp}{2\pi} \sim \frac{1}{2\pi} f(q) \log \frac{\epsilon_1}{\epsilon_1}. \qquad (2.71)$$

This can take any value as we let ϵ_1 and ϵ_1 to go to zero. Requiring that $\epsilon_1 = \epsilon_2$ removes this ambiguity. Indeed there are many other rules that could have been chosen: the particular one we choose must be justified by physical considerations: rather like the choice of boundary conditions in the solution of differential equations. In problems of interest to us there is a symmetry $p \to -p$ (ultimately due to charge conjugation invariance) which selects out the Principal Value (or the Finite Part we will define soon) as the correct prescription.

We already saw a use of the Cauchy principal value in the definition of the operator ϵ :

$$\epsilon \psi(x) = \frac{i}{\pi} \mathcal{P} \int \frac{\psi(y)}{x-y} dy. \qquad (2.72)$$

The symmetric choice of regularization implicit in the Principal value is required by the condition that ϵ is hermitean.

---

[3] A function f is Hölder continuous of order ν (or $f \in C^\nu$) if $\lim_{p \to q} \frac{|f(p)-f(q)|}{|p-q|^\nu}$ exists for all q .

We will often have to deal with integrals that have a worse singularity:

$$\int_a^b \frac{f(p)}{(p-q)^2} \frac{dp}{2\pi}. \tag{2.73}$$

The 'Hadamard Finite Part' ('part finie') of such an integral is defined in terms of the Cauchy Principal value:

$$\mathcal{FP} \int \frac{f(p)}{(p-q)^2} \frac{dp}{2\pi} = \mathcal{P} \int \frac{f(p) - f(q)}{(p-q)^2} \frac{dp}{2\pi}. \tag{2.74}$$

As long as $f \in C^1$, this will make sense: the vanishing of the numerator on the r.h.s. makes the integral exist as a Cauchy principal value.

Although we wont need this yet, we note for completenes the definition of the finite part integral for a finite range of integration:

$$\mathcal{FP} \int_a^b \frac{f(p)}{(p-q)^2} \frac{dp}{2\pi} = \mathcal{P} \int_a^b \frac{f(p) - f(q)}{(p-q)^2} \frac{dp}{2\pi} + \frac{f(q)}{2\pi} \left[\frac{1}{a-q} - \frac{1}{b-q} \right]. \tag{2.75}$$

We need the notion of a finite part integral because

$$\mathcal{FP} \int \frac{1}{p^2} e^{ipx} \frac{dp}{2\pi} = -\frac{1}{2}|x-y|. \tag{2.76}$$

The quantity on the right hand side is a Green's function of the Laplace operator on the real line:

$$\frac{d^2}{dx^2} \frac{1}{2}|x-y| = \delta(x-y). \tag{2.77}$$

The choice of boundary conditions in this Green's function corresponds to the choice of the definition of the singular integral above.

As an aside we note that

$$\mathcal{FP} \int \frac{g(p)}{p^2} \frac{dp}{2\pi} = \lim_{\epsilon \to 0+} \left[\int_{-\infty}^{-\epsilon} + \int_{\epsilon}^{\infty} \right] \frac{g(p) - g(0)}{p^2} \frac{dp}{2\pi} \geq 0 \tag{2.78}$$

if

$$g(p) \geq g(0) \tag{2.79}$$

for all p. This is useful in checking that the hamiltonian is positive.

2.12 Hamiltonian in Momentum Space

Now,

$$\frac{1}{2}\int M_b^a(x,y)M_a^b(y,x)|x-y|dxdy = \tag{2.80}$$

$$-\mathcal{FP}\int \frac{1}{r^2}\tilde{M}_b^a(p,q)\tilde{M}_a^b(p',q')\frac{dpdp'dqdq'dr}{(2\pi)^5}$$

$$\int e^{ir(x-y)+ipx-iqy+ip'y-iq'x}dxdy$$

$$= -\mathcal{FP}\int \frac{1}{r^2}\tilde{M}_b^a(p,q)\tilde{M}_a^b(q+r,p+r)\frac{dpdqdr}{(2\pi)^3}$$

Similarly,

$$\frac{1}{2}\int M_a^a(x,x)M_b^b(y,y)|x-y|dxdy \tag{2.81}$$

$$-\mathcal{FP}\int \frac{1}{r^2}\tilde{M}_b^a(p,q)\tilde{M}_a^b(p',q')\frac{dpdp'dqdq'dr}{(2\pi)^5}$$

$$\int e^{ir(x-y)+ipx-iqx+ip'y-iq'y}dxdy$$

$$= -\mathcal{FP}\int \frac{1}{r^2}\tilde{M}_a^a(p,p+r)\tilde{M}_b^b(p',p'-r)\frac{dpdp'dr}{(2\pi)^3}$$

Thus the hamiltonian becomes

$$E[M] = -\frac{1}{4}\int [p+\frac{\mu^2}{p}]\tilde{M}(p,p)\frac{dp}{2\pi}$$

$$-\frac{\alpha_1}{4}\mathcal{FP}\int \frac{1}{r^2}\tilde{M}_b^a(p,q)\tilde{M}_a^b(q+r,p+r)\frac{dpdqdr}{(2\pi)^3}$$

$$-\frac{\alpha_2}{4}\mathcal{FP}\int \frac{1}{r^2}\tilde{M}_a^a(p,p+r)\tilde{M}_b^b(p',p'-r)\frac{dpdp'dr}{(2\pi)^3}$$

In spite of the minus signs on the r.h.s., this is in fact a positive function of $\tilde{M}$. We already saw that the kinetic energy is positive whenever the constraint is satisfied.

We can rewrite the second term in a more symmetric form,

$$E[M] = -\frac{1}{4}\int [p+\frac{\mu^2}{p}]\tilde{M}(p,p)\frac{dp}{2\pi}$$

$$-\frac{\alpha_1}{4}\mathcal{FP}\int \frac{1}{r^2}\tilde{M}_b^a(p-\frac{r}{2},q-\frac{r}{2})\tilde{M}_a^b(q+\frac{r}{2},p+\frac{r}{2})\frac{dpdqdr}{(2\pi)^3}$$

$$-\frac{\alpha_2}{4}\mathcal{FP}\int\frac{1}{r^2}\tilde{M}^a_a(p,p+r)\tilde{M}^b_b(p',p'-r)\frac{dp\,dp'\,dr}{(2\pi)^3}$$

Now,

$$f(r) = \int\tilde{M}^a_b(p-\frac{r}{2},q-\frac{r}{2})\tilde{M}^b_a(q+\frac{r}{2},p+\frac{r}{2})\frac{dp\,dq}{(2\pi)^2}$$

satisfies

$$f(r) = f^*(-r) \tag{2.82}$$

so that the second term in the hamiltonian can be written as

$$-\frac{\alpha_1}{4}\mathcal{FP}\int\frac{1}{r^2}\operatorname{Re} f(r)\frac{dr}{2\pi}. \tag{2.83}$$

Moreover, by a simple use of the Schwarz inequality, we see that

$$\operatorname{Re} f(r) \le f(0). \tag{2.84}$$

Thus this term in the hamiltonian is positive. In the same way the last term can also be proved to be positive.

2.13   The Equations of Motion

From our Poisson brackets and the above hamiltonian we can derive the equations of motion. We can regard the derivative $E'(M) = \frac{\partial E}{\partial M}$ of the energy with respect to M as an operator valued function of M:

$$dE = \operatorname{tr} E'(M)dM = \int \tilde{E}''^b_a(M;q,p)d\tilde{M}^a_b(p,q)\frac{dp\,dq}{(2\pi)^2}. \tag{2.85}$$

Explicitly,

$$
\begin{aligned}
dE = \ & -\frac{1}{4}\int[p+\frac{\mu^2}{p}]d\tilde{M}^a_a(p,p)\frac{dp}{2\pi} \\
& -\frac{\alpha_1}{2}\mathcal{FP}\int\frac{1}{r^2}\tilde{M}^b_a(q+r,p+r)\frac{dr}{(2\pi)}d\tilde{M}^a_b(p,q)\frac{dp\,dq}{(2\pi)^2} \\
& -\frac{\alpha_2}{2}\mathcal{FP}\int\frac{1}{(q-p)^2}\tilde{M}^b_b(p',p'+p-q)\frac{dp'}{2\pi}d\tilde{M}^a_a(p,q)\frac{dp\,dq}{(2\pi)^2}.
\end{aligned}
$$

Thus

$$\tilde{E}_a^{\prime b}(M; q, p) = -\frac{1}{4}[p + \frac{\mu^2}{p}]2\pi\delta(p - q)\delta_a^b$$
$$-\frac{\alpha_1}{2}\mathcal{FP}\int\frac{1}{r^2}\tilde{M}_a^b(q + r, p + r)\frac{dr}{(2\pi)}$$
$$-\frac{\alpha_2}{2}\delta_a^b\mathcal{FP}\frac{1}{(q-p)^2}\int\tilde{M}_c^c(p', p' + p - q)\frac{dp'}{2\pi}.$$

Now the equations of motion implied by our Poisson brackets are, in operator notation,

$$k\frac{dM}{dt} = k\{E(M), M\} = [E'(M), \epsilon + M]. \tag{2.86}$$

(Here, k is the constant we fixed earlier to be $\frac{i}{2}$.) Substituting the above formula for $E'(M)$ will give a rather complicated system of integral equations as our equations of motion.

In particular, static solutions are given by

$$[E'(M), \epsilon + M] = 0 \tag{2.87}$$

The obvious solution to this equation is the vacuum,

$$M = 0. \tag{2.88}$$

This is just the point where the hamiltonian is a minimum. Actually each connected component will have a minimum for the hamiltonian; the absolute minimum is the one with renormalized dimension zero. We will return to the study of the minimum in the components of non-zero renormalized dimension.

2.14 Linear Approximation

If M is infinitesimally close to the vacuum value $M = 0$ we can linearize the equations of motion:

$$\frac{i}{2}\frac{d\tilde{M}_b^a(p, q)}{dt} = -\frac{1}{2}[\tilde{K}(p) - \tilde{K}(q)]\tilde{M}_b^a(p, q)$$
$$-\frac{\alpha_1}{2}\mathcal{FP}\int\frac{1}{s^2}[\tilde{M}_b^a(p + s, r + s)2\pi\delta(r - q)\,\mathrm{sgn}\,(q)$$

$$-2\pi\delta(p-r)\,\text{sgn}\,(p)M_b^a(r+s,q+s)]\frac{drds}{(2\pi)^2}$$

$$-\frac{\alpha_2}{2}\delta_b^a\mathcal{FP}\int\Big[\frac{1}{(p-r)^2}2\pi\delta(r-q)\,\text{sgn}\,(q)\tilde{M}_c^c(p',p'+r-p)$$

$$-\frac{1}{(r-q)^2}2\pi\delta(p-r)\,\text{sgn}\,(p)\tilde{M}_c^c(p',p'+q-r)\Big]\frac{dp'}{2\pi}\frac{dr}{(2\pi)}$$

Simplifying,

$$\frac{i}{2}\frac{d\tilde{M}_b^a(p,q)}{dt} =$$

$$-\frac{1}{2}[\tilde{K}(p)-\tilde{K}(q)]\tilde{M}_b^a(p,q)$$

$$-\frac{\alpha_1}{2}\mathcal{FP}\int\frac{1}{s^2}[\tilde{M}_b^a(p+s,q+s)\,\text{sgn}\,(q)-\text{sgn}\,(p)M_b^a(p+s,q+s)]\frac{ds}{2\pi}$$

$$-\frac{\alpha_2}{2}\delta_b^a\mathcal{FP}\int\Big[\frac{1}{(p-q)^2}\,\text{sgn}\,(q)\tilde{M}_c^c(p',p'+q-p)$$

$$-\frac{1}{(p-q)^2}\,\text{sgn}\,(p)\tilde{M}_c^c(p',p'+q-p)\Big]\frac{dp'}{2\pi}$$

Here,

$$\tilde{K}(p)=\frac{1}{2}[p+\frac{\mu^2}{p}]. \qquad (2.89)$$

Recall that the constraint on M becomes, in this linear approximation,

$$[\,\text{sgn}\,(p)+\text{sgn}\,(q)]\tilde{M}(p,q)=0; \qquad (2.90)$$

i.e., that p and q have opposite signs. We can assume that $p>0, q<0$ since the opposite case is determined by the hermiticity condition

$$\tilde{M}(p,q)=\tilde{M}^*(q,p). \qquad (2.91)$$

Thus,

$$\frac{i}{2}\frac{d\tilde{M}_b^a(p,q)}{dt} = -\frac{1}{2}[\tilde{K}(p)-\tilde{K}(q)]\tilde{M}_b^a(p,q)$$

$$+\alpha_1\mathcal{FP}\int\frac{1}{s^2}\tilde{M}_b^a(p+s,q+s)\frac{ds}{2\pi}$$

$$+\alpha_2 \delta_b^a \mathcal{FP} \int \frac{1}{(p-q)^2} \tilde{M}_c^c(p', p' + q - p) \frac{dp'}{2\pi}$$

Now, under translation, $M(x, y) \rightarrow M(x + a, y + a)$ and $\tilde{M}(p, q) \mapsto e^{i(p-q)a} \tilde{M}(p, q)$. Thus the total momentum, which is a conserved quantity, is $P = p - q$. We can use as independent variables

$$x_B = p/P, \quad P = p - q. \tag{2.92}$$

Clearly $0 \le x_B \le 1$ and $P \ge 0$. For a stationary solution,

$$\tilde{M}(p, q; t) = e^{i\omega t} \chi(x_B). \tag{2.93}$$

We get thus

$$\frac{1}{2}\omega\chi(x_B) =$$

$$\frac{1}{4}[2P + \frac{\mu^2}{Px_B} + \frac{\mu^2}{P(1 - x_B)}]\chi(x_B) - \frac{\alpha_1}{2\pi P}\mathcal{FP}\int_0^1 \frac{\chi(y)}{(x - y)^2}dy \tag{2.94}$$

$$+\frac{\alpha_2}{2\pi P}\int_0^1 \chi(y)dy$$

Recalling that the mass of the meson is given in light cone coomponents by,

$$2\omega P - P^2 = \mu^2 \tag{2.95}$$

we get

$$\mu^2\chi(x_B) =$$

$$[\frac{\mu^2}{x_B} + \frac{\mu^2}{(1 - x_B)}]\chi(x_B) - \frac{2\alpha_1}{\pi}\mathcal{FP}\int_0^1 \frac{\chi(y)}{(x_B - y)^2}dy - \frac{\alpha_2}{\pi}\int_0^1 \chi(y)dy$$

If we set $\alpha_2 = 0$ this is exactly the equation that 't Hooft obtained for the meson spectrum of two dimensional QCD.

't Hooft obtained this result by summing over an infinite class of Feynman diagrams. He showed that in the usual perturbaton series of two dimensional QCD, the diagrams of planar topology dominate in the large N limit. He then found a set of integral equations that describe the sum over these planar diagrams.

't Hooft's approach is limited by its origins in perturbations theory. It does not in fact reproduce the full large N limit of two dimensional QCD; only the *linear approximation* to the large N limit is obtained this way. The complete large N limit that we have constructed, by very different methods, is a nonlinear theory, reproduces the earlier theory of 't Hooft as just the linear approximation.

Our theory of course contains much more information: it describes all the interactions among the mesons, since the hamiltonian is a non-linear function of the co-ordinates on the Grassmannian. Indeed it is not even a polynomial which means that there are interaction vertices at every order in perturbation theory. Some of these have been determined, but our analysis gives all the infinite number of vertices in terms of a single constant: the geometry of the Grassmannian fixes all these uniquely in terms of α_1. The first such vertex was obtained by a summation of planar diagrams as well; but this approach will quickly get bogged dpwn by combinatorial complications. Before our work there was no indication that it was even possible to get a closed formula for all the interactions among the mesons in the large N limit of two dimensional QCD.

The solution of this linear integral equation will give the structure functions of mesons. Experimental data exists on the structure functiosn of mesons as well. It is not from Deep Inelastic Scattering, but from Drell-Yan scattering. Preliminary results on the comparison of these structure functions with data are encouraging. However the focus of these lectures is the much more conceptually deep problem of obtaining the structure function of the proton.

In fact our quantum hadrondynamics is not only a theory of mesons but also of baryons. We will show in the next chapter that there are topological solitons in our theory that describe baryons.

2.15   Quantization of HadronDynamics

We have so far described a classical theory, whose phase space is the Grassmannian. The Poisson bracket relations for the $M(x, y)$ play a role analogous to the canonical commutation relations. There are two distinct approaches to quatization of this theory:
(i) Find a representation of the Lie algebra in terms of operators ona complex Hilbert space ('canonical quantization', pursued in Ref. [11]; and, (ii)

Realize the wavefunctions as holomorphic sections of a line bundle on the Grassmannians ('geometrical quantization' pursued in Ref. [13].

In the first approach, we can use the complete classification of highest weight unitary representation of the (central extension of the) infinite-dimensional unitary Lie algebra, due to Kac and Peterson [27]. These turn out to be described by Young tableaux just as in the finite dimensional case, except they may have infinite depth. The question arises which irreducible representation to choose. The quadratic constraint selects out one class of representations, whose Young Tableaux are rectangular, of width N and infinite depth. Such a representation can be realized as

$$\hat{M}(x,y) = \frac{1}{N} : q^{i\dagger}(x)q_i(y) : \qquad (2.96)$$

where $q, q^\dagger$ satisfy the canonical anti-commutation relations. The representation is carried by the vector space of color invariant states in the fermionic Fock space. Thus we recover exactly the light-cone gauged fixed version of two dimensional QCD with N colors as the quantum theory of hadrondynamics. We have shown that two dimensional QCD and two dimensional hadrondynamics are *exactly* equivalent at all energies and for all values of the number of colors. We have already given the details of this argument in Ref.[11].

In the geometric approach, we construct a line bundle of Chern number N on the Grassmannian. There is a connection on this line budle whose curvature is the natural symplectic form of the Grassmannian. The infinite dimensional Grassmannian is a complex manifold, just like the finite dimensional ones. Holomorphic sections of the line bundle exist when the Chern number is a positive integer. An inner product can be established on the vector space of these holomorphic sections by using ideas of Segal [28] or the measure of Pickrell [29]. The observables $M(x,y)$ can be represented as operators on this Hilbert space. Again the formulae we get are just the same as that of light-cone gauge fixed QCD. See [13] for details.

In addition it is sometimes convenient to quantize a restricted version of the theory where the phase space is some symplectic submanifold of the Grassmannian. This will give us some insight into the connection between our theory and the parton model. We will return to this in the next chapter.

To summarize,we have a theory that doesnt just describe mesons to all orders in perturbation theory; we even have a theory of baryons. The topologically non-trivial solutions of the theory- the solitons- describe baryons.

In the next chapter we will study these solitons in more detail and obtain a theory of the structure functions of the baryon.

Chapter 3

Solitons

We saw that our phase space is a disconnected manifold, each connected component being labelled by an integer

$$B = -\frac{1}{2} \operatorname{tr}_\epsilon M. \tag{3.1}$$

Thus each connected component will have a minimum for the energy. These will describe stable static solutions of the equations of motion. They are called topological solitons.

The minimum of the energy in each connected component with a given value of B desribes a stable particle, as the only particles with lower energy would be in a sector with a different value of B. It cannot decay into those states since B is a conserved quantity.

We saw that hadron dynamics describes, in the linearized approximation, the spectrum of mesons of two dimensional QCD. What does a topological soliton describe? It is an old idea of Skyrme that the topological solitons of a theory of hadrons are baryons. This idea was revived in the mid eighties [3, 4] and shown to be consistent with QCD in four dimensions.

With this in mind, we should expect that the topological solitons of our two-dimensional theory describe baryons in the Deep Inelastic region. In the center of mass frame, a baryon will (due to Lorentz contraction) have a thin flat shape: one that can be described within a two dimensional theory. This offers us the possibility of solving one of the long-standing problems of particle physics: to explain the structure functions of hadrons as measured in Deep Inelastic Scattering. By solving the static equations of motion, we will be able to determine the 'shape' of the soliton, which will then give us the dependence of the hadronic structure function on the Bjorken x_B variable. The equations we need to solve are certain singular nonlinear integral equations. They are exactly solvable, so we will have to resort to numerical techniques. Indeed some ingenuity is needed even in the numerical part of this project: 'off the shelf' methods do not work due to the singularities in the integrals. These singularities do not pose any new conceptual problems however: we are dealing with a finite quantum field theory. The basic definitions of these singular integrals (as dealt with an earlir section) go back to the nineteenth century work of Cauchy and Hadamard.

The static equations of motion are

$$[E'(M), \epsilon + M] = 0 \tag{3.2}$$

where $E'(M)$ is the derivative of the energy, computed above. It will be most convenient to view them as integral equations for $\tilde{M}(p.q)$, the integral kernel in momentum space. Of course we must solve these equations subject to the constraints

$$[\epsilon, M]_+ + M^2 = 0; \quad \mathrm{tr}_\epsilon M = -2. \tag{3.3}$$

The last condition picks out the sector with baryon number one.

3.1 Quark Distribution Functions from QHD

If we split the operator M into submatrices according to the splitting $\mathcal{H} = \mathcal{H}_- \oplus \mathcal{H}_+$,

$$M = \begin{pmatrix} a & b \\ b^\dagger & d \end{pmatrix}, \tag{3.4}$$

the operator $a : \mathcal{H}_- \to \mathcal{H}_-$ is positive while $d : \mathcal{H}_+ \to \mathcal{H}_+$ is negative. The baryon number $B = -\frac{1}{2} \, \mathrm{tr}_\epsilon M$ is the sum of two terms: one from positive momenta and one from negative momenta:

$$B = -\frac{1}{2} \sum_a \int_0^\infty [\tilde{a}_{aa}(-p, -p) + \tilde{d}_{aa}(p, p)] \frac{dp}{2\pi}. \tag{3.5}$$

The first term is always negative and the second always positive. Thus, $\phi_{\bar{a}}(p) = \frac{1}{2} \tilde{a}_{aa}(-p, -p)$ can be thought of as the distribution function of antiquarks in a hadron while $\phi_{\bar{a}}(p) = -\frac{1}{2} d_{aa}(p, p)$ is the distribution function for quarks. The quantity

$$\phi_a^V(p) = \phi_a(p) - \bar{\phi}_{\bar{a}}(p) \tag{3.6}$$

corresponds to what is usually called the valence parton distribution. Its integral over p and sum over a is equal to the baryon number.

So the quark distribution function can be thought of as the sum of the 'valence' and 'Sea' contributions:

$$\phi_a(p) = \phi_a^V(p) + \phi_a^S(p). \tag{3.7}$$

The 'Sea' quark distribution function is thus just the same as the anti-quark distribution function:

$$\phi_a^S(p) = \phi_{\bar{a}}(p). \tag{3.8}$$

We find this terminology a bit confusing and mention it only for purposes of comparison. The point is that there is no physical meaning to the splitting of the quark distribution function into a valence and a sea quark contribution: only the sum has a measurable, physical significance. The anti-quark distribution does in fact have a physical significance. The above definition of a Sea quark distribution is thsu completely arbitrary and is *not* imposed on us by any symmetry such as charge conjugation invariance. In fact we will find later that there is another splitting which is more natural, where the Sea quarks have a wavefunctions are orthogonal to the valence quark wavefunction. This at least respects the Pauli principle for quarks. Even that spliting is merely a matter of convenience of interpretation: all comparisons with experiment should be in terms of the measurable quantities $\phi_a(p)$ and $\phi\bar{a}(p)$.

Actually the distribution functions are usually thought of as functions of a dimensionless variable x_B . If P is the total momentun (in the null direction) of the hadron, the momentum of the quark can be measured as a fraction of P . So we should actually write

$$\phi_a(x_B) = -\frac{1}{2}\sum_a \tilde{d}_{aa}(x_B P), \quad \phi_{\bar{a}}(x_B) = \frac{1}{2}\sum_a \tilde{a}_{aa}(-x_B P). \tag{3.9}$$

These identifications of the parton distribution functions in terms of the diagional matrix elements of $\tilde{M}_{ab}(p, p')$ can also be seen in terms of the formula we derived in terms of QCD.

The structure functions are supposed to vanish for $x_B > 1$. Why should the r.h.s. of the above equation vanish for such x_B? For large N, the structure functions will be of order $e^{-cN^{\frac{3}{2}}}$ at $x_B = 1$. So even though the wavefunction doesnt vanish at $x_B = 1$, it is exponentially small. A more proper analysis of the large N limit, taking into account semi-classical corrections will in fact reproduce wavefunctions that vanish at $x_B = 1$. The difference between these two versions of the large N limit is like the difference between the microcanonical ensemble and the canonical ensemble in statistical mechanics.

3.2 Rank One Ansatz

Before we set out to solve the above integral equations, it is useful to consider a simpler, variational approximation to them. A solution to the above constraints on M is the rank one (or separable) ansatz,

$$M = -2\psi \otimes \psi^{\dagger}. \tag{3.10}$$

We showed earlier that this is a solution to the constraints with baryon number one if ψ is a positive energy wavefunction of length one:

$$\epsilon\psi = \psi, \quad \sum_a \int |\tilde{\psi}_a(p)|^2 \frac{dp}{2\pi} = 1. \tag{3.11}$$

We are not claiming that the exact solution of the static equations of motion are of this form: only that this is a reasonable variational ansatz for it. The physical meaning of this ansatz will be made clear in the next section: it is equivalent to the valence quark model of the baryon in the large N limit.

The technical advantage of this ansatz is that it solves the constraint; more precisely it reduces it to the much simpler constraint that ψ is of length one. Geometrically, the set of states we are considering forms a projective space: changing ψ by a complex number of modulus one does not change M. The separable ansatz is an embedding of the complex projective space $CP(\mathcal{H}_+)$ of the positive energy states into the component of the Grassmannian with renormalized dimension one. Instead of minimizing the energy on the whole Grassmannian, in this approximation, we minimize it on this submanifold.

48

The parton structure functions have a simple meaning within this approximation: the structure function $\phi_a(x_B)$ is just the square of the wavefunction $\tilde{\psi}(p)$ in momentum space:

$$\phi_a(x_B) = |\tilde{\psi}_a(x_B P)|^2. \tag{3.12}$$

The anti-quark structure function is just zero in this separable approximation.

The energy on this submanifold can be easily calculated:

$$E_1(\psi) = \sum_a \int_0^\infty \frac{1}{2}[p + \frac{\mu^2}{p}]|\tilde{\psi}_a(p)|^2 \frac{dp}{2\pi} \tag{3.13}$$

$$+ \frac{\alpha_1 + \alpha_2}{2} \sum_{ab} \int |\psi_a(x)|^2 |\psi_b(y)|^2 |x - y| dx dy \tag{3.14}$$

where as usual,

$$\psi_a(x) = \int_0^\infty \tilde{\psi}_a(p) e^{ipx} \frac{dp}{2\pi}. \tag{3.15}$$

[1] Such a 'positive momentum' function has an analytic continuation into the upper half plane in the x variable. A simple choice would be

$$\dot{\psi}_a(x) = \frac{C_a}{(x + ib)^2} \tag{3.17}$$

with the location of the double pole serving as a variational parameter. C_a is constrained by the normalization condition that the length of ψ is one. (A simple pole would have infinite kinetic energy, so we consider the next simplest possibility.) It is easy to see that there is a term in energy that scales like b^{-1} (coming from $p|\tilde{\psi}(p)|^2$) and the remaining terms scale like b. Thus there is a minimum. This encourages us to proceed to a more accurate determination of the wavefunction ψ that minimizes the energy.

The integral equations we have to solve are

$$\frac{1}{2}[p + \frac{\mu^2}{p}]\tilde{\psi}_a(p) + \mathcal{F}P\tilde{g}^2 \int \tilde{V}(p - q)\tilde{\psi}_a(q)\frac{dq}{2\pi} = \lambda \tilde{\psi}_a(p)$$

---

[1] We will find it convenient to denote

$$\alpha_1 + \alpha_2 = \tilde{g}^2 \tag{3.16}$$

to agree with previous papers.

$$\tilde{V}(p) = -\frac{1}{p^2}\int_0^\infty \tilde{\psi}^{*a}(p+q)\tilde{\psi}_a(q)\frac{dq}{2\pi}.$$

subject of course to $\tilde{\psi}(p) = 0$ for $p \leq 0$ and $\int_0^\infty |\tilde{\psi}(p)|^2 \frac{dp}{2\pi} = 1$. The second of the above integral equations is just the momentum space version of Poisson's equation of electrostatics:

$$V''(x) = -\psi^{a*}(x)\psi_a(x). \tag{3.18}$$

From Gauss' law,

$$V(x) \sim \frac{1}{2}|x| \tag{3.19}$$

as $|x| \to \infty$: it is just the electrostatic potential of a unit charge located near the origin. Thus

$$\tilde{V}(q) \sim -\frac{1}{q^2} \tag{3.20}$$

as $|q| \to 0$. This is why the integrals are singular: the finite part prescription is just a way of imposing the boundary condition that $V(x) \sim \frac{1}{2}|x|$ at infinity in position space. We also note that for a solution centered at the origin,

$$\psi(x) = \psi^*(-x) \tag{3.21}$$

and $V(x) = V(-x)$. The boundary condition at the origin we impose is

$$V(0) = 0 \tag{3.22}$$

which translates to

$$\mathcal{FP}\int \tilde{V}(q)\frac{dq}{2\pi} = 0. \tag{3.23}$$

3.3   Approximate Analytic Solution for $\mu = 0$

It will be useful to have an approximate analytical solution; even if it only works for a physically uninteresting region it will help us to validate our numerical method.

Consider the singular integral (setting $\tilde{W}(q) = q^2 \tilde{V}(q)$):

$$\mathcal{FP} \int \tilde{V}(q)\tilde{\psi}_a(p+q)\frac{dq}{2\pi} = \mathcal{P} \int \frac{\tilde{W}(q)\psi_a(p+q) - \tilde{W}(0)\psi_a(p)}{q^2}\frac{dq}{2\pi}$$

$$= \frac{1}{2}\mathcal{P} \int \frac{\tilde{W}(q)[\tilde{\psi}_a(p+q) + \tilde{\psi}_a(p-q)] - 2\tilde{W}(0)\tilde{\psi}_a(p)}{q^2}\frac{dq}{2\pi}$$

$$= \frac{1}{2}\mathcal{P} \int \tilde{W}(q)\frac{[\tilde{\psi}_a(p+q) + \tilde{\psi}_a(p-q) - 2\tilde{\psi}_a(p)]}{q^2}\frac{dq}{2\pi}$$

$$+ \frac{1}{2}\left[\mathcal{P} \int \frac{2[\tilde{W}(p) - \tilde{W}(0)]}{q^2}\frac{dq}{2\pi}\right]\tilde{\psi}_a(p)$$

The first integral is not singular any more. The integral in the square brackets in the last term is in fact zero by our boundary condition:

$$\mathcal{P} \int \frac{2[\tilde{W}(p) - \tilde{W}(0)]}{q^2}\frac{dq}{2\pi} = \mathcal{FP} \int \tilde{V}(q)\frac{dq}{2\pi} = V(0) = 0. \qquad (3.24)$$

Thus

$$\mathcal{FP} \int \tilde{V}(q)\tilde{\psi}_a(p+q)\frac{dq}{2\pi} = \frac{1}{2} \int \tilde{W}(q)\frac{[\tilde{\psi}_a(p+q) + \tilde{\psi}_a(p-q) - 2\tilde{\psi}_a(p)]}{q^2}\frac{dq}{2\pi} \qquad (3.25)$$

So far we havent made any approximations on this integral: just rewritten the singular integral in a better way. The main contribution to this integral ought to come from the neighbourhood of the point $q = 0$. So we should be able to approximate the quatity in the square brackes by its leading Taylor series approximation

$$\mathcal{FP} \int \tilde{V}(q)\tilde{\psi}_a(p+q)\frac{dq}{2\pi} \sim \left[\int q^2\tilde{V}(q)\frac{dq}{2\pi}\right]\tilde{\psi}_a''(p). \qquad (3.26)$$

Then our nonlinear singular integral equation reduces to an ordinary differential equation:

$$-b\tilde{\psi}_a''(p) + \frac{1}{2}[p + \frac{\mu^2}{p}]\tilde{\psi}_a(p) = \lambda\tilde{\psi}_a(p) \qquad (3.27)$$

where b is determined by the self-consistency constraint

$$b = -\int q^2\tilde{V}(q)\frac{dq}{2\pi} = \sum_a\left|\int_0^\infty \tilde{\psi}^a(q)\frac{dq}{2\pi}\right|^2. \qquad (3.28)$$

This equation can be thought of as the nonrelativistic Schrödinger equation for a particle of mass b^{-1} in a linear plus Coloumb potential. The only additional complications are that b is determined by the above self-consistency relation and also, we have the boundary condition $\tilde{\psi}(p) = 0$ for $p \leq 0$. The only non-linearity in the problem is in the equation determining b.

Such non-relativistic models with linear plus Coloumb potentials have been used to describe mesons made of heavy quarks [32]. But the physical origin of this Schrödinger equation is completely different in our case: we get this equation in momentum space and not position space. The eigenvalue λ does not have the physical meaning of energy for us: energy is to be determined by substituting the solution into the formula for the hamiltonian. Moreover, we are studying the fully relativistic bound state problem. And of course we are studying baryons not mesons. Still it is encouraging that it is possible to reduce a fully relativistic bound state problem to a mathematical problem that is no more complicated than the non-relativistic case.

The special case $\mu = 0$ is particularly simple. In this extreme relativistic case, the solution is an Airy function:

$$\tilde{\psi}(p) = C \text{Ai}\left(\frac{p - 2\lambda}{[2b(\alpha_1 + \alpha_2)]^{\frac{1}{3}}}\right) \tag{3.29}$$

C is fixed by normalization: $\int |\tilde{\psi}(p)|^2 \frac{dp}{2\pi} = 1$. The eigenvalue is fixed by the continuity of the wavefunction: since it must vanish for $p \leq 0$ we require it to vanish at $p = 0$ as well by continuity. Then

$$\lambda = -\frac{1}{2}\xi_1[2b(\alpha_1 + \alpha_2)]^{\frac{1}{3}} \tag{3.30}$$

where $\xi = -2.33811$ is the zero of the Airy function closest to the origin. Now we fix the constant b by putting this back into the nonlinear self-consistency condition. We get

$$2b = \tilde{g}\frac{1}{(2\pi)^{\frac{3}{2}}}\frac{|\int_{\xi_1}^{\infty} \text{Ai}(\xi)d\xi|^3}{[\int_{\xi_1}^{\infty} \text{Ai}^2(\xi)d\xi]^{\frac{3}{2}}}$$

$$\lambda = \frac{|\xi_1|}{2\sqrt{(2\pi)}}\frac{\int_{\xi_1}^{\infty} \text{Ai}(\xi)d\xi}{[\int_{\xi_1}^{\infty} \text{Ai}^2(\xi)d\xi]^{\frac{1}{2}}}\tilde{g} \sim 0.847589\tilde{g}.$$

Moreover

$$\tilde{\psi}(p) = C \text{Ai}(\xi_1(\frac{p}{\tilde{g}\lambda} - 1)). \tag{3.31}$$

The Analytic Approximation to the Wave Function with m=0

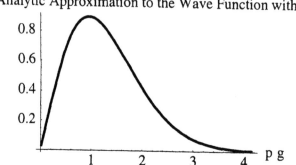

Figure 3.1: The approximate analytic solution for $\mu = 0$.

We plot the solution so obtained below.

3.4 Numerical Solution with the Rank One Ansatz

We now describe how to solve our integral equation numerically. First we replace the nonlinear integral into a recursion relation.

$$\frac{1}{2}[p + \frac{\mu^2}{p} - 2\lambda_s]\tilde{\psi}_{s+1}(p) + \tilde{\alpha}_1 \mathcal{P} \int_0^\infty \tilde{V}_s(p - q)\tilde{\psi}_{s+1}(q)\frac{dq}{2\pi} = 0 \qquad (3.32)$$

and

$$\tilde{V}_s(p) = -\frac{1}{p^2} \int_0^\infty \tilde{\psi}_s^*(p + q)\tilde{\psi}(q)\frac{dq}{2\pi}. \qquad (3.33)$$

We start with an initial guess $\tilde{V}_0(p)$, solve the linear integral equation to get a solution $\tilde{\psi}_1(p)$. Among all the solutions of this linear integral equation we pick the one without a node; this happens to be the one with the smallest eigenvalue λ . Then we calculate $\tilde{V}_1(p)$ as above and then again solve the integral equation to get $\tilde{\psi}_2$ and so on till our iteration converges.

We pick the nodeless eigenfunction at each stage since we expect our final answer for the ground state eigenfunction to have this property. It is just a coincidence that this happens to have the smallest eiganvalue λ : in any case λ does not have the meaning of energy.

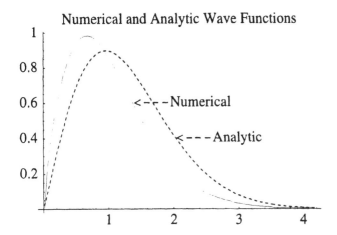

Figure 3.2: Comparison of Numerical and Approximate Analytic Solutions for $\mu = 0$.

Of course to carry out this algorithm, we need to convert the above integral equations into matrix equations. That is done by the quadrature method (described in the appendix C) for singular integrals. The whole procedure can be implemented in Mathamatica quite well.

We plot the numerical solution so obtained against the approximate analytical solution we got earlier for the special case $\mu = 0$.

The curves are qualitatively the same: the difference has to do with the approximations we had to make in order to get an analytic solution. The physicaly interesting case in fact does not have $\mu = 0$: indeed we will see that the value of μ^2 is in fact negative. For example when $m = 0$, we have, $\mu^2 = -\frac{g^2}{\pi}$. It is of much interest to see if there is a variant of the analytic approximatin method above that applies to this more realistic case.

3.5   Quantization of the Rank One Ansatz

We have considered elesewhere [11, 13] the quantization of our hadron dynamics, to recover QCD. It is also instructive to study the quantization of a simpler version of our theory, corresponding to the rank one ansatz. The quantization of this truncated theory gives a sort of approximation to QHD (hence QCD) which is of some interest in itself. We will see in the next chapter that this is just the valence parton model with interactions between the partons as predicted by QCD.

Given a vector satisfying

$$\epsilon\psi = \psi, \quad ||\psi||^2 = 1 \tag{3.34}$$

we have an element of the Grassmannian of renormalized dimension one:

$$M = -2\psi \otimes \psi^\dagger. \tag{3.35}$$

But this element remains the same if we change ψ by a complex number of modulus one: $\psi \to e^{i\theta}\psi$. Thus we have an embedding of the projective space $\mathcal{P}(\mathcal{H}_+)$ into the Grassmannian. This is a sympletic embedding: the symplectic form on the projective space induced by this embedding is the same as the natural sympletic form on the projective space. Thus we can regard the separable operators as forming a 'reduced phase space' describing part of the degrees of freedom of our theory.

A way to understand this is to consider $\tilde{\psi}(p)$ as a complex-valued observable on the effective phase space. In order to reproduce the Poisson brackets of M , these must satisfy

$$\{\tilde{\psi}_a(p), \tilde{\psi}_b(p')\} = 0 = \{\tilde{\psi}^{*a}(p), \tilde{\psi}^{*b}(p')\}, \tag{3.36}$$

$$\{\tilde{\psi}_a(p), \tilde{\psi}^{*b}(p')\} = -i2\pi\delta(p-q)\delta_a^b. \tag{3.37}$$

They satisfy the constraints,

$$\tilde{\psi}(p) = 0; \text{for } p < 0, \quad \int_0^\infty |\psi(p)|^2 \frac{dp}{2\pi} = 1. \tag{3.38}$$

We should regard the function $\tilde{\psi}(p)$ on momentum space to be the fundametal variable from which other observables such as

$$\psi(x) = \int_0^\infty \tilde{\psi}(p)e^{ipx}\frac{dp}{2\pi} \tag{3.39}$$

can be obtained.

Both the Poisson brackets and the constraint are simpler than the full theory, which is why we consider this case first.

The Hamiltonian of our reduced dynamical system is obtained by putting the ansatz into $E(M)$:

$$E_1(\psi) = \sum_a \int_0^\infty \frac{1}{2}[p + \frac{\mu^2}{p}]|\tilde\psi_a(p)|^2 \frac{dp}{2\pi} \tag{3.40}$$
$$+\frac{\alpha_1 + \alpha_2}{2} \sum_{ab} \int |\psi_a(x)|^2|\psi_b(y)|^2|x - y|dxdy.$$

These Poisson brackets and hamiltonian form a perfectly well–defined dynamical system on its own right. We will now consider how to quantize this theory, and obtain the rules for the semi-classical approximation.

We can quantize the theory by looking for operators satisfying canonical commutation relations:

$$[\hat{\tilde\psi}_a(p), \hat{\tilde\psi}_b(p')] = 0 = [\hat{\tilde\psi}^{\dagger a}(p), \hat{\tilde\psi}^{\dagger b}(p')], \quad [\hat{\tilde\psi}_a(p), \hat{\tilde\psi}^{\dagger b}(p')] = \frac{1}{N}2\pi\delta(p - p')\delta_a^b \tag{3.41}$$

As usual classical Poisson brackets go over to quantum commutation relations:

$$\{A, B\} \to -i\hbar[\hat A, \hat B]. \tag{3.42}$$

In our case we will denote the parameter that measures the quantum correction, analogous to $\hbar$, by $\frac{1}{N}$. In a minute we will see that this number N must actually be an integer. Th elimit $N \to \infty$ is the classical limit. The constraint on the observables can be implemented by restricting attention to those states satisfying

$$\int_0^\infty \hat{\tilde\psi}^{*a}(p)\hat{\tilde\psi}_a(p)\frac{dp}{2\pi}| >= 1. \tag{3.43}$$

Now it is obvious that a representation for our commutation relations is provided by bosonic creation annihilation operators:

$$[\hat{\tilde a}_a(p), \hat{\tilde a}_b(p')] = 0 = [\hat{\tilde a}^{\dagger a}(p), \hat{\tilde a}^{\dagger b}(p')], \quad [\hat{\tilde a}_a(p), \hat{\tilde a}^{\dagger b}(p')] = 2\pi\delta(p - q)\delta_a^b. \tag{3.44}$$

with

$$\psi_a(x) = \frac{1}{\sqrt N}a_a(x), \quad \psi^{\dagger a}(x) = \frac{1}{\sqrt N}a^{\dagger a}(x). \tag{3.45}$$

Then the constraint becomes just the condition that we restrict to states containing N particles:

$$\int_0^\infty \tilde{a}^{\dagger a}(p)\tilde{a}_p(x)\frac{dp}{2\pi} = N. \tag{3.46}$$

This is why N must be an integer!.

Now we know that we are dealing with a system of N bosons interacting with each other under the hamiltonian

$$\frac{1}{N}\hat{E}_1(\psi) = \sum_a \int_0^\infty \frac{1}{2}[p + \frac{\mu^2}{p}]\tilde{a}^{\dagger a}(p)\tilde{a}_a(p)\frac{dp}{2\pi} \tag{3.47}$$

$$+\frac{\alpha_1 + \alpha_2}{2}N\int a^{\dagger a}(x)a^{\dagger b}(y)a_b(y)a_a(x)|x - y|dxdy.$$

The classical (or large N) limit we have been discussing so far is just the mean field approximation to this many-body problem. The semi-classical approximation will give us the leading corrections in the case of finite N. This basic insight is due to Witten, in a by now classic paper [8].

What are these bosons? We will see in the next chapter that these bosons are just the valence quarks of the parton model, sripped of their color!. Quarks are of course fermions. However the wavefunction of the system must be totally anti-symmetric in the color indices beacuse of the condition that the state be invariant under $SU(N)$. Thus in the remaining indices the wavefunction must be symmetric: if we ignore color the valence quarks behave like bosons.

Note that the momentum of the particles created by $a^{\dagger}(p)$ is always positive. Thus the total momentum

$$\hat{P} = \int_0^\infty p\, a^{\dagger a}(p)a_a(p)\frac{dp}{2\pi} \tag{3.48}$$

is a positive operator. Indeed on a state containing N particles,

$$|a_1, p_1; a_2, p_2; \cdots a_N, p_N> = a^{\dagger a_1}(p_1)a^{\dagger a_2}(p_2)\cdots a^{\dagger a_N}(p_N)|0> \tag{3.49}$$

P is just the sum of individual momenta, each of which is positive:

$$\hat{P}|a_1, p_1; a_2, p_2; \cdots a_N, p_N> = [p_1 + p_2 + \cdots p_N]|a_1, p_1; a_2, p_2; \cdots a_N, p_N> \tag{3.50}$$

A general state will be, in this basis described a wavefunction $\tilde{\phi}$

$$|\phi> = \sum_{a_1, \cdots a_N} \int_0^\infty \frac{dp_1}{2\pi} \cdots \frac{dp_N}{2\pi}\tilde{\phi}(a_1, p_1; \cdots a_N, p_N)|a_1, p_1; \cdots a_N, p_N> . \tag{3.51}$$

An eigenstate of $\hat{P}$ with eigenvalue P will satisfy

$$[p_1 + \cdots p_N]\,\tilde{\phi}(a_1, p_1; \cdots a_N, p_N) = P\tilde{\phi}(a_1, p_1; \cdots a_N, p_N). \tag{3.52}$$

Since each of the momenta p_i are positive, it follows that they must each be less than the total momentum P:

$$0 \leq p_i \leq P, \text{ for } i = 1, \cdots N. \tag{3.53}$$

We will see in the next chapter that the p_i are the momenta of the valence partons: the proton is a simultaneous eignstate of $\hat{H}$ and $\hat{P}$. We have just seen a very important point: our theory at finite N (but within the approximation of the valence parton model) predicts that the wavefunction must vanish unless each of the parton momenta p_i are less than P. But there is no N in this inequality; so it must hold even in the large N limit!.

We will be finding an approximate eigenstate of $\hat{H}$, by a variational principle: a sort of mean field theory, the large N limit.The naive choice is a wavefunction which is a product of single particle wavefunctions. But such a naive version of mean field theory will violate the exact inequality we just established on momentum eigenstates. We should find our variational approximation to the eigenstate of the hamiltonian, within the space of momentum eigenfunctions.

Thus we assume that the wavefunction is approximated by a wavefunction that is just a product *except* for the constraint that the momenta add up to P:

$$\tilde{\psi}(a_1, p_1; a_2, p_2; \cdots a_N, p_N) = 2\pi\delta(\sum_i p_i - P)\tilde{\psi}(a_1, p_1)\tilde{\psi}(a_2, p_2) \cdots \tilde{\psi}(a_N, p_N) \tag{3.54}$$

Thus the fraction of the momentum carried by each particle is less than one:

$$\psi(p) = 0, \quad \text{unless } 0 \leq \frac{p}{P} \leq 1. \tag{3.55}$$

That is how we recover the fact that the quark distribution function must vanish when the Bjorken variable is greater than one. This is a sort of semi-classical correction in the $\frac{1}{N}$ approximation. We must solve the classical equations of motion for $\tilde{\psi}(p)$ subject to the boundary condition that it vanish outside of the interval $0 \leq p \leq P$. This requires a modification of the variational ansatz but it it is possible to do that.

This variant of mean field theory is rather like the micro-canonical ensemble. The naive mean field theory where only the expectation value of momentum is required to be P is like the canonical ensemble.

3.6   Valence Quark Distributions

Now we come to the point of comparison of the calculated valence parton distribuion function againt the data. It is not in fact necessary to make a direct comparison with the data on Deep Inelastic Scattering. A generation of phenemenologists have extracted the valance parton structure functions from the data. More precisely they have assumed a parametric form such as

$$\phi(x_B) = Ax_B^{\nu_1}(1 - x_B)^{\nu_2}[1 + a_1x_B + a_2x_B^2] \qquad (3.56)$$

for the parton distribution functions and fit to all known data points. Since the data consists of measurement of the DIS (Deep Inelastic Scattering) cross-section at different values of Q^2 this fit uses the convolution of this distribution function with the structure functions of the quarks calculated in perturbation theory. There are about 6 structure functions (corresponding to $u, \bar{u}, d, \bar{d}, s$ quarks and the and gluons) so altogether there are about 30 parameters in addition to the fundamental parameters of perturbative QCD. Altogether there are about a thousand data points, coming from measurement of the cross-section for $e - p$, $e - n$, $\nu - p$ or $\nu - n$ scattering at various energies. Thus the mere extraction of these distribution functions has itself become a subfield of particle physics [30, 17, 31]. The major groups seem to be. in general agreement with each other, although a detailed analyzis of the errors in their parameters is not yet available. There are recent attempts to estimate the systematic and statistical errors, but they seem incomplete.

The comparison should be made with the weighted average over the u and d quark distributions,weighted so as to get an isospin invariant combination. This is because we have not yet done a collective variable variable quantization of the isospin degrees of freedom of the proton, so the distribution function we are computing is the isospin invariant one.

Another complication in the comparison is that not all the momentum in the baryon is carried by the valence quarks: it is known that only about half of the momentum is in the valence quarks, the rest being in the anti-quarks (or 'sea' quarks) or gluons. This affects our momentum sum rule. Essentially what it does is that the N in the sumrule gets replaced by $N_{\text{eff}} = \frac{N}{f}$ where f is the fraction of the momentum carried by the valence quarks. Until we do a complete calculation allowing for the contributions of the sea quarks and the gluons (i.e., without assuming the factorizable ansatz for M and without ignoring the scalar fields ϕ_A) we must treat f as a parameter

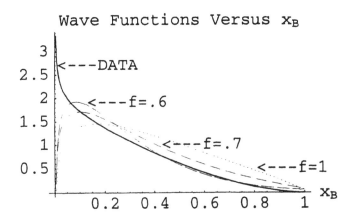

Figure 3.3: Comparison of parton wavefunctions $\sqrt{\phi(x)}$. The MRST distribution agrees with our prediction best when the fraction f of the momentum carried by the valence quarks is about 0.6 .

and choose the value that gives the best fit. The good news is that it is the only parameter: m^2 is fixed to be zero[2], since the up and down quarks are known to have masses that are smal compared to the QCD scale α_1 . ($m_u' \sim 5\ MeV$ and $m_d \sim 10\ Mev$ while $\alpha_1 \sim 100\ MeV$.)

The only other parameter of the theory is $\bar{g}^2$; but since the distribution functions are dimensionless function of a dimensionless variable, the dependence on α_1 cancels out.

In comparison with experimental data we should keep in mind that we are in fact ignoring some of the constituents of the baryon: the valence quarks are knwon to carry only about half of the momentum of the baryon. The rest is in the sea quarks and the gluons. Thus we introduce a parameter f , the fraction of the momentum carried by the valence quarks. We find that for a value of $f = 0.6$ the wavenfunction predicted by us agrees quite well with experiment. (The value of being about a half is consistent with other ways of looking at the situation.) Thus we have solved the problem of deriving the quark structure functions of the baryon from QCD.

It still remains to calculate the anti-quark and gluon structure functions.

---

[2]This by the way means that $\mu^2 = m^2 - \frac{\bar{g}^2}{\pi}$ is negative. There is no contradiction here, since the quark is not a true particle: the mass of the baryon is still predicted to be real.

We will turn to that in later publications. In a later chapter we will give some ideas that help in solving this problem.

In the next chapter we will see how to reconcile the soliton model with the parton model.

Chapter 4

The Parton Model

This chapter is based on the paper [14] which in turn was an expansion of some short statements made in Ref. [11].

We have been describing the idea that the baryon is a soliton, essentially made up of an infinite number of mesons. But this seems to contradict the simple and successful picture of the baryon as a bound state of quarks. At the time the Skyrme model was revived, it was shown that the static (low energy properties) of the baryon that are so well explained by the quark model (magnetic moments, mass differences, flavor multiplet structure) can all be rederived in the soliton model[5]. But the only explanation for the structure of the proton in deep inelastic scattering still appeared to be the quark-parton model. In these lectures, we are describing how to explain the structure functions within the soliton model as well. In fact we will be able to go beyond the simple minded parton model.

It is an old idea in the study of the structure of the proton that the constituents-the partons- can be thought of as free particles. This then raises the question of how they bind to form the proton in the first place. It is the binding mechanism that determines the wavefunction of the partons and hence the structure functions of the hadron. In this section we will present a model of interacting partons. Perhaps surprsingly, it will turn out to be equivalent to the soliton model in the factorizable ansatz (and in the large N limit). Thus the parton model is merely an approximation to the soliton model. Later, by a deeper analysis of the soliton model, we will derive the sea quark and gluon distributions as well.

It is at the moment impossible to derive the particle spectrum of four dimensional QCD directly. There are many attempts to do compute the spectrum by direct numerical simulation of QCD, but they have to yet to surmount many obstructions. For example, the asymptotic freedom of QCD implies that the energy of all physical states are exponentially small compared to the cut-off. Thus extra-ordinary accuracy is required of all direct numerical computations of the energy, a problem that is only exacerbated by the large (in principle infinite) number of degrees of freedom in the system. However, it may be possible to understand the spectrum of two dimensional QCD by diagonalizing its hamiltonian numerically. Being a finite (rather than asymptotically free) theory numerical calculations are much more reliable.

4.1   Parton Wavefunctions

We will follow largely the discussion in [14] The valence parton model assumes that the proton is made of N partons (quarks) which are fermions transforming under the fundamental representation of color. The idea of color of course was not present in the original version of this model, due to Bjorken and Feynman. Also it was not known then that the partons were spin half particles and hence fermions. But these were straightened out soon with the identification of the valence partons with the quarks. footnoteIt is known that only about half of the momentum of the baryon is accounted for by these valence partons: the rest must be carried by the 'sea quarks' and by the gluons.

  We will work think of the parton wavefunctions as functions of the null momentum $p = p_0 + p_1$. In addition, the partons carries the quantum numbers of flavor and spin (together denoted by a) and color α . Thus the wavefunction of a single parton will be $\tilde{\psi}(a, \alpha, p)$. Since $p \geq 0$ for the null component of momomentum, we require that ψ vanish for negative p

  A baryon is made of N such partons so its wavefunction is a completely antisymmetric function $\tilde{\psi}(a_1, \alpha_1, p_1; a_2, \alpha_2, p_2; \cdots; a_N, \alpha_N, p_N)$. However, the baryon must be invariant under color: transform under the trivial representation of color $SU(N)$. This means that the wavefunction is completely antisymmetric in color alone:

$$\tilde{\psi}(a_1, \alpha_1, p_1; a_2, \alpha_2, p_2; \cdots, a_N, \alpha_N, p_N) = \epsilon_{\alpha_1, \alpha_2, \cdots \alpha_N} \tilde{\psi}(\alpha_1, p_1; \alpha_2, p_2; \cdots \alpha_N, p_N) \quad (4.1)$$

In other words, the wavefunction is completely *symmetric* in spin, flavor and momentum quantum numbers. If we factor out color from the wavefunction, the partons behave as if they are *bosons*.

4.2   Hamiltonian

The kinetic energy of a single parton is

$$\frac{1}{2}[p + \frac{\mu^2}{p}].$$

(4.2)

Hence the kinetic energy of the partons making up a baryon is

$$\sum_{a_1 \cdots a_N} \int_0^\infty \sum_{i=1}^N \frac{1}{2}[p_i + \frac{m_{a_i}^2}{p_i}]|\tilde{\psi}(a_1, p_1; \cdots; a_N, p_N)|^2 \frac{dp_1 \cdots dp_N}{(2\pi)^N}.$$

(4.3)

In the early versions of the quark-parton model, they were treated as if they are free particles. That cannot be quite correct, since they must after all bind to form the baryon. The simplest idea would be to let them interact through a pairwise potential $\alpha_1 v(x)$:

$$\frac{1}{2}\alpha_1 \sum_{a_1 \cdots a_N} \int_0^\infty \sum_{i \neq j} v(x_i - x_j)|\psi(a_1, x_1; \cdots; a_N, x_N)|^2 dx_1 \cdots dx_N.$$

(4.4)

Now we must allow for a self-energy term as well in the hamiltonian. The point is that the effective mass of the partons may not be the same as their bare mass. Just as the electrons in a metal have a different effective mass from the free electron, the effective mass of the parton inside a baryon may be different from that of a 'bare' parton: there can be a finite renormalization of the mass. We can allow for this by replacing the μ^2 in the kinetic energy by an effective mass μ^2 .

Since the partons (quarks) cannot be separated out to infinity, one could wonder what the 'bare' parton mass m means: it has no meaning as the mass of any particle that can exist in an asymptotic state. However, it is possible to give the bare parton mass a meaning in terms of high energy processes that do not involve asymptotic states. For example the weak interaction can be used as a probe of the quark masses: the weak decay rates of the quarks are sensitive to the bare masses. This gives a way to make sense of the bare parameter m . But for our purposes, what matters is the effective mass μ .

Thus the hamiltonian of the valence quark model is

$$\mathcal{E}_N(\tilde{\psi}) = \sum_{a_1 \cdots a_N} \int_0^\infty \sum_{i=1}^N \frac{1}{2}[p_i + \frac{\mu_{a_i}^2}{p_i}]|\tilde{\psi}(a_1, p_1; \cdots a_N, p_N)|^2 \frac{dp_1 \cdots dp_N}{(2\pi)^N}$$
$$+ \frac{1}{2}\alpha_1 \sum_{a_1 \cdots a_N} \int_0^\infty \sum_{i \neq j} v(x_i - x_j)|\psi(a_1, x_1; \cdots a_N, x_N)|^2 dx_1 \cdots dx_N.$$

The ground state of this hamiltonian is the baryon of the valence quark model.

We havent yet decided what potential $v(x)$ to use. There is ample evidence that the quark-quark potential is linear [32]. In any case as in QHD, Lorentz invariance will lead to this choice n our lightcone co-ordinates. So we choose $v(x) = \frac{1}{2}|x|$.

With this choice we get exactly the same hamiltonian we had in the last chapter, from the quantization of the rank one ansatz with $\frac{1}{N}$ playing the role of $\hbar$. Thus the rank one approximation to the topological soliton model is equivalent to this interacting valence parton model.

4.3   Hartree Ansatz

The ground state of a many boson system can often be described by mean field theory: each boson moves in the field created by all the others. Moreover, all the bosons can be assumed to occupy the same single particle state in this ground state. After the color is factored out, the partons in our model behave just like bosons. Hence we should be able to simplify the problem by making this mean field approximation. More specifically,

$$\tilde{\psi}(a_1, p_1; \cdots a_N, p_N) = 2\pi\delta(\sum_i p_i - P) \prod_{i=1}^N \tilde{\psi}(a_i, p_i). \tag{4.5}$$

Here P is the total momentum of the system. Since $p \geq 0$ we must impose

$$\tilde{\psi}(p) = 0 \tag{4.6}$$

for $p \leq 0$. It then follows that p is less than the total momentum P .
The wavefunction satisfies the normalization condition

$$\int_0^P |\tilde{\psi}(p)|^2 \frac{dp}{2\pi} = 1 \tag{4.7}$$

and the momentum sum rule:

$$N \int_0^P p|\tilde{\psi}(p)|^2 \frac{dp}{2\pi} = P. \tag{4.8}$$

The energy is calculated by putting this ansatz into the earlier formula:

$$
\begin{aligned}
E &= \sum_a \int_0^\infty \frac{1}{2}[p + \frac{\mu_a^2}{p}]|\tilde{\psi}(a,p)|^2 \frac{dp}{2\pi} + \\
&\quad \frac{1}{2}\tilde{g}^2 \int_{-\infty}^\infty v(x-y) \sum_a |\psi(a,x)|^2 \sum_b |\psi(b,y)|^2 dx dy.
\end{aligned}
$$

At this point we notice that we have *exactly* the same problem as in the last chapter. The energy of the rank one ansatz for $\tilde{M}(p,q)$ is exactly the same as that of the parton model in the Hartree approximation. Thus the mean field theory of the valence parton model is our classical theory on the projective space $CP(\mathcal{H}_+)$ .. This interpretation of the rank one ansatz in terms of the parton model was already noted in Ref. [11].

Chapter 5

Beyond the Valence Parton Model

We saw that the separable ansatz in the soliton theory corresponds to the valence parton approximation. Now we see how to go beyond that and get the true minimum of the energy. We can improve on the valence parton model by adding Sea quarks and anti-quarks; this requires us to invent an improved variational ansatz. In another direction we can minimize the energy in the Grassmannian by numerical methods. We will describe ideas in both directions.

5.1   A Co-ordinate System

Recall that the equations to be solved are

$$[E'(M), \epsilon + M] = 0, \tag{5.1}$$

along with the constraints:

$$[\epsilon + M]^2 = 1, \quad \text{tr}_\epsilon M = -2. \tag{5.2}$$

Define

$$\epsilon_1 = \epsilon - 2\psi \otimes \psi^\dagger, \tag{5.3}$$

where ψ is the minimum in the subset of separable kernels; or at least a good approximation to it. Rather than parametrize the soliton by the

69

deviation from the vacuum ϵ it makes more sense to use the deviation from our approximate solution ϵ_1 which is in the same connected component. So we define a new variable M_1 by

$$\epsilon + M = \epsilon_1 + M_1. \tag{5.4}$$

Then the constraints on M_1 are

$$[\epsilon_1, M_1]_+ + M_1^2 = 0, \qquad \mathrm{tr}_\epsilon M_1 = 0. \tag{5.5}$$

The trace of M_1 is zero since the baryon number is now already carried by ϵ_1 :

$$-\frac{1}{2} \mathrm{tr}[\epsilon_1 - \epsilon] = 1. \tag{5.6}$$

Let W be the subspace where ϵ_1 takes eigenvalue -1 :

$$W = \mathcal{H}_- \oplus < \psi > \tag{5.7}$$

where $< \psi >$ denotes the linear span of ψ . We have an orthogonal splitting of $\mathcal{H}$ according to this splitting. We can now represent M_1 in terms of a co-ordinate system centered at the point ϵ_1 :

$$M_1 = -2 \begin{pmatrix} [1 + ZZ^\dagger]^{-1} - 1 & [1 + ZZ^\dagger]^{-1}Z \\ Z^\dagger[1 + ZZ^\dagger]^{-1} & Z^\dagger[1 + ZZ^\dagger]^{-1}Z \end{pmatrix} \tag{5.8}$$

Here, $Z : W^\perp \to W$ is an arbitrary Hilbert-Schmidt operator. It is straightforward to check that this is a solution to the constraint equations. Moreover, in some finite neighborhood of ϵ_1 , all points of the Grassmannian can be represented this way. Of course this co-ordinate system will break down if we go too far away from ϵ_1 : we are assuming that the true minimum lies close enough to the approximate solution to be in this co-ordinate chart.

Given a fixed ψ , the operator Z above provides a co-ordinate system on an neighborhood of the Grassmannian. This co-ordinate system only covers a part of the connected component with baryon number one. It takes a countably infinite number of such charts (corresponding to different choices of ψ) to cover the whole space of baryon number one configurations.

70

5.2   The Method of Steepest Descent

We need to minimize the function $E(M)$ subject to the constraints

$$[\epsilon, M]_+ + M^2 = 0, \quad -\frac{1}{2} \, \text{tr} M = 1, \quad \text{tr} M^2 < \infty. \tag{5.9}$$

A simple method for minimizing functions of several variables is steepest descent: we start at some initial point and move along the straight line opposite to the gradient of the function at that point a small distance. Then we recompute the gradient at the new point and repeat the procedure. If the topography of the constant energy surfaces is not too complicated we will eventualy arrive at the minimum.

The main complication in our case is of course the constraint: the gradient vector $E'(M)$ is not tangential to the Grassmannian so if we move opposite to it we will leave the constraint surface. Even if we project out the tangential component of the vector, we will still leave the surface if we move along the straightline in that direction. The proper geometric solution to this problem is to move a small distance along the geodesic at M tangential to the gradient vector. Then we will recompute the gradient and find the geodesic at the new point. This is feasible beacuse the Grassmannian is a homogenous manifold and we can find the geodesic on it easily using its high degree of symmetry.

More explicitly, recall that the tangential projection of the gradient vector is

$$T = \frac{1}{4}[\epsilon + M, [\epsilon + M, E'(M)]]. \tag{5.10}$$

The vector

$$Y = \frac{1}{2}[\epsilon + M, E'(M)] \tag{5.11}$$

is at right angles to the tangential part of the gradient: it is in fact obtained by multplying T by the complex structure of the Grassmannian. Just as in the case of the sphere, the geodesic is obtained by rotating the point about an axis orthogonal to the tangent vector. More explicitly, the geodesic starting at M tangential to T is

$$\gamma(\tau) = e^{\tau Y}[\epsilon + M]e^{-\tau Y} - \epsilon. \tag{5.12}$$

71

Thus the steepest descent algorithm is

1. Choose an initial configuration M_0 and small parameter τ.
2. Given the kth configuration M_k, calculate the gradient $E'(M_k)$ and $Y_k = \frac{1}{2}[\epsilon + M_k, E'(M_k)]$.
3. Set $M_{k+1} = e^{\tau Y_k}[\epsilon + M_k]e^{-\tau Y_k} - \epsilon$ and repeat the previous step.

The value of τ has to be chosen by some trial and error. Too small a value will produce changes in the configuration within the noise due to numerical errors. Too large a choice will not give a convergent sequence: we will bounce around all over the Grassmannian. But we found that in practice, a proper value of τ can be found quickly.

How do we choose the initial configuration M_0? Since we believe that the valence approximation is good, we could use as the starting point a separable configuration, minimizing the energy within that subspace.

5.3   Sea Quarks and Anti-quarks

We saw that if we restrict the dynamics of our theory to the rank one ansatz, $M = -2\psi \otimes \psi^\dagger$, we get the valence parton model. To get a more general picture that includes Sea quarks and anti-quarks (but is still not the total picture) we must use the ansatz with rank three. In fact by going to ansatzes of larger and larger rank we can get better and better approximations to the baryon structure functions.

The rank three ansatz is

$$-\frac{1}{2}M = \psi_0 \otimes \psi_0^\dagger + \zeta_-\{\zeta_-[\psi_+ \otimes \psi_+^\dagger - \psi_- \otimes \psi_-^\dagger] \\ -\sqrt{[1 - \zeta_-^2]}[\psi_- \otimes \psi_+^\dagger + \psi_+ \otimes \psi_-^\dagger]\}$$

where ψ_-, ψ_0, ψ_+ are three vectors in $\mathcal{H}$ satisfying

$$\epsilon\psi_- = -\psi_-, \quad \epsilon\psi_0 = \psi_0, \quad \epsilon\psi_+ = \psi_+, \tag{5.13}$$

$$||\psi_-||^2 = ||\psi_0||^2 = ||\psi_+||^2 = 1, \quad \psi_0^\dagger\psi_+ = 0. \tag{5.14}$$

Moreover,

$$0 \le \zeta_-^2 \le 1. \tag{5.15}$$

We saw that ζ_-^2 is the probability of finding an anti-quark inside the baryon.

As noted earlier, the anti-quark distribution function is just $\zeta_-^2 |\tilde{\psi}_a(-p)|^2$. The baryon number is

$$B = \sum_a \int_0^\infty \{|\tilde{\psi}_{0a}(p)|^2 + \zeta_-^2 [|\tilde{\psi}_{+a}(p)|^2 - \tilde{\psi}_{-a}(-p)|^2]\} \frac{dp}{2\pi}. \qquad (5.16)$$

The total momentum is, similarly,

$$P = \sum_a \int_0^\infty p\{|\tilde{\psi}_{0a}(p)|^2| + \zeta_-^2 [|\tilde{\psi}_{-a}(-p)|^2 + |\tilde{\psi}_{+a}(p)|^2]\} \frac{dp}{2\pi} \qquad (5.17)$$

These confirm the interpretation of ψ_0 as the valence quark wavefunction and $\psi_+(p)$ as the Sea quark wavefunction.

Also, the kinetic energy is

$$K = \sum_a \int_0^\infty \frac{1}{2}[p + \frac{\mu^2}{p}]\{|\tilde{\psi}_{0a}(p)|^2| + \zeta_-^2 [|\tilde{\psi}_{-a}(-p)|^2 + |\tilde{\psi}_{+a}(p)|^2]\} \frac{dp}{2\pi} \qquad (5.18)$$

The potential energy is more complicated. Recall that the potential energy is simpler in position space while the kinetic energy is simpler in momentum space. Our wavefunctions now depend on a discrete variable in addition to momentum, and there is a unitary transformation in these discrete variables that is the counterpart to Fourier transformation. This is the transformation to a basis in which μ is diagonal; the wavefunctions will no longer be eigenstates of ϵ.

$$-\frac{1}{2}M = \psi_0 \otimes \psi_0^\dagger + \zeta_-\{\psi_1 \otimes \psi_1^\dagger - \psi_2 \otimes \psi_2^\dagger\} \qquad (5.19)$$

where

$$\psi_1 = \frac{1}{\sqrt{2}}\left\{\sqrt{[1-\zeta_-]}\psi_- - \sqrt{[1+\zeta_-]}\psi_+\right\} \qquad (5.20)$$

$$\psi_2 = \frac{1}{\sqrt{2}}\left\{\sqrt{[1+\zeta_-]}\psi_- + \sqrt{[1-\zeta_-]}\psi_+\right\} \qquad (5.21)$$

The potential energy is then,

$$U = \frac{\alpha_1}{2} \sum_a \int \left\{|\psi_{0a}(x)|^2 V_{00}(x) + \right.$$

$$\zeta_-^2 |\psi_{1a}(x)|^2 V_{11}(x) + \zeta_-^2 |\psi_{2a}(x)|^2 V_{22}(x) \Big\} dx$$
$$+ \frac{\alpha_1}{2} 2 \operatorname{Re} \sum_a \int \Big\{ \zeta_- \psi_{0a}(x) \psi_{1a}^*(x) V_{10}(x) - $$
$$\zeta_- \psi_{0a}(x) \psi_{2a}^*(x) V_{20}(x) - \zeta_-^2 \psi_{1a}(x) \psi_{2a}^*(x) V_{21}(x) \Big\} dx.$$

The mean fields are determined by solving the differential equations

$$V''_{\alpha\beta}(x) = \psi_{\alpha a}(x) \psi_{\beta a}^*(x). \tag{5.22}$$

with the boundary conditions

$$V_{\alpha\beta}(x) \to \delta_{\alpha\beta} \frac{1}{2}|x|, \text{ for } |x| \to \infty. \tag{5.23}$$

Equivalently,

$$V_{\alpha\alpha}(0) = V'_{\alpha\alpha}(0) = 0 \tag{5.24}$$

for the diagonal components, and,

$$V_{\alpha\beta}(\infty) = V'_{\alpha\beta}(\infty) = 0 \text{ for } \alpha \neq \beta \tag{5.25}$$

for the off diagonal components.

By choosing appropriate variational ansatzes we can estimate the anti-quark content of the proton. Or we can derive integral equations for the functions $\psi_{\pm,0}$ and solve them numerically. Both methods are being pursued. Initial results are encouraging. Detailed results will appear in a separate publication.

Acknowledgement

I thank the students and postdoctoral fellows at Rochester who have worked withme at various stages in implementing this research program: P. Bedaque, K. S. Gupta, S. Guruswamy, I. Horvath,V. John, G. Krishnaswami, C-W. H. Lee, O. T. Turgut. I have also benefitted from discussions with A. P. Balachandran, A. Bodek, J. Mickelsson, S. Okubo, D. Pickrell and J. Schechter. I am especially grateful to O. T. Turgut and V. John for proof-reading an early version of these notes. Also I thank C. Saclioglu and J. Mickelsson for invitations to Istanbul and Stockholm and making my visits there so enjoyable. This work is supported by the Department of Energy.

Appendix A

A Null Co-ordinate System

A.1 Kinematics

It will be convenient to use a co-ordinate system that combines the advantages of the null and Cartesian co-ordinate systems.

If x^0 and x^1 are the usual Cartesian co-ordinates in Minkowski space, the metric is

$$ds^2 = [dx^0]^2 - [dx^1]^2. \tag{A.1}$$

We define now

$$t = x^0 - x^1, \quad x = x^1 \tag{A.2}$$

so that

$$ds^2 = dt[dt + 2dx]. \tag{A.3}$$

The Minkowski metric is,

$$\eta_{\mu\nu} = \begin{pmatrix} 1 & 1 \\ 1 & 0 \end{pmatrix} \tag{A.4}$$

Thus the vector $\frac{\partial}{\partial t}$ is time-like while $\frac{\partial}{\partial x}$ is null. The initial values of fields will be given on a surface of constant t, which is a null line.

Momentum $p = p_\mu dx^\mu = p_0 dt + p dx$ is a 1-form (co-vector); we will use the same letter to denote the momentum 1-form as well as its null componenet, but it should be clear from the context which one we mean. To find its magnitude we must use the inverse of the above metric tensor:

$$\eta^{\mu\nu} = \begin{pmatrix} 0 & 1 \\ 1 & -1 \end{pmatrix}, \quad p^2 = p_\mu p_\nu \eta^{\mu\nu} = 2p_0 p - p^2. \tag{A.5}$$

Thus the mass shell condition becomes

$$p_0 = \frac{1}{2}\left[p + \frac{\mu^2}{p}\right]. \tag{A.6}$$

Here, μ is the rest mass of the particle.

We see now one of the main technical advantages of using the null-time co-ordinate system: energy and momentum have the same sign. In the usual space-time co-ordinates, $p_o = \pm\sqrt{[p_1^2 + \mu^2]}$ and therefore no such simple relationship exists. In the Dirac theory of fermions the states of negative energy are occupied; this becomes merely the condtition that the negative momentum states be occupied.

A.2   Dirac Matrices

The Dirac matrices are best thought of as matrix-valued vectors, since they appear in the form $\gamma^\mu \nabla_\mu$ in the action. Thus the Dirac algebra $\gamma^\mu \gamma^\nu + \gamma^\nu \gamma^\mu = 2\eta^{\mu\nu}$ becomes

$$[\gamma^t]^2 = 0, \quad \gamma^t\gamma + \gamma\gamma^t = 2, \quad \gamma^2 = -1. \tag{A.7}$$

We will choose the explicit representation

$$\gamma^t = \begin{pmatrix} 0 & 2 \\ 0 & 0 \end{pmatrix}, \quad \gamma = \begin{pmatrix} 0 & -1 \\ 1 & 0 \end{pmatrix}. \tag{A.8}$$

In any representation of Dirac matrices there is a 'charge-conjugation' matrix C satisfying

$$C\gamma^\mu C^{-1} = (\gamma^\mu)^T. \tag{A.9}$$

In the usual space-time formalism this matrix is often γ^0 itself, but that is a representation-dependent fact. In our representation,

$$C = \begin{pmatrix} 0 & 1 \\ 1 & 0 \end{pmatrix}. \tag{A.10}$$

Given the Dirac spinor $q = \begin{pmatrix} q_1 \\ q_2 \end{pmatrix}$, the conjugate spinor is

$$\bar{q} = q^\dagger C = (\, q_2^\dagger \quad q_1^\dagger \,).\qquad\qquad(A.11)$$

A.3 Free Fermions

The Lagrangian of a free Dirac fermion becomes

$$
\begin{aligned}
L_D \;=\;& \bar{q}\gamma^\mu[-i\partial_\mu]q + m\bar{q}q \\
& +2q_2^\dagger[-i\partial_t]q_2 + q_1^\dagger(-i\partial_x)q_1 - q_2^\dagger(-i\partial_x)q_2 + m[q_2^\dagger q_1 + q_1^\dagger q_2].
\end{aligned}
$$

We see that the field q_1 has no dynamical degrees of freedom: it has no time derivative in the action. Hence it can be eliminated by its equation of motion:

$$q_1 = -\frac{m}{\hat{p}}q_2 \qquad\qquad (A.12)$$

where $\hat{p} = -i\partial_x$. Putting this back into the action and changing variables

$$\chi = \sqrt{2}\, q_2 \qquad\qquad (A.13)$$

gives us the effective action for the propagating field:

$$L_\chi = \chi^\dagger(-i\partial_t)\chi - \chi^\dagger\frac{1}{2}[\hat{p} + \frac{m^2}{\hat{p}}]\chi. \qquad\qquad (A.14)$$

A.4 Gauge Fields

The lagrangian of two dimensional QCD is

$$L \;=\; \frac{N}{4\alpha_1}\int \mathrm{tr}F_{\mu\nu}F^{\mu\nu} + \sum_{a=1}^{N_f}\int \bar{q}^{a\alpha}[-i\gamma\cdot\nabla + m_a]q_{a\alpha}.$$

We have added in the flavor indices a, b and the color indices i, j .

The freedom of gauge transformations can be utilized partially to impose the null gauge condition $A_x = 0$. Then, the fermionic part of the lagrangian becomes

$$L_D \;=\; \bar{q}^{ai}\gamma^\mu[-i(\partial_\mu\delta_i^j + A_{\mu i}^j)]q_{aj} + m\bar{q}^{ai}q_{ai}$$

$$+ \quad \chi^{\dagger ai}[-i(\partial_t + A^j_{ti})]\chi_j$$
$$+q_1^{\dagger ai}(-i\partial_x)q_{1ai} - \frac{1}{2}\chi^{\dagger ai}(-i\partial_x)\chi_{ai} + \frac{m}{\sqrt{2}}[\chi^{\dagger ai}q_{1ai} + q_1^{\dagger ai}\chi_{ai}].$$

which becomes

$$L_\chi = \chi^{\dagger a}(-i\partial_t)\chi_a - \chi^{\dagger a}\left\{\frac{1}{2}[\hat{p} + \frac{m^2}{\hat{p}}] - iA_t\right\}\chi_a \tag{A.15}$$

upon eliminating q_1.

To this we must add the action of the Yang–Mills field itself, which looks quite simple in this co-ordinate system and gauge:

$$L_{YM} = \frac{N}{4\alpha_1} \text{tr} F_{\mu\nu} F_{\rho\sigma} \eta^{\mu\rho}\eta^{\nu\sigma} = \frac{N}{2\alpha_1} \text{tr}[\partial_x A_t]^2. \tag{A.16}$$

The field A_t does not propagate and can be eliminated. Thus the action of two-dimensional QCD can be written entirely in terms of the field χ.

A.5   The Dirac Vacuum

Let us return to the free fermion theory with Lagrangian

$$L = \chi^{\dagger}(-i\partial_t)\chi - \chi\frac{1}{2}[\hat{p} + \frac{m^2}{\hat{p}}]\chi \tag{A.17}$$

Upon quantization, the field χ becomes an operator satisfying the fermionic anti-commutation relations

$$[\chi(x), \chi(y)]_+ = \delta(x - y), \quad [\chi(x), \chi(y)]_+ = 0. \tag{A.18}$$

If there were only a finite number of operators, such canonical anti-commutation relations would have a unique representation. In the infinite dimensional case physical ideas have to be brought in to choose the right representation. Dirac showed that the correct choice is to assume that all the negative energy states are filled even in the vacuum. Since energy and momentum have the same sign in our co-ordinate system, this condition is easy to implement. We define the vacuum by

$$\tilde{\chi}^{\dagger}(p)|0> = 0 \text{ for } p < 0, \quad \tilde{\chi}(p)|0> = 0 \text{ for } p > 0. \tag{A.19}$$

We have defined the Fourier transforms

$$\chi(x) = \int \tilde{\chi}(p) e^{ipx} \frac{dp}{2\pi} \tag{A.20}$$

etc. Then the normal ordered product of a pair of operators is defined as

$$: \chi^\dagger(p)\chi(p') : \, = \chi^\dagger(p)\chi(p') \tag{A.21}$$

unless both p and p' are negative, in which case it is

$$: \chi^\dagger(p)\chi(p') : \, = -\chi(p')\chi^\dagger(p). \tag{A.22}$$

The point is that then the expectation value of normal ordered current operators are zero in the Dirac vacuum. Indeed,

$$
\begin{aligned}
< 0|\chi^\dagger(x)\chi(y)|0 > \; &= \; \int \frac{dp\,dq}{2\pi\,2\pi} e^{-ipx+iqy} < 0|\tilde{\chi}^\dagger(p)\tilde{\chi}(q)|0 > \\
&= \; \int_{-\infty}^{0} \frac{dp}{2\pi} e^{ip(y-x)} = \frac{1}{2}[\delta(x-y) + \epsilon(x-y)].
\end{aligned}
$$

Here,

$$\epsilon(x-y) = \int \text{sgn}\,(p) e^{ip(x-y)} \frac{dp}{2\pi} = \frac{1}{\pi i}\mathcal{P}\frac{1}{x-y} \tag{A.23}$$

is to be thought of as a distribution. It is (upto a factor of i) the kernel of a well-known integral transform, the Hilbert transform.

Thus we should regard the current operators of the fermionic theory as defined with the normal ordering. For example, the equation of motion of the A_t will be, in the quantum theory,

$$-\partial_x^2 A_{tj}^i(x) = \frac{1}{N}\alpha_1 : \chi^{\dagger ai}(x)\chi_{aj}(x) : . \tag{A.24}$$

We can use this to eliminate the gauge field from the theory completely. Thus two dimensional QCD can be written as a theory of fermions interacting with each other through a Coulomb-like potential. The further analysis of this theory is carried out in the text, towards the end of the first chapter.

Appendix B

Operator Ideals

Here we give the basic definitions of the operator ideals we use in the text. A deeper discussion may be found in Ref. [25, 33, 34]; we give only a rough outline of the theory. It is the author's fervent hope that experts in functional analysis will *not* read this appendix.

B.1 Compact and Hilbert-Schimdt Operators

The *rank* of an operator $A : \mathcal{H} \to \mathcal{H}$ on a complex Hilbert space is the dimension of its range; i.e., the dimension of the subspace of all vectors that can be written as Au for some $u \in \mathcal{H}$. When the rank of A is finite, it can be thought of as a sort of 'rectangular matrix' with (possibly) an infinite number of columns but only a finite number of rows; at least there is a basis in which it has this form.

For such an operator, we can define several measures of its size (norm). For example (the operator norm),

$$|A| = \sup_u \frac{||Au||}{||u||}. \tag{B.1}$$

Another (the Hilber-Schmidt or H-S norm) is

$$|A|_2 = \mathrm{tr}[AA^\dagger]^{\frac{1}{2}}. \tag{B.2}$$

$|A|_2^2$ is also the sum of the absolute squares of all the matrix elements in any basis.

81

The completion of the space of finite rank operators in the operator norm is the space of *compact* operators. In other words, a compact operator is one that can be approximated arbitrarily closely by finite rank operators, distance between operators being measured with the operator norm. If we instead complete in the H-S norm, we get the space of Hilbert-Schimdt operators.

A bounded operator is one which has finite operator norm; i.e.,

$$|A| = \sup_{u \in \mathcal{H}} \frac{||Au||}{|u|} \tag{B.3}$$

exists. The set of bounded operators on $\mathcal{H}$ is an algebra $\mathcal{B}(\mathcal{H})$. Not all bounded operators are compact; for example the identity is bounded yet not compact.

A compact operator has the expansion, with $\mu_n > 0$,

$$A = \sum_{n=1} \mu_n |\psi_n ><\phi_n| \tag{B.4}$$

the sum being either finite (when A is of finite rank) or infinite (more generally). The numbers μ_n are called singular values; if A is positive they are its eigenvalues. A and its adjoint $A^\dagger$ have the same singular values.

"Roughly speaking, the operator norm of A is the largest of its singular values, while the Hilbert-Schmidt norm is the sum of the squares of the singular values. The singular values of a compact operator form a sequence that converges to zero; in fact zero is the only limit point of the sequence. Thus, if the H-S norm is finite, the eigenvalues must be tend to zero. In fact all H-S operators are compact.

B.2   Schatten Ideals

Many other norms can be defined in terms of the singular values. For example the trace class operators are those for which the sum of singular values is convergent. This is stronger than the requirement that the diagonal elements in some basis form a summable sequence: being trace-class reqires a sort of absolute convergence. The product of two Hilbert-Schimdt operators in trace-class.

More generally, for $p \geq 1$, we have the class $\mathcal{I}_p(\mathcal{H})$ of operators for which the sum $\sum \mu_n^p$ converges. Of course $\mathcal{I}_2$ is the space of H-S operators and $\mathcal{I}_1$ that of trace class operators. The space of compact operators can be thought of as the limiting case $\mathcal{I}_\infty$ and that of finite rank rank operators as the opposite limit, $\mathcal{I}_0$. We have the inclusions

$$\mathcal{I}_0 \subset \mathcal{I}_1 \subset \mathcal{I}_2 \cdots \mathcal{I}_\infty \subset \mathcal{B}. \tag{B.5}$$

It is very important for us that the $\mathcal{I}_p$ are *two-sided ideals* in the algebra of bounded operators; i.e., that $A \in \mathcal{B}$ and $B \in \mathcal{I}_p$ implies that both $AB \in \mathcal{I}_p$ and $BA \in \mathcal{I}_p$. These are called the Schatten ideals. We are especially interested, of course, in the cases $p = 1, 2$.

B.3   The Restricted Unitary Group and its Grassmannian

In the text we are interested in the case of a Hilbert space $\mathcal{H}$ with a given orthogonal splitting into two infinite dimensional orthogonal subspaces: $\mathcal{H} = \mathcal{H}_- \oplus \mathcal{H}_+$. Recall that the operator ϵ is defined to have eigenvalues ± 1 on $\mathcal{H}_\pm$. The restricted Unitary group is the subset of all unitaries satisfying a convergence condition:

$$U_1(\mathcal{H}, \epsilon) = \{g | gg^\dagger = g^\dagger g = 1; [\epsilon, g] \in \mathcal{I}_2\}. \tag{B.6}$$

It is vital for this definition to make sense that $\mathcal{I}_2$ is an ideal; that is why the product of two elements still satisfies the convergence conditon. (Any unitary operator is bounded). If we split $g \in U_1(\mathcal{H}, \epsilon)$ into submatrices according to the splitting $g = \begin{pmatrix} a & b \\ c & d \end{pmatrix}$ where $a : \mathcal{H}_- \to \mathcal{H}_-$ etc., the submatrices b, c are H-S. The matrices a, d may not be invertible in general, but they are Fredholm (see below) since g^{-1} exists.

The restricted Grassmannian is the set of all self-adjoint operators of square one whose distance (in the H-S sense) from ϵ is finite.

$$Gr_1(\mathcal{H}, \epsilon) = \{\Phi | \Phi^\dagger = \Phi; \Phi^2 = 1, \Phi - \epsilon \in \mathcal{I}_2\}. \tag{B.7}$$

In the text we often use $M = \Phi - \epsilon$ as the variable that describes a point in the Grassmannian. We showed that each such operator corresponds to a subspace of $\mathcal{H}$ (the negative eigenspace of Φ) which is at a finite distance from the standard subspace $\mathcal{H}_-$.

The fact that $\mathcal{I}_2$ is an ideal ensures that the action of the restricted unitary group on this Grassmannian well-defined:

$$\Phi \mapsto g\Phi g^\dagger, \quad M \to gMg^\dagger + g[\epsilon, g^\dagger]. \tag{B.8}$$

B.4 Fredholm Index

The material in this section is contained in Ref. [25] to which we refer for proofs and more precise statements. A bounded operator A is *Fredholm* if it is invertible modulo a compact operator; i.e., if there is a compact operator K such that $A + K$ has an inverse. The set of Fredhom operators is a toplogical space, with the toplogy induced by the operator norm. It is disconnected, each connected component being labelled by an integer called the Fredholm index.

To understand the Fredholm index, consider the kernel of an operator; i.e., the subspace of all u such that $Au = 0$. Even in the finite dimensional case the kernel can change change discontinuosly under a small change in A .; for example we may cross an eigenvalue. In the finite dimensional case, the dimensions of the kernels of A and $A^\dagger$ are the same. In the case of Fredholm operators on infinite dimensional Hilbert spaces, the difference

$$\text{index } (A) = \dim \ker A - \dim \ker A^\dagger \tag{B.9}$$

is always finite but need not vanish. It however does not change under continuos changes in A : it is constant in each connected component of the space of Freholm operators. In fact it is the only such function: any two Fredholm operators of the same index are connected by a continuos curve.

Now $g \in U_1(\mathcal{H}, \epsilon)$ is of course Fredholm of index zero: it is invertible as is its adjoint. But the submatrices a, d (where $a : \mathcal{H}_- \to \mathcal{H}_-$ etc.) defined above are not invertible in general. Yet they are invertible modulo some compact operators (products such as $bb^\dagger$), so they are Fredholm. Since g as a whole invertible, they must have opposite Fredholm indexes. The index of a (for example) is a topological invariant of g . The group U_1 is the union of connected components labelled by this index, which can take any integer value. The connected component of the identity, of course, has index zero.

We can understand the renormalized dimension of a subspace (called virtual rank in [25]) in terms of the Fredhom index. Every self-adjoint operator can be diagonalized; an operator $\Phi \in Gr_1$ can be brought to the standard form ϵ by an element in U_1 : $\Phi = g\epsilon g^{\dagger}$. The index of $a(g)$ is then the topological invariant associated to Φ .

Another point of view is in terms of the subspace W of $\mathcal{H}$ corresponding to Φ . If Φ is at a finite distance from ϵ , W will not differ 'too much' from $\mathcal{H}_-$. More precisely, the projection operator $\pi_+ : W \to \mathcal{H}_+$ will be compact and $\pi_- : W \to \mathcal{H}_-$ will be Fredholm. The index of π_- measures the 'difference in dimensions' between W and $\mathcal{H}_-$; this is the renormalized dimension of W . We saw in the text that this has the physical meaning of baryon number.

It si crucial for all this that we allow only subspaces at a finite distance from the standard one in Gr_1 . If we had allowed for all subspaces, the Grassmannian would have been contractible!. The convergence condition we must impose is required for the Poisson brackets to make sense: the symplectic form of the Grassmannian would not make sense otherwise. It is gratifying that as a consequence, we get a topological invariant which has the physical meaning of baryon number.

Although the symplectic form makes sense on all of the phase space Gr_1 , we should not expect the hamiltonian to make sense on all of it. There should be some dense domain in which the hamiltonian does make sense however. We leave such questions as challenges for the analyst who is interested in solving problems of relevance to physics.

Appendix C

Quadrature of Singular Integrals

In this appendix we summarize some ideas on the numerical methods that are used in the chapter on solitons. Some originality is needed even in this part of the problem.

C.1   Quadrature Formulas

The equations we have are just too hard to be solved analytically. We have to resort to numerical methods. The basic idea is to convert the integral equation into a matrix equation by allowing p to take just a finite set of values: we must find a way to approximate the integral by a finite sum. Then we will solve the resulting nonlinear matrix equations by iteration.

Quadrature is the approximation of integrals by finite sums. There are standard methods for quadrature, (method of moments) going back to the days of Gauss. But we have to modify these methods since our integrals are singular. The basic idea is well-known in the literature on quadrature at least for the case of the Cauchy principal value[35]. Our integrals are one step harder (Hadamard Finite Part) but the idea is the same. See S. Chandrashekhar,'s classic book [36] for a clear discussion of numerical integration.

Let $\rho(x)$ be a continuous positive function on the close interval $[a, b]$. We are interested in evaluating integrals such as

$$\int_a^b f(x)\rho(x)dx \qquad (C.1)$$

by numerical approximations. Here $f(x)$ is some continuous function.

Let $x_j, j = 1, \cdots \nu$ be a set of points in the interval $[a, b]$. We expect a weighted sum such as

$$\sum_j w_j f(x_j) \tag{C.2}$$

to be good approximation for the integral, provided that (i) the number points ν is large enough and (ii) the points x_j are distributed roughly uniformly.

Every function can be approximated by a polynomial within the interval; as the order $\nu - 1$ of the polynomials grows the approximation gets better. We can thus approximate the integral of $f(x)$ by that of its polynomial approximation of order $\nu - 1$. The weights w_j are determined (once the points x_j are chosen) by this requirement: the above formula should in particular be exact for polynomials of order $\nu - 1$.

This is the same as the condition

$$\int_a^b x^k \rho(x)dx = \sum_j w_j x_j^k, \text{ for } k = 0, 1, \cdots k - 1. \tag{C.3}$$

The left hand side are the moments of the distribution $\rho(x)dx$, and are assumed to be known. Then the above set of linear equations determine the weights in terms of x_j.

If the function $\rho(x)dx$ is not too rapidly varying, a simple choice such as equally spaced points $x_j = a + (b - a)\frac{j-1}{\nu}$ should give reasonable approximation to the integral: certainly for polynomials upto order $\nu - 1$ we will get the exact answer anyway. But it is possible to do better by choosing the ν points x_j cleverly, as pointed out by Gauss: we can ensure that the answer is exact for polynomials of order $2\nu - 1$. But we wont be using this idea: we will just use equal spacing for the points, which turns out to be more convenient. This because our integrands involveterms such $\tilde{\psi}(p+r)$, so it is convenient if the sum two points $x_j + x_k$ is also a point at which we evaluate the integrand. The loss of precision in quadrature can be made up because the simplicity of equal spacing allows us to choose a larger number of points.

C.2 Singular Measures

Now consider integral[35] $\mathcal{FP} \int_0^b f(x) \frac{dx}{x^2}$. The symbol $\mathcal{FP}$ indicates as before the Hadamard 'finite part' of the integral. For such singular integrals we can also find a numerical approximation as above. But the system of moments is no longer positive. This is because the integation measure is no longer positive: $\mathcal{FP} \int_0^b f(x) \frac{dx}{x^2}$ can be negative even when $f(x)$ is positive. The moments of the measure are given by

$$\mathcal{FP} \int_0^b x^k \frac{dx}{x^2} = \frac{b^{k-1}}{k-1} \text{ for } k \neq 1 \tag{C.4}$$

and

$$\mathcal{FP} \int_0^b x^k \frac{dx}{x^2} = \log b \text{ for } k = 1. \tag{C.5}$$

Note that the zeroth moment is negative. Also the second moment violates scale invariance and is negative if $b < 1$.

Given a system of points x_j we can approximate the singular integral by a sum

$$\mathcal{FP} \int_0^b f(x) \frac{dx}{x^2} = \sum_{j=1}^{\nu} w_j f(x_j) \tag{C.6}$$

where the weights are determined as above by solving the system

$$\mu_k = \sum_{j=1}^{\nu} w_j x_j^k \text{ for } k = 0, \cdots \nu - 1. \tag{C.7}$$

The choice of equally spaced points gives good answers in many cases.

These methods are used in the text (towards the end of the second chapter) to solve the integrals equations for the wavefunction of the baryon.

Bibliography

[1] J.D. Bjorken, *Partons*, Invited talk Presented at the Int. Conf. on Duality and Symmetry in Hadron Physics, Tel-Aviv, 1971

[2] R. P. Feynman, *Photon-Hadron Interacions* Benjamin, Reading (1972).

[3] A. P. Balachandran, V. P. Nair, S. G. Rajeev and A. Stern, Phys. Rev. Lett., 49,1124 (1982); Phys. Rev. D27,1153 (1983); S. G. Rajeev, Phys. Rev. D29, 2844 (1984).

[4] S. G. Rajeev, *Baryons as Solitons* Ph. D. Thesis Syracuse University (1984).

[5] A. P. Balachandran, in *Proceedings of the Yale Theoretical Advanced Study Institute, High Energy Physics 1985* ed. M. J. Bowick and F. Gursey, World Scientific, Singapore (1986).

[6] E. Witten, Nucl. Phys. B223, 422 (1983); B223, 433 (1983).

[7] G. 't Hooft Nucl. Phys. B72, 461 (1974);B75 461 (1974).

[8] E. Witten, Nucl. Phys. B160, 57 (1979).

[9] S.G. Rajeev, 'In 1991 Summer School on High Energy Physics and Cosmology', ed. E. Gava, et. al. World Scientific, Singapore, (1992)

[10] P.F. Bedaque, I. Horvath, S.G. Rajeev, Mod.Phys.Lett.A7:3347-3356,1992;hep-th/9209027.

[11] S.G. Rajeev, Int.J.Mod.Phys.A9:5583,1994;hep-th/9401115.

[12] K. S. Gupta and S. Guruswamy and S. G. Rajeev,Phys. Rev. D48, 3354 (1993);hep-ph/9301208.

[13] S.G. Rajeev and O.T. Turgut, Comm. Math. Phys. 192, 493-517, (1998); hep-th/9705103.

[14] G. Krishnaswami and S. G. Rajeev, Phys. Lett. 441, 429 (1998).

[15] V. John, G. Krishnaswami and S. G. Rajeev *Derivation of the Anti-QuarkDistribution function of the Proton from Quantum ChromoDynamics*, in preparation.

[16] C. W. H. Lee and S. G. Rajeev, Phys. Rev. Lett. 80, 2285-2288(1998); Nucl. Phys. B 529, 656,(1998); J. Math. Phys. 39, 5199 (1998); J. Math. Phys 40, 1870 (1999); Phys. Lett. B 436, 91(1998).

[17] R. Brock *et al.* [CTEQ Collaboration], "Handbook of perturbative QCD: Version 1.0," Rev. Mod. Phys. **67**, 157 (1995).

[18] G. ' Hooft, Unpublished lectures at the Les Houches Summer School (1973)

[19] D. Gross and F. Wilczek, Phys. Rev. Lett. 30, 1343 (1973).

[20] D. Politzer, Phys. Rev. Lett. 30, 1346 (1973).

[21] D. Politzer, Phys. Rep. 14,129 (1974).

[22] A. H. Mueller, Phys. Rep. 73,237-368 (1981); J. C. Collins and D. E. Soper Nucl. Phys. B194 445 (1982)

[23] S. S. Chern, *Complex Manifolds without Potential Theory*, Springer-Verlag, New York (1979).

[24] J. Mickelsson, *Current Algebras and Groups*, Plenum (1989).

[25] A. Pressley and G. Segal, *Loop Groups* Clarendon Press, Oxford (1986).

[26] W. Hackbush, *Integral Equations: Theory and Numerical Treatment* Birkhauser, Boston 1995. Verlag

[27] V. Kac and D. H. Peterson, Proc. Natl. Acad. Sci. USA 78, 3308 (1981).

[28] G. Segal, Comm. Math. Phys. 80, 301 9(1981).

[29] D. Pickrell, Journ. Funct. Anal. **70** (1987) 323.

[30] A. D. Martin, R. G. Roberts, W. J. Stirling and R. S. Thorne, hepph/9803445.

[31] M. Glück, E. Reya, Vogt, Z. Phys. C67(1995) 433.

[32] W. Kwong, J. L. Rosner and C. Quigg, Ann. Rev. Nucl. Part. Sci. 37, 325 (1987).

[33] I. C. Gohberg and M. G. Krein, *Introduction to the theory of Linear Non-Self Adjoint Operators*, Translations of the Amer. Math. Soc., 18 (1969).

[34] B. Simon, *Trace Ideals and Their Applications*, Cambridge University Press, Cambridge, (1979).

[35] P. J. Davis and P. Rabinowitz *Methods of Numerical Integration*, Academic Press, Orlando Florida (1984).

[36] S. Chandrashekhar, *Radiative Transfer* Dover, (1960).

Monstrous Moonshine
and the Classification of CFT

(16 lectures given in Istanbul, August 1998)

Terry Gannon

Department of Mathematical Sciences, University of Alberta,
Edmonton, Alberta, Canada, T6G 2G1
e-mail: tgannon@math.ualberta.ca

Abstract

In these notes we give an introduction both to Monstrous Moonshine and to the classification of rational conformal field theories, using this as an excuse to explore several related structures and go on a little tour of modern math. We will discuss Lie algebras, modular functions, the finite simple group classification, vertex operator algebras, Fermat's Last Theorem, category theory, (generalised) Kac-Moody algebras, denominator identities, the A-D-E meta-pattern, representations of affine algebras, Galois theory, etc. This work is informal and pedagogical, and aimed mostly at grad students in math or math phys, but I hope that many interested nonexperts will find something of value here — like any good Walt Disney movie I try not to completely ignore the 'grown-ups'. My emphasis is on ideas and motivations, so these notes are intended to complement other papers and books where this material is presented with more technical detail. The level of difficulty varies significantly from topic to topic. The two parts — in fact any of the sections — can be read independently.

Table of Contents

Part 1. The classification of conformal field theories

Part 2. Monstrous Moonshine

Glossary

Part 1. The classification of conformal field theories

l.1. INFORMAL MOTIVATION

In this section we will sketch a very informal and 'hand-wavy' motivation to what we shall call the classification problem for rational conformal field theory (RCFT). Much of this material is more carefully treated in e.g. [18].

A CFT is a quantum field theory (QFT), usually with a two-dimensional space-time, whose symmetries include the conformal transformations. There are different approaches to CFT — for one of these see [26,27]. Another formulation which has been deeply influential s due to Graeme Segal [52]. It is motivated by string theory and is phrased in an important mathematical language called *category theory*.

A *category* consists of two types of things. One are called *objects*, and the other are called *arrows* (or *morphisms*). An arrow, written $f : A \to B$, has an initial and a final object (A and B respectively). Arrows f, g can be composed to yield a new arrow $f \circ g$, f the final object of g equals the initial object of f. Maps between categories are called *functors* if they take the objects (resp., arrows) of one to the objects (resp., arrows) of the other, and preserve composition.

The only difficulty people can have in understanding categories is in realising that there is no real content to them. It's just a language, highly abstract like the more familiar set theory, but in many contexts (a great example is the theory of knot invariants [58]) one which is both natural and suggestive. It tries to deflect some of our instinctive infatuation with objects (nouns), to the mathematically more fruitful one with structure-preserving maps between objects (verbs). A gentle introduction to the mathematics and philosophy of categories is [43]; we'll give a taste of this shortly.

The standard example of a category is called **Set**, where the 'objects' are sets, and the arrows from A to B are functions $A \to B$. A related example that Segal uses is **Vect**, where the objects are complex vector spaces and the arrows are linear maps. A rather trivial example of a functor $\mathcal{F} :$ **Vect** $\to$ **Set** sends a vector space V to its underlying set, also called V — i.e. $\mathcal{F}$ simply 'forgets' the vector space structure on V and ignores the fact that the arrows f in **Vect** are linear. The other category Segal uses he calls **C**; its objects are disjoint unions of (parametrised) circles S^1, and the arrows are (conformal equivalence classes of) *cobordisms*, i.e. (Riemann) surfaces whose boundaries are those circles. Composition of arrows in **C** is accomplished by gluing the surfaces along the appropriate boundary circles.

Consider the usual definition of a one-to-one function: $f(x) = f(y)$ only when $x = y$. Category theory replaces this with the following. The arrow $f : A \to B$ is called 'monic' if for any arrow $g : C \to B$, there exists a unique arrow $h : C \to A$ such that $f \circ h = g$. So it's a sort of factorisation property. You can easily verify that in **Set** the notions of 'one-to-one' and 'monic' coincide. What does this redefinition gain us? It certainly doesn't seem any simpler. But it does change the focus from the *argument* of f, to the *global* functional behaviour of f, and a change of perspective can never be bad. And it allows us to transport the idea of one-to-one-ness to arbitrary categories. For instance, in the Riemann surface category **C**, the 'one-to-one functions' are the genus-0 cobordisms.

1

Or consider the notion of *product*. In category theory, we say that the triple (P, a, b) is a product of objects A, B if $a : P \to A$ and $b : P \to B$ are arrows, and if for any $f : C \to A$, $g : C \to B$, there is a unique arrow $h : C \to P$ such that $f = a \circ h$ and $g = b \circ h$. This notion unifies several constructions (each of which is the 'product' in an appropriately chosen category): Cartesian product of sets; intersection of sets; multiplication of numbers; the logical operator 'and'; direct product; infimum in a partially ordered set; etc. *Sum* can be defined similarly, unifying the constructions of disjoint union, 'or', addition, tensor product, direct sum, supremum, etc.

This generality of course comes with a price: it can wash away all of the endearing special features of a favourite theory or structure. There certainly are contexts where e.g. all human beings should be thought of as equal, but there are other contexts where the given human is none other than your mother and must be treated as such. It turns out that category theory provides a beautiful framework for understanding topological invariants such as the Jones-Reshetikhin-Turaev-Witten knot invariants (see e.g. [58]). And it seems to be a natural language for formulating CFT axiomatically, as we'll now see.

According to Segal, a CFT is a functor $\mathcal{T}$ from **C** to **Vect**, which obeys various properties. The picture comes from string theory: the fundamental object is a 'vibrating' loop; a state is given by a collection of these loops; each classical Feynman path from the initial to the final states is a world-sheet, i.e. a surface Σ whose boundary is all the loops in the initial and final states. QFT assigns a complex number I (the action) to each of these world-sheets, and the quantum amplitude, written $\langle \text{final}|\text{initial}\rangle$, will then be the integral over all worldsheets of $e^{iI/\hbar}$. (The quantum amplitude is how the theory makes contact with experiment, as it tells us the probability of the given process $|\text{initial}\rangle \mapsto |\text{final}\rangle$ happening.) This is what Segal is trying to capture formally. The vector spaces in **Vect** come in in order to handle uniformly and simultaneously the various 'vibrations' of the strings. In particular there is one basic vector space H (a Hilbert space of quantum states), and the functor will take n copies of S^1 to $H^n = H \otimes \cdots \otimes H$.

The simplest interesting example here is the 'tree-level creation of a string from the vacuum'. In this case the world-sheet looks like a bowl, i.e. *topologically* is a disk D (if we imagine the bowl to be made of rubber, we could grab its rim and stretch it down flat onto the table, so we say a bowl and a disk are topologically equivalent — see also §2.3). Segal's functor gives us a linear map $\mathcal{T}(D) : \mathbb{C} \to H$ (H^0 is just $\mathbb{C}$), which we can think of equivalently as the assignment of the vector $\mathcal{T}(D)(1)$ in H to D. In the case of the standard unit disk (i.e. where the parametrisation of the boundary S^1 is simply $\theta \mapsto e^{2\pi i\theta}$), this vector is cal0led the *vacuum state* $\Omega = |0\rangle$.

For another example, consider the 'vacuum-to-vacuum expectation value'. The initial and final states (objects) here are both the empty set, so the world-sheets (arrows) are closed Riemann surfaces. As usual in QFT, we can organise these by how many internal 'loops' are involved (this number is called the *genus* of the surface): topologically, 0-loop (i.e. 'tree-level') world-sheets are spheres, 1-loop world-sheets are tori, etc. These closed Riemann surfaces are discussed in more detail in §2.3. The 0-loop contribution isn't very interesting (there is only one conformal equivalence class of spheres), so let us look at the 1-loop contribution. It will be of the form $\int \mathcal{Z}([torus]) \, d[torus]$, where $\mathcal{Z}$ is a complex-valued function called the partition function, and $[torus]$ is a conformal equivalence class of tori. In the Segal formalism we recover this in the following way: the functor takes [*torus*] to a linear function from $H^0 = \mathbb{C}$ to $H^0 = \mathbb{C}$. Any such linear function is simply a 1×1 matrix, i.e. a complex number, which we call $\mathcal{Z}([torus])$.

Now there is a nice parametrisation of conformal equivalence classes of tori, as we will see more explicitly in §2.3. Namely, a representative for each class can be chosen to be of the form $\mathbb{C}/(\mathbb{Z} + \mathbb{Z}\tau)$ where τ is in the upper half plane $\mathcal{H}$. Thus we can write $\mathcal{Z}$ as a function of a complex variable τ. However, different τ correspond to the same equivalence class of tori: the redundancy is exactly captured by the modular group $\mathrm{PSL}_2(\mathbb{Z})$. Namely, τ and $\frac{a\tau+b}{c\tau+d}$ are equivalent, whenever $a, b, c, d \in \mathbb{Z}$ and $ad - bc = 1$. Thus $\mathcal{Z}(\tau) = \mathcal{Z}(\frac{a\tau+b}{c\tau+d})$. In other words, the partition function is modular invariant![1]

There are two sectors in CFT, a holomorphic one and an antiholomorphic one, corresponding to the two directions ('left-' and 'right-moving') of motion on a string, or the two components of the group $\mathrm{Diff}(S^1)$ of diffeomorphisms of the circle. This means that many of the quantities (e.g. the partition function) factorise into parts depending holomorphically and anti-holomorphically on the modular parameters (e.g. τ in genus 1). In a *rational* CFT there are finitely many 'primary fields' $a \in \Phi$ — the precise meaning of this is not important here, but it says that the space of states for the theory decomposes into a finite sum[2] $H = \oplus_{a,b \in \Phi} M_{ab} H_a \otimes \overline{H}_b$, where M_{ab} are nonnegative integers which count the multiplicity of $H_a \otimes \overline{H}_b$ in H. The linear maps $\mathcal{T}(\Sigma) : H^m \to H^n$ in an RCFT will factorise similarly; this 'chiral factorisation' is captured by what Segal calls the 'modular functor' [52]. The partition function becomes

$$\mathcal{Z}(\tau) = \sum_{a,b \in \Phi} M_{ab}\, \chi_a(\tau)\, \chi_b(\tau)^* \tag{1.1.1}$$

for certain holomorphic functions χ_a. One of the primary fields (we'll denote it '0') corresponds to the vacuum Ω, and uniqueness of the vacuum means that $M_{00} = 1$.

H_0 is called a *chiral algebra*; in the language of §2.6, H_0 will be a vertex operator algebra (VOA). Φ parametrises the irreducible H_0-modules and the χ's are their characters; in an RCFT we require this number to be finite. For example, for the Moonshine VOA $V^\natural$ discussed in Part 2, Φ consists of only one element.

The higher-genus behaviour of an RCFT is determined from the lower-genus behaviour, by composition of 'arrows' (i.e. the gluing together of surfaces) in **C**. See Figure 3 of [30] to find how a genus-2 surface is built up from genus-0 ones. In fact, it's generally believed that an RCFT will be uniquely determined by: (i) the choice of chiral algebra; (ii) the partition function (which tells you the spectrum of the theory, i.e. how the two sectors link up); and (iii) the structure constants C_{ab}^c, which in the Segal formalism correspond to the surfaces called 'pairs-of-pants', equivalently disks with two interior disks removed. Our approach will be to start with a chiral algebra, and find all possible partition functions. We will thus ignore the important question of existence and uniqueness of the structure

---

[1] In higher-dimensional string theories, a similar argument shows more generally that *automorphic forms* will appear naturally.

[2] It seems though that 'rational' logarithmic CFT is trying to teach us the lesson that this familiar requirement can and should be weakened. See Gaberdiel-Kausch (1999).

constants, though at least for our chiral algebras, it seems to be generally believed that the structure constants will be unique.

Perhaps all chiral algebras come from standard constructions (e.g. orbifolds and the Goddard-Kent-Olive (GKO) coset construction — see e.g. [18]) involving lattices and affine Kac-Moody algebras. For instance a $\mathbb{Z}_2$-orbifold of the VOA of the Leech lattice gives us the Moonshine module $V^\natural$, and the so-called minimal models come from GKO cosets involving $A_1^{(1)}$. This is in line with the spirit of Tannaka-Krein duality (and its generalisations by Deligne and Doplicher-Roberts), which roughly says that if a bunch of things *act like* they're the set of representations of a Lie group, then they *are* the set of representations of a Lie group.

In any case, one of the simplest, best understood, and important classes (called Wess-Zumino-Witten (WZW) models — see for instance [30,59] in this volume) of RCFTs correspond to affine Kac-Moody algebras at a positive integer level k. We will have much more to say later about these algebras, but for now let us remark that Φ here will be the (finite) set P_+^k of integrable level k highest weights λ. Their chiral algebras were constructed by Frenkel and Zhu. The following sections concern the attempt to classify the partition functions corresponding to Kac-Moody algebras — see especially §1.5. I will use this theme as an excuse to describe many other things, e.g. the A-D-E meta-pattern, Lie theory, Galois, fusion rings, ... I dedicate these notes to the profound friendship developing in recent years between mathematics and physics. As Victor Kac said in his 1996 Wigner medal acceptance speech, "Some of the best ideas come to my field from the physicists. And on top of this they award me a medal. One couldn't hope for a better deal."

1.2. LIE ALGEBRAS

Lie algebras (and their nonlinear partners *Lie groups*) appear in numerous places throughout math and mathematical physics. A nice introduction is [9]; Lie theory is presented with more of a physics flavour in [24], as well as [59].

An *algebra* is a vector space with a way to multiply vectors which is compatible with the vector space structure (i.e. the vector-valued product is *bilinear*: $(a\vec{u}+a'\vec{u}')(b\vec{v}+b'\vec{v}') = ab\,\vec{u}\vec{v} + ab'\,\vec{u}\vec{v}' + a'b\,\vec{u}'\vec{v} + a'b'\,\vec{u}'\vec{v}'$). For example, the complex numbers $\mathbb{C}$ can be thought of as a 2-dimensional algebra over $\mathbb{R}$ (a basis is 1 and i $= \sqrt{-1}$; the *scalars* here are real numbers and the *vectors* are complex numbers). The quaternions are 4-dimensional over $\mathbb{R}$ and the octonions are 8-dimensional over $\mathbb{R}$. Incidentally, these are the only finite-dimensional algebras over $\mathbb{R}$ which obey the cancellation law: $\vec{u} \neq 0$ and $\vec{u}\vec{v} = 0$ implies $\vec{v} = 0$ (the reader should try to convince himself why the familiar vector product on $\mathbb{R}^3$ fails the cancellation law). This important little fact makes several unexpected appearances in math. For instance, it is trivially possible to 'comb the hair' on the circle S^1 without 'cheating' (i.e. needing a hair-part or exploiting a bald spot): just comb the hair clockwise for example. However it is not possible to comb the hair on the sphere S^2 (e.g. your own head) without cheating. The only other k-spheres S^k which can be combed (i.e. for which there exist k linearly independent continuous vector fields) are $k = 3$ and 7. This is intimately connected with the existence of $\mathbb{C}$, the quaternions, and octonions (namely,

S^1, S^3, S^7 can be thought of as the length 1 complex numbers, quaternions, and octonions, resp.).

In a *Lie* algebra $\mathfrak{g}$, the product is usually called a 'bracket' and is written $[xy]$. It is required to be 'anti-commutative' and 'anti-associative':

$$[xy] + [yx] = 0 \qquad (1.2.1a)$$
$$[x[yz]] + [y[zx]] + [z[xy]] = 0 \qquad (1.2.1b)$$

(like most other equalities in math, (1.2.1b) is usually called the *Jacobi identity*). Usually we will consider Lie algebras over $\mathbb{C}$, but sometimes over $\mathbb{R}$. Note that (1.2.1a) says $[xx] = 0$.

One important consequence of bilinearity is that it is enough to know the values of all the brackets $[x_i x_j]$ for $i < j$, for any basis $\{x_1, x_2, \ldots\}$ of $\mathfrak{g}$. (The reader should convince himself of this before proceeding.)

The simplest example of a Lie algebra is $\mathfrak{g} = \mathbb{C}$ (or $\mathfrak{g} = \mathbb{R}$), with the bracket $[xy]$ identically 0. In fact, this is the only 1-dimensional Lie algebra. It is a straightforward exercise for the reader to find all 2- and 3-dimensional Lie algebras (over $\mathbb{C}$) up to isomorphism (i.e. change of basis): there are precisely 2 and 6 of them, respectively (though one of the 6 depends on a complex parameter). Over $\mathbb{R}$, there are 2 and 9 (with 2 depending on real parameters). This exercise cannot be continued much further — e.g. not all 7-dimensional Lie algebras (over $\mathbb{C}$) are known. Nor is it obvious that this would be an interesting or valuable exercise. We should suspect that our definition of Lie algebra is probably a little too general for anything obeying it to be automatically an interesting structure. More often than not, a classification turns out to be a stale and useless list.

Two of the 3-dimensional Lie algebras are important in what follows. One of them is well-known to the reader: consider the vector-product (also called cross-product) in $\mathbb{C}^3$. Taking the standard basis $\{e_1, e_2, e_3\}$ of $\mathbb{C}^3$, the bracket can be defined by the relations

$$[e_1 e_2] = e_3 , \qquad [e_1 e_3] = -e_2 , \qquad [e_2 e_3] = e_1 . \qquad (1.2.2a)$$

This Lie algebra, denoted A_1 or $\mathrm{sl}_2(\mathbb{C})$, can be called the 'mother of all (semi-simple) Lie algebras'. A more familiar realisation of A_1 uses a basis $\{e, f, h\}$ with relations

$$[ef] = h , \qquad [he] = 2e , \qquad [hf] = -2f . \qquad (1.2.2b)$$

The reader can find the change-of-basis (valid over $\mathbb{C}$ but not $\mathbb{R}$) showing that (1.2.2) define isomorphic *complex* (but not *real*) Lie algebras.

Another important 3-dimensional Lie algebra is called the Heisenberg algebra[3] and is the algebra of the canonical commutation relations in quantum mechanics: choosing a basis x, p, h, it is defined by

$$[xp] = h , \qquad [xh] = [ph] = 0 . \qquad (1.2.3)$$

---

[3] Actually, 'Heisenberg algebra' refers to a family of Lie algebras, with (1.2.3) being the one of lowest dimension.

From our definition, it is far from clear that Lie algebras, as a class, should be natural and worth studying. After all, there are infinitely many possible axiomatic systems: why should the one defining a Lie algebra be anything special *a priori*? Perhaps this could have been anticipated by the following line of reasoning.

Axiom. Groups are important and interesting.

Axiom. Manifolds are important and interesting.

Manifolds are structures where calculus is possible; locally, a manifold looks like a piece of $\mathbb{R}^n$ (or $\mathbb{C}^n$), but these pieces can be bent and stitched together to create more interesting shapes. For instance a circle is a 1-dimensional manifold, while Einstein claimed space-time is a curved 4-dimensional one.

Definition. A Lie group is a manifold with a compatible group structure.

This means that 'multiplication' and 'inverse' are differentiable maps. $\mathbb{R}$ is a Lie group, under addition: obviously, $\mu : \mathbb{R}^2 \to \mathbb{R}$ and $\iota : \mathbb{R} \to \mathbb{R}$ defined by $\mu(a,b) = a + b$ and $\iota(a) = -a$ are both differentiable. (Why isn't $\mathbb{R}$ a Lie group under multiplication?) A circle is also a Lie group: parametrise the points with the angle θ defined mod 2π (or mod 360 if you prefer); the 'product' of the point at angle θ_1 with the point at angle θ_2 will be the point at angle $\theta_1 + \theta_2$. Surprisingly, the only other k-sphere which is a Lie group is S^3 (the product can be defined using quaternions of unit length[4], or using the matrix group $\text{SU}_2(\mathbb{C})$). Many but not all Lie groups can be expressed as matrix groups. Two other examples are GL_n (invertible $n \times n$ matrices) and SL_n (ones with determinant 1).

A consequence of the above axioms is then surely:

Corollary. Lie groups should be important and interesting.

Lie group structure theory can be thought of as a major generalisation of linear algebra. The basic constructions familiar to undergraduates have important analogues valid in many Lie groups. For instance, years ago we were taught to solve linear equations and invert matrices by using elementary row operations to reduce a matrix to row-echelon form. What this says is that any matrix $A \in \text{GL}_n(\mathbb{C})$ can be factorised $A = BPN$, where N is uppertriangular with 1's on the diagonal, P is a permutation matrix, and B is an uppertriangular matrix. This is essentially what is called the Bruhat decomposition of the Lie group $\text{GL}_n(\mathbb{C})$. More generally (where it applies to any 'reductive' Lie group G), P will be an element of the so-called 'Weyl group' of G (of which we'll have much more to say later), and B will be in a 'Borel subgroup'.

Lie groups appear throughout physics. E.g. the orthogonal group $\text{SO}_3(\mathbb{R})$ is the configuration space of a rigid body centred at the origin, while $\text{SU}_2(\mathbb{C})$ is the set of states of an electron at rest. The gauge group of the Standard Model of particle physics is $\text{SU}_3(\mathbb{C}) \times \text{SU}_2(\mathbb{C}) \times \text{U}_1(\mathbb{C})$, while the Lorentz group of special relativity is $\text{SO}_{3,1}(\mathbb{R})$.

There is an important relation between Lie groups and Lie algebras.

Fact. The tangent space of a Lie group is a Lie algebra. Any (finite-dimensional real or complex) Lie algebra is the tangent space to some Lie group.

---

[4] Similarly, the 7-sphere inherits from the octonions a *nonassociative* (hence nongroup) product, compatible with its manifold structure.

More precisely, the tangent space at 1 (i.e. the set of all infinitesimal generators of the Lie group) can be given a natural Lie algebra structure. A Lie algebra, being a linearised Lie group, is much simpler and easier to handle. The Lie algebra preserves the local properties of the Lie group, though it loses global topological properties (like boundedness). A Lie group has a single Lie algebra, but a Lie algebra can correspond to many different Lie groups. The Lie algebra corresponding to both $\mathbb{R}$ and S^1 is $\mathfrak{g} = \mathbb{R}$ with trivial bracket. The Lie algebra corresponding to both $S^3 = SU_2(\mathbb{C})$ and $SO_3(\mathbb{R})$ is the cross-product algebra on $\mathbb{R}^3$ (usually called $\mathfrak{so}_3(\mathbb{R})$). Given the above fact, a safe guess would be:

Conjecture. Lie algebras are important and interesting.

·From this line of reasoning, it should be expected that historically Lie groups arose first. Indeed that is the case: the Norwegian Sophus Lie introduced them in 1873 to try to develop a Galois theory for ordinary differential equations. As the reader may be aware, Galois theory is used for instance to show that not all 5th degree (or higher) polynomials can be explicitly 'solved' using radicals — we will meet Galois theory in §1.8. Lie wanted to study the explicit solvability (integrability) of differential equations, and this led him to develop what we now call Lie theory. The importance of Lie groups however have grown well beyond this initial motivation.

An important class of Lie algebras are the so-called finite-dimensional *simple* ones. Their definition and motivation will be studied in §2.7 below, but in a certain sense they serve as building blocks for all other finite-dimensional Lie algebras.

The classification of simple finite-dimensional Lie algebras over $\mathbb{C}$ is quite important and was accomplished at the turn of the century by Killing and Cartan. There are 4 infinite families A_ℓ ($\ell \geq 1$), B_ℓ ($\ell \geq 3$), C_ℓ ($\ell \geq 2$), and D_ℓ ($\ell \geq 4$), and 5 exceptionals E_6, E_7, E_8, F_4 and G_2. A_ℓ can be thought of[5] as $\mathfrak{sl}_{\ell+1}(\mathbb{C})$, the $(\ell+1) \times (\ell+1)$ matrices with trace 0. The orthogonal algebras B_ℓ and D_ℓ can be identified with $\mathfrak{so}_{2\ell+1}(\mathbb{C})$ and $\mathfrak{so}_{2\ell}(\mathbb{C})$, resp., where $\mathfrak{so}_n(\mathbb{C})$ is all $n \times n$ anti-symmetric matrices $A^t = -A$. The symplectic algebra C_ℓ is $\mathfrak{sp}_{2\ell}(\mathbb{C})$, i.e. all $2\ell \times 2\ell$ matrices A obeying $A\Omega = -\Omega A^t$, where $\Omega = \begin{pmatrix} 0 & I_\ell \\ -I_\ell & 0 \end{pmatrix}$ and I_ℓ is the identity matrix. The exceptionals can be constructed e.g. using the octonions. In all these cases the bracket is given by the commutator

$$[AB] = [A, B] := AB - BA \tag{1.2.4}$$

(it is a good exercise for the reader to confirm that the commutator satisfies (1.2.1), and that e.g. $\mathfrak{sl}_n(\mathbb{C})$ is indeed closed under it). To see that (1.2.2b) truly is $\mathfrak{sl}_2(\mathbb{C})$, put

$$e = \begin{pmatrix} 0 & 1 \\ 0 & 0 \end{pmatrix}, \qquad f = \begin{pmatrix} 0 & 0 \\ 1 & 0 \end{pmatrix}, \qquad h = \begin{pmatrix} 1 & 0 \\ 0 & -1 \end{pmatrix}. \tag{1.2.5}$$

Incidentally the names A, B, C, D have no significance: since the 4 series start at $\ell = 1, 2, 3, 4$, it seemed natural to call these A, B, C, D, resp. Unfortunately a bit of bad luck happened: B_2 and C_2 are isomorphic and so at random that algebra was placed in the orthogonal series; however affine Dynkin diagrams make it clear that it really is a

---

[5] Strictly speaking these are *representations* (see next section).

symplectic algebra which accidentally looks orthogonal; hence in hindsight the names of the B- and C-series really should have been switched.

This classification changes if the *field* — the choice of scalars=numbers — is changed. By a field, we mean we can add, subtract, multiply and divide, such that all the usual properties like commutativity and distributivity are obeyed. Fields will make a few different appearances in these notes. $\mathbb{C}$, $\mathbb{R}$, and $\mathbb{Q}$ are fields, while $\mathbb{Z}$ is not (you can't always divide an integer by e.g. 3, and remain in $\mathbb{Z}$). The integers mod n, which we will write $\mathbb{Z}_n$, are a field iff n is prime (the reader can verify that in e.g. $\mathbb{Z}_4$, it is not possible to divide by the field element $[2] \in \mathbb{Z}_4$ even though $[2] \neq [0]$ there). $\mathbb{C}$ and $\mathbb{R}$ are examples of fields of characteristic 0 — this means that 0 is the only integer k with the property that $kx = 0$ for all x in the field. $\mathbb{Z}_p$ is the simplest example of a field with nonzero characteristic: in $\mathbb{Z}_p$, multiplying by the integer p has the same effect as multiplying by 0, and so we say $\mathbb{Z}_p$ has characteristic p. Strange fields have important applications in e.g. coding theory and, ironically, in number theory itself — see e.g. §1.8.

As always, $\mathbb{C}$ is better behaved than e.g. $\mathbb{R}$ because every polynomial can be factorised over $\mathbb{C}$ (we say $\mathbb{C}$ is *algebraically closed*) — this implies for example that every matrix has an eigenvector over $\mathbb{C}$ but not necessarily over $\mathbb{R}$. Over $\mathbb{R}$, the difference in the simple Lie algebra classification is that each symbol $X_\ell \in \{A_\ell, \ldots, G_2\}$ corresponds to a number of inequivalent algebras (over $\mathbb{C}$, each algebra has its own symbol). For example, 'A_1' corresponds to 3 different real simple Lie algebras, namely the matrix algebras $\mathrm{sl}_2(\mathbb{R})$, $\mathrm{sl}_2(\mathbb{C})$ (interpreted as a *real* vector space), and $\mathrm{so}_3(\mathbb{R})$. The simple Lie algebra classification has recently been done in any characteristic $p > 7$. It is surprising but very common that the smaller primes behave very poorly, and the classification for characteristic 2 is probably completely hopeless.

Associated with each simple algebra X_ℓ is a Weyl group, and a (Coxeter-)Dynkin diagram. The Weyl group is a finite reflection group, e.g. for A_ℓ it is the symmetric group $\mathfrak{S}_{\ell+1}$. See Figure 7 of [59] for the Weyl group of A_2. The Dynkin diagram of X_ℓ (see e.g. [24,36,38] or Figure 6 in [59]) is a graph with ℓ nodes, and with possibly some double and triple edges. It says how to construct X_ℓ abstractly using generators and relations — see §2.7. We will keep meeting both throughout these notes.

Another source of Lie algebras are the vector fields Vect(M) on a manifold M. A vector field v is a choice (in a smooth way) of a tangent vector $v(p) \in T_p M$ at each point of M. It can be thought of as a (1st order) differential operator, acting on functions $f : M \to \mathbb{R}$ (or $f : M \to \mathbb{C}$); at each point $p \in M$ take the directional derivative of f in the direction $v(p)$. For example the vector fields on the circle, Vect(S^1), can be thought of as anything of the form $g(\theta)\frac{d}{d\theta}$ where $g(\theta)$ can be any function with period 1. We can compose vector fields $u \circ v$, but this will result in a 2nd order differential operator: e.g.

$$(f(\theta)\frac{d}{d\theta}) \circ (g(\theta)\frac{d}{d\theta}) = f(\theta)\,g(\theta)\,\frac{d^2}{d\theta^2} + f(\theta)\,g'(\theta)\,\frac{d}{d\theta}\ .$$

Instead, the natural 'product' of vector fields is given by their commutator $[u, v] = u \circ v - v \circ u$, as it always results in a vector field: e.g.

$$[f(\theta)\frac{d}{d\theta}, g(\theta)\frac{d}{d\theta}] = (f(\theta)\,g'(\theta) - f'(\theta)\,g(\theta))\frac{d}{d\theta}$$

8

in Vect(S^1). Vect(M) with this bracket is an infinite-dimensional Lie algebra. In the case where M is a Lie group G, the Lie algebra of G can be interpreted as a certain finite-dimensional subalgebra of Vect(G) given by the 'left-invariant vector fields'.

Simple algebras need not be finite-dimensional. An example of an infinite-dimensional one is the *Witt algebra* $\mathcal{W}$, which can be defined (over $\mathbb{C}$) by the basis[6] L_n, $n \in \mathbb{Z}$, and the relations

$$[L_m L_n] = (m - n)L_{m+n} . \tag{1.2.6}$$

Using the realisation $L_n = -ie^{-in\theta}\frac{d}{d\theta}$, the Witt algebra can also be interpreted as the polynomial subalgebra of the complexification $\mathbb{C} \otimes \text{Vect}(S^1)$ — i.e. change the scalar field of Vect(S^1) from $\mathbb{R}$ to $\mathbb{C}$. Incidentally, infinite-dimensional Lie algebras need not have a corresponding Lie group: e.g. the real algebra Vect(S^1) is the Lie algebra of the Lie group $\text{Diff}^+(S^1)$ of orientation-preserving diffeomorphisms $S^1 \to S^1$, but $\mathbb{C}\otimes\text{Vect}(S^1)$ has no Lie group. $\text{Diff}^+(S^1)$ plays a large role in CFT, by acting on the objects of Segal's category C.

The Witt algebra appears naturally in CFT: e.g. using the realisation $L_n = -z^{n+1}\frac{d}{dz}$ it is the polynomial subalgebra of the Lie algebra Vect($\mathbb{C}/\{0\}$). Very carelessly, Vect($\mathbb{C}/\{0\}$) is often thought of as the infinitesimal conformal transformations on a suitable neighbourhood of 0 (yet clearly L_{-2}, L_{-3}, ... are singular at 0!). Indeed the CFT literature is very sloppy when discussing the conformal group in 2-dimensions. The unfortunate fact is that, contrary to claims, *there is no infinite-dimensional conformal group* for $\mathbb{C} \cong \mathbb{R}^2$. The best we can do is the 3-dimensional group $\text{PSL}_2(\mathbb{C})$ of Möbius transformations $z \mapsto \frac{az+b}{cz+d}$, which are orientation-preserving conformal transformations for the Riemann sphere $\mathbb{C} \cup \{\infty\}$. There seem to be 2 ways out of this rather embarrassing predicament. One is to argue that we are really interested in 'infinitesimal conformal invariance' in some meromorphic sense, so the full Witt algebra can appear. The other way is to argue that it is the conformal group of 'Minkowski space' $\mathbb{R}^{1,1}$ (or better, its compactification $S^1 \times S^1$) rather than $\mathbb{R}^2 \cong \mathbb{C}$ (or its compactification S^2) which is relevant for CFT. That conformal group *is* infinite-dimensional; for $S^1 \times S^1$ it consists of 2 copies of $\text{Diff}^+(S^1) \times \text{Diff}^+(S^1)$. For a more careful treatment of this point, see [51].

For reasons we will discuss in §1.4, we are more interested in the *Virasoro algebra* $\mathcal{V}$ rather than the Witt algebra $\mathcal{W}$. This is a '1-dimensional central extension' of $\mathcal{W}$; as a vector space $\mathcal{V} = \mathcal{W} \oplus \mathbb{C}C$ with relations given by

$$[L_m L_n] = (m - n)L_{m+n} + \delta_{n,-m}\frac{m(m^2 - 1)}{12}C \tag{1.2.7a}$$

$$[L_m C] = 0 . \tag{1.2.7b}$$

'1-dimensional central extension' means $\mathcal{V}$ has one extra basis vector C, which lies in the *centre* of $\mathcal{V}$ (i.e. $[xC] = 0$ for all $x \in \mathcal{V}$), and sending $C \to 0$ recovers $\mathcal{W}$ (i.e. takes (1.2.7a) to (1.2.6)). A common mistake in the physics literature is to regard C as a number: it is in fact a vector, though in many (but not all) representations it is mapped to a scalar multiple of the identity.

---

[6] In infinite dimensions, to avoid convergence complications, only finite linear combinations of basis vectors are generally permitted. Infinite linear combinations would involve taking some 'completion'.

The reason for the strange-looking (1.2.7a) is that we have little choice: $\mathcal{V}$ is the unique nontrivial 1-dimensional central extension of $\mathcal{W}$. The factor $\frac{1}{12}$ there is conventional but standard, and has to do with 'zeta-function regularisation' in string theory — i.e. the divergent sum $\sum_{n=1}^{\infty} n$ is 'reinterpreted' as $\zeta(-1) = \frac{-1}{12}$, where $\zeta(s) = \sum_{n=1}^{\infty} n^{-s}$ is the Riemann zeta function. Incidentally $\zeta(s)$ can be written in terms of the product $\prod_{p}(1 - p^{-s})^{-1}$ over all primes $p = 2, 3, 5, \ldots$ (try to see why); hence $\zeta(s)$ has a lot to do with primes, in particular their distribution. In fact the most famous unsolved problem in math today is the Riemann conjecture, which says that $\zeta(s) \neq 0$ whenever $\mathrm{Re}(s) \neq \frac{1}{2}$. One researcher recently described this conjecture as saying that the primes have music in them.

In CFT, L_0 is the energy operator. For example the partition function is given by $\mathcal{Z}(\tau) = \mathrm{Tr}_H(q^{L_0-c/24}q^{*L_0-c/24})$ and the (normalised) character χ_a equals $\mathrm{Tr}_{H_a}(q^{L_0-c/24})$ for $q = e^{2\pi i \tau}$. cI is the scalar multiple of the identity to which C gets sent; it has a physical interpretation [18] involving Casimir (vacuum) energy, which depends on space-time topology, and the strange shift by $c/24$ is due to an implicit mapping from the cylinder to the plane.

1.3. Representations of finite-dimensional simple Lie algebras

The representation theory of the simple Lie algebras[7] can perhaps be regarded as an enormous generalisation of trigonometry. For instance the facts that $\frac{\sin(nx)}{\sin(x)}$ can be written as a polynomial in $\cos(x)$ for any $n \in \mathbb{Z}$, and that

$$\frac{\sin(mx)\sin(nx)}{\sin(x)} = \sin((m+n)x) + \sin((m+n-2)x) + \cdots + \sin((m-n)x)$$

for any $m, n \in \mathbb{Z}_{>}$, are both easy special cases of the theory.

The classic example of an algebraic structure are the numbers, and they prejudice us into thinking that commutativity and associativity are the ideal. We have learned over the past couple of centuries that commutativity can often be dropped without losing depth and usefulness, but most interesting structures seem to obey some form of associativity. Moreover, true associativity (as opposed to e.g. anti-associativity) really simplifies the arithmetic. Given the happy 'accident' that the commutator $[x, y] := xy - yx$ in any associative algebra obeys anti-associativity, it would seem to be both tempting and natural to study the ways (if any) in which associative algebras $\mathfrak{A}$ can 'model' or represent a given Lie algebra. Precisely, we are looking for a map $\rho : \mathfrak{g} \to \mathfrak{A}$ which preserves the linear structure (i.e. ρ is a linear function), and which sends the bracket $[xy]$ in $\mathfrak{g}$ to the commutator $[\rho(x), \rho(y)]$ in $\mathfrak{A}$.

In practice groups (resp., algebras) often appear as symmetries (resp., infinitesimal generators of symmetries). These symmetries often act linearly. In other words, in practise the preferred associative algebras will usually be matrix algebras, and this is the usual

---

[7] See e.g. [25] for more details. Historically, representations of Lie algebras were considered even before representations of finite groups.

form for a representation and the only kind we will consider. The *dimension* of these representations is the size of the matrices.

Finding all possible representations, even for the simple Lie algebras, is probably hopeless. However, it is possible to find all *finite-dimensional* representations of the simple Lie algebras, and the answer is easy to describe. Given a simple Lie algebra X_ℓ, there is a representation L_λ for each ℓ-tuple $\lambda = (\lambda_1, \ldots, \lambda_\ell)$ of nonnegative integers. λ is called a *highest-weight*. Moreover, we can take direct sums $\oplus_i L_{\lambda^{(i)}}$ of finitely many of these representations. The matrices in such a direct sum will be in block form. It turns out that, up to change-of-basis, this exhausts all finite-dimensional representations of X_ℓ.

It is common to replace 'representation ρ of $\mathfrak{g}$' with the equivalent notion of '$\mathfrak{g}$-module M' — i.e. we think of the matrices $\rho(x)$ as linear maps $M \to M$. A $\mathfrak{g}$-*module* is a vector space on which $\mathfrak{g}$ acts (on the left). Instead of considering the matrix $\rho(x)$, we consider 'products' xv (think of this as the matrix $\rho(x)$ times the column vector v) for $v \in M$. This product must be bilinear, and must obey $[xy]v = x(yv) - y(xv)$.

To get an idea of what L_λ looks like, consider A_1. Recall its generators e, f, h and relations (1.2.2b). Choose any $\lambda \in \mathbb{C}$. Define $x_0 \neq 0$ to formally obey $hx_0 = \lambda x_0$ and $ex_0 = 0$. Define inductively $x_{i+1} := fx_i$ for $i = 0, 1, \ldots$. Define M_λ to be the span of all x_i — we will see shortly that they are linearly independent (so M_λ is infinite-dimensional). M_λ is a module of A_1: the calculations $hx_{i+1} = hfx_i = ([hf] + fh)x_i = (-2f + fh)x_i$ and $ex_{i+1} = efx_i = ([ef] + fe)x_i = (h + fe)x_i$ show inductively that $hx_m = (\lambda - 2m)x_m$ and $ex_m = (\lambda - m + 1)m\,x_{m-1}$. From these the reader can show that the x_i are linearly independent. M_λ is called a *Verma module*; λ is called its highest-weight, and x_0 is called a highest-weight vector.

Now specialise to $\lambda = n \in \mathbb{Z}_\geq := \{0, 1, 2, \ldots\}$. Note that $ex_{n+1} = 0$ and $hx_{n+1} = (-n-2)x_{n+1}$. This means that, for these n, M_n contains a *submodule* with highest-weight vector x_{n+1}, isomorphic to M_{-n-2}. x_{n+1} is called a *null vector*. In other words, we could set $x_{n+1} := 0$ and still have an A_1-module. We would then get a *finite-dimensional* module which we'll call $L_n := M_n/M_{-n-2}$ (not to be confused with the Virasoro generator in (1.2.7)). Its basis is $\{x_0, x_1, \ldots, x_n\}$ and so it has dimension $n + 1$.

For example, take $n = 1$. Note that what we get in terms of the basis $\{x_0, x_1\}$ is the familiar representation $\mathrm{sl}_2(\mathbb{C})$ given in (1.2.5).

The situation for the other simple Lie algebras X_ℓ is similar.

It turns out to be hard to compare representations: ρ and ρ' could be equivalent (i.e. differ merely by a change-of-basis) but look very different. Or if we are given a representation, we may want to decompose it into the direct sum of some $L_{\lambda^{(i)}}$. When working with representations, it is often very useful to avoid much of the extraneous basis-dependent detail present in the function ρ. Finite group theory suggests how to do this: we should use *characters*. The character of an A_1-module M is given by Weyl: write M as a direct sum of eigenspaces $M(m)$ of h; then define

$$\mathrm{ch}_M(z) := \sum_m \dim M(m)\, e^{mz}\,, \tag{1.3.1}$$

for any $z \in \mathbb{C}$. The m are called *weights* and the $M(m)$ *weight-spaces*. For example, for

L_n the weights are $m = n, n - 2, \ldots, -n$, the weight-spaces $L_n(m)$ are $\mathbb{C}x_{(n-m)/2}$, and

$$\mathrm{ch}_n(z) = \sum_{i=0}^{n} e^{(n-2i)z} = \frac{\sin((n + 1)z)}{\sin(z)} . \tag{1.3.2}$$

Analogous formulas apply to any algebra X_ℓ: the character will then be a function of an ℓ-dimensional subspace $\mathfrak{h}$ called the *Cartan subalgebra*, spanned by all the h_i (see §2.7), so can be thought of as a complex-valued function of ℓ complex variables. The weights m will lie in the dual space to $\mathfrak{h}$ — i.e. are linear maps $\mathfrak{h} \to \mathbb{C}$ — so will have ℓ components. See for instance Figure 8 in [59]. Incidentally, ℓ is called the *rank* of X_ℓ.

Weyl's definition works: two representations are equivalent iff their characters are identical, and $M = \oplus_i L_{\lambda^{(i)}}$ iff $\mathrm{ch}_M(z) = \sum_i \mathrm{ch}_{\lambda^{(i)}}(z)$. It also is enormously simpler: e.g. the smallest nontrivial representation of E_8 is a map from $\mathbb{C}^{248}$ to the space of 248×248 matrices, while its character is a function $\mathbb{C}^8 \to \mathbb{C}$. But why is Weyl's definition natural? How did he come up with it?

To answer that question, we must remind ourselves of the characters of finite groups[8]. A representation of a finite group G is a structure-preserving map ρ (i.e. a group homomorphism) from G to matrices. The group's product becomes matrix product. In these notes we will be exclusively interested in group representations over $\mathbb{C}$. Two representations ρ, ρ' are called *equivalent* if there exists a matrix (change-of-basis) U such that $\rho'(g) = U\rho(g)U^{-1}$ for all g. The character ch_ρ is the map $G \to \mathbb{C}$ given by the trace: $\mathrm{ch}_\rho(g) = \mathrm{tr}(\rho(g))$. We see that equivalent representations will have the same character, because of the fundamental identity $\mathrm{tr}(AB) = \mathrm{tr}(BA)$. This identity also tells us that the character is a 'class function', i.e. $\mathrm{ch}_\rho(hgh^{-1}) = \mathrm{tr}(\rho(h)\rho(g)\rho(h)^{-1}) = \mathrm{ch}_\rho(g)$ so ch_ρ is constant on each 'conjugacy class'. Group characters are also enormously simpler than representations: e.g. the smallest nontrivial representation of the Monster $\mathbb{M}$ (see Part 2) consists of almost 10^{54} matrices, each of size 196883×196883, while its character consists of 194 complex numbers. Incidentally, finite group representations behave analogously to the representations of X_ℓ: the role of the modules L_λ is played by the irreducible representations ρ_i, and any finite-dimensional representation of G can be decomposed uniquely into a direct sum of various ρ_i. The difference is that there are only finitely many ρ_i — their number equals the number of conjugacy classes of G.

We can use this group intuition here. In particular, given any Lie algebra X_ℓ and representation ρ, we can think of the map $e^x \mapsto e^{\rho(x)}$ as a representation of a Lie group $G(X_\ell)$ corresponding to X_ℓ (the exponential e^A of a matrix is defined by the usual power series; it will always converge). The trace of the matrix $e^{\rho(x)}$ will be the *group* character value at $e^x \in G(X_\ell)$, so we'll define it to be the *algebra* character value at $x \in X_\ell$. Again, it suffices to consider only representatives of each conjugacy class of $G(X_\ell)$, because the character will be a class function. Now, almost every matrix is diagonalisable (since almost any $n \times n$ matrix has n distinct eigenvalues), and so it would seem we aren't losing much by restricting $x \in X_\ell$ to *diagonalisable* matrices. Hence we may take our conjugacy class

---

[8] Surprisingly, what we now call the characters of group representations were invented almost a decade before group representations were.

representatives to be *diagonal* matrices $x \in X_\ell$, i.e. (for $X_\ell = A_1$) to $x = zh$ for $z \in \mathbb{C}$ (h is diagonal in the x_i basis of L_λ). So the algebra character can be chosen to be a function of z. Finally, the trace of $e^{\rho(x)} = e^{z\rho(h)}$ will be given by (1.3.1). This completes the motivation for Weyl's character formula.

There is one other important observation we can make. Different diagonal matrices can belong to the same conjugacy class. For instance,

$$\begin{pmatrix} 0 & -1 \\ 1 & 0 \end{pmatrix} \begin{pmatrix} a & 0 \\ 0 & b \end{pmatrix} \begin{pmatrix} 0 & -1 \\ 1 & 0 \end{pmatrix}^{-1} = \begin{pmatrix} b & 0 \\ 0 & a \end{pmatrix},$$

so e^{zh} and e^{-zh} lie in the same $G(A_1) = SL_2(\mathbb{C})$ conjugacy class. Hence $\mathrm{ch}_M(z) = \mathrm{ch}_M(-z)$. This symmetry $z \mapsto -z$ belongs to the *Weyl group* for A_1. Each X_ℓ has similar symmetries, and the Weyl group plays an important role in the whole theory, sort of analogous to the modular group for modular functions we'll discuss in §2.3.

Weyl found a generalisation of the right-side of (1.3.2), valid for all X_ℓ. The character of L_λ can be written as a fraction (2.8.1): the numerator will be a alternating sum over the Weyl group, and the denominator will be a product over 'positive roots'. This formula and its generalisations have profound consequences, as we'll see in §2.8.

Incidentally, the trigonometric identities given at the beginning of this section are the tensor product formula of representations (interpreted as the product and sum of characters), and the fact that an arbitrary character can be written as a polynomial in the fundamental characters, both specialised to A_1 (see (1.3.2) for the A_1 characters).

1.4. Affine Algebras and the Kac-Peterson Matrices

The theory of nontwisted affine Kac-Moody algebras (usually called *affine algebras* or *current algebras*) is extremely analogous to that of the finite-dimensional simple Lie algebras. Nothing infinite-dimensional tries harder to be finite-dimensional than affine algebras. Standard references for the following material are [38,41,24].

Let X_ℓ be any simple finite-dimensional Lie algebra. The affine algebra $X_\ell^{(1)}$ is essentially the *loop algebra* $\mathcal{L}(X_\ell)$, defined to be all possible 'Laurent polynomials' $\sum_{n \in \mathbb{Z}} a_n t^n$ where each $a_n \in X_\ell$ and all but finitely many $a_n = 0$. t here is an indeterminant. The bracket in $\mathcal{L}(X_\ell)$ is the obvious one: e.g. $[at^n, bt^m] = [ab]t^{n+m}$. Geometrically, $\mathcal{L}(X_\ell)$ is the Lie algebra of polynomial maps $S^1 \to X_\ell$ — hence the name (for that realisation, think of $t = e^{2\pi i\theta}$). Hence there are many generalisations of the loop algebra (e.g. any manifold in place of S^1 will do), closely related ones called *toroidal algebras* being the Lie algebra of maps $S^1 \times \cdots \times S^1 \to X_\ell$. But the loop algebra is simplest and best understood, and the only one we'll consider. Note that $\mathcal{L}(X_\ell)$ is infinite-dimensional. Its Lie groups are the *loop groups*, consisting of all loops $S^1 \to G(X_\ell)$ in a Lie group for X_ℓ.

We saw S^1 before, in the discussion of the Witt algebra. Thus the Virasoro and affine algebras should be related. In fact, the Virasoro algebra acts on the affine algebras as 'derivations', and this connection plays an important technical role in the theory.

$X_\ell^{(1)}$ is in the same relation to the loop algebra, that the Virasoro $\mathcal{V}$ is to the Witt $\mathcal{W}$. Namely, it is its (unique nontrivial 1-dimensional) central extension — see e.g. (7.7.1) of [38] for the analogue of (1.2.7a) here. In addition, for more technical reasons, a further (noncentral) 1-dimensional extension is usually made: the derivation $t\frac{d}{dt}$ is included (see footnote 33). $X_\ell^{(1)}$ is the simplest of the infinite-dimensional Kac-Moody algebras. The superscript '(1)' denotes the fact that the loop algebra was twisted by an order-1 automorphism — i.e. that it is untwisted. It is called 'affine' because of its Weyl group, as we shall see.

Central extensions are a common theme in today's infinite-dimensional Lie theory[9]. Their *raison d'être* is always the same: a richer supply of representations. For example, $\mathcal{W}$ has several representations, but no nontrivial one is an 'irreducible unitary positive-energy representation' — the kind of greatest interest in math phys. On the other hand, its central extension $\mathcal{V}$ has a rich supply of those representations (e.g. there's one for each choice of $c > 1, h > 0$, namely the Verma module $V_{c,h}$ corresponding to $L_0 x_0 = h x_0, C x_0 = c x_0$). At the level of groups, central extensions allow *projective* representations (i.e. representations up to a scalar factor) to become true representations. Projective representations (hence central extensions) appear naturally in QFT because a quantum state vector $|v\rangle$ is physically indistinguishable from any nonzero scalar multiple $\alpha|v\rangle$.

All of the quantities associated to X_ℓ have an analogue here: Dynkin diagram, Weyl group, weights,... For instance, the affine Dynkin diagram is obtained from the Dynkin diagram for X_ℓ by adding one node. See for example Figure 9 of [59]. The extra node is always labelled by a '0'. The Cartan subalgebra $\mathfrak{h}$ here will be $(\ell + 2)$-dimensional. Many of these details will be discussed in more detail in §2.7 below.

The construction of $X_\ell^{(1)}$ is so trivial that it seems surprising anything interesting and new can happen here. But a certain 'miracle' happens...

No interesting representation of $X_\ell^{(1)}$ is finite-dimensional. The analogue for $X_\ell^{(1)}$ of the finite-dimensional representations of X_ℓ are called the *integrable highest-weight representations*, and will be denoted L_λ. The highest-weight λ here will be an $(\ell + 1)$-tuple $(\lambda_0, \lambda_1, \ldots, \lambda_\ell)$, $\lambda_i \in \mathbb{Z}_{\geq}$ (strictly speaking, it will be an $(\ell + 2)$-tuple, but the extra component is not important and is usually ignored). As for X_ℓ, the highest-weights can be thought of as the assignment of a nonnegative integer to each node of the Dynkin diagram. The construction of L_λ is as in the finite-dimensional case. They are called *integrable* because they are precisely those highest-weight representations which can be 'integrated' to a projective representation of the corresponding loop group, and hence a representation of a central extension of the loop group.

We define the character χ_λ as in (1.3.1), though now the weights m will be $(\ell + 2)$-tuples, and there will be infinitely many of them. χ_λ will be a complex-valued function of $\ell + 2$ complex variables $(\vec{z}, \tau, u)$ (see (1.4.1a) below). It can be written as an alternating sum over the Weyl group W, over a 'nice' denominator. The difference here is that W is now infinite.

Perhaps most of the interest in affine algebras can be traced to the 'miracle' that their Weyl groups are a semidirect product $Q^\vee \rtimes \overline{W}$ of translations in a lattice $\mathbb{Q}^\vee$ (the ℓ-dimensional 'co-root lattice' of X_ℓ — see §1.6) with the (finite) Weyl group $\overline{W}$ of X_ℓ.

---

[9] Incidentally the *finite-dimensional* simple Lie algebras do not have nontrivial central extensions.

See Figure 10 of [59] for the Weyl group of $A_2^{(1)}$. 'Semidirect product'[10] means that any element of W can be written uniquely as (t, w) for some translation t and some $w \in \overline{W}$, and $(t, w) \circ (t', w') = (\text{stuff}, w \circ w')$.

One thing this implies is that χ_λ will be of the form 'theta function'/denominator. Theta functions are classically-studied modular forms (we will discuss these terms in §2.3), and thus the modular group $SL_2(\mathbb{Z})$ will make an appearance! To make this more precise, consider the highest-weight $\lambda = (\lambda_0, \lambda_1)$ of $A_1^{(1)}$, and write $k = \lambda_0 + \lambda_1$. Then

$$\chi_\lambda = \frac{\Theta_{\lambda_1+1}^{(k+2)} - \Theta_{-\lambda_1-1}^{(k+2)}}{\Theta_1^{(2)} - \Theta_{-1}^{(2)}} \tag{1.4.1a}$$

where these functions all depend on 3 complex variables z, τ, u, and

$$\Theta_m^{(n)}(z, \tau, u) := e^{-2\pi i n u} \sum_{\ell \in \mathbb{Z} + \frac{m}{2n}} \exp[\pi i n \tau \ell^2 - 2\sqrt{2}\pi i n \ell z] . \tag{1.4.1b}$$

In (1.4.1a) we can see the alternating sum over the Weyl group of A_1 in the numerator (and denominator, since we've used the $A_1^{(1)}$ denominator identity in writing (1.4.1a)). For general $X_\ell^{(1)}$, the denominator will always be independent of λ, and the theta function (1.4.1b) will become a multidimensional one involving a sum over $Q^\vee$ shifted by some weight and appropriately rescaled. The (co-)root lattice of A_1 is $\sqrt{2}\mathbb{Z}$. The key variable in (1.4.1a) is the modular one τ, which will lie in the upper half complex plane $\mathcal{H}$ (in order to have convergence). In the applications to CFT, the other variables are often set to 0.

The number k introduced in (1.4.1a) plays an important role in the general theory. In the representation L_λ, the central term C will get sent to some multiple of the identity — the multiplier is labelled k and is called the *level* of the representation. For any $X_\ell^{(1)}$ there is a simple formula expressing the level k in terms of the highest-weight λ; e.g. for $A_\ell^{(1)}$ and $C_\ell^{(1)}$ it is given by $k = \lambda_0 + \lambda_1 + \cdots + \lambda_\ell$. Write P_+^k for the (finite) set of level k highest-weights (so the size of P_+^k for $A_\ell^{(1)}$ is $\binom{k+\ell}{\ell}$). An important weight in P_+^k is $(k, 0, \ldots, 0)$. We will denote this '0'. In RCFT it corresponds to the vacuum.

The modular group $SL_2(\mathbb{Z})$ acts on the Cartan subalgebra $\mathfrak{h}$ of $X_\ell^{(1)}$ in the following way:

$$\begin{pmatrix} a & b \\ c & d \end{pmatrix} (\vec{z}, \tau, u) = (\frac{\vec{z}}{c\tau + d}, \frac{a\tau + b}{c\tau + d}, u - \frac{c\vec{z} \cdot \vec{z}}{2(c\tau + d)})$$

Under this action, the characters χ_λ also transform nicely: in particular we find for any level k weight λ

$$\chi_\lambda(\frac{\vec{z}}{\tau}, \frac{-1}{\tau}, u - \frac{\vec{z} \cdot \vec{z}}{2\tau}) = \sum_{\mu \in P_+^k} S_{\lambda\mu} \chi_\mu(\vec{z}, \tau, u) \tag{1.4.2a}$$

$$\chi_\lambda(\vec{z}, \tau + 1, u) = \sum_{\mu \in P_+^k} T_{\lambda\mu} \chi_\mu(\vec{z}, \tau, u) \tag{1.4.2b}$$

---

[10] This is also discussed briefly in section 2.2.

where S and T are complex matrices called the *Kac-Peterson matrices*. S will always be symmetric and unitary, and has many remarkable properties as we shall see. Its entries are related to Lie group characters at elements of finite order (see (1.4.5) below). T is diagonal and unitary; its entries are related to the eigenvalues of the quadratic Casimir.

For example, consider $A_1^{(1)}$ at level k. Then S and T will be $(k+1) \times (k+1)$ matrices given by

$$S_{\lambda\mu} = \sqrt{\frac{2}{k+2}} \sin(\pi \frac{(\lambda_1 + 1)(\mu_1 + 1)}{k+2}) , \qquad T_{\lambda\mu} = \exp[\pi i \frac{(\lambda_1 + 1)^2}{2(k+2)} - \frac{\pi i}{4}] \delta_{\lambda,\mu} . \quad (1.4.3)$$

One important place S appears is the famous *Verlinde formula*

$$N_{\lambda\mu}^{\nu} = \sum_{\kappa \in P_+^k} \frac{S_{\lambda\kappa} S_{\mu\kappa} S_{\nu\kappa}^*}{S_{0,\kappa}} \quad (1.4.4)$$

for the fusion coefficients $N_{\lambda\mu}^{\nu}$ of the corresponding RCFT. We will investigate some consequences of this formula in a later section. The fusion coefficients for the affine algebras are well-understood; see e.g. Section 4 of [59] for their interpretation (usually called the Kac-Walton formula) as 'folded tensor product coefficients'.

We will see in §1.7 that symmetries of the extended Dynkin diagram have consequences for S and T (simple-currents, charge-conjugation). There is a 'Galois action' on S which we will discuss in §1.8. There is a strange property of S and T called *rank-level duality* (see e.g. [45]): the matrices for $A_\ell^{(1)}$ at level k are closely related to those of $A_{k-1}^{(1)}$ at level $\ell + 1$, and similar statements hold for $B_\ell^{(1)}$, $C_\ell^{(1)}$ and $D_\ell^{(1)}$. Another reason S is mathematically interesting is the formula

$$\frac{S_{\lambda\mu}}{S_{0\mu}} = \mathrm{ch}_\lambda(-2\pi i \, \overline{\frac{\mu + \rho}{k + h^\vee}}) . \quad (1.4.5)$$

The right-side is a character of X_ℓ, and $\bar{\lambda} = (\lambda_1, \ldots, \lambda_\ell)$ means ignore the extended node. ρ is the 'Weyl vector' $(1, 1, \ldots, 1)$ and $h^\vee$ is called the *dual Coxeter number* and is the level of ρ. For $A_\ell^{(1)}$, $h^\vee = \ell + 1$. Of course the right-side can also be regarded as a character for a Lie group associated to X_ℓ, in which case the argument would have to be exponentiated and would correspond to an element of finite order in the group. These numbers (1.4.5) have been studied by many people (most extensively by Pianzola) and have some nice properties. For instance Moody-Patera (1984) have argued that exploiting them leads to some quick algorithms for computing e.g. tensor product coefficients. Kac [37] found a curious application for them: a Lie theoretic proof of 'quadratic reciprocity'!

Quadratic reciprocity is one of the gems of classical number theory. It tells us that the equations

$$x^2 \equiv a \pmod{b}$$
$$y^2 \equiv b \pmod{a}$$

are related; more precisely, for fixed a and b (for simplicity take them to both be primes $\neq 2$) the questions of whether there is a solution x to the first equation and a solution y to the second, are related. They will both have the same yes or no answer, unless $a \equiv b \equiv 3$ (mod 4), in which case they will have opposite answers. E.g. take $a = 23$ and $b = 3$, then we know the first equation does not have a solution (since $a \equiv 2$ (mod 3) and $x^2 \equiv 2$ doesn't have a solution mod 3), and hence the second equation must have a solution (indeed, $y = 7$ works). There are now many proofs for quadratic reciprocity, and Kac used Lie characters at elements of finite order to find another one.

What is interesting here is that Kac's proof uses only certain special weights for A_ℓ. The natural question is: is it possible to find any generalisations of quadratic reciprocity using other weights and algebras? Many generalisations of quadratic reciprocity are known; will generalising Kac's argument recover them, or will they perhaps yield new reciprocity laws? It seems no one knows.

The relation (1.4.5) is important because it connects finite-dimensional Lie data with infinite-dimensional Lie data. The 'conceptual arrow' can be exploited both ways: in the generalisations of the arguments of §1.9 to other algebras, (1.4.5) allows us to use our extensive knowledge of finite-dimensional algebras to squeeze out some information in the affine setting; but also it is possible to use the richer symmetries of the affine data to see 'hidden' symmetries in finite-dimensional data. For example it can be used (Gannon-Walton 1995) to find a sort of Galois symmetry of dominant weight multiplicities in X_ℓ, which would be difficult or impossible to anticipate without (1.4.5).

1.5. THE CLASSIFICATION OF PHYSICAL INVARIANTS

We are interested in the following classification problem. Choose any affine algebra $X_\ell^{(1)}$ and level $k \in \mathbb{Z}_{\geq}$. Find all matrices $M = (M_{\lambda\mu})_{\lambda,\mu \in P_+^k}$ such that

P1) $MS = SM$ and $MT = TM$, where S, T are the Kac-Peterson matrices (1.4.2);
P2) each entry $M_{\lambda\mu} \in \mathbb{Z}_{\geq}$;
P3) $M_{00} = 1$.

Any such M, or equivalently the corresponding partition function $\mathcal{Z} = \sum_{\lambda,\mu} M_{\lambda\mu} \chi_\lambda \chi_\mu^*$, is called a *physical invariant*.

The first and most important classification of physical invariants was the Cappelli-Itzykson-Zuber A-D-E classification for $A_1^{(1)}$ at all levels k [8]. We will give their result shortly. This implies for instance the minimal model RCFT classification, as well as the $N = 1$ super(symmetric)conformal minimal models. The other classifications of comparable magnitude are $A_2^{(1)}$ for all k; $A_\ell^{(1)}$, $B_\ell^{(1)}$ and $D_\ell^{(1)}$ for all $k \leq 3$; $(A_1 \oplus A_1)^{(1)}$ for all levels (k_1, k_2); and $(u(1) \oplus \cdots \oplus u(1))^{(1)}$ for all (matrix-valued) levels k. See e.g. [29] for references. The most difficult of these classifications is for $A_2^{(1)}$, done by Gannon (1994).

In other words, very little in this direction has been accomplished in the 15 or so years this problem has existed. But this is not really a good measure of progress. The effort instead has been directed primarily towards the full classification; most of these partial results are merely easy spin-offs from that more serious and ambitious assault.

The proof in [8] was very complicated and followed the following lines. First, an explicit basis was found for the vector space (called the 'commutant') of all matrices obeying (P1). Then (P2) and (P3) were imposed. Unfortunately their proof was long and formidable. Others tried to apply their approach to $A_2^{(1)}$, but without success. The eventual proof for $A_2^{(1)}$ was completely independent of the [8] argument, and exploited more of the structure implicit in the problem. As the $A_2^{(1)}$ argument became more refined, it became the model for the general assault. In §1.9 we sketch this new approach.

From this more general perspective, of these completed classifications only the level 2 $B_\ell^{(1)}$ and $D_\ell^{(1)}$ ones will have any lasting value (the orthogonal algebras at level 2 behave very peculiarly, possess large numbers of exceptional physical invariants, and must be treated separately). The others behave more generically and will fall out as special cases once the more general classifications are concluded. Other classifications which should be straightforward with our present understanding are $C_2^{(1)}$ at all k; $G_2^{(1)}$ at all k; and $B_\ell^{(1)}$ and $D_\ell^{(1)}$ at $k = 4$. The $C_2^{(1)}$ should be easiest and would imply the $C_\ell^{(1)}$ level 2, as well as the $B_\ell^{(1)}$ and $D_\ell^{(1)}$ level 5, classifications. A very safe conjecture is that the only *exceptional* physical invariants (we define this term in §1.7) for $C_2^{(1)}$ occur at $k = 3, 7, 8, 12$ — this is known to be true for all $k \leq 500$. $G_2^{(1)}$ would be more difficult but also much more valuable; its only known exceptionals occur at $k = 3, 4$, and these are the only exceptionals for $k \leq 500$, and a very safe conjecture is that there are no other $G_2^{(1)}$ exceptionals. $B_\ell^{(1)}$ and $D_\ell^{(1)}$ at level 4 will also be more difficult, but also would be valuable; less is understood about its physical invariants and there is a good chance new exceptionals exist there.

The most surprising thing about the known physical invariant classifications is that there so few surprises: almost every physical invariant is 'generic'. We will see that the symmetries of the extended Dynkin diagram give rise to general families of physical invariants. We will call any physical invariants which do not arise in these generic ways (i.e. using what are called simple-currents or conjugations), *exceptional*. Many exceptionals have been found, and now we are almost at the point where we can safely conjecture the complete list of physical invariants for $X_\ell^{(1)}$ at any k, for X_ℓ a simple algebra.

Unfortunately the classification for semi-simple algebras $X_{\ell_1} \oplus \cdots \oplus X_{\ell_s}$ does not reduce to the one for simple ones. In fact, *any explicit classification of the physical invariants for $X^{(1)}$, for all semi-simple X, would easily be one of the greatest accomplishments in the history of math*, for it would include as a small part such monumental things as an explicit classification of all positive-definite integral lattices. Thus we unfortunately cannot expect an *explicit* classification for the *semi*-simple algebras.

To make this discussion more concrete and explicit, consider $A_1^{(1)}$. For convenience drop λ_0, so $P_+^k = \{0, 1, \ldots, k\}$. Write J for the permutation (called a simple-current) $Ja := k - a$. Then the complete list of physical invariants for $A_1^{(1)}$ is

$$\mathcal{A}_{k+1} = \sum_{a=0}^{k} |\chi_a|^2 , \qquad \text{for all } k \geq 1$$

$$\mathcal{D}_{\frac{k}{2}+2} = \sum_{a=0}^{k} \chi_a \chi_{J^a a}^* , \qquad \text{whenever } \frac{k}{2} \text{ is odd}$$

18

$$\mathcal{D}_{\frac{k}{2}+2} = |\chi_0 + \chi_{J0}|^2 + |\chi_2 + \chi_{J2}|^2 + \cdots + 2|\chi_{\frac{k}{2}}|^2 , \qquad \text{whenever } \frac{k}{2} \text{ is even}$$

$$\mathcal{E}_6 = |\chi_0 + \chi_6|^2 + |\chi_3 + \chi_7|^2 + |\chi_4 + \chi_{10}|^2 , \qquad \text{for } k = 10$$

$$\mathcal{E}_7 = |\chi_0 + \chi_{16}|^2 + |\chi_4 + \chi_{12}|^2 + |\chi_6 + \chi_{10}|^2$$
$$+ \chi_8 (\chi_2 + \chi_{14})^* + (\chi_2 + \chi_{14}) \chi_8^* + |\chi_8|^2 , \qquad \text{for } k = 16$$

$$\mathcal{E}_8 = |\chi_0 + \chi_{10} + \chi_{18} + \chi_{28}|^2 + |\chi_6 + \chi_{12} + \chi_{16} + \chi_{22}|^2 , \qquad \text{for } k = 28 .$$

The physical invariants $\mathcal{A}_n$ and $\mathcal{D}_n$ are generic, corresponding respectively to the order 1 (i.e. identity) and order 2 (i.e. the simple-current J) Dynkin diagram symmetries, as we shall see in §1.7. Physically, they are the partition functions of WZW models on $\mathrm{SU}_2(\mathbb{C})$ and $\mathrm{SO}_3(\mathbb{R})$ group manifolds, resp. The exceptionals $\mathcal{E}_6$ and $\mathcal{E}_8$ are best interpreted as due to the $C_{2,1} \supset A_{1,10}$ and $G_{2,1} \supset A_{1,28}$ conformal embeddings (see §1.7; standard notation is to write '$X_{\ell,k}$' for '$X_\ell^{(1)}$ and level k'). The $\mathcal{E}_7$ exceptional is harder to interpret, but can be thought of as the first in an infinite series of exceptionals involving rank-level duality and D_4 triality.

Around Christmas 1985, Zuber wrote Kac about the $A_1^{(1)}$ physical invariant problem, and mentioned the physical invariants he and Itzykson knew at that point (what we now call $\mathcal{A}_*$ and $\mathcal{D}_{even}$). A few weeks later, Kac wrote back saying he found one more invariant, and jokingly pointed out that it must be indeed quite exceptional as the exponents of E_6 appeared in it. "I must confess that I didn't pay much attention to that last remark (I hardly knew what Coxeter exponents were, at the time!)" [63]. By spring 1986, Cappelli arrived in Paris and got things moving again; together Cappelli-Itzykson-Zuber found $\mathcal{E}_7$, $\mathcal{D}_{odd}$, and then $\mathcal{E}_8$, and struggled to find more. "And it is only in August [1986], during a conversation with Pasquier, in which he was showing me his construction of lattice models based on Dynkin diagrams, that I suddenly remembered this cryptic but crucial! observation of Victor, rushed to the library to find a list of the exponents of the other algebras... and found with the delight that you can imagine that they were matching our list" [63]. Thus the A-D-E pattern to these physical invariants was discovered.

1.6. The A-D-E meta-pattern

Before we discuss *meta-patterns* in math, let's introduce the notion of *lattice*[11], a simple geometric structure we'll keep returning to in these notes. The standard reference for lattice theory is [13].

Consider the real vector space $\mathbb{R}^{m,n}$: its vectors look like $\vec{x} = (\vec{x}_+; \vec{x}_-)$ where $\vec{x}_+$ and $\vec{x}_-$ are m- and n-component vectors respectively, and dot products are given by $\vec{x} \cdot \vec{y} = \vec{x}_+ \cdot \vec{y}_+ - \vec{x}_- \cdot \vec{y}_-$. The dot products $\vec{x}_\pm \cdot \vec{y}_\pm$ are given by the usual product and sum of components. For example, the familiar Euclidean (positive-definite) space is $\mathbb{R}^n = \mathbb{R}^{n,0}$, while Minkowski space is $\mathbb{R}^{3,1}$.

---

[11] There are many words in math which have several incompatible meanings. For example, there are *vector fields* and *number fields*, and modular *forms* and modular *representations*. 'Lattice' is another of these words. Aside from the geometric meaning we will use, it also refers to a 'partially ordered set'.

Now choose any basis $B = \{\vec{x}_1, \ldots, \vec{x}_{m+n}\}$ in $\mathbb{R}^{m,n}$. So $\mathbb{R}^{m,n} = \mathbb{R}\vec{x}_1 + \cdots + \mathbb{R}\vec{x}_{m+n}$. Define the set $\Lambda(B) := \mathbb{Z}\vec{x}_1 + \cdots + \mathbb{Z}\vec{x}_{m+n}$. This is a *lattice*, and all lattices can be formed in this way[12]. So a lattice is discrete and is closed under sums and integer multiples. For example, $\mathbb{Z}^{m,n}$ is a lattice (take the standard basis in $\mathbb{R}^{m,n}$). A more interesting lattice is the hexagonal lattice (also called A_2), given by the basis $B = \{(\frac{\sqrt{2}}{2}, \frac{\sqrt{6}}{2}), (\sqrt{2}, 0)\}$ of $\mathbb{R}^2$ — try to plot several points. If you wanted to slide a bunch of coins on a table together as tightly as possible, their centres would form this hexagonal lattice. Another important lattice is $II_{1,1} \subset \mathbb{R}^{1,1}$, given by $B = \{(\frac{1}{\sqrt{2}}; \frac{1}{\sqrt{2}}), (\frac{1}{\sqrt{2}}; \frac{-1}{\sqrt{2}})\}$; equivalently it can be thought of as the set of all pairs $(a, b) \in \mathbb{Z}^2$ with dot product

$$(a, b) \cdot (c, d) = ad + bc . \tag{1.6.1}$$

It is important to note that different choices of basis may or may not result in a different lattice. For a trivial example, consider $B = \{1\}$ and $B' = \{-1\}$ in $\mathbb{R} = \mathbb{R}^{1,0}$: they both give the lattice $\mathbb{Z} = \mathbb{Z}^{1,0}$. Two lattices are called *equivalent* if they only differ by a change-of-basis. E.g. $B = \{(\frac{1}{\sqrt{2}}, \frac{1}{\sqrt{2}}), (\frac{1}{\sqrt{2}}, \frac{-1}{\sqrt{2}})\}$ in $\mathbb{R}^2$ yields a lattice equivalent to $\mathbb{Z}^2$.

The *dimension* of the lattice is $m + n$. The lattice is called *positive-definite* if it lies in some $\mathbb{R}^m$ (i.e. $n = 0$). The lattice is called *integral* if all dot products $\vec{x} \cdot \vec{y}$ are integers, for $\vec{x}, \vec{y} \in \Lambda$. A lattice Λ is called *even* if it is integral and in addition all norms $\vec{x} \cdot \vec{x}$ are *even* integers. For example, $\mathbb{Z}^{m,n}$ is integral but not even, while A_2 and $II_{1,1}$ are even. The *dual* Λ^* of a lattice Λ consists of all vectors $\vec{x} \in \mathbb{R}^{m,n}$ such that $\vec{x} \cdot \Lambda \subset \mathbb{Z}$. So a lattice is integral iff $\Lambda \subseteq \Lambda^*$. A lattice is called *self-dual* if $\Lambda = \Lambda^*$. $\mathbb{Z}^{m,n}$ and $II_{1,1}$ are self-dual but A_2 is not.

There are lots of 'meta-patterns' in math, i.e. collections of seemingly different problems which have similar answers. Once one of these meta-patterns is identified it is always helpful to understand what is responsible for it. For example, while I was writing up my PhD thesis I noticed in several places the numbers 1, 2, 3, 4, and 6. For instance $\cos(2\pi r) \in \mathbb{Q}$ for $r \in \mathbb{Q}$ iff the denominator of r is 1, 2, 3, 4, or 6. This pattern was easy to explain: they are precisely those positive integers n with Euler totient $\phi(n) \leq 2$, i.e. there are at most 2 positive numbers less than n coprime[13] to n. The other incidences of these numbers can usually be reduced to this $\phi(n) \leq 2$ property (e.g. the dimension of the number field $\mathbb{Q}[\cos(2\pi \frac{a}{b})]$ (see §1.8) considered as a vector space over $\mathbb{Q}$ will be $\phi(b)/2$).

A more interesting meta-pattern involves the number 24 and its divisors. One sees 24 wherever modular forms naturally appear. For instance, we see it in the critical dimensions in string theory: $24 + 2$ and $8 + 2$. Another example: the dimensions of even self-dual positive-definite lattices must be a multiple of 8 (e.g. the E_8 root lattice defined shortly has dimension 8, while the Leech lattice discussed in §2.4 has dimension 24). The meta-pattern 24 is also understood: the fundamental problem for which it is the answer is the following one. Fix n, and consider the congruence $x^2 \equiv 1 \pmod{n}$. Certainly in order to have a chance of satisfying this, x and n must be coprime. The extreme situation is when *every*

---

[12] In most presentations a lattice is permitted to have smaller dimension than its ambient space, however that freedom gains no real generality.

[13] We say m, n are *coprime* if any prime p which divides m does not divide n, and vice versa.

number x coprime to n satisfies this congruence:

$$\gcd(x,n) = 1 \qquad \Longleftrightarrow \qquad x^2 \equiv 1 \pmod{n} . \qquad (1.6.2)$$

The reader can try to verify the following simple fact: n obeys this extreme situation (1.6.2) iff n divides 24.

What does this congruence property have to do with these other occurrences of 24? Let Λ be an even self-dual positive-definite lattice of dimension n. Then an elementary argument shows that there will exist an n-tuple $\vec{a} = (a_1, \ldots, a_n)$ of odd integers with the property that 8 must divide $\vec{a} \cdot \vec{a} = \sum_i a_i^2$. But $a_i^2 \equiv 1 \pmod{8}$, and so we get $8|n$.

A much deeper and still not-completely-understood meta-pattern is called A-D-E (see [1] for a discussion and examples). The name comes from the so-called *simply-laced algebras*, i.e. the simple finite-dimensional Lie algebras whose Dynkin diagrams — see Figure 6 in [59] — contain only single edges (i.e. no arrows). These are the A_*- and D_*-series, along with the E_6, E_7 and E_8 exceptionals. The claim is that many other problems, which don't seem to have anything directly in common with simple Lie algebras, have a solution which falls into this A-D-E pattern (for an object to be meaningfully labelled X_ℓ, some of the data associated to the algebra X_ℓ should reappear in some form in that object). Let's look at some examples.

Consider even positive-definite lattices Λ. The smallest possible nonzero norm in Λ will be 2, and the vectors of norm 2 are special and are called *roots*. The reason they are special is that reflecting through them will always be an automorphism of Λ. That is, the reflection $\vec{u} \mapsto \vec{u} - 2\frac{\vec{u} \cdot \vec{\alpha}}{\vec{\alpha} \cdot \vec{\alpha}} \vec{\alpha}$ through $\vec{\alpha} \neq \vec{0}$ won't in general map Λ to itself, unless $\vec{\alpha}$ is a root of Λ. It is important in lattice theory to know the lattices which are spanned by their roots; it turns out these are precisely the orthogonal direct sums of lattices called A_n, D_n, and E_6, E_7 and E_8. They carry those names for a number of reasons. For example, the lattice called X_n will have a basis $\{\vec{\alpha}_1, \ldots, \vec{\alpha}_n\}$ with the property that the matrix $A_{ij} := \vec{\alpha}_i \cdot \vec{\alpha}_j$ is the Cartan matrix (see §2.7) for the Lie algebra X_n! Also, the reflection group generated by reflections in the roots of the lattice X_n will be isomorphic to the Weyl group of the Lie algebra X_n. Finally, to any simple Lie algebra there is canonically associated a lattice called the root lattice; for the simply-laced algebras, these will equal the corresponding lattice of the same name. Incidentally, the root lattices for the non-simply-laced simple algebras will (up to rescalings) be direct sums of the simply-laced root lattices.

We have already met the A_2 lattice: it is the densest packing of circles in the plane. It has long been believed that the obvious pyramidal way to pack oranges is also the densest possible way — the centres of the oranges form the A_3 root lattice. A controversial proof for this famous conjecture has been offered by W.-Y. Hsiang in 1991; in 1998 a new proof by Hale *et al* has been proposed. The densest known packings in dimensions 4,5,6,7,8 are D_4, D_5, E_6, E_7, E_8, resp. E_8 is the smallest even self-dual positive-definite lattice.

· A famous A-D-E example is called the McKay[14] correspondence. Consider any finite subgroup G of the Lie group $SU_2(\mathbb{C})$ (i.e. the 2×2 unitary matrices with determinant 1). For example, there is the cyclic group $\mathbb{Z}_n$ of n elements generated by the matrix

$$M_n = \begin{pmatrix} \exp[2\pi i/n] & 0 \\ 0 & \exp[-2\pi i/n] \end{pmatrix}$$

---

[14] He is the same John McKay we will celebrate in section 2.1.

Let R_i be the irreducible representations of G. For instance, for $\mathbb{Z}_n$, there are precisely n of these, all 1-dimensional, given by sending the generator M_n to $\exp[2k\pi i/n]$ for each $k = 1, 2, \ldots, n$. Now consider the tensor product $G \otimes R_i$, where we interpret $G \subset SU_2(\mathbb{C})$ here as a 2-dimensional representation. We can decompose that product into a direct sum $\oplus_j m_{ij} R_j$ of irreducibles (the m_{ij} here are multiplicities). Now create a graph with one node for each R_i, and with the ith and jth nodes ($i \neq j$) connected with precisely m_{ij} directed edges $i \to j$. If $m_{ij} = m_{ji}$, we agree to erase the double arrows from the m_{ij} edges. Then McKay observed that the graph of any G will be a distinct extended Dynkin diagram of A-D-E type! For instance, the cyclic group with n elements corresponds to the extended graph of A_{n-1}.

How was McKay led to his remarkable correspondence? He knew that the sum of the 'marks' $a_i = 1, 2, 3, 4, 5, 6, 4, 2, 3$ associated to each node of the extended E_8 Dynkin diagram equaled 30, the Coxeter number of E_8. So what did their *squares* add to? 120, which he recognised as the cardinality of one of the exceptional finite subgroups of $SU_2(\mathbb{C})$, and that got him thinking...

Another famous example of A-D-E, due to Arnol'd, are the 'simple critical points' of smooth complex-valued functions f, on e.g. $\mathbb{C}^3$. For example, both $x^2 + y^2 + z^{n+1}$ and $x^2 + y^3 + z^5$ have singularities at $(0, 0, 0)$ (i.e. their first partial derivatives all vanish there), and they are assigned to A_n and E_8, respectively. The $SU_2(\mathbb{C})$ subgroups can be related to singularities as follows. The group $SU_2(\mathbb{C})$ acts on $\mathbb{C}^2$ in the obvious way (matrix multiplication). If G is a discrete subgroup, then consider the (ring of) polynomials in 2 variables w_1, w_2 invariant under G. It turns out it will have 3 generators $x(w_1, w_2)$, $y(w_1, w_2)$, $z(w_1, w_2)$, which are connected by 1 polynomial relation (syzygy). For instance, take G to be the cyclic group $\mathbb{Z}_n$, then we're interested in polynomials $p(w_1, w_2)$ invariant under $w_1 \mapsto \exp[2\pi i/n]w_1$, $w_2 \mapsto \exp[-2\pi i/n]w_2$. Any such invariant $p(w_1, w_2)$ is clearly generated by (i.e. can be written as a polynomial in) $w_1 w_2$, w_1^n and w_2^n. Choosing instead the generators $x = \frac{w_1^n - w_2^n}{2}$, $y = i\frac{w_1^n + w_2^n}{2}$, $z = w_1 w_2$, we get the syzygy $z^n = -(x^2 + y^2)$. For any G, generators x, y, z can always be found so that the syzygy will be one of the polynomials associated to a simple singularity, and in fact will give the equation of the algebraic surface $\mathbb{C}^2/G$ as a 2-dimensional complex surface in $\mathbb{C}^3$ (e.g. the complex surfaces $\mathbb{C}^2/\mathbb{Z}_n$ and $\{(x, y, z) \in \mathbb{C}^3 \mid x^2 + y^2 + z^n = 0\}$ are equivalent).

Arguably the first A-D-E classification goes back to Theaetetus, around 400 B.C. He classified the regular solids. For instance the tetrahedron can be associated to E_6 while the cube is matched with E_7. This A-D-E is only partial, as there are no regular solids assigned to the A-series, and to get the D-series one must look at 'degenerate regular solids'.

The closest thing to an explanation of the A-D-E meta-pattern would seem to be the notion of 'additive assignments' on graphs (which is a picturesque way of describing the corresponding eigenvalue problem). Consider any graph $\mathcal{G}$ with undirected edges, and none of the edges run from a node to itself. We can also assume without loss of generality that $\mathcal{G}$ is connected. Assign a positive number a_i to each node. If this assignment has the property that for each i, $2a_i = \sum a_j$ where the sum is over all nodes j adjacent to i (counting multiplicities of edges), then we call it 'additive'. For instance, for the graph o=o, the assignment $a_1 = 1 = a_2$ is additive, but the assignment $a_1 = 1, a_2 = 2$ is not. The question is, which graphs have an additive assignment? The answer is: precisely the extended Dynkin diagrams of A-D-E type! And their additive assignments are unique (up to constant proportionality) and are given by the *marks* a_i of the algebra (see e.g. the Table on p.54 of [38]). For example the extended A_n graph consists of $n + 1$ nodes arranged in a circle, and its marks a_i all equal 1.

What do additive assignments have to do with the other A-D-E classifications? Consider a finite subgroup G of $SU_2(\mathbb{C})$. Take the dimension of the equation $G \otimes R_i = \oplus_j m_{ij} R_j$: we get $2d_i = \sum_j m_{ij} d_j$ where $d_j = \dim(R_j)$. Hence the dimensions of the irreducible representations define an additive assignment for each of McKay's graphs, and hence those graphs must be of A-D-E type (provided we know $m_{ij} = m_{ji}$).

As Cappelli-Itzykson-Zuber observed, the physical invariants for $A_1^{(1)}$ also realise the A-D-E pattern, in the following sense. The *Coxeter number* h of the name $\mathcal{X}_\ell$ (i.e. the sum $\sum_i a_i$ of the marks) equals $k + 2$, and the *exponents* m_i of X_ℓ equal those $a \in P_+^k$ for which $M_{aa} \neq 0$ (for the simply-laced algebras, the m_i are defined by writing the eigenvalues of the corresponding Cartan matrix (see §2.7) as $4 \sin^2(\frac{\pi m_i}{2h})$ — the m_i are integers and the smallest is always 1). Probably what first led Kac to his observation about the E_6 exponents was that $k + 2$ (this is how k enters most formulas) for his exceptional equalled the Coxeter number 12 for E_6. More recently, the operator algebraists Ocneanu [48] and independently Böckenhauer-Evans [4] found an A-D-E interpretation for the off-diagonal entries M_{ab} of the $A_1^{(1)}$ physical invariants, using subfactor theory.

We are not claiming that this $A_1^{(1)}$ classification is 'equivalent' to any other A-D-E one — that would miss the point of meta-patterns. What we really want to do is to identify some critical combinatorial part of an $A_1^{(1)}$ proof with critical parts in other A-D-E classifications — this is what we did with the other meta-patterns. A considerably simplified proof of the $A_1^{(1)}$ classification is now available [29], so hopefully this task will now be easier.

There has been some progress at understanding this $A_1^{(1)}$ A-D-E. Nahm [46] constructed the invariant $\mathcal{X}_\ell$ in terms of the compact simply-connected Lie group of type X_ℓ, and in this way could interpret the $k + 2 = h$ and $M_{m_i m_i} \neq 0$ coincidences. A very general explanation for A-D-E has been suggested by Ocneanu [48] using his theory of path algebras on graphs; although his work has never been published, others are now rediscovering (and publishing!) similar work (see e.g. [4]). Nevertheless, the A-D-E in CFT remains almost as mysterious now as it did a dozen years ago — for example it still isn't clear how it directly relates to additive assignments.

There are 4 other claims for A-D-E classifications of families of RCFT physical invariants, and all of them inherit their (approximate) A-D-E pattern from the more fundamental $A_1^{(1)}$ one. One is the $c < 1$ minimal models, also proven in [8], and another is the $N = 1$ superconformal minimal models, proved by Cappelli (1987). In both cases the physical invariants are parametrised by pairs of A-D-E diagrams. The list of known $c = 1$ RCFTs also looks like A-D-E (two series parametrised by $\mathbb{Q}_+$, and three exceptionals), but the completeness of that list has never been rigourously established.

The fourth classification often quoted as A-D-E, is the $N = 2$ superconformal minimal models. Their classification was done by Gannon (1997). The connection here with A-D-E turns out to be rather weak: e.g. 20, 30, and 24 distinct invariants would have an equal

23

right to be called $\mathcal{E}_6$, $\mathcal{E}_7$, and $\mathcal{E}_8$ respectively. It would appear that the frequent claims that the $N = 2$ minimal models fall into an A-D-E pattern are rather dubious.

Hanany-He [35] suggest that the $A_1^{(1)}$ A-D-E pattern can be related to subgroups $G \subset \mathrm{SU}_2(\mathbb{C})$ by orbifolding 4-dimensional $N = 4$ supersymmetric gauge theory by G, resulting in an $N = 2$ superCFT whose 'matter matrix' can be read off from the Dynkin diagram corresponding to G. The same game can be played with finite subgroups of $\mathrm{SU}_3(\mathbb{C})$, resulting in $N = 1$ superCFTs whose matter matrices correspond to graphs very reminiscent of the 'fusion graphs' of Di Francesco-Petkova-Zuber (see e.g. [62]) corresponding to $A_2^{(1)}$ physical invariants. [35] use this to conjecture a McKay-type correspondence between singularities of type $\mathbb{C}^n/G$, for $G \subset \mathrm{SU}_n(\mathbb{C})$, and the physical invariants of $A_{n-1}^{(1)}$. This in their view would be the form A-D-E takes for higher rank physical invariants. Their actual conjecture though is still somewhat too vague.

For a final example of meta-pattern, consider 'modular function' (see §2.3). After all, they appear in a surprising variety of places and disguises. Maybe we shouldn't regard their ubiquity as fortuitous, instead perhaps there's a deeper common 'situation' which is the source for that ubiquity. Just as 'symmetry' yields 'group', or 'rain-followed-by-heat' breeds mosquitos. Math is not above metaphysics; like any area it grows by asking questions, and changing your perspective — even to a metaphysical one — should suggest new questions.

1.7. SIMPLE-CURRENTS AND CHARGE-CONJUGATION

The key properties[15] of the matrix S are that it's unitary and symmetric (so M in §1.5 equals SMS^*),

$$S_{0\mu} > 0 \text{ for all } \mu \in P_+^k , \qquad (1.7.1)$$

and that the numbers $N_{\lambda\mu}^{\nu}$ defined by Verlinde's formula (1.4.4) are nonnegative integers. These are obeyed by the matrix S in any (unitary) RCFT. From these basic properties, we will obtain here some elementary consequences which have important applications.

But first, let's make an observation which isn't difficult to prove, but doesn't appear to be generally known.

Verlinde's formula looks strange, but it is quite generic,

and we can see it throughout math and mathematical physics. Consider the following.

Let $\mathfrak{A}$ be a commutative associative algebra, over $\mathbb{R}$ say. Suppose $\mathfrak{A}$ has a finite basis Φ (over $\mathbb{R}$) containing the unit 1. Define the 'structure constants' $N_{ab}^c \in \mathbb{R}$, for $a, b, c \in \Phi$, by $ab = \sum_{c \in \Phi} N_{ab}^c c$. Suppose there is an algebra homomorphism $*$ (so $*$ is linear, and $(xy)^* = x^* y^*$) which permutes the basis vectors (so $\Phi^* = \Phi$), and we have the relation $N_{ab}^1 = \delta_{b,a^*}$. We call any such algebra $\mathfrak{A}$ a *fusion algebra*. Then any fusion algebra will necessarily have a unitary matrix S with $S_{1a} > 0$ and with the structure constants given by Verlinde's formula. Algebraically, the relation $S = S^t$ holds if $\mathfrak{A}$ is 'self-dual' in a certain natural sense.

---

[15] A good exercise for the reader is to prove that if S is unitary and symmetric, and obeys (1.7.1), then there will be at most finitely many physical invariants M for that S,T.

Define the 'fusion matrices' N_λ by $(N_\lambda)_{\mu\nu} = N^\nu_{\lambda\mu}$. Then Verlinde's formula says that the μth column $S_{\uparrow\mu}$ of S is an eigenvector of each fusion matrix N_λ, with eigenvalue $\frac{S_{\lambda\mu}}{S_{0\mu}}$.

Useful Fact. If v is a simultaneous eigenvector of each fusion matrix N_λ, then there exists a constant $c \in \mathbb{C}$ and a $\lambda \in P^k_+$ such that $v = c\, S_{\uparrow\lambda}$.

For one consequence, take the complex conjugate of the eigenvector equation $N_\lambda S_{\uparrow\mu} = \frac{S_{\lambda\mu}}{S_{0\mu}} S_{\uparrow\mu}$: we get that the vector $S^*_{\uparrow\mu}$ is a simultaneous eigenvector of all N_λ, and hence must equal $c\, S_{\uparrow\gamma}$ for some number c and weight $\gamma \in P^k_+$, both depending on μ. Write $\gamma = C\mu$; then C defines a permutation of P^k_+. The reader can verify that unitarity of S forces $|c| = 1$, while (1.7.1) forces $c > 0$. Thus $c = 1$ and we obtain the formula

$$S^*_{\lambda\mu} = S_{\lambda,C\mu} = S_{C\lambda,\mu} . \tag{1.7.2}$$

Also, unitarity and symmetry of S forces $C = S^2$, while conjugating twice shows $C^2 = id$. C is an important matrix in RCFT, and is called *charge-conjugation*. When $C = id.$, then the matrix S is real.

Note that (1.7.1) now implies $C0 = 0$. Also $CT = TC$. Hence $M = C$ will always define a physical invariant, and if M is any other physical invariant, the matrix product $MC = CM$ will define another physical invariant. Also, $N^{C\nu}_{C\lambda,C\mu} = N^\nu_{\lambda\mu}$ and $(N_\lambda)^t = CN_\lambda = N_\lambda C = N_{C\lambda}$.

For the WZW (=affine) case, C has a special meaning: $C\lambda$ is the highest-weight 'con-tragredient' to λ. C corresponds to an order 2 (or 1) symmetry of the (unextended) Dynkin diagram. For example, for $A^{(1)}_\ell$, we have $C(\lambda_0, \lambda_1, \ldots, \lambda_{\ell-1}, \lambda_\ell) = (\lambda_0, \lambda_\ell, \lambda_{\ell-1}, \ldots, \lambda_1)$. For $A^{(1)}_1$ then, $C = id.$, which can also be read off from (1.4.3).

The algebras $D^{(1)}_{even}$ all have at least one nontrivial symmetry of the (unextended) Dynkin diagram which isn't the charge-conjugation. The most interesting example is $D^{(1)}_4$, which has 5 of these. By a *conjugation*, we will mean any symmetry of the unextended Dynkin diagram.

To go much further, we need a fascinating tool called *Perron-Frobenius theory* — a collection of results concerning the eigenvalues and eigenvectors of nonnegative matrices (i.e. matrices in which every entry is a nonnegative real number). Whenever you have such matrices in your problem, and it is natural to multiply them, then there is a good chance Perron-Frobenius theory will tell you something interesting. The basic result here is that if A is a nonnegative matrix, then there will be a nonnegative eigenvector $x \geq 0$ with eigenvalue $\rho \geq 0$, such that if λ is any other eigenvalue of A, then $|\lambda| \leq \rho$. There are lots of other results (see e.g. [44]), e.g. ρ must be at least as large as any diagonal entry of A, and there must be a row-sum of A no bigger than ρ, and another row-sum no smaller than ρ.

For instance, consider

$$A = \begin{pmatrix} 1 & 1 & 1 \\ 1 & 1 & 1 \\ 1 & 1 & 1 \end{pmatrix}, \quad B = \begin{pmatrix} 1 & 1 & 1 \\ 1 & 0 & 0 \\ 1 & 0 & 0 \end{pmatrix} .$$

Perron-Frobenius eigenvectors for A and B are $\begin{pmatrix} 1 \\ 1 \\ 1 \end{pmatrix}$ and $\begin{pmatrix} 2 \\ 1 \\ 1 \end{pmatrix}$, with eigenvalues 3 and 2 resp. The other eigenvalue of A is 0 (multiplicity 2), while those of B are 0 and -1.

Fusion matrices N_λ are nonnegative, and it is indeed natural to multiply them:

$$N_\lambda N_\mu = \sum_{\nu \in P_+^k} N_{\lambda\mu}^\nu N_\nu \ .$$

So we can expect Perron-Frobenius to tell us something interesting. This is the case, and we obtain the curious-looking inequalities

$$S_{\lambda 0}\, S_{0\mu} \geq |S_{\lambda\mu}|\, S_{00} \ . \tag{1.7.3}$$

Squaring both sides, summing over μ and using unitarity, we get that $S_{\lambda 0} \geq S_{00}$. In other words, the ratio $\frac{S_{\lambda 0}}{S_{00}}$, called the *quantum-dimension* of λ, will necessarily be ≥ 1.

The term 'quantum-dimension' comes from quantum groups, where $\frac{S_{\lambda 0}}{S_{00}}$ is the quantum-dimension of the module labelled by λ of the quantum group $U_q(X_\ell)$.

The borderline case then [16] is when a quantum-dimension *equals* 1. Any such weight is called a *simple-current*. The theory of simple-currents was developed most extensively by Schellekens and collaborators (see e.g. [50]). The simple-currents for the affine algebras were classified by J. Fuchs (1991), and the result is that (with one unimportant exception: $E_8^{(1)}$ at level 2) they all correspond to symmetries of the extended Dynkin diagrams. In particular, applying any such symmetry to the vacuum $0 = (k, 0, \ldots, 0)$ gives the list of simple-currents. For instance, the $\ell + 1$ weights of the form $(0, \ldots, k, \ldots, 0)$ (k in the ith spot) are the simple-currents for $A_\ell^{(1)}$. There are 2 simple-currents for $B_\ell^{(1)}$, $C_\ell^{(1)}$ and $E_7^{(1)}$, 3 for $E_6^{(1)}$, and 4 for $D_\ell^{(1)}$. Simple-currents play a large role in RCFT, as we shall see.

Let j be any simple-current. Then (1.7.3) becomes $S_{0\mu} \geq |S_{j\mu}|$ for all μ, so unitarity forces $S_{0\mu} = |S_{j\mu}|$, that is

$$S_{j\mu} = \exp[2\pi\mathrm{i}\, Q_j(\mu)]\, S_{0\mu} \qquad \forall \mu \in P_+^k \tag{1.7.4}$$

for some rational numbers $0 \leq Q_j(\mu) < 1$. Hence by diagonalising, we get $N_j N_{Cj} = I$. But the inverse of a nonnegative matrix A is itself nonnegative, only if A is a 'generalised permutation matrix', i.e. a permutation matrix except the 1's can be replaced by any positive numbers. But N_j and N_{Cj} are also integral, and so they must in fact be permutation matrices. Write $(N_j)_{\lambda\mu} = \delta_{\mu, J\lambda}$ for some permutation J of P_+^k. So $j = J0$. Then

$$\delta_{\lambda,\mu} = N_{j,\lambda}^{J\mu} = \sum_\nu \exp[2\pi\mathrm{i}\, Q_j(\nu)]\, S_{\lambda\nu}\, S_{J\mu,\nu}^* \ ,$$

---

[16] This seems to be a standard trick in math: when some sort of bound is established, look at the extremal cases which realise that bound. If your bound is a good one, it should be possible to say something about those extremal cases, and having something to say is always of paramount importance. This trick is used for instance in the definition of 24 last section, and the definition of normal subgroup in section 2.2.

so taking absolute values and using the triangle inequality and unitarity of S, we find that (1.7.4) generalises:

$$S_{J\lambda,\mu} = \exp[2\pi i \, Q_j(\mu)] \, S_{\lambda\mu} . \tag{1.7.5}$$

The simple-currents form a finite abelian group, corresponding to the composition of the permutations J. For any simple-currents J, J', we get the symmetry $N^\nu_{J\lambda, J'\mu} = N^{JJ'\nu}_{\lambda\mu}$. The $\mathbb{Q}/\mathbb{Z}$-valued functions Q_j define gradings on the fusion rings, and conversely any grading corresponds to a simple-current in this way.

For example, the simple-current $j = (0, k)$ of $A_1^{(1)}$ at level k corresponds to $Q_j(\lambda) = \lambda_1/2$ and the permutation $J\lambda = (\lambda_1, \lambda_0)$. We can see this directly from (1.4.3). For $A_2^{(1)}$ level k, there are 2 nontrivial simple-currents, $(0, k, 0)$ and $(0, 0, k)$. The first of these corresponds to triality $Q(\lambda) = (\lambda_1 + 2\lambda_2)/3$ and $\lambda \mapsto (\lambda_2, \lambda_0, \lambda_1)$, while the second to $(2\lambda_1 + \lambda_2)/3$ and $\lambda \mapsto (\lambda_1, \lambda_2, \lambda_0)$. Similar statements hold for all affine algebras: e.g. for $B_\ell^{(1)}$ level k, the nontrivial simple-current has $Q_j(\lambda) = \lambda_\ell/2$ and $J\lambda = (\lambda_1, \lambda_0, \lambda_2, \ldots, \lambda_\ell)$.

One of the applications of simple-currents is that physical invariants can be built from them in generic ways. These physical invariants all obey the selection rule

$$M_{\lambda\mu} \neq 0 \quad \Longrightarrow \quad \mu = J\lambda \text{ for some simple-current } J = J(\lambda, \mu) . \tag{1.7.6}$$

We will call any such physical invariant M a *simple-current invariant*. A special case is the $\mathcal{D}_{\frac{k}{2}+2}$ physical invariant for $A_1^{(1)}$ at even level k. Up to a fairly mild assumption, all simple-current invariants have been classified for any RCFT by Schellekens and collaborators; given that assumption, they can all be constructed by generic methods. The basic construction is due to Bernard [3], though it has been generalised by others. In the WZW case, all simple-current invariants (except some for $D_\ell^{(1)}$) correspond to strings on nonsimply-connected Lie groups.

By a *generic physical invariant* of $X_\ell^{(1)}$ we mean one of the form $M = C'M'C''$ where C', C'' are (charge-)conjugations, and M' is a simple-current invariant. In other words, M is constructed in generic ways from symmetries of the extended Dynkin diagram of X_ℓ. Any other M are called *exceptional*.

All known results point to the validity of the following guess:

Conjecture. Choose a simple algebra X_ℓ. Then for all sufficiently large k, all physical invariants of $X_\ell^{(1)}$ at level k will be generic.

In other words, any given $X_\ell^{(1)}$ will have only finitely many exceptionals. For instance, for $A_1^{(1)}$ and $A_2^{(1)}$ at any $k > 28$ and $k > 21$ resp., all physical invariants are generic. For $C_2^{(1)}$ and $G_2^{(1)}$, $k > 12$ and $k > 4$ resp. should work.

The richest source of exceptionals are *conformal embeddings*. In some cases the affine representations L_λ for some algebra $X_\ell^{(1)}$ (necessarily at level 1) can be decomposed into *finite* direct sums of representations of some affine subalgebra $Y_m^{(1)}$ (at some level k). In this case, a physical invariant for $X_\ell^{(1)}$ level 1 will yield a physical invariant for $Y_m^{(1)}$ level k, obtained by replacing every $X_\ell^{(1)}$ level 1 character χ_λ by the appropriate finite sum of

$Y_m^{(1)}$ level k characters. An example will demonstrate this simple idea: $A_1^{(1)}$ level 28 is a conformal subalgebra of $G_2^{(1)}$ level 1, and we have the character decompositions

$$\chi_{(1,0,0)} = \chi'_{(28,0)} + \chi'_{(18,10)} + \chi'_{(10,18)} + \chi'_{(0,28)}$$
$$\chi_{(0,0,1)} = \chi'_{(22,6)} + \chi'_{(16,12)} + \chi'_{(12,16)} + \chi'_{(6,22)} .$$

Thus the unique level 1 $G_2^{(1)}$ physical invariant $|\chi_0|^2 + |\chi_{(0,0,1)}|^2$ yields what we call the $\mathcal{E}_8$ physical invariant of $A_1^{(1)}$. All level 1 physical invariants are known, as are all conformal embeddings and the corresponding character decompositions (branching rules).

1.8. GALOIS THEORY

Evariste Galois was a brilliantly original French mathematician. Born shortly before Napoleon's ill-fated invasion of Russia, he died shortly before the ill-fated 1832 uprising in Paris. His last words: "Don't cry, I need all my courage to die at 20".

Galois grew up in a time and place confused and excited by revolution. He was known to say "if I were only sure that a body would be enough to incite the people to revolt, I would offer mine". On May 2 1832, after frustration over failure in love and failure to convince the Paris math establishment of the depth of his ideas, he made his decision. A duel was arranged with a friend, but only his friend's gun would be loaded. Galois died the day after a bullet perforated his intestine. At his funeral it was discovered that a famous general had also just died, and the revolutionaries decided to use the general's death rather than Galois' as a pretext for an armed uprising. A few days later the streets of Paris were blocked by barricades, but not because of Galois' sacrifice: his death had been pointless [56].

Galois theory in its most general form is the study of relations between objects defined implicitly by some conditions. For example, the objects could be the solutions to a given differential equation. In the incarnation of Galois we are interested in here, the objects are numbers, namely the zeros of certain polynomials. We will sketch this theory below, but see e.g. the article by Stark in [60] for more details.

Gauss seems to have been the first to show that 'weird' (complex) numbers could tell us about the integers. For instance, suppose we are interested in the equation $n = a^2 + b^2$. Consider $5 = 2^2 + 1^2$. We can write this as $5 = (2 + i)(2 - i)$, so we are led to consider complex numbers of the form $a + bi$, for $a, b \in \mathbb{Z}$. These are now called 'Gaussian integers'. Suppose we know the following theorem:

Fact. Let $p \in \mathbb{Z}$ be any prime number. Then p factorises over the Gaussian integers iff $p = 2$ or $p \equiv 1 \pmod 4$.

By 'factorise' there, we mean $p = zw$ where neither z nor w is a 'unit': ± 1, $\pm i$.

Now suppose p is a prime, $= 2$ or $\equiv 1 \pmod 4$, and we write $p = (a + bi)(c + di)$. Then $p^2 = (a^2 + b^2)(c^2 + d^2)$, so $a^2 + b^2 = c^2 + d^2 = p$. Conversely, suppose $p = a^2 + b^2$, then $p = (a + bi)(a - bi)$. Thus:

Consequence. [17] Let $p \in \mathbb{Z}$ be any prime number. Then $p = a^2 + b^2$ for $a, b \in \mathbb{Z}$ iff $p = 2$ or $p \equiv 1 \pmod 4$.

Now we can answer the question: can a given n be written as a sum of 2 squares $n = a^2 + b^2$? Write out the prime decomposition $n = \prod p^{a_p}$. Then $n = a^2 + b^2$ has a solution iff a_p is *even* for every $p \equiv 3 \pmod 4$. For instance $60 = 2^2 \cdot 3^1 \cdot 5^1$ cannot be written as the sum of 2 squares, but $90 = 2^1 \cdot 3^2 \cdot 5^1$ can. We can also find (and count) all solutions: e.g. $90 = 2 \cdot 3^2 \cdot 5 = \{(1 + i)3(1 + 2i)\}\{(1 - i)3(1 - 2i)\}$, giving $90 = (-3)^2 + 9^2$.

This problem should give the reader a small appreciation for the power of using non-integers to study integers. Nonintegers often lurk in the shadows, secretly watching their more arrogant brethren the integers strut. One of the consequences of their presence can be the existence of certain 'Galois' symmetries. Such happens in RCFT, as we will show below.

Look at complex conjugation: $(wz)^* = w^* z^*$ and $(w + z)^* = w^* + z^*$. Also, $r^* = r$ for any $r \in \mathbb{R}$. So we can say that $*$ is a structure-preserving map $\mathbb{C} \to \mathbb{C}$ (called an *automorphism* of $\mathbb{C}$) fixing the reals. We will write this $* \in \mathrm{Gal}(\mathbb{C}/\mathbb{R})$. '$\mathrm{Gal}(\mathbb{C}/\mathbb{R})$' is the Galois group of $\mathbb{C}$ over $\mathbb{R}$; it turns out to contain only $*$ and the identity.

A way of thinking about the automorphism $*$ is that it says that, as far as the real numbers are concerned, i and $-$i are identical twins.

Let $\mathbb{F}$ be any *field* containing $\mathbb{Q}$ (we defined 'field' in §1.2). The *Galois group* $\mathrm{Gal}(\mathbb{F}/\mathbb{Q})$ then will be the set of all automorphisms=symmetries of $\mathbb{F}$ which fix all rationals.

For example, take $\mathbb{F}$ to be the set of all numbers of the form $a + b\sqrt{5}$, where $a, b \in \mathbb{Q}$. Then $\mathbb{F}$ will be a field, which is commonly denoted $\mathbb{Q}[\sqrt{5}]$ because it is generated by $\mathbb{Q}$ and $\sqrt{5}$. Let's try to find its Galois group. Let $\sigma \in \mathrm{Gal}(\mathbb{F}/\mathbb{Q})$. Then $\sigma(a + b\sqrt{5}) = \sigma(a) + \sigma(b)\sigma(\sqrt{5}) = a + b\sigma(\sqrt{5})$, so once we know what σ does to $\sqrt{5}$, we know everything about σ. But $5 = \sigma(5) = \sigma(\sqrt{5}^2) = (\sigma(\sqrt{5}))^2$, so $\sigma(\sqrt{5}) = \pm\sqrt{5}$ and there are precisely 2 possible Galois automorphisms here (one is the identity). As far as $\mathbb{Q}$ is concerned, $\pm\sqrt{5}$ are interchangeable: it cannot see the difference.

For a more important example, consider the *cyclotomic field* $\mathbb{F} = \mathbb{Q}[\xi_n]$, where $\xi_n := \exp[2\pi i/n]$ is an nth root of 1. So $\mathbb{Q}[\xi_n]$ consists of all complex numbers which can be expressed as polynomials $a_m \xi_n^m + a_{m-1}\xi_n^{m-1} + \cdots + a_0$ in ξ_n with rational coefficients a_i. Once again, to find the Galois group $\mathrm{Gal}(\mathbb{Q}[\xi_n]/\mathbb{Q})$, it is enough to see what an automorphism σ does to the generator ξ_n. Since $\xi_n^n = 1$, we see that it must send it to another nth root of 1, ξ_n^ℓ say; in fact it is easy to see that $\sigma(\xi_n)$ must be another 'primitive' nth root of 1, i.e. ℓ must be coprime to n. So $\mathrm{Gal}(\mathbb{Q}[\xi_n]/\mathbb{Q})$ will be isomorphic to the multiplicative group $\mathbb{Z}_n^\times$ of numbers between 1 and n coprime to n. The rationals can't see any difference between the primitive nth roots of 1 — for instance $\mathbb{Q}$ can't tell that $\xi_n^{\pm 1}$ are 'closer to 1' than the other primitive roots. So any $\sigma \in \mathrm{Gal}(\mathbb{Q}[\xi_n]/\mathbb{Q})$ will correspond to some $\ell \in \mathbb{Z}_n^\times$, and to see what σ does to some $z \in \mathbb{Q}[\xi_n]$ what we do is write z as a polynomial in ξ_n and then replace each occurrence of ξ_n with ξ_n^ℓ. For example,

$$\sigma\big(\cos(2\pi a/n)\big) = \sigma\left(\frac{\xi_n^a + \xi_n^{-a}}{2}\right) = \frac{\xi_n^{a\ell} + \xi_n^{-a\ell}}{2} = \cos(2\pi a\ell/n) \ .$$

---

[17] This result was first stated by Fermat in one of his infamous margin notes (another is discussed in Section 2.3), and was finally proved a century later by Euler. A remarkable 1-line proof was found by Zagier [61].

So to summarise, Galois automorphisms are a massive generalisation of the idea of complex conjugation. If in your problem complex conjugation seems interesting, then there is a good chance more general Galois automorphisms will play an interesting role. This is what happens in RCFT, as we now show.

Fact. [14] Suppose S is unitary and symmetric and each $S_{0a} > 0$.
  (a) If in addition the numbers N_{ab}^c given by Verlinde's formula (1.4.4) are rational, then the entries S_{ab} of S must lie in a cyclotomic field.
  (b) The numbers N_{ab}^c will be rational iff for any $\sigma \in \mathrm{Gal}(\mathbb{Q}[S]/\mathbb{Q})$, there is a permutation $a \mapsto a^\sigma$, and a choice of signs $\epsilon_\sigma(a) \in \{\pm 1\}$, such that

$$\sigma(S_{ab}) = \epsilon_\sigma(a)\, S_{a^\sigma,b} = \epsilon_\sigma(b)\, S_{a,b^\sigma} \ . \tag{1.8.1}$$

'$\mathbb{Q}[S]$' in part (b) denotes the field generated by $\mathbb{Q}$ and all matrix entries S_{ab}. The argument follows the one given for the charge-conjugation C at the beginning of the last section. The kinds of complex numbers which lie in cyclotomic fields are $\sin(\pi r)$, $\cos(\pi r)$, $\sqrt{r}$ and ri for any $r \in \mathbb{Q}$. Almost all complex numbers fail to lie in any cyclotomic field: e.g. generic cube roots, 4th roots, ..., of rationals, as well as transcendental numbers like e, π and e^π.

Of course the affine algebras satisfy the conditions of the Fact, as does more generally the modular matrix S for any unitary RCFT, and so these will possess the Galois action. For the affine algebras this action has a geometric interpretation in terms of multiplying weights by an integer ℓ and applying Weyl group elements — see [14] for a description.

This Fact is useful in both directions: as a way of testing whether a conjectured matrix S has a chance of producing the integral fusions we want it to yield; and more importantly as a source of a symmetry of the RCFT which generalises charge-conjugation. Any statement about charge-conjugation seems to have an analogue for any of these Galois symmetries, although it is usually more complicated.

As an example, consider $A_1^{(1)}$: (1.4.3) shows explicitly that $S_{\lambda\mu}$ lies in the cyclotomic field $\mathbb{Q}[\xi_{4(k+2)}]$. Write $\{x\}$ for the number congruent to $x \bmod 2(k+2)$ satisfying $0 \le \{x\} < 2(k+2)$. Choose any Galois automorphism σ, and let $\ell \in \mathbb{Z}_{4(k+2)}^\times$ be the corresponding integer. Then if $\{\ell(a+1)\} < k+2$, we will have $a^\sigma = \{\ell(a+1)\}-1$, while if $\{\ell(a+1)\} > k+2$, we'll have $a^\sigma = 2(k+2) - \{\ell(a+1)\} - 1$. The sign $\epsilon_\sigma(a)$ will depend on a contribution from $\sqrt{\frac{2}{k+2}}$ (which for most purposes can be ignored), as well as the sign $+1$ or -1, resp., depending on whether or not $\{\ell(a+1)\} < k+2$.

Consider specifically $k = 10$, and the Galois automorphism σ_5 corresponding to $\ell = 5$. Then the permutation is $0 \leftrightarrow (6,4)$, $(9,1) \leftrightarrow (1,9)$, $(8,2) \leftrightarrow (2,8)$, $(4,6) \leftrightarrow (0,10)$, while $(7,3)$ and $(5,5)$ are fixed.

This Galois symmetry has been used to find certain exceptional physical invariants, but its greatest use so far is as a powerful selection rule we will describe next section.

1.9. THE MODERN APPROACH TO CLASSIFYING PHYSICAL INVARIANTS

In this final section we include some of the basic tools belonging to the 'modern' classifications of physical invariants, and we give a flavour of their proofs. We will state them for the $A_1^{(1)}$ level k problem given above, but everything generalises without effort. See [29] and references therein for more details. Recall the matrices S, T in (1.4.3).

First note that commutation of M with T implies the selection rule

$$M_{\lambda\mu} \neq 0 \implies (\lambda_1 + 1)^2 \equiv (\mu_1 + 1)^2 \pmod{4(k+2)} . \tag{1.9.1}$$

It is much harder to squeeze information out of the commutation with S, but the resulting information turns out surprisingly to be much more useful. In fact, commutation with S is almost incompatible with the constraint $M_{\lambda\mu} \in \mathbb{Z}_{\geq}$.

Note that the vacuum $0 \in P_+^k$ is both physically and mathematically special; our strategy will be to find all possible 0th rows and columns of M, and then for each of these possibilities to find the remaining entries of M.

The easiest result follows by evaluating $MS = SM$ at $(0, \lambda)$ for any $\lambda \in P_+^k$:

$$\sum_{\mu \in P_+^k} M_{0\mu} S_{\mu\lambda} \geq 0 , \tag{1.9.2}$$

with equality iff the λth column of M is identically 0. (1.9.2) has two uses: it severely constrains the values of $M_{0\mu}$ (similarly $M_{\mu 0}$), and it says precisely which columns (and rows) are nonzero.

Next, let's apply the triangle inequality to sums involving (1.7.5). Choose any $i, j \in \{0, 1\}$. Then

$$M_{J^i 0, J^j 0} = \sum_{\lambda, \mu} (-1)^{\lambda_1 i} S_{0\lambda} M_{\lambda\mu} (-1)^{\mu_1 j} S_{0\mu} .$$

Taking absolute values, we obtain

$$M_{J^i 0, J^j 0} \leq \sum_{\lambda, \mu} S_{0\lambda} M_{\lambda\mu} S_{0\mu} = M_{00} = 1 .$$

Thus $M_{J^i 0, J^j 0}$ can equal only 0 or 1. If it equals 1, then we obtain the selection rule:

$$\lambda_1 i \equiv \mu_1 j \pmod 2 \text{ whenever } M_{\lambda\mu} \neq 0 ;$$

this implies the symmetry $M_{J^i \lambda, J^j \mu} = M_{\lambda\mu}$ for all $\lambda, \mu \in P_+^k$. We can see both of these in the list of physical invariants for $A_1^{(1)}$ level k. This explains a lot of the properties of those invariants. For instance, try to use this selection rule to explain why no χ_{odd} appears in the exceptional called $\mathcal{E}_8$.

Our M is nonnegative, and although multiplying M's may not give us back a physical invariant, it will give us a matrix commuting with S and T. In other words, the commutant is much more than merely a vector space, it is in fact an algebra. Thus we should expect Perron-Frobenius to tell us something here. A first application is the following.

Suppose $M_{\lambda 0} = \delta_{\lambda,0}$ — i.e. the 0th column of M is all zeros except for $M_{00} = 1$. Then Perron-Frobenius implies (with a little work) that M will be a permutation matrix — i.e. there is some permutation π of P_+^k such that $M_{\lambda\mu} = \delta_{\mu,\pi\lambda}$, and $S_{\pi\lambda,\pi\mu} = S_{\lambda\mu}$. This nice fact applies directly to the $\mathcal{A}_*$ and $\mathcal{D}_{odd}$ physical invariants of $A_1^{(1)}$.

This is proved by studying the powers $(M^t M)^L$ as L goes to infinity: its diagonal entries will grow exponentially with L, unless there is at most one nonzero entry on each row of M, and that entry equals 1.

More careful reasoning along those lines tells us about the other generic situation here. Namely, suppose $M_{\lambda 0} \neq 0$ only for $\lambda = 0$ and $\lambda = J0$, and similarly for $M_{0\lambda}$ — i.e. the 0th row and column of M are all zeros except for $M_{J^i 0, J^i 0} = 1$. Then the λth row (or column) of M will be identically 0 iff λ_1 is odd. Moreover, let λ, μ be any non-fixed-points of J, and suppose $M_{\lambda\mu} \neq 0$. Then

$$M_{\lambda\nu} = \begin{cases} 1 & \text{if } \nu = \mu \text{ or } \nu = J\mu \\ 0 & \text{otherwise} \end{cases}$$

with a similar formula for $M_{\nu\mu}$. This applies to the $\mathcal{D}_{even}$ and $\mathcal{E}_7$ invariants of $A_1^{(1)}$.

Our final ingredient is the Galois symmetry (1.8.1) obeyed by S. Choose any Galois automorphism σ. It will correspond to some integer ℓ coprime to $2(k+2)$. From (1.8.1) and $M = SMS^*$ we get, for all λ, μ, the important relation

$$M_{\lambda\mu} = \epsilon_\sigma(\lambda)\,\epsilon_\sigma(\mu)\,M_{\lambda^\sigma,\mu^\sigma} . \tag{1.9.3}$$

From (1.9.3) and the positivity of M, we obtain the powerful *Galois selection rule*

$$M_{\lambda\mu} \neq 0 \quad \Longrightarrow \quad \epsilon_\sigma(\lambda) = \epsilon_\sigma(\mu) . \tag{1.9.4}$$

Next let us quickly sketch how these tools are used to obtain the $A_1^{(1)}$ classification. For details the reader should consult [29].

The first step will be to find all possible values of λ such that $M_{0\lambda} \neq 0$ or $M_{\lambda 0} \neq 0$. These λ are severely constrained. We know two generic possibilities: $\lambda_1 = 0$ (good for all k), and $\lambda = J0$ (good when $\frac{k}{2}$ is even). We now ask the question, what other possibilities for λ are there? Our goal is to prove (1.9.7). Assume $\lambda \neq 0, J0$, and write $a = \lambda_1 + 1$ and $n = k + 2$.

There are only two constraints on λ which we will need. One is (1.9.1):

$$(a-1)(a+1) \equiv 0 \pmod{4n} . \tag{1.9.5}$$

More useful is the Galois selection rule (1.9.4), which we can write as $\sin(\pi\ell\frac{a}{n})\sin(\pi\ell\frac{1}{n}) > 0$, for all those ℓ. But a product of sines can be rewritten as a difference of cosines, so

$$\cos(\pi\ell\,\frac{a-1}{n}) > \cos(\pi\ell\,\frac{a+1}{n}) . \tag{1.9.6}$$

(1.9.6) is strong and easy to solve; the reader should try to find her own argument. What we get is that, provided $n \neq 12, 30$, M obeys the strong condition

$$M_{\lambda 0} \neq 0 \text{ or } M_{0\lambda} \neq 0 \qquad \Longrightarrow \qquad \lambda \in \{0, J0\} . \qquad (1.9.7)$$

Consider first **case 1**: $M_{\lambda 0} = \delta_{\lambda, 0}$. From above, we know $M_{\lambda\mu} = \delta_{\mu, \pi\lambda}$ for some permutation π of P_+^k obeying $S_{\lambda\mu} = S_{\pi\lambda, \pi\mu}$. We know $\pi 0 = 0$; put $\mu := \pi(k-1, 1)$. Then $\sin(\pi \frac{2}{n}) = \sin(\pi \frac{\mu_1 + 1}{n})$, and so we get either $\mu = (k-1, 1)$ or $\mu = J(k-1, 1)$. By T-invariance (1.9.1), the second possibility can only occur if $4 \equiv (n-2)^2 \pmod{4n}$, i.e. 4 divides n. But for those n, $\mathcal{D}_{\frac{n}{2}+1}$ is also a permutation matrix, so replacing M if necessary with the matrix product $M \mathcal{D}_{\frac{n}{2}+1}$, we can always require $\mu = (k-1, 1)$, i.e. π also fixes $(k-1, 1)$. It is now easy to show π must fix any λ, i.e. that M is the identity matrix $\mathcal{A}_{n-1}$.

The other possibility, **case 2**, is that both $M_{0, J0} \neq 0$ and $M_{J0, 0} \neq 0$. (1.9.1) says $1 \equiv (n-1)^2 \pmod{4n}$, i.e. $\frac{n}{2}$ is odd. The argument here is similar to that of **case 1**, but with $(k-2, 2)$ playing the role of $(k-1, 1)$. We can show that $M_{(k-2, 2), (k-2, 2)} \neq 0$, except possibly for $k = 16$, where we find the exceptional $\mathcal{E}_7$. Otherwise we get $M = \mathcal{D}_{\frac{n}{2}+1}$.

For more general $X_\ell^{(1)}$ level k, the approach is

(i) to look at all the constraints on the $\lambda \in P_+^k$ for which $M_{0\lambda} \neq 0$ or $M_{\lambda 0} \neq 0$. Most important here are $TM = MT$ (which will always be some sort of norm selection rule) and the Galois selection rule (1.9.4). Generically, what we will find is that such a λ must equal $J0$ for some simple-current J, as in (1.9.7) for $A_1^{(1)}$.

(ii) Solve this generic case (in the $A_1^{(1)}$ classification, these were the physical invariants $\mathcal{A}_*$, $\mathcal{D}_*$ and $\mathcal{E}_7$).

(iii) Solve the nongeneric case. The worst of these are the orthogonal algebras at $k = 2$, as well as the places where conformal embeddings (see §1.7) occur.

(ii) has recently been completed for all simple X_ℓ, as has the $k = 2$ part of (iii). (i) is the main remaining task in the physical invariant classification for simple X_ℓ.

A natural question to ask is whether A-D-E has been observed in e.g. the $A_2^{(1)}$ classification. The answer is no, although the fusion graph theory of Di Francesco-Petkova-Zuber [62] is an attempt to assign to these physical invariants graphs reminiscent of the A-D-E Dynkin diagrams. Also, there is related work trying to understand the $A_2^{(1)}$ classification in terms of subgroups of $SU_3(\mathbb{C})$ (as opposed to $SU_2(\mathbb{C})$ for $A_1^{(1)}$) — see e.g. [35]. Finding the $A_3^{(1)}$, $A_4^{(1)}$,... classifications would permit the clarification and testing of this vaguely conjectured relation between the $A_n^{(1)}$ physical invariants, and singularities $\mathbb{C}^{n+1}/G$ for G a finite subgroup of $SU_{n+1}(\mathbb{C})$.

However, a few years ago Philippe Ruelle was walking in a library in Dublin. He spotted a yellow book in the math section, called *Complex Multiplication* by Lang. A strange title for a book by Lang! After all, there can't be all that much even Lang could really say about complex multiplication! Ruelle flipped it to a random page, which turned out to be p.26. On there he found what we would call the Galois selection rule for $A_2^{(1)}$, analysed and solved for the cases where $k + 3$ is coprime to 6. Lang however didn't know about physical invariants; he was reporting on work by Koblitz and Rohrlich on decomposing the Jacobians of the Fermat curve $x^n + y^n = z^n$ into their prime pieces, called 'simple factors' in algebraic geometry. n here corresponds to $k + 3$. Similarly, Itzykson discovered traces of the $A_2^{(1)}$ exceptionals — these occur when $k + 3 = 8, 12, 24$ — in the Jacobian of $x^{24} + y^{24} = z^{24}$. See [2] for further observations along these lines. These 'coincidences' are still far from understood. Nor is it known if, more generally, the $A_\ell^{(1)}$ level k classification will somehow be related to the hypersurface $x_1^n + \cdots + x_\ell^n = z^n$, for $n = k + \ell + 1$.

The $(u(1) \oplus \cdots \oplus u(1))^{(1)}$ classification has connections to rational points on Grassmannians. The Grassmannian is (essentially) the moduli space for the Narain compactifications of the (classical) lattice string. It would be very interesting to interpret other large families of physical invariants as special points on other moduli spaces.

These new connections relating various physical invariant classifications to other areas of math seem to indicate that although the physical invariant classifications are difficult, they could be well worth the effort and be of interest outside RCFT. Once the physical invariant lists are obtained, we will still have the fascinating task of explaining and developing all these mysterious connections. These thoughts keep me going!

Another motivation for completing these lists comes from their relation to subfactor theory in von Neumann algebras[18]. These algebras (see e.g. [22]) can be thought of as symmetries of a (generally infinite) group. Their building blocks are called *factors*. Jones initiated the combinatorial study of *subfactors* N of M (i.e. inclusions $N \subseteq M$ where M, N are factors), relating it to e.g. knots, and for this won a Fields medal in 1990. Jones assigned to each subfactor $N \subseteq M$ a numerical invariant called an 'index', a sort of (generally irrational) ratio of dimensions. Graphs (called principal and dual principal) are also associated to subfactors. A much more refined subfactor invariant, called a 'paragroup', has been introduced by Ocneanu. It is essentially equivalent to a $(2+1)$-dimensional topological field theory. Moreover, any RCFT can be assigned a paragroup, and any paragroup (via a process called asymptotic inclusion which is akin to Drinfeld's quantum doubling of Hopf algebras) yields an RCFT. See [22] for details.

Böckenhauer-Evans [4] have recently developed this much further, and have clarified the fusion graph $\leftrightarrow$ physical invariant relation. The fusion graphs will correspond to subfactor principal graphs. In the work of Di Francesco-Petkova-Zuber, that relation seems to be only empirical (i.e. nonconceptual).

Subfactor theory together with singularity theory is our best hope at present for understanding and generalising the A-D-E meta-pattern.

---

[18] For reasons of necessity, in the following discussion I'll take more liberties than usual in the presentation.

Part 2. Monstrous Moonshine

2.1. INTRODUCTION

In 1978, John McKay made a very curious observation. One of the well-known[19] functions of classical number theory is the j-function[20], given by

$$j(\tau) := \frac{(1 + 240 \sum_{n=1}^{\infty} \sigma_3(n) q^n)^3}{q \prod_{n=1}^{\infty} (1 - q^n)^{24}} = \frac{\Theta_{E_8}(\tau)^3}{\eta(\tau)^{24}}$$

$$= q^{-1} + 744 + 196\,884\,q + 21\,493\,760\,q^2 + 864\,299\,970\,q^3 + \cdots \qquad (2.1.1)$$

Here as elsewhere in this paper, $q = \exp[2\pi i\,\tau]$. Also, $\sigma_3(n) = \sum_{d|n} d^3$, Θ_{E_8} is the theta function of the E_8 root lattice, and η is the Dedekind eta. What is important here are the values of the first few coefficients. What McKay noticed was that $196\,884 \approx 196\,883$. Closer inspection shows $21\,493\,760 \approx 21\,296\,876$, and $864\,299\,970 \approx 842\,609\,326$. In fact,

$$196\,884 = 196\,883 + 1 \qquad (2.1.2a)$$

$$21\,493\,760 = 21\,296\,876 + 196\,883 + 1 \qquad (2.1.2b)$$

$$864\,299\,970 = 842\,609\,326 + 21\,296\,876 + 2 \cdot 196\,883 + 2 \cdot 1 \qquad (2.1.2c)$$

The numbers on the right-side are the dimensions of the smallest irreducible representations of the Monster finite simple group $\mathbb{M}$ (in 1978 it still wasn't certain that $\mathbb{M}$ even existed so back then these numbers were merely conjectural). The same game could be played with other coefficients of the j-function. With numbers so large, it seemed to him doubtful that this numerology was merely a coincidence. On the other hand, it was hard to imagine any deep conceptual connection between the Monster and the j-function: they seem completely unrelated.

In November 1978 he mailed the 'McKay equation' (2.1.2a) to John Thompson. At first Thompson dismissed this as nonsense, but after checking the next few coefficients he became convinced. He then added a vital piece to the puzzle. It should be well-known that when one sees a nonnegative integer, it often helps to try to interpret it as the dimension of some vector space. Essentially, that is what McKay was proposing here. (2.1.2) are really hinting that there is a 'graded' representation V of $\mathbb{M}$:

$$V = V_{-1} \oplus V_1 \oplus V_2 \oplus V_3 \oplus \cdots$$

where $V_{-1} = \rho_0$, $V_1 = \rho_1 \oplus \rho_0$, $V_2 = \rho_2 \oplus \rho_1 \oplus \rho_0$, $V_3 = \rho_3 \oplus \rho_2 \oplus \rho_1 \oplus \rho_1 \oplus \rho_0 \oplus \rho_0$, etc, where ρ_i are the irreducible representations of $\mathbb{M}$ (ordered by dimension), and that

$$j(\tau) - 744 = \dim_q(V) := \dim(V_{-1})\,q^{-1} + \sum_{i=1}^{\infty} \dim(V_i)\,q^i \,, \qquad (2.1.3)$$

---

[19] 'Well-known' is math euphemism for 'a basic result of which until recently we were utterly ignorant.' As Conway later said, "the j-function was 'well-known' to other people, but not 'well-known' to me."

[20] This and other technical terms used in this introduction will be carefully explained in the following subsections. This section is merely offered as a quick overview.

the graded dimension of V.

Thompson suggested that we twist $\dim_q(V)$, i.e. that more generally we consider the series (now called the *McKay-Thompson series*)

$$T_g(\tau) := \mathrm{ch}_{V,q}(g) = \mathrm{ch}_{V_{-1}}(g)\, q^{-1} + \sum_{i=1}^{\infty} \mathrm{ch}_{V_i}(g)\, q^i \,, \qquad (2.1.4)$$

for each element $g \in \mathbb{M}$. The point is that, for any group representation ρ, the character value $\mathrm{ch}_\rho(id.)$ equals the dimension of ρ, and so $T_{id.}(\tau) = j(\tau) - 744$ and we recover (2.1.2) as special cases. But there are many other possible choices of $g \in \mathbb{M}$. Thompson couldn't guess what these functions T_g would be, but he suggested that they too might be interesting. This is a nice thought: when we see a positive integer, we should try to interpret it as a dimension of a vector space; if there is a symmetry present, then it may act on the vector space — i.e. our vector space may carry a representation of that symmetry group — in which case we can apply the Thompson trick and see what if any significance the other character values have in our context.

Conway and Norton [12] did precisely what Thompson asked. Conway called it "one of the most exciting moments in my life" [11] when he opened Jacobi's foundational (but 150 year old!) book on elliptic and modular functions and found that the first few terms of the McKay-Thompson series agreed perfectly with the first few terms of certain special functions, namely the Hauptmoduls of various genus 0 modular groups. Monstrous Moonshine was officially born.

The word 'moonshine' here is English slang for 'unsubstantial or unreal'. It was chosen by Conway to convey as well the feeling that things here are dimly lit, and that Conway-Norton were 'distilling information illegally' from the Monster character table.

In fact the first incarnation of Moonshine goes back to Andrew Ogg in 1975. He was in France describing his result that the primes p for which the group $\Gamma_0(p)+$ has genus 0, are $\{2, 3, 5, 11, 13, 17, 19, 23, 29, 31, 41, 47, 59, 71\}$. $\Gamma_0(p)+$ is the group generated by $\begin{pmatrix} 0 & 1 \\ -p & 0 \end{pmatrix}$ and $\Gamma_0(p)$, and is the normaliser of $\Gamma_0(p)$ in $\mathrm{SL}_2(\mathbb{R})$ (this sentence will make a little more sense after §2.3, but it isn't important here to understand it). He also attended a lecture by Jacques Tits, who was describing a newly conjectured simple group. When Tits wrote down the prime decomposition of the order of that group (see (2.2.1) below), Ogg noticed its prime factors precisely equalled his list of primes. Presumably as a joke, he offered a bottle of Jack Daniels' whisky to the first person to explain the coincidence.

The next step was accomplished by Griess in 1980, with the construction of the Monster[21] $\mathbb{M}$, and with it the proof that the conjectured character table for $\mathbb{M}$ was correct. Griess did this by explicitly constructing the 196883-dimensional representation ρ_1; it turns out to have a (commutative nonassociative) algebra structure, now called the *Griess algebra*. Though this paper was clearly important, the construction was artificial and 100 pages long: since the Monster is presumably a natural mathematical object (see §2.2), an elegant construction for it should exist. This was ultimately accomplished in the mid 1980s with the construction by Frenkel-Lepowsky-Meurman [23] of the Moonshine module

---

[21] Griess also came up with the symbol for the Monster; Conway came up with the name.

$V^\natural$ and its interpretation by Borcherds as a *vertex operator algebra*. The Griess algebra appears naturally in $V^\natural$, as we shall see. $V^\natural$ does indeed seem to be a 'natural' mathematical structure, and $\mathbb{M}$ is its automorphism group: in fact $V^\natural$ is the graded representation V of $\mathbb{M}$ conjectured by McKay and Thompson.

Connections with physics (CFT) go back to Dixon-Ginsparg-Harvey [19] in 1988, in a paper titled "Beauty and the beast: Superconformal symmetry in a Monster module". The Moonshine module $V^\natural$ can be interpreted as the string theory for a $\mathbb{Z}_2$-orbifold of free bosons compactified on the torus $\mathbb{R}^{24}/\Lambda_{24}$ (Λ_{24} is the Leech lattice). Many aspects of Moonshine make complete sense within CFT, but some (e.g. the genus zero property) remain more obscure. (Though in 1987 Moore speculated that the 0-genus of $\Gamma_0(a)+$ could be related to the vanishing of the cosmological constant in certain string theories related to $\mathbb{M}$, and Tuite [57] related genus-zero with the conjectured uniqueness of $V^\natural$.) Nevertheless this helps make the words of Dyson ring prophetic: "I have a sneaking hope, a hope unsupported by any facts or any evidence, that sometime in the twenty-first century physicists will stumble upon the Monster group, built in some unsuspected way into the structure of the universe" [21].

Finally, in 1992 Borcherds [5] completed the proof of the Conway-Norton conjectures by showing $V^\natural$ is the desired representation V. The full conceptual relationship between the Monster and the Hauptmoduls (like j) seems to remain 'dimly lit', although much progress has been realised. This is a subject where it is much easier to conjecture than to prove, and we are still awash in unresolved conjectures.

McKay also noticed in 1978 that similar coincidences hold if $\mathbb{M}$ and $j(\tau)$ are replaced with the Lie group $E_8(\mathbb{C})$ and $(qj(q))^{\frac{1}{3}} = 1 + 248q + \cdots$. This turns out to be much easier to explain, and in 1980 both Kac and Lepowsky remarked that the unique level 1 highest-weight representation of the affine algebra $E_8^{(1)}$ has graded dimension $(qj(q))^{\frac{1}{3}}$.

Moonshiners have a little chip on their shoulders. Modern math, they say, tends to be a little too infatuated with the pursuit of generalisations for generalisations' sake. Surely a noble goal for math is to find interesting and fundamentally new theorems. It can be argued that both history and common-sense suggest that to this end it is most profitable to look simultaneously at both exceptional structures and generic structures, to understand the special features of the former in the context of the latter, and to be led in this way to a new generation of exceptional and generic structures. Moonshiners would sympathise with those biologists who study the duck-billed platypus and lungfish rather than hide them in the closet as monsters: BECAUSE they appear to be unique, those animals presumably have much to teach us about our general understanding of evolution, etc.

It often seems to people that Moonshine can't be very deep: the Conway-Norton conjectures seem to be so finite[22] and specialised. There only are 171 distinct McKay-Thompson series T_g in Monstrous Moonshine, after all. The whole point though is to try to understand *why* the Monster and the Hauptmoduls are so related, and then to try to extend and apply this understanding to other contexts. Moonshine is still young, and our

---

[22] Indeed the Moonshine conjectures *are* finite (it is enough to check the first 1200 coefficients), and a slightly weaker form was quickly proved on a computer by Atkin, Fong and Smith [54]. However this sort of argument adds no light to Moonshine, and tells us nothing of V except that it exists.

understanding remains incomplete. But already math has benefitted: e.g. we now have a natural definition of $\mathbb{M}$ (as the automorphism group of $V^\natural$), and Moonshine helped lead us to the rich structures of generalised Kac-Moody algebras and vertex operator algebras.

We will see that Moonshine involves the interplay between *exceptional structures* such as the number 24, the Leech lattice Λ_{24}, the Monster group $\mathbb{M}$, and the Moonshine module $V^\natural$, and *generic structures* such as modular functions, vertex operator algebras, generalised Kac-Moody algebras, and conformal field theories. The following sections will introduce the reader to many of these structures, as we use Moonshine as another happy excuse to take a second little tour through modern mathematics.

2.2. INGREDIENT #1: FINITE SIMPLE GROUPS AND THE MONSTER

A readable introduction to the basics of finite group representation theory is [25]. The finite simple groups are described in [33]; see also [11]. Group representations were introduced in §1.3.

A *normal* subgroup H of a group is one obeying $gHg^{-1} = H$ for all $g \in G$. These are important because the set G/H of 'cosets' gH has a natural group structure precisely when H is normal. Every group has two trivial normal subgroups: itself and $\{1\}$. If these are the only normal subgroups, the group is called *simple*. It is conventional to regard the trivial group $\{1\}$ as not simple (just as 1 is conventionally regarded as not prime). An alternate definition of a (finite) simple group G is that if $\varphi : G \to H$ is any group homomorphism (i.e. structure-preserving map: $\varphi(gg') = \varphi(g)\varphi(g')$), then φ is either constant (i.e. $\varphi(G) = \{1\}$), or φ is one-to-one.

The importance of simple groups is provided by the *Jordan-Hölder Theorem*. By a 'composition series' for a group G, we mean a nested sequence

$$G = H_0 \supset H_1 \supset H_2 \supset \cdots \supset H_k \supset H_{k+1} = \{1\}$$

of groups such that H_i is normal in H_{i-1}, and H_{i-1}/H_i (called a 'composition factor') is simple. Any finite group G has at least one composition series. If $H'_0 \supset \cdots \supset H'_{\ell+1} = \{1\}$ is a second composition series for G, then Jordan-Hölder says that $k = \ell$ and, up to a reordering π, the simple groups H_{i-1}/H_i and $H'_{\pi j-1}/H'_{\pi j}$ are isomorphic.

For example, the cyclic group $\mathbb{Z}_n$ of order($=$size) n — you can think of it as the integers modulo n under addition — is simple iff n is prime. Consider the group $\mathbb{Z}_{12} = \langle 1 \rangle$. Two composition series are

$$\mathbb{Z}_{12} \supset \langle 2 \rangle \supset \langle 4 \rangle \supset \langle 0 \rangle$$
$$\mathbb{Z}_{12} \supset \langle 3 \rangle \supset \langle 6 \rangle \supset \langle 0 \rangle$$

corresponding to composition factors $\mathbb{Z}_2$, $\mathbb{Z}_2$, $\mathbb{Z}_3$, and $\mathbb{Z}_3$, $\mathbb{Z}_2$, $\mathbb{Z}_2$. Of course this is consistent with Jordan-Hölder. This is reminiscent of the fact that $2 \cdot 2 \cdot 3 = 3 \cdot 2 \cdot 2$ are both prime factorisations of 12.

There is some value to regarding finite groups as a massive generalisation of the notion of number. The number n can be identified with the cyclic group $\mathbb{Z}_n$. The divisor of a number corresponds to a normal subgroup, so a prime number corresponds to a simple group. The Jordan-Hölder Theorem generalises the uniqueness of prime factorisations. That you can build up any number by multiplying primes, is generalised to building up a group by semi-direct products (more generally, by group extensions): if H is a normal subgroup of G, then G will be an extension of H by the quotient group G/H.

Note however that $\mathbb{Z}_6 \times \mathbb{Z}_2$ and $\mathfrak{S}_3 \times \mathbb{Z}_2$ — both different from $\mathbb{Z}_{12}$ — will also have $\mathbb{Z}_2, \mathbb{Z}_2, \mathbb{Z}_3$ as composition factors: unlike for numbers, 'multiplication' here does not give a unique answer. The semidirect product $\mathbb{Z}_2 \rtimes \mathbb{Z}_2$ can equal either $\mathbb{Z}_4$ or $\mathbb{Z}_2 \times \mathbb{Z}_2$, depending on how the product is taken. More precisely, the notation $G \rtimes G'$ means a group where every element can be written uniquely as a pair (g, g'), for $g \in G$ and $g' \in G'$, and where the group operation is $(g, g')(h, h') = (\text{stuff}, gh)$.

Thus simple groups have an importance for group theory approximating what primes have for number theory. One of the greatest accomplishments of twentieth century math is surely the classification of the finite simple groups. (On the other hand, group extensions turn out to be technically quite difficult and leads one into group cohomology.) This work, completed in the early 1980s (although gaps are continually being discovered and filled in the arguments), runs to approximately 15 000 journal pages, spread over 500 individual papers, and is the work of a whole generation of group theorists. A modern revision is currently underway (see e.g. [34]) to simplify the proof and find and fill all gaps, but the final proof is still expected to be around 4000 pages long. The resulting list is:

- the cyclic groups $\mathbb{Z}_p$ (p a prime);
- the alternating groups $\mathfrak{A}_n$ for $n \geq 5$;
- 16 families of Lie type;
- 26 sporadics.

We've already met the cyclic groups. The alternating group $\mathfrak{A}_n$ consists of the even permutations in the symmetric group $\mathfrak{S}_n$, and so has order(=size) $\frac{1}{2} n!$. The groups of Lie type are essentially Lie groups defined over finite fields[23] $\mathbb{F}_q$ (such as $\mathbb{Z}_p$), sometimes 'twisted' in certain senses. The simplest example is $\mathrm{PSL}_n(\mathbb{F}_q)$, which consists of the $n \times n$ matrices with entries in $\mathbb{F}_q$, with determinant 1, quotiented out by the centre of $\mathrm{SL}_n(\mathbb{F}_q)$ (namely the scalar matrices $\mathrm{diag}(a, a, \ldots, a)$ for $a^n = 1$) (except for $\mathrm{PSL}_2(\mathbb{Z}_2)$ and $\mathrm{PSL}_2(\mathbb{Z}_3)$, which aren't simple).

Note that the determinant $|\rho(g)|$ for any representation ρ of any (noncyclic) simple group must be 1, otherwise we would violate the homomorphism definition of simple group (try to see why). Also, the centre of any (noncyclic) simple group must be trivial (why?). The smallest noncyclic simple group is $\mathfrak{A}_5$, with order 60.[24] It is the same as (isomorphic to) $\mathrm{PSL}_2(\mathbb{Z}_5)$ and $\mathrm{PSL}_2(\mathbb{F}_4)$, and can also be expressed as the group of all rotations (reflections have determinant -1 and so cannot belong to any simple group) of $\mathbb{R}^3$ that bring a regular icosahedron back to itself.

---

[23] There is a finite field with q elements, iff q is a power of a prime. For each such q, there is only 1 field of that size. The field with prime p elements is the integers taken mod p.

[24] This implies, incidentally, that if G and H are any two groups with the same order below 60, then they will have the same composition factors.

The smallest sporadic group is the Mathieu group M_{11}, order 7920, discovered in 1861[25]. The largest is the Monster $\mathbb{M}$, conjectured by Fischer and Griess in 1973 and finally proved to exist by Griess in 1980. Its order is[26]

$$\|\mathbb{M}\| = 2^{46} \cdot 3^{20} \cdot 5^9 \cdot 7^6 \cdot 11^2 \cdot 13^3 \cdot 17 \cdot 19 \cdot 23 \cdot 29 \cdot 31 \cdot 41 \cdot 47 \cdot 59 \cdot 71 \approx 8 \times 10^{53} . \quad (2.2.1)$$

20 of the 26 sporadics are involved in (i.e. are quotients of subgroups of) the Monster. Some relations among $\mathbb{M}$, the Leech lattice Λ_{24} and the largest Mathieu group M_{24} are given in Chapters 10 and 29 of [13].

Moonshine hints at a tantalising connection between the classification of finite simple groups, and the classification of RCFTs discussed in Part 1. Speculates [23] (page xli): "One can certainly hope for a uniform description of the finite simple groups as automorphism groups of certain vertex operator algebras — or conformal quantum field theories. If such a quantum field theory could somehow be attached a priori to a finite simple group, the classification of such theories, a problem of great current interest among string theorists, might some day be part of a new approach to the classification of the finite simple groups. On the other hand, can the known classification of the finite simple groups help in the classification of conformal field theories?"

2.3. INGREDIENT #2: MODULAR FUNCTIONS AND HAUPTMODULS

A readable introduction to some of the topics discussed in this section is [16,42,60].

We know from complex analysis that the group $SL_2(\mathbb{R})$ of 2×2 matrices with real entries and determinant 1, acts on the upper-half plane $\mathcal{H} = \{\tau \in \mathbb{C} \,|\, \mathrm{Im}(\tau) > 0\}$ by fractional linear (or *Möbius*) transformations:

$$\begin{pmatrix} a & b \\ c & d \end{pmatrix} \cdot \tau := \frac{a\tau + b}{c\tau + d} . \quad (2.3.1)$$

For example the matrix $S := \begin{pmatrix} 0 & -1 \\ 1 & 0 \end{pmatrix}$ corresponds to the function $\tau \mapsto -1/\tau$, while the matrix $T := \begin{pmatrix} 1 & 1 \\ 0 & 1 \end{pmatrix}$ corresponds to the translation $\tau \mapsto \tau + 1$. Since $\pm \begin{pmatrix} 1 & 0 \\ 0 & 1 \end{pmatrix}$ correspond to the same Möbius transformation, strictly speaking our group here is $PSL_2(\mathbb{R}) = SL_2(\mathbb{R})/\{\pm I\}$.

The only reason this action (2.3.1) of the 2×2 matrices on complex numbers (or more precisely the Riemann sphere $\mathbb{C} \cup \{\infty\}$) might not look strange to us, is because familiarity breeds numbness. What we really have is a natural action of $n \times n$ matrices on $\mathbb{C}^n$, and this

---

[25] Although his arguments apparently weren't very convincing. In fact some people, including the Jordan of Jordan-Hölder fame, argued in later papers that the largest of Mathieu's sporadic groups couldn't exist.

[26] The inquisitive reader, hungry for more 'coincidences', may have noticed that 196883 and 21296876 — see (2.1.2) — exactly divide the order of the Monster. Indeed this will hold for any finite group: the dimensions of the irreducible representations of a finite group will always divide its order.

induces their action on $\mathbb{C}^{n-1}$ (together with a codimension-2 set of 'points at infinity') by interpreting $\mathbb{C}^n$ as projective coordinates for $\mathbb{C}^{n-1}$. Specialising to $n = 2$ gives us (2.3.1). In projective geometry, 'parallel lines' intersect at ∞. Projective coordinates allow one to treat 'finite' and 'infinite' points on an equal footing.

Consider $\Gamma := \mathrm{SL}_2(\mathbb{Z})$, the subgroup of $\mathrm{SL}_2(\mathbb{R})$ consisting of the matrices with integer entries. It can be shown that it is generated by S and T (in other words, every matrix $\alpha \in \Gamma$ can be expressed as a monomial in S and T). For reasons that will be clear shortly, consider the extended upper-half plane $\overline{\mathcal{H}} := \mathcal{H} \cup \{i\infty\} \cup \mathbb{Q}$ — the extra points $\{i\infty\} \cup \mathbb{Q}$ are called *cusps*. Γ acts on $\overline{\mathcal{H}}$ (e.g. S interchanges 0 and $i\infty$). By a *modular function* for Γ, we mean a meromorphic function $f : \overline{\mathcal{H}} \to \mathbb{C}$, symmetric with respect to Γ: i.e. $f(\alpha(\tau)) = f(\tau)$ for all $\alpha \in \Gamma$. Note that we require f to be meromorphic at the cusps (e.g. polynomials are meromorphic at $i\infty$, but e^z is not).

It is not obvious why modular functions should be interesting, but in fact they are one of the most fundamental notions in modern number theory (see the last paragraph of §1.6). For example, consider the question of writing numbers as sums of squares. We can write $5 = 1^2 + (-2)^2 = (-1)^2 + 1^2 + 0^2 + 1^2 + (-1)^2$, to give a couple of trivial examples. Let $N_n(k)$ be the number of ways we can write the integer n as a sum of k squares, counting order and signs. For example $N_5(1) = 0$ (since 5 is not a perfect square), $N_5(2) = 8$ (since $5 = (\pm 1)^2 + (\pm 2)^2 = (\pm 2)^2 + (\pm 1)^2$), $N_5(3) = 24$, etc. Their generating functions are:[27]

$$\sum_{n=0}^{\infty} N_n(k)\, q^n = (\theta_3(q))^k \ ,$$

where

$$\theta_3(q) = 1 + 2q + 2q^4 + \cdots = \sum_{n \in \mathbb{Z}} q^{n^2}$$

is called a *theta function*. It turns out that θ_3 transforms nicely with respect to Γ, once we make the change-of-variables $q = \exp[\pi i \tau]$. This takes work to show. For example, θ_3 is clearly invariant under the action of $\begin{pmatrix} 1 & 2 \\ 0 & 1 \end{pmatrix}$, and a little work (from e.g. Poisson summation) shows that $\begin{pmatrix} 0 & -1 \\ 1 & 0 \end{pmatrix}$ takes $\theta_3(\tau)$ to $\sqrt{\frac{\tau}{i}}\,\theta_3(\tau)$. θ_3 is not precisely a modular function (it is a 'modular form of weight $\frac{1}{2}$' for $\Gamma_0(4)$), but this simple example illustrates the point that Γ (and related groups) appear throughout number theory. More on this shortly.

That important change-of-variables $q = \exp[\pi i \tau]$ was introduced by Jacobi early last century, in his analysis of 'elliptic integrals'. The theory is beautiful and poorly remembered today, which is very disappointing considering how much of modern math was touched by it. I strongly recommend the book [10], written over a century ago; the style and motivation of math in our century is different from that in Jacobi's, and we've lost a little in motivation what we've gained in power. I'll briefly sketch Jacobi's theory.

---

[27] A fundamental principle in math is: whenever you have a subscript with an infinite range, make a power series (called a *generating function*) out of it.

Just as we could develop a theory of 'circular functions' (i.e. sine etc.) starting from the integral $s(a) = \int_0^a \frac{dx}{\sqrt{1-x^2}}$, so can we develop a theory of 'elliptic functions' starting from the 'elliptic integral' $F(k,a) = \int_0^a \frac{dx}{\sqrt{(1-x^2)(1-k^2x^2)}}$. Inverting $s(a)$ gives a function both more useful and with nicer properties than $s(a)$: we call it $\sin(u)$. Similarly, for any k the elliptic function $\mathrm{sn}(k,u)$ is defined by $u = F(k, \mathrm{sn}(k,u))$. Just as we can define a numerical constant π by $\sin(\frac{1}{2}\pi) = 1$ (i.e. $\frac{1}{2}\pi = \int_0^1 \frac{dx}{\sqrt{1-x^2}}$), we get a function $K(k) = \int_0^1 \frac{dx}{\sqrt{(1-x^2)(1-k^2x^2)}}$. Just as $\sin(u)$ has period $4(\frac{1}{2}\pi)$, so has sn u-period $4K(k)$. sn also turns out to have u-period $4\mathrm{i}\,K(k')$ where $k' = \sqrt{1-k^2}$ — today we take this as the starting point and define an elliptic function to be doubly periodic (see [42] or Cohen in [60]).

The theta functions aren't elliptic functions, but they are closely related, and e.g. sn can be written as a quotient of them. In Jacobi's language, we have

$$\theta_3\Big(\frac{\mathrm{i}\,K(k')}{K(k)}\Big) = \sqrt{\frac{2K(k)}{\pi}}\ .$$

The 'modular transformation' $\tau \mapsto \frac{-1}{\tau}$ corresponds to interchanging the 'modulus' k with the 'complementary modulus' k', and thus is completely natural in Jacobi's theory. The important formula $\theta_3(\frac{-1}{\tau}) = \sqrt{\frac{\tau}{\mathrm{i}}}\,\theta_3(\tau)$ is trivial here.

A certain interpretation of modular functions also indicates their usefulness, and played an important role in Part 1. A *torus* is something that looks like the surface of a bagel, at least as far as its topology is concerned. For example, the Cartesian product $S^1 \times S^1$ of circles is a torus (think of one circle being the contact-circle of the bagel with the table on which it rests, then from each point on that horizontal circle imagine placing a vertical circle perpendicular to it, like a rib; together all these ribs fill out the bagel's surface). A more sophisticated example of a torus is an elliptic curve (a complex curve of the form $y^2 = ax^3 + bx^2 + cx + d$ and a special point on it playing the role of 0). A final example is the quotient $\mathbb{C}/\Lambda$ of the complex plane $\mathbb{C}$ with a 2-dimensional lattice Λ (we saw lattices in §1.6; Λ here will be a discrete doubly-periodic set of points in $\mathbb{C}$, containing 0). It turns out that certain equivalence classes of tori (e.g. with respect to conformal or complex-analytic equivalence) always contain a representative torus of the form $\mathbb{C}/\Lambda$, where Λ consists of all points $\mathbb{Z} + \mathbb{Z}\,\tau$, for some $\tau \in \mathcal{H}$. (Incidentally, the cusps correspond to degenerate tori.) In other words, these equivalence classes are parametrized by complex numbers τ in $\mathcal{H}$. So if we have a complex-valued function F on the set of all tori, which is e.g. conformally invariant (an example is the genus-one partition function $\mathcal{Z}$ in conformal field theories — see §1.1), then we can consider F as a well-defined function $F : \mathcal{H} \to \mathbb{C}$. However, it turns out that different points τ in $\mathcal{H}$ correspond to the same equivalence class of tori: e.g. the lattice for τ is the same as that for $\tau + 1$, and these are a rescaling of that for $-1/\tau$. Thus $F(\tau) = F(\tau + 1) = F(-1/\tau)$, because $\tau, \tau + 1, -1/\tau$ all represent equivalent tori. Since $\tau \mapsto \tau + 1$ and $\tau \mapsto -1/\tau$ generate $\mathrm{PSL}_2(\mathbb{Z})$, what in fact we find is that F has Γ as its group of symmetries. One often says that Γ is the 'modular group of the torus', and that the orbit space $\Gamma\backslash\mathcal{H}$ is the 'moduli space' of (conformal equivalence classes of) tori. $\mathcal{H}$ is called its 'Teichmüller space' or 'universal cover'. This is exactly analogous to $S^1 = \mathbb{R}/\mathbb{Z}$: $\mathbb{R}$ is its universal cover and $\mathbb{Z}$ is its 'modular group' (or 'mapping class group'). Another example: the Teichmüller space for (conformal equivalence classes of) 'pair-of-pants', or equivalently a disc minus two open interior disks, is $\mathbb{R}^3$ (an ordered triple), while its modular group is the symmetric group $\mathfrak{S}_3$ and its moduli space consists of unordered triples. Incidentally, we write $\Gamma \backslash \mathcal{H}$ instead of $\mathcal{H}/\Gamma$ because the group Γ acts on $\mathcal{H}$ 'on the left'. A good introduction to the geometry here is [55].

In any case, a surprising number of innocent-looking questions in number theory can be dragged (usually with effort) into the richly developed realm of elliptic curves and modular functions, and it is there they are often solved. For instance, we all know the ancient Greeks were interested in Pythagorean triples: find all integer solutions a, b, c to $a^2 + b^2 = c^2$, i.e. find all integer (or if you prefer, rational) right-angle triangles. They solved this by elementary means: choose any integers (or rationals) x, y and put $u = \frac{x^2 - y^2}{x^2 + y^2}$, $v = \frac{2xy}{x^2 + y^2}$; then $u^2 + v^2 = 1$ and (multiplying by the denominator) this gives all Pythagorean triples.

There are two ways of extending this problem. One is to ask which $n \in \mathbb{Z}$ can arise as areas of these rational right-angle triangles. It turns out $n = 5$ is the smallest one: $a = \frac{3}{2}$, $b = \frac{20}{3}$, $c = \frac{41}{6}$ works ($5 = \frac{1}{2}(\frac{3}{2})(\frac{20}{3})$ and $(\frac{3}{2})^2 + (\frac{20}{3})^2 = (\frac{41}{6})^2$). This is a hard problem — just try to show $n = 1$ cannot work. $n = 157$ turns out to work: the simplest triangle has a and b as quotients of integers of size around 10^{25}, and c as the quotient of integers around 10^{47}. Although this problem was studied by the ancient Greeks and also by the Arabs in the 10th century, it was finally cracked in the 1980s. It was solved by first translating it into the question of whether the elliptic curve $y^2 = x^3 - n^2 x$ has infinitely-many rational points, and then applying all the rich 20th century machinery to answering that question.

The other continuation of the Pythagorean triples question is more famous: find all integer solutions to $a^n + b^n = c^n$ (or equivalently all rational solutions to $a^n + b^n = 1$). 350 years ago Fermat wrote in the margin of the book he was reading (the book was describing the Greek solution to Pythagorean triples) that he had found a "truly marvelous" proof that for $n > 2$ there are no nontrivial solutions, but that the margin was too narrow to contain it. This result came to be known as 'Fermat's Last Theorem'[28] and despite considerable effort no one has succeeded in rediscovering his proof. Most people today believe that Fermat soon realised his 'proof' wasn't valid, otherwise he would have alluded to it in later letters. In any case, a very long and complicated proof was finally achieved in the 1990s: the 'Taniyama conjecture' says that a certain function associated to any elliptic curve over $\mathbb{Q}$ will be modular; if $a^n + b^n = c^n$ for some $n > 2$, then the elliptic curve $y^2 = x^3 + (a^n - b^n)x^2 - a^n b^n$ will violate the Taniyama conjecture; finally Wiles proved the Taniyama conjecture is true.

To most mathematicians, the 'area-n problem' and 'Fermat's Last Theorem' are interesting only because they can be related to elliptic curves and modular forms — it's easy to ask hard questions in math, but most questions tend to be stale. Number theory is infatuated with modular stuff because (in increasing order of significance) (a) it's exceedingly rich, with lots of connections to other areas of math and math phys; (b) it's

---

[28] It was called his 'Last Theorem' because it was the last of his 48 margin notes to be proved by other mathematicians — another one is discussed in Section 1.8. The story of Fermat's Last Theorem is a fascinating one, but alas this footnote is too small to do it credit. See for instance the excellent book [53].

a battleground on which many innocent-looking but hard-to-crack problems can be slain; and (c) last generation's number theorists also worked on modular stuff.

In any case, modular functions turn out to be important for math (and mathematical physics) even though they may at first glance look artificial. Poincaré explained how to study them. He said to look at the orbits of $\overline{\mathcal{H}}$ with respect to Γ. For example, one orbit, hence one point in $\Gamma\backslash\overline{\mathcal{H}}$, contains all cusps. We write this as $\Gamma\backslash\overline{\mathcal{H}}$, and give it the natural topological structure (i.e. 2 points $[\tau], [\tau'] \in \Gamma\backslash\overline{\mathcal{H}}$ are considered 'close' if the 2 sets $\Gamma\tau, \Gamma\tau'$ nearly overlap). Note first that by applying T repeatedly, every point in $\mathcal{H}$ corresponds to a point in the vertical strip $-\frac{1}{2} \leq \text{Re}(\tau) \leq \frac{1}{2}$ — in fact to a unique point in that strip, if we avoid the two edges. S is an inversion through the unit circle, so it permits us to restrict to those points in the vertical strip which are distance at least 1 from the origin. The resulting region R is called a fundamental region for Γ. Apart from the boundary of R, every Γ-orbit will intersect R in one and only one point.

What should we do about the boundary? Well, the edge $\text{Re}(\tau) = -\frac{1}{2}$ gets mapped by T to the edge $\text{Re}(\tau) = \frac{1}{2}$, so we should identify (=glue together) these. The result is a cylinder running off to infinity, with a strange lip at the bottom. S tells us how we should close that lip: identify $ie^{i\theta}$ and $ie^{-i\theta}$. This seals the bottom of the cylinder, so we get an infinitely tall cup with a strangely puckered base. In fact the top of this cup is also capped off, by the cusp i∞. So what we have (topologically speaking) is a *sphere*. It does not look like a smooth sphere, but in fact it inherits the smoothness of $\mathcal{H}$.

Incidentally, topological manifolds of dimension ≤ 3 always have a unique compatible smooth structure. 'Topological structure' means you can speak of continuity or closeness, 'smooth structure' means you can also do calculus. On the other hand, $\mathbb{R}^4$ has infinitely many smooth structures compatible with its topological structure; mysteriously, all other Euclidean spaces $\mathbb{R}^n$ have a unique smooth structure! Thus both mathematics and physics single out 4-space. Coincidence???

So anyways, what this construction of $\Gamma\backslash\overline{\mathcal{H}}$ means is that a modular function can be reinterpreted as a meromorphic complex-valued function on this sphere. This is very useful, because our undergraduate complex variables class taught us all about meromorphic complex-valued functions f on the Riemann sphere $\mathbb{C} \cup \infty$. There are many meromorphic functions on $\mathbb{C}$, but to also be meromorphic at ∞ forces f to be *rational*, i.e. $f(w) = \frac{\text{some polynomial } P(w)}{\text{some polynomial } Q(w)}$, where w is the complex parameter on the Riemann sphere. So our modular function $f(\tau)$ will simply be some rational function P/Q evaluated at the change-of-variables function $w = c(\tau)$ which maps us from our sphere $\Gamma\backslash\overline{\mathcal{H}}$ to the Riemann sphere. There are many different choices for this function $c(\tau)$, but the standard one is $c(\tau) = j(\tau)$, the j-function of $(2.1.1)$[29]. Thus, any modular function can be written as a rational function $f(\tau) = P(j(\tau))/Q(j(\tau))$ in the j-function. Conversely, any such function will be modular.

This is analogous to saying that any function $g(x)$ periodic under $x \mapsto x + 1$ can be thought of as a function on the unit circle $S^1 \subset \mathbb{C}$ evaluated at the change-of-variables function $x \mapsto e^{2\pi i x}$, and hence has a Fourier expansion $\sum_n g_n \exp[2\pi i n x]$.

---

[29] Historically, j was the standard choice, but in Moonshine the preferred choice would be the function $J = j$ − 744 with zero constant term.

We can generalise this argument. Consider a subgroup G of $SL_2(\mathbb{R})$ which is both not too big, and not too small. 'Not too big' means it should be *discrete*, i.e. the matrices in G can only get so close to the identity matrix $\begin{pmatrix} 1 & 0 \\ 0 & 1 \end{pmatrix}$. To make sure G is 'not too small', it is enough to require that G contains some subgroup of the form

$$\Gamma_0(N) := \{ \begin{pmatrix} a & b \\ c & d \end{pmatrix} \in SL_2(\mathbb{Z}) \,|\, c \equiv 0 \ (\mathrm{mod}\ N) \} \ , \tag{2.3.2a}$$

i.e. G must contain all matrices in Γ whose bottom-left entry is a multiple of N. So G must contain T, for example. We will also be interested only in those G which obey

$$\begin{pmatrix} 1 & t \\ 0 & 1 \end{pmatrix} \in G \Rightarrow t \in \mathbb{Z} \tag{2.3.2b}$$

i.e. the only translations in G are by integers. We will call a function $f : \overline{\mathcal{H}} \to \mathbb{C}$ a *modular function for G* if it is meromorphic (including at the cusps $\mathbb{Q} \cup \{i\infty\}$), and if also f is symmetric with respect to G: $f \circ \alpha = f$ for all $\alpha \in G$. This implies we will be able to expand f as a Laurent series in q. We analyse this as before: look at the orbit space $\Sigma = G\backslash\overline{\mathcal{H}}$; because G is not too big, Σ will be a (Riemann) surface; because G is not too small, Σ will be compact.

The compact Riemann surfaces have been classified (up to homeomorphism — i.e. considering only topology as relevant), and are characterised by a number called the *genus*. Genus 0 is a sphere, genus 1 is a torus, genus 2 is like two tori resting side-by-side, etc. For example, the surface of a wine glass, or a fork, is topologically a sphere, while a coffee cup and a key will (usually) be tori. Eye glasses with the lenses popped out is a 2-torus, while a ladder with n rungs on it has genus $n-1$.

We will call G 'genus g' if its surface Σ has genus g. For example, $G = \Gamma_0(2)$ and $G = \Gamma_0(25)$ are both genus 0, while $\Gamma_0(50)$ is genus 2 and $\Gamma_0(24)$ is genus 3. Once again, we are interested here in the genus 0 case. As before, this means that there is a change-of-variables function we'll denote J_G which has the property that it's a modular function for G, and all other modular functions for G can be written as a rational function in it. Because of (2.3.2), we can choose J_G to look like

$$J_G(\tau) = q^{-1} + a_1(G)\,q + a_2(G)\,q^2 + \cdots$$

So J_G plays exactly the same role for G that $J := j - 744$ plays for Γ. J_G is called the Hauptmodul for G. (Incidentally for genus > 0, two generators, not one, are needed.)

For example, $\Gamma_0(2)$, $\Gamma_0(13)$ and $\Gamma_0(25)$ are all genus 0, with Hauptmoduls

$$J_2(\tau) = q^{-1} + 276\,q - 2048\,q^2 + 11202\,q^3 - 49152\,q^4 + 184024\,q^5 + \cdots \tag{2.3.3}$$
$$J_{13}(\tau) = q^{-1} - q + 2\,q^2 + q^3 + 2\,q^4 - 2\,q^5 - 2\,q^7 - 2\,q^8 + q^9 + \cdots \tag{2.3.4}$$
$$J_{25}(\tau) = q^{-1} - q + q^4 + q^6 - q^{11} - q^{14} + q^{21} + \cdots \tag{2.3.5}$$

The smaller the modular group, the smaller the coefficients of the Hauptmodul. In this sense, the j-function is optimally bad among the Hauptmoduls: e.g. for it $a_{23} \approx 10^{25}$.

An obvious question is, how many genus 0 groups (equivalently, how many Hauptmoduls) are there? It turns out that $\Gamma_0(p)$ is genus 0, for a prime p, iff $p - 1$ divides 24. Thompson in 1980 proved that for any g, there are only finitely many genus g groups obeying our two conditions (2.3.2). In particular this means there are only finitely many Hauptmoduls. Over 600 Hauptmoduls with integer coefficients $a_i(G)$ are presently known.

2.4. THE MONSTROUS MOONSHINE CONJECTURES

We are now ready to make precise the main conjecture of Conway and Norton [12]. (We should emphasise though that there have been several other conjectures, some of which turned out to be partially wrong.)

They conjectured that for each element g of the Monster $\mathbb{M}$, there is a Hauptmodul

$$J_g(\tau) = q^{-1} + \sum_{n=1}^{\infty} a_n(g)\, q^n \qquad (2.4.1)$$

for a genus 0 group G_g such that each coefficient $a_n(g)$ is an integer, and for each n the map $g \mapsto a_n(g)$ is a character of $\mathbb{M}$. They also conjectured that G_g contains $\Gamma_0(N)$ as a normal subgroup, for some N depending on the order of g.

Another way of saying this is that there exists an infinite-dimensional graded representation $V = V_{-1} \oplus \bigoplus_{n=1}^{\infty} V_n$ of $\mathbb{M}$ such that the McKay-Thompson series $T_g(\tau)$ in (2.1.4) is a Hauptmodul.

There are around 8×10^{53} elements to the Monster, so naively we may expect around 8×10^{53} different Hauptmoduls $J_g = T_g$. However the character of a representation evaluated at g and at hgh^{-1} will always be the same, so $J_g = J_{hgh^{-1}}$. Hence the relevant quantity is the number of conjugacy classes, which for $\mathbb{M}$ is only 194. Moreover, a character evaluated at g^{-1} will always be the complex conjugate of its value at g, but here all character values $\chi_{V_n}(g)$ are integers (according to the conjecture). Thus $J_g = J_{g^{-1}}$. The total number of distinct Hauptmoduls J_g arising in Monstrous Moonshine turns out to be only 171.

For example, if we choose g to be the identity, we recover $T_{id.} = J$. It turns out that there are precisely 2 different conjugacy classes of order 2 elements, one of them giving the Hauptmodul J_2 in (2.3.3). Similarly for 13, but J_{25} doesn't correspond to any conjugacy class of $\mathbb{M}$.

Moonshine provides an explanation for a forgotten mystery of classical mathematics: why are the coefficients of the j-function *positive* integers? On the other hand, that they are *integers* has long been important to number theory (complex multiplication, class field theory — see e.g. [16]).

There are lots of other less important conjectures. One which played a role in ultimately proving the main conjecture involves the *replication formulae*. Conway-Norton want to think of the Hauptmoduls J_g as being intimately connected with $\mathbb{M}$; if so, then the group structure of $\mathbb{M}$ should somehow directly relate different J_g. In particular, consider the power map $g \mapsto g^p$. Now, it was well-known that $j(\tau)$ has the property that $j(p\tau) + j(\frac{\tau}{p}) + j(\frac{\tau+1}{p}) + \cdots + j(\frac{\tau+p-1}{p})$ equals a polynomial in j, for any prime p (*sketch of proof:* it's a modular function for Γ, and hence equals a rational function of j; since its only poles will be at the cusps, the denominator polynomial must be trivial). Hence the same will hold for J. Explicitly we get

$$J(2\tau) + J(\frac{\tau}{2}) + J(\frac{\tau+1}{2}) = J^2(\tau) - 2a_1 \qquad (2.4.2a)$$

$$J(3\tau) + J(\frac{\tau}{3}) + J(\frac{\tau+1}{3}) + J(\frac{\tau+2}{3}) = J^3(\tau) - 3a_1 J(\tau) - 3a_2 \qquad (2.4.2b)$$

where $J(\tau) = \sum_k a_k q^k$. Slightly more complicated formulas hold in fact for any composite n. Conway and Norton conjectured that these formulas have an analogue for the Moonshine functions J_g in (2.4.1). In particular, (2.4.2) become for any $g \in \mathbb{M}$

$$J_{g^2}(2\tau) + J_g(\frac{\tau}{2}) + J_g(\frac{\tau+1}{2}) = J_g^2(\tau) - 2a_1(g) \qquad (2.4.3a)$$

$$J_{g^3}(3\tau) + J_g(\frac{\tau}{3}) + J_g(\frac{\tau+1}{3}) + J_g(\frac{\tau+2}{3}) = J_g^3(\tau) - 3a_1(g) J_g(\tau) - 3a_2(g). (2.4.3b)$$

These are examples of the replication formulae.

'Replication' concerns the power map $g \mapsto g^n$ in $\mathbb{M}$. Can Moonshine see more of the group structure of $\mathbb{M}$? A step in this direction was made by Norton [47], who associated a Hauptmodul to commuting elements g, h in $\mathbb{M}$. Physically [19], this corresponds to orbifold traces, i.e. the $V^\natural$ RCFT with boundary conditions twisted by g and h in the 'time' and 'space' directions. Still, we would like to see more of $\mathbb{M}$ in Moonshine.

An important part of the Monstrous Moonshine conjectures came a few years after [12]. Frenkel-Lepowsky-Meurman [23] constructed a graded infinite-dimensional representation $V^\natural$ of $\mathbb{M}$ and conjectured (correctly) that it is the representation in (2.1.4). $V^\natural$ has a very rich algebraic structure, which will be discussed in §2.6.

A major claim of [23] was that $V^\natural$ is a 'natural' structure (hence their notation). To see what they mean by that, it's best to view another simpler example of a natural construction: that of the *Leech lattice* Λ_{24}. Recall the discussion of (root) lattices in §1.6.

Λ_{24} is one of the most interesting lattices, and is related to Moonshine. It can be defined using 'laminated lattices'. Start with the 0-dimensional lattice $\Lambda_0 = \{0\}$, which consists of just a single point. Use it to construct a 1-dimensional lattice, with minimal (nonzero) norm 4, built out of infinitely many copies of Λ_0 laid side by side. The result of course is simply the even integers $2\mathbb{Z}$, which we will call here Λ_1. Now construct a 2-dimensional lattice, of minimum norm 4, built out of infinitely many copies of Λ_1 laid next to each other. There are lots of ways to do this, but choose the densest lattice possible. The result is unique: it is the hexagonal lattice A_2 scaled by a factor of $\sqrt{2}$: call it Λ_2. Continue in this way: $\Lambda_3, \Lambda_4, \Lambda_5, \Lambda_6, \Lambda_7$, and Λ_8 will be the root lattices A_3, D_4, D_5, E_6, E_7 and E_8, respectively, all scaled by $\sqrt{2}$. See [13] chapter 6 for a more complete treatment of laminated lattices.

The 24th repetition of this construction yields the Leech lattice. It is the unique 24-dimensional self-dual lattice with no norm-2 vectors, and provides among other things the densest known packing of 23-dimensional spheres in $\mathbb{R}^{24}$. Many of its properties are discussed throughout [13]. So lamination provides us with a sort of no-input construction of the Leech lattice, and a good example of the mathematical meaning of 'natural'. After dimension 24, it seems chaos results from the lamination procedure (there are 23 different 25-dimensional lattices that have an equal right to be called Λ_{25}, and over 75 000 are expected for Λ_{26}).

It is natural to ask about Moonshine for other groups. There is a partial Moonshine for the Mathieu groups M_{24} and M_{12} (which have about 2×10^8 and 10^5 elements resp.), the automorphism group .0 of Λ_{24} (which has about 8×10^{18} elements), and a few others — see e.g. [49]. These groups are either simple or almost simple (e.g. .0 is the direct product of $\mathbb{Z}_2$ with the simple group .1). More generally, there will be some sort of Moonshine for any group which is the automorphism group of a vertex operator algebra; the finite simple groups of Lie type should be automorphism groups of VOAs closely related to the affine algebras except defined over fields like $\mathbb{Z}_p$.

There is a geometric side to Moonshine, associated to names like Lian-Yau and Hirzebruch. In particular, Hirzebruch's 'prize question' asks for the construction of a 24-dimensional manifold on which $\mathbb{M}$ acts, whose twisted elliptic genus are the McKay-Thompson series. This is still open.

It should be emphasised that Monstrous Moonshine is a completely unexpected connection between finite groups and modular functions. Although there has been enormous progress in our understanding of this connection (so much so, that Richard Borcherds won the 1998 Fields medal for his work on this), there still is mystery at its heart. In particular, that $\mathbb{M}$ is associated with *modular functions* can be explained mathematically by it being the automorphism group of the Moonshine VOA $V^\natural$, and physically by the associated RCFT, but what is so special about $\mathbb{M}$ that these modular functions should be genus 0? We will come back to this in §2.9.

2.5. FORMAL POWER SERIES

Vertex (operator) algebras (VOAs) are a mathematically precise formulation of the notion of *W-algebra* or *chiral algebra*[30] which is so central to conformal field theory (see §1.1). VOAs were first defined by Borcherds, and their theory has since been developed by a number of people (Frenkel, Lepowsky, Meurman, Zhu, Dong, Li, Mason, Huang, ...). Because our primary motivation here is Moonshine, I will only focus on one aspect of their theory (the connection with Lie algebras). Useful to consult while reading this review are the notes [27] — they take a more analytic approach to many of the things we discuss, and their approach (namely that of CFT) motivates beautifully much of VOA theory.

In quantum field theory the basic object is the quantum field, which roughly speaking is a choice of operator $\hat{A}(x)$ at each space-time point x. 'Operator' means something that 'operates on' functions or vectors. E.g. an indefinite integral is an operator, as is a derivative. The operators in the QFT act on the space spanned by the states $|\star\rangle$, and together form an infinite-dimensional vector space (e.g. a C^* algebra) — this infinite-dimensionality of QFT is a major source of its mathematical difficulties, and QFT still has not been put on completely satisfactory mathematical grounds.

---

[30] An alternate (and much more complicated) mathematical formulation of chiral algebra is due to Beilinson and Drinfeld, and belongs to algebraic geometry. See [28] for a good — but still difficult — review.

But another difficulty is that the quantum field $\hat{A}$ really isn't an operator-valued *function* of space-time. 'Function' is too narrow a concept. For example, one of the most familiar 'functions' in quantum mechanics is the Dirac delta $\delta(x)$. You see it for example in the canonical commutation relations: e.g. for a scalar field $\hat{\varphi}$, we have $[\hat{\varphi}(\vec{x},t), \frac{\partial}{\partial t}\hat{\varphi}(\vec{y},t)] = i\hbar\delta^3(\vec{x} - \vec{y})$. $\delta(x)$ has the property that for any other smooth function f,

$$\int_{-1}^{1} f(y)\,\delta(y)\,\mathrm{d}y = f(0) \ , \qquad \int_{-1}^{1} f(y)\,\delta'(y)\,\mathrm{d}y = f'(0) \ ,$$

etc. The problem is that $\delta(x)$ isn't a function — no function could possibly have those properties.

One way to make sense of 'functions' like the Dirac delta and its derivatives is distribution theory. Although it was first informally used in physics, it was rigourously developed around 1950 by Laurent Schwartz, and uses the idea of test functions. See e.g. [15].

What I will describe now is an alternate approach, *algebraic* as opposed to *analytic*. These two approaches are not equivalent: you can do some things in one approach which you can't do in the other. But the algebraic approach is considerably simpler technically — no calculus or convergence to worry about — and it is remarkable how much can still be captured. This approach is the starting point for the VOA story described next section, and was first created around 1980 by Garland and Date-Kashiwara-Miwa. Keep in mind that what we are trying to capture is an operator-valued 'function' on space-time. Space-time in CFT is 2-dimensional, and so we can think of it (at least locally) as being on the complex plane $\mathbb{C}$ (more precisely, we will usually associate the space-time point (x,t) with the complex number $z = e^{t+ix}$). Good introductions to the material in this section are [23,39,31].

Let W be any vector space. We are most interested in it being an infinite-dimensional space of matrices (i.e. operators on an infinite-dimensional space), but forget that for now. Define $W[[z,z^{-1}]]$ to be the set of all formal series $\sum_{n=-\infty}^{\infty} w_n z^n$, where the coefficients w_n lie in our space W. We don't ask here whether a given series converges or diverges — z is merely a formal variable. We will also be interested in $W[z,z^{-1}]$ (Laurent polynomials). We can add these formal series in the usual way, and multiply them by numbers (scalars) in the usual way.

Remember our ultimate aim here: we want to capture quantum fields. So we want our formal series to be operator-valued. The way to accomplish this is to choose W to be a vector space of operators, or matrices if you prefer. A fancy way to say this is '$W = \mathrm{End}(V)$', which means the things in W operate on vectors in V. If we take $V = \mathbb{C}^m$, then we can think of W as being the space of all $m \times m$ complex matrices. We are ultimately interested in the case $m = \infty$, but we won't lose much now by taking $m = 1$, which would mean formal power series with numerical coefficients.

Because our coefficients w_n are operators, we can multiply our formal series. We define multiplication in the usual way. For example, consider $W = V = \mathbb{C}$, and take $c(z) = z^{21} - 5z^{100}$ and $d(z) = \sum_{n=-\infty}^{\infty} z^n$. Then

$$c(z)\,d(z) = \sum_{n=-\infty}^{\infty} z^{n+21} - 5\sum_{n=-\infty}^{\infty} z^{n+100} = \sum_{n=-\infty}^{\infty} z^n - 5\sum_{n=-\infty}^{\infty} z^n = -4d(z) \ .$$

So far so good. Now try to compute the square $d(z)^2$. You get infinity. So the lesson is: you can't always multiply in $W[[z, z^{-1}]]$. We'll come back to this later.

But first, look again at that first product: $c(z)\,d(z) = -4d(z)$. One thing it tells us is that *we can't always divide* (certainly $c(z)$ and -4 are two very different power series!). But there's another lesson here: if you work out a few more multiplications of this kind, what you'll find is that $f(z)\,d(z) = f(1)\,d(z)$ for any f, at least for those f for which $f(1)$ exists (e.g. any $f \in W[z, z^{-1}]$). Thus $d(z)$ is what we would call the Dirac delta $\delta(z - 1)$! (You can think of it as the Fourier expansion of the Dirac delta, followed by a change of variables). Unfortunately, the standard notation here is to write it without the '-1':

$$\delta(z) := \sum_{n=-\infty}^{\infty} z^n$$

and that is the notation we will also adopt. Similarly, $\delta(az)$ and $\delta'(z)$ etc (which are the formal series defined in the obvious way) act on $W[z, z^{-1}]$ in the way one would expect: $f(z)\,\delta(az) = f(\frac{1}{a})\,\delta(az)$ and $f(z)\,\delta'(z) = f'(1)\,\delta'(z)$. So of course it makes perfect sense that we couldn't work out $d(z)^2$: we were trying to square the Dirac delta, which we know is impossible!

A similar theory can be developed for several variables z_i, with identities such as $f(z_1, z_2)\,\delta(z_1/z_2) = f(z_2, z_2)\,\delta(z_1/z_2) = f(z_1, z_1)\,\delta(z_1/z_2)$.

But we must not get too overconfident:

Paradox 1. Consider the following product:

$$\delta(z) = \left[\left(\sum_{n=0}^{\infty} z^n \right)(1 - z) \right] \delta(z) = \left(\sum_{n=0}^{\infty} z^n \right) \left[(1 - z)\,\delta(z) \right] = \left(\sum_{n=0}^{\infty} z^n \right) \left[0\,\delta(z) \right] = 0 \ .$$

When physicists are confronted with 'paradoxes' such as this, they tend to respond by keeping them in the back of their mind, by treading with care when they are involved in a calculation which reminds them of one of the paradoxes, and otherwise trusting their instincts. Mathematicians typically over-react: they kick themselves for getting overconfident and walking head-first into a 'paradox', and then they devise some rule which will absolutely guarantee that that paradox will always be safely avoided in the future. We will follow the mathematicians' approach, and in the next few paragraphs will describe their rule for avoiding Paradox 1: to forbid certain innocent-looking products.

Remember that we are actually interested in the vector space $W = \text{End}(V)$. Suppose we have infinitely many matrices $w_i \in \text{End}(V)$. We will call them *summable* if for every column vector $v \in V$, only finitely many products $w_i(v) \in V$ are different from 0. In other words, only finitely many of the matrices w_i have a nonzero first column, only finitely many have a nonzero second column, ...

We will certainly have a well-defined sum $\sum_i w_i(z)$ if for each fixed n, the set $\{w_i(n)\}$ (as i varies) of matrices is summable. All other sums are forbidden. We will certainly have a well-defined product[31] $\prod_{i=1}^{m} w_i(z)$ if for each n, the set $\{w_1(n_1)\,w_2(n_2) \cdots w_m(n_m)\}_{\sum n_i = n}$

---

[31] m here will be finite: we permit infinite sums but only finite products.

(vary the n_i subject to $\sum_i n_i = n$) is summable. All other products are forbidden. This is reasonable because the sum of those matrix products $w_1(n_1) \cdots w_m(n_m)$ will precisely equal the nth coefficient of the product $\prod_{i=1}^m w_i(z)$.

Note that there are certainly more general ways to have a well-defined product (or sum). For example, according to our rule, we cannot even add $\sum_n 2^{-n}$! This way has the advantage of not touching the more complicated realm of convergence issues. We are doing *algebra* here, not *analysis*. The way out of Paradox 1 is that $(\sum z^n)(1 - z)$ doesn't equal 1 — rather, it's a forbidden product.

An interesting consequence of the fact that we are doing algebra instead of analysis is that the product $z^{\frac{1}{2}} \delta(z)$ here does not and cannot equal $1^{\frac{1}{2}} \delta(z) = \delta(z)$ — their formal power series are very different. In hindsight this 'failing' is understandable: algebraically, it seems artificial to prefer the positive root of 1 over the negative root.

Paradox 2. Expand $\frac{1}{1-z}$ in a formal power series in z to get $\sum_{n\geq 0} z^n$. Next, expand $\frac{1}{1-z} = \frac{-z^{-1}}{1-z^{-1}}$ in a formal power series in z^{-1} to get $-\sum_{n<0} z^n$. Subtract these; we presumably should get 0, but we actually get $\delta(z)$!

The analytic explanation is that the first expression converges only for $|z| < 1$, and the second for $|z| > 1$, so it would be naive to expect their difference to be 0. We see from this 'paradox' that *it really matters in which variable we expand rational functions*. For instance, at first glance the identity

$$z_0^{-1} \delta\left(\frac{z_1 - z_2}{z_0}\right) - z_0^{-1} \delta\left(\frac{z_2 - z_1}{-z_0}\right) = z_2^{-1} \delta\left(\frac{z_1 - z_0}{z_2}\right)$$

is nonsense; it only holds if you expand the terms in positive powers of z_2, z_1, and z_0 respectively. The procedure of expanding a function in positive and negative powers of a variable and then subtracting the results, yields what are called *expansions of zero*; it is possible to show that expansions of zero will always be linear combinations of Dirac deltas $\delta(az)$ and their various derivatives $\delta^{(k)}(az)$, as we saw in Paradox 2.

2.6. Ingredient #3: Vertex Operator Algebras

We are now prepared to introduce the important new structure called vertex operator algebras (VOAs). They are essentially the chiral algebras of RCFTs — see [26,27] for excellent motivation of the 7 axioms below. A more detailed treatment of the basic theory of VOAs is provided by e.g. [23,39,31]. Although VOAs are natural from the CFT perspective and appear to be an important and rapidly developing area in math, their definition is not easy: Borcherds is known to have said that you either know what they are, or you don't want to know.

A VOA is a (infinite-dimensional) graded vector space $V = \oplus_{n\in\mathbb{Z}} V_n$ with infinitely many bilinear products $u *_n v$ respecting the grading (in particular $V_k *_n V_\ell \subseteq V_{k+\ell-n-1}$), which obey infinitely many constraints. 'Bilinear' means that for any $a, a', b, b' \in \mathbb{C}$ and $u, u', v, v' \in V$, $(au+a'u') *_n (bv+b'v') = ab \, u *_n v + ab' \, u *_n v' + a'b \, u' *_n v + a'b' \, u' *_n v'$ — i.e. that the products are compatible with the vector space structure of V. The subspaces V_n

must all be finite-dimensional, and they must be trivial (i.e. $V_n = \{0\}$) for all sufficiently small n (i.e. for $n \approx -\infty$). Note that we can collect all these products into one generating function: a linear map $Y : V \to (\text{End}V)[[z, z^{-1}]]$. That is, to each vector $u \in V$ we associate the formal power series (called a *vertex operator*) $Y(u, z) = \sum_{n \in \mathbb{Z}} u_n z^{-n-1}$. For each u, the coefficients u_n will be functions from V to V. The idea is that the product $u *_n v$ will now be written $u_n v := u_n(v)$. The bilinearity of $*_n$ translates into two things in this new language: that $Y(*, z)$ is linear, and that each function u_n is itself linear (i.e. they are endomorphisms).

The constraints are:

VOA 1. (*regularity*) $u_n v = 0$ for all $n > N(u, v)$;

VOA 2. (*vacuum*) there is a vector $1 \in V$ such that $Y(1, z)$ is the identity (i.e. $1_n v = \delta_{n,-1} v$);

VOA 3. (*state-field correspondence*) $Y(u, 0)1 = u$;

VOA 4. (*conformal*) there is a vector $\omega \in V$, called the *conformal vector*, such that $L_n := \omega_{n+1}$ gives us a representation of the Virasoro algebra $\mathcal{V}$, with central term $C \mapsto cI$ for some $c \in \mathbb{C}$;

VOA 5. (*translation generator*) $Y(L_{-1}u, z) = \frac{d}{dz}Y(u, z)$;

VOA 6. (*conformal weight*) $L_0 u = nu$ whenever $u \in V_n$;

VOA 7. (*locality*) $(z - w)^M [Y(u, z), Y(v, w)] = 0$ for some integer $M = M(u, v)$.

We saw the Virasoro algebra in Part 1 (see (1.2.7)). The number c in **VOA 4** is called the *central charge*, and is an important invariant of V. The peculiar-looking **VOA 7** simply says that the commutator $[Y(u, z), Y(v, w)]$ of two vertex operators will be a finite linear combination of derivatives of various orders of the Dirac delta centred at $z - w$. A recommended exercise for the reader is to show that $M = 4$ works in **VOA 7** for $u = v = \omega$. Note that in a VOA, any $Y(u, z)v$ will be a finite sum — i.e. the series $Y(u, z)$ is summable (defined last section). It is a consequence of the axioms that $1 \in V_0$ and $\omega \in V_2$: for instance, **VOA 7** says all $u_n 1 = 0$ for any $n \geq 0$, so $L_0 1 = \omega_1 1 = 0$ and hence $1 \in V_0$.

In RCFT, V would be the 'Hilbert space of states' (more carefully, V will be a dense subspace of it), and $z = e^{t+ix}$ would be a local complex coordinate on a Riemann surface. L_0 generates time translations, and so its eigenvalues (the *conformal weights*) can be identified with energy. Physically, the requirement that $V_n \to 0$ for $n \to -\infty$ corresponds to the energy of the RCFT being bounded from below. Also, $z = 0$ in **VOA 3** corresponds to the time limit $t \to -\infty$. For each state u, the vertex operator $Y(u, z)$ is a holomorphic (chiral) quantum field. The vector 1 is the vacuum $|0\rangle$, and $Y(\omega, z)$ is the stress-energy tensor T. The most important axiom, **VOA 7**, says that vertex operators commute up to a possible pole at $z = w$, and so are local quantum fields. It is equivalent to the duality axiom of many treatments of CFT. In the physics literature, there is a minor notational difference: for $u \in V_k$, $Y(u, z) = \sum u_n z^{-n-1}$ is written $\sum u_{(n)} z^{-n-k}$. (Physicists prefer this because it cleans up some formulas a little; mathematicians abhor it because it artificially prefers the 'homogeneous' vectors $u \in V_k$.)

In Segal's language (see §1.1), $Y(u, z)$ appears quite naturally. Consider the physical event of two strings combining to form a third. To first order (i.e. the tree-level Feynman diagram), this would correspond in Segal's language to a 'pair-of-pants', or a sphere with three punctures, two of which are positively oriented (corresponding to the incoming strings) and the other being negatively oriented. We can think of the sphere as the Riemann sphere $\mathbb{C} \cup \{\infty\}$; put the punctures at ∞ (outgoing) and z and 0 (the incomings). Segal's functor $\mathcal{T}$ will associate to this a z-dependent homomorphism $\varphi_z : V \times V \to V$. We write $\varphi_z(u, v) \in V$ as $Y(u, z)v$. Incidentally, the symbol 'Y' was chosen because of this 'pair-of-pants' picture (time flows from the top of the 'Y' to the bottom), as was the name 'vertex operator'.

The original axioms by Borcherds were a little more complicated and general: he didn't require $\dim(V_n) < \infty$ nor the $V_n \to 0$ condition, and he only considered L_0 and L_{-1} rather than the full Virasoro algebra. The resulting generalisation is called a *vertex algebra*.

VOA 7 can be rewritten in the form (usually called the *Jacobi identity* for the VOA)

$$z_0^{-1}\delta\left(\frac{z_1 - z_2}{z_0}\right) Y(u, z_1) Y(v, z_2) - z_0^{-1}\delta\left(\frac{z_2 - z_1}{-z_0}\right) Y(v, z_2) Y(u, z_1) \tag{2.6.1}$$

$$= z_2^{-1}\delta\left(\frac{z_1 - z_0}{z_2}\right) Y(Y(u, z_0)v, z_2) ,$$

where the formal series are expanded in the appropriate way. This is the embodiment of commutativity and associativity in the VOA, as we will see. To bring it into a more useful form, hit it with $t \in V$ and expand out into $z_0^\ell z_1^m z_2^n$: we obtain

$$\sum_{i \geq 0} (-1)^i \binom{\ell}{i} \left(u_{\ell+m-i} \circ v_{n+i} - (-1)^\ell v_{\ell+n-i} \circ u_{m+i}\right) = \sum_{i \geq 0} \binom{m}{i} (u_{\ell+i}v)_{m+n-i} , \tag{2.6.2}$$

where for any $k \in \mathbb{Z}$, $j \in \mathbb{Z}_\geq$, $\binom{k}{j} := \frac{k(k-1)\cdots(k-j+1)}{j!}$. For instance, specialising (2.6.2) to $\ell = 0$ and $m = 0$, resp., gives us

$$[u_m, v_n] = \sum_{i \geq 0} \binom{m}{i} (u_i v)_{m+n-i} \tag{2.6.3}$$

$$(u_\ell v)_n = \sum_{i \geq 0} (-1)^i \binom{\ell}{i} \left(u_{\ell-i} \circ v_{n+i} - (-1)^\ell v_{\ell+n-i} \circ u_i\right) . \tag{2.6.4}$$

Why is (2.6.1) called the Jacobi identity? Put $\ell = m = n = 0$ in (2.6.2): we get $u_0(v_0 t) - v_0(u_0 t) = (u_0 v)_0 t$. If we now formally write $[xy] := x_0 y$, then this becomes $[u[vt]] - [v[ut]] = [[uv]t]$, which is one of the forms of the Lie algebra Jacobi identity (1.2.1b). Even though $[xy] \neq -[yx]$ here, this formal little trick will turn out to be quite important next section.

The simplest examples of VOAs correspond to any even positive-definite lattice Λ; for their construction see e.g. [27,39]. Physically, they correspond to a bosonic string compactified on the torus $\mathbb{R}^n/\Lambda \cong S^1 \times \cdots \times S^1$ (where n is the dimension of Λ); the central charge $c = n$. Other important examples, first constructed by Frenkel-Zhu (again see e.g. [27,39]), correspond to affine Kac-Moody algebras $X_\ell^{(1)}$ at level $k \in \mathbb{Z}_>$, and physically to WZW theories on simply-connected compact group manifolds. (We discussed affine algebras in §1.4.) These have central charge $c = \frac{k \dim(X_\ell)}{k+h^\vee}$.

In 1984 Frenkel-Lepowsky-Meurman [23] constructed the *Moonshine module* $V^{\natural}$. It is a VOA with $c = 24$, with $V^{\natural} = V_0^{\natural} \oplus V_1^{\natural} \oplus V_2^{\natural} \oplus \cdots$, where $V_0^{\natural} = \mathbb{C}1$ is 1-dimensional, $V_1^{\natural} = \{0\}$ is trivial, and $V_2^{\natural} = (\mathbb{C}\omega) \oplus$ (Griess algebra) is $(1 + 196883)$-dimensional. Its automorphism group (=symmetry group) is precisely the Monster $\mathbb{M}$. Each graded piece $V_n^{\natural}$ is a finite-dimensional representation of $\mathbb{M}$; Borcherds proved that in fact $V^{\natural}$ is the McKay-Thompson infinite-dimensional representation of $\mathbb{M}$. It can be regarded as the most natural representation of $\mathbb{M}$ — it is rather surprising that important aspects of a finite group need to be studied via an infinite-dimensional representation.

$V^{\natural}$ has an elegant physical interpretation. First construct the bosonic string on $\mathbb{R}^{24}/\Lambda_{24}$ (recall that Λ_{24} is the Leech lattice). The resulting $c = 24$ VOA has partition function (=graded dimension) $J(\tau) + 24$, but although its graded pieces (at least for $n > 0$) have the right dimensions, they don't carry a natural representation of $\mathbb{M}$ and so can't qualify for the McKay-Thompson representation. To get $V^{\natural}$, orbifold this Λ_{24} VOA by the order-2 automorphism of Λ_{24} sending $\vec{x} \mapsto -\vec{x}$. $V^{\natural}$ thus corresponds to a holomorphic $c = 24$ RCFT, and Moonshine is related to physics. Most of Moonshine can be interpreted physically, except perhaps the genus 0 property of the McKay-Thompson series T_g.

There is a formal parallel between e.g. lattices and VOAs. For example, the Leech lattice Λ_{24} and the Moonshine module $V^{\natural}$ play analogous roles: Λ_{24} is the unique even lattice which (i) is self-dual, (ii) contains no norm 2 vectors, and (iii) has dimension 24; $V^{\natural}$ is believed to be the unique VOA which (i) possesses only one irreducible representation (namely itself), (ii) contains no conformal weight 1 elements, and (iii) has central charge $c = 24$. Analogies of these kinds are always useful as they suggest new directions to explore, and the history of math blooms with them. The battlecry 'Why invent what can be profitably copied' is not only heard in Hollywood.

We will end this section on a more speculative note. Witten (1986) said that to understand string theory conceptually, we need a new analogue of Riemannian geometry. Huang (1997) has pushed this thought a little further, saying that there is a more classical 'particle-math' and a more modern 'string-math'. According to Huang we have the real numbers (particle physics) vrs the complex numbers (string theory); Lie algebras vrs VOAs; and the representation theory of Lie algebras vrs RCFT, etc. What are the stringy analogues of calculus, ordinary differential equations, Riemannian manifolds, the Atiyah-Singer Index theorem,...? At present these are all unknown. However, Huang suggests that just as we could imagine Moonshine as a mystery which is explained in some way by RCFT, perhaps the stringy version of calculus would similarly explain the mystery of 2-dimensional gravity, stringy ODEs would explain the mystery of infinite-dimensional integrable systems, stringy Riemannian manifolds would help explain the mystery of mirror symmetry, and the stringy index theorem would help explain the elliptic genus.

2.7. INGREDIENT #4: GENERALISED KAC-MOODY ALGEBRAS

In this section we investigate Lie algebras arising from VOAs. These Lie algebras are an interesting generalisation of Kac-Moody algebras. See e.g. [5,6,38 Chapter 11.13, 29].

Much of Lie theory (indeed much of algebra) is developed by analogy with simple properties of integers. In §2.2 I invited you to think of a finite group as a massive generalisation of the concept of whole number. Specifically, the number n can be identified with the cyclic group $\mathbb{Z}_n$ with n elements. A divisor d of n generalises to a normal subgroup of a group. A prime number then corresponds to a simple group. Multiplying numbers corresponds to taking the semidirect product of groups (more generally, taking extensions of groups). Then we find that every group has a unique set of simple building blocks (although unlike numbers, different groups can have the same list of building blocks).

For a finite-dimensional Lie algebra, a divisor is called an ideal; a prime is called simple; and multiplying corresponds to semidirect sum. Lie algebras behave simpler than groups but not as simple as numbers, and the analogy sketched above is a reasonably satisfactory one. In particular, simple Lie algebras are important for similar reasons that simple groups are, and as mentioned in §1.2 can also be classified (with *much* less effort). A good treatment of this important classification (over $\mathbb{C}$) is provided by [36]. The proof is now reaching the state of perfection of the formulation of classical mechanics. One unobvious discovery is that the best way to capture the structure of a simple Lie algebra is by an integer matrix, called the *Cartan matrix*, or equivalently but more effectively (since most entries in the Cartan matrix are 0's) by using a graph called the *(Coxeter-)Dynkin diagram*. For instance the Dynkin diagram for A_ℓ consists of ℓ nodes connected sequentially in a line. See Figure 6 in [59].

More precisely, define a *symmetrised Cartan matrix* to be a symmetric real matrix $A = (a_{ij})_{i,j \leq \ell}$ such that $a_{ij} \leq 0$ if $i \neq j$, $a_{ii} > 0$, each $2\frac{a_{ii}}{a_{ii}} \in \mathbb{Z}$, and A is positive-definite. Examples of 2×2 symmetrised Cartan matrices are[32]

$$\begin{pmatrix} 2 & -1 \\ -1 & 2 \end{pmatrix}, \quad \begin{pmatrix} 2 & 0 \\ 0 & 2 \end{pmatrix}, \quad \begin{pmatrix} 1 & -1 \\ -1 & 2 \end{pmatrix}, \quad \begin{pmatrix} 2 & -3 \\ -3 & 6 \end{pmatrix}$$

The Dynkin diagram corresponding to A consists of ℓ nodes; the ith and jth nodes are connected with $4a_{ij}^2/a_{ii}a_{jj}$ lines, and if $a_{ii} \neq a_{jj}$, then we put an arrow over those lines pointing to i if $a_{ii} < a_{jj}$. The Dynkin diagrams corresponding to those four Cartan matrices are respectively

$$\text{o—o} \quad , \quad \text{o o} \quad , \quad \text{o⇐o} \quad , \quad \text{o⇚o}$$

We may without loss of generality require A to be *indecomposable*, or equivalently that the Dynkin diagram be connected. Of the 4 given above, only the second is decomposable.

To any $\ell \times \ell$ symmetrisable Cartan matrix, we can construct the corresponding Lie algebra $\mathfrak{g}$ in the following way. For each i, create 3 generators e_i, f_i, h_i (so there are a total of 3ℓ generators). The relations these generators obey are given by the following brackets: $[e_i f_j] = \delta_{ij} h_i$, $[h_i e_j] = a_{ij} e_j$, $[h_i f_j] = -a_{ij} f_j$, and for $i \neq j$ $\text{ad}(e_i)^n e_j = \text{ad}(f_i)^n f_j = 0$ where $n = 1 - 2\frac{a_{ij}}{a_{ii}}$. By 'ad$(e)$' here I mean the function $\mathfrak{g} \to \mathfrak{g}$ defined by $\text{ad}(e)f = [ef]$. So $\text{ad}(e)^2 f = [e[ef]]$, $\text{ad}(e)^3 f = [e[e[ef]]]$, etc.

---

[32] Note that our Cartan matrices differ from the usual definition, in which every diagonal entry equals 2.

To get a better feeling for these relations, consider a fixed i. The generators $e = \sqrt{\frac{2}{a_{ii}}}\, e_i, f = \sqrt{\frac{2}{a_{ii}}}\, f_i, h = \frac{2}{a_{ii}} h_i$ obey the relations (1.2.2b). In other words, every node in the Dynkin diagram corresponds to a copy of the A_1 Lie algebra. The lines connecting these nodes tells how these ℓ copies of A_1 intertwine.

For instance consider the first Cartan matrix given above. It corresponds to the Lie algebra A_2, or $sl_3(\mathbb{C})$. The two A_1 subalgebras which generate it (corresponding to the 2 nodes of the Dynkin diagram) can be chosen to be the trace-zero matrices of the form

$$\begin{pmatrix} \star & \star & 0 \\ \star & \star & 0 \\ 0 & 0 & 0 \end{pmatrix}, \quad \begin{pmatrix} 0 & 0 & 0 \\ 0 & \star & \star \\ 0 & \star & \star \end{pmatrix} .$$

It can be shown that the Lie algebra corresponding to an indecomposable symmetrised Cartan matrix will be finite-dimensional and simple, and conversely that any finite-dimensional simple Lie algebra corresponds to an indecomposable symmetrisable Cartan matrix in this way.

A confusion sometimes arises between the terms 'generators' and 'basis'. Both generators and basis vectors build up the whole algebra; the difference lies in which operations you are permitted to use. For a basis, you are only allowed to use linear combinations (i.e. addition of vectors and multiplication by numbers), while for generators you are also permitted multiplication of vectors (or the bracket, in the Lie case). 'Dimension' refers to basis, while 'rank' usually refers in some way to generators. For instance the (commutative associative) algebra of polynomials in one variable x is infinite-dimensional — any basis needs infinitely many vectors. However, the single polynomial x is enough to generate it (so we could say that its rank is 1). Although those Lie algebras have 3ℓ generators, their dimensions in general will be greater.

From the point of view of generators and relations, the step from 'finite-dimensional simple' to 'symmetrisable Kac-Moody' is rather easy: the only difference is that we drop the 'positive-definite' condition (which was responsible for finite-dimensionality). Kac-Moody (KM) algebras are also generated by (finitely many) A_1 subalgebras, and their theory is quite parallel to that of the simple algebras. Compare Figures 6 and 9 in [59].

Now, it is easy to generalise something; the challenge is to generalise it in a rich and interesting direction. One natural and appealing strategy for generalisation was followed instinctively by a grad student named Robert Moody. Moody's original motivation for developing the theory of Kac-Moody algebras was the Weyl group. If there were Lie algebras for finite Coxeter groups, he asked, why not also for the Euclidean (=affine) ones? For another example of this style of generalisation, consider the question: What is the analogue of calculus (or manifolds) over weird fields — fields (like $\mathbb{Z}_p$) for which the usual limit definitions make no sense? This question leads to the riches of algebraic geometry. Nevertheless this generalisation strategy, even in the hands of a master, will not always be successful. For instance, consider all the trouble the following metaphor has caused: my watch has a maker, so so should the Universe.

Recently Borcherds produced a further generalisation of finite-dimensional simple Lie algebras, which is rather less obvious than that of Kac-Moody algebras. It is easy to associate a Lie algebra to a matrix A, but which class of matrices will yield a deep theory?

Borcherds found such a class by holding in his hand a single algebra (the fake Monster Lie algebra, see §2.9) which acted a lot like a KM algebra, even though it had 'imaginary simple roots'.

By a *generalised symmetrised Cartan matrix* $A = (a_{ij})$ we will mean a symmetric real matrix (possibly infinite), such that $a_{ij} \leq 0$ if $i \neq j$, and if $a_{ii} > 0$ then $2\frac{a_{ij}}{a_{ii}} \in \mathbb{Z}$ for all j. By a *universal generalised Kac-Moody algebra* (universal GKM) or *universal Borcherds-Kac-Moody algebra* $\mathfrak{g}$ we mean the algebra[33] with generators e_i, f_i, h_{ij}, and with relations: $[e_i f_j] = h_{ij}$; $[h_{ij} e_k] = \delta_{ij} a_{ik} e_k$; $[h_{ij} f_k] = -\delta_{ij} a_{ik} f_k$; if $a_{ii} > 0$ and $i \neq j$ then $\mathrm{ad}(e_i)^n e_j = \mathrm{ad}(f_i)^n f_j = 0$ for $n = 1 - 2\frac{a_{ij}}{a_{ii}}$; and if $a_{ij} = 0$ then $[e_i e_j] = [f_i f_j] = 0$.

For example the Heisenberg algebra (1.2.3) corresponds to the choice $A = (0)$, while any other 1×1 $A = (a)$ corresponds to A_1. A universal GKM algebra differs from a KM algebra in that it is built up from Heisenberg algebras as well as A_1, and these subalgebras intertwine in more complicated ways. Nevertheless *much of the theory for finite-dimensional simple Lie algebras continues to find an analogue in this much more general setting* (e.g. root-space decomposition, Weyl group, character formula,...). This unexpected fact is the point of GKM algebras.

To get a feel for these algebras, let us prove a few simple results concerning the h_{ij}. Note first that, using the above relations together with the Jacobi identity, we obtain $[h_{ij} h_{k\ell}] = \delta_{ij}(a_{jk} - a_{j\ell}) h_{k\ell}$. Comparing this with $[h_{k\ell} h_{ij}] = -[h_{ij} h_{k\ell}]$, we see that bracket must always equal 0. Hence all h's pairwise commute, and $h_{ij} = 0$ unless the ith and jth columns of A are identical. An easy exercise now is to show that when $i \neq j$, h_{ij} will lie in the centre of the algebra (i.e. h_{ij} will commute with all other generators).

Although the definition of universal GKM algebra is more natural, it turns out that an equivalent ·form can be more useful in practice. (It's simpler to describe over $\mathbb{R}$, so in most expositions the reals are used, but alas it's far too late for us to switch loyalties now.) By a *generalised Kac-Moody algebra* (or *Borcherds-Kac-Moody algebra*) $\mathfrak{g}$, we mean a (complex!) Lie algebra which is:

- $\mathbb{Z}$-graded, i.e. $\mathfrak{g} = \oplus_i \mathfrak{g}_i$, where $[\mathfrak{g}_i \mathfrak{g}_j] \subseteq \mathfrak{g}_{i+j}$;
- $\mathfrak{g}_i$ is finite-dimensional for $i \neq 0$;
- $\mathfrak{g}$ has an antilinear involution ω (i.e. $\omega(kx + y) = k^* \omega(x) + \omega(y)$, $[\omega(x)\omega(y)] = \omega([xy])$, and $\omega \circ \omega = id.$) which maps $\mathfrak{g}_i$ to $\mathfrak{g}_{-i}$ and acts as multiplication by -1 on some basis of $\mathfrak{g}_0$;
- $\mathfrak{g}$ has an invariant symmetric bilinear form $(\star, \star)$ (i.e. $([xy], z) = (x, [yz])$ and $(y, z) = (z, y) \in \mathbb{C}$), obeying $(\omega(x), \omega(y)) = (x, y)^*$, such that $(\mathfrak{g}_i, \mathfrak{g}_j) = 0$ if $i \neq -j$;
- the Hermitian form defined by $(x|y) := -(\omega(x), y)$ is positive-definite on $\mathfrak{g}_{i \neq 0}$.

Note that for some basis x_i of $\mathfrak{g}_0$, the third condition tells us $-[x_i x_j] = [(-x_i)(-x_j)] = [x_i x_j]$, i.e. $\mathfrak{g}_0$ has a trivial bracket. It plays the role of the Cartan subalgebra $\mathfrak{h}$ in the theory.

For example, let $\mathfrak{g} = sl_2(\mathbb{C})$ and recall (1.2.5). Then $\mathfrak{g}_1 = \mathbb{C}e$, $\mathfrak{g}_0 = \mathbb{C}h$, $\mathfrak{g}_{-1} = \mathbb{C}f$ is the root-space decomposition. $\omega(x) = -x^\dagger$ is the Cartan involution. $(x, y) = \mathrm{tr}(xy)$ is the Killing form.

---

[33] As with KM algebras, usually we want to extend it by some derivations; enough derivations are added so that the simple roots are linearly independent.

It turns out [5] that any universal GKM algebra is a GKM algebra, and any GKM algebra can be constructed from a unique universal GKM algebra (by quotienting out part of the centre and adding derivations), so in that sense the two structures are equivalent. This theorem is important, because it tells us that *GKM algebras are the ultimate generalisation of simple Lie algebras*, in the sense that any further generalisation will lose some basic structural ingredient.

We know simple Lie algebras (and groups) arise in both classical and quantum physics, and the affine KM algebras are important in CFT, as we saw in Part 1. GKM algebras have recently appeared in the physics literature (see Harvey-Moore) in the context of BPS states in string theory.

How do GKM algebras arise in VOAs? If we define $[xy] := x_0 y$, then as mentioned in §2.6 we get from the VOA Jacobi identity the equation $[x[yz]] - [y[xz]] = [[xy]z]$. Thus our bracket will be anti-associative if it is anti-commutative. But is it anti-commutative? It can be shown

$$u_n v = \sum_{i=0}^{\infty} \frac{1}{i!}(-1)^{1+n+1}(L_{-1})^i(v_{n+i}u) \tag{2.7.1}$$

so $u_0 v \equiv -v_0 u$ if we look at things mod $L_{-1}V$.

Since our bracket is clearly bilinear, we thus get a Lie algebra structure on $V/L_{-1}V$. Similarly, we get a symmetric bilinear product on $V/L_{-1}V$, given by $\langle u, v \rangle := u_1 v$.

We would like $\langle \star, \star \rangle$ to respect the Lie algebra structure, i.e. be $[\star\star]$-invariant. We compute from (2.6.4)

$$\langle [uv], t \rangle = -[v\langle u, t \rangle] + \langle u, [vt] \rangle . \tag{2.7.2}$$

Since we would like to identify $\langle \star, \star \rangle$ with the bilinear form in the GKM algebra definition, we also would like it to be number-valued (i.e. have 1-dimensional range).

There is a simple way to satisfy both of these. First, restrict attention to V_1, i.e. the conformal weight 1 vectors: $V_1 \cap (V/L_{-1}V) = V_1/L_{-1}V_0$. Then $\langle u, v \rangle \in V_0$. Assume V_0 is 1-dimensional: i.e. $V_0 = \mathbb{C}1$. Then $\langle u, v \rangle$ will equal a number times 1, so call (u, v) that number. Also, $[\langle u, t \rangle v] = (u, t)1_0 v = 0$, so $\langle \star, \star \rangle$ and hence $(\star, \star)$, will be invariant. Of course, when $V_0 = \mathbb{C}1$, $L_{-1}V_0 = \{0\}$.

$V_1/L_{-1}V_0$ is generally too large in practice to be useful; a subalgebra can be defined as follows. Let P_n be the 'primary states with conformal weight n', i.e. the $u \in V_n$ killed by L_m for all $m > 0$. Then $\mathfrak{g}(V) := P_1/L_{-1}P_0$ will be a subalgebra of $V_1/L_{-1}V_0$. Through the assignment $u \mapsto u_0$, $\mathfrak{g}(V)$ acts on V and this action commutes with that of L_1. This association of a Lie algebra to a VOA is due to Borcherds (1986).

Similar arguments show that when V_0 is 1-dimensional and V_1 is 0-dimensional, then V_2 will necessarily be a commutative nonassociative algebra with product $u \times v := u_1 v \in V_2$ and identity element $\frac{1}{2}\omega$ (*proof:* $\omega \times u = L_0 u = 2u$). Now, those conditions on V_0, V_1 are satisfied by the Moonshine module $V^\natural$. We find that $V_2^\natural$ is none other than the Griess algebra extended by an identity element.

2.8. INGREDIENT #5: DENOMINATOR IDENTITIES

In §1.3, we discussed the representation theory of Lie algebras. An important invariant of a representation is its *character*. Simple Lie algebras possess a very useful formula for their characters, due to Weyl:

$$\mathrm{ch}_\lambda(z) := \sum_\mu \dim(L_\lambda(\mu))\, e^{\mu\cdot z} = e^{-\rho\cdot z}\, \frac{\sum_{w\in W} \pm e^{w(\lambda+\rho)\cdot z}}{\prod_{\alpha\in\Delta_+}(1-e^{-\alpha\cdot z})}\,, \qquad (2.8.1)$$

where W is the Weyl group, Δ_+ the positive roots, and where $\oplus_\mu L_\lambda(\mu)$ is the weight-space decomposition of L_λ — i.e. the simultaneous eigenspaces of the h_i. Here z belongs to the Cartan subalgebra $\mathfrak{h}$; the character is complex-valued. Analogous statements hold for all GKM algebras.

It is rare indeed when a trivial special case of a theorem or formula produces something interesting. But that is what happens here. Consider the trivial representation: i.e. $x \mapsto 0$ for all $x \in X_\ell$. Then the character is identically 1, by definition: $\mathrm{ch}_0 \equiv 1$. Thus the character formula tells us that a certain alternating sum over a Weyl group, equals a certain product over positive roots. These formulas, called *denominator formulas*, are nontrivial even in the finite-dimensional cases.

Consider for instance the smallest simple algebra, A_1. Here the identity indeed is too trivial: it reads $e^{z/2} - e^{-z/2} = e^{z/2}(1 - e^{-z})$. For A_2 we get a sum of 6 terms equalling a product of 3 terms, and the complexity continues to rise from there.

Around 1970 Macdonald tried to generalise these finite denominator identities to infinite identities, corresponding to the extended Dynkin diagrams. These were later reinterpreted by Kac, Moody and others as denominator identities for affine nontwisted KM algebras. The simplest one was known classically as the Jacobi triple product identity:

$$\sum_{n=-\infty}^{\infty} (-1)^n x^{n^2} y^n = \prod_{m=1}^{\infty} (1 - x^{2m})(1 - x^{2m-1}y)(1 - x^{2m-1}y^{-1})$$

We now know it to be the denominator identity for the simplest infinite-dimensional KM algebra, $A_1^{(1)}$.

Freeman Dyson is famous for his work in quantum field theory, but he started as an undergraduate in number theory and still enjoys it as a hobby. Dyson [20] found a curious formula for the Ramanujan τ-function, which can be defined by the generating function $\sum_{n=1}^{\infty} \tau(n)x^n = \eta^{24}(x) := x \prod_{m=1}^{\infty}(1 - x^m)^{24}$. Dyson found the remarkable formula

$$\tau(n) = \sum \frac{(a-b)(a-c)(a-d)(a-e)(b-c)(b-d)(b-e)(c-d)(c-e)(d-e)}{1!\,2!\,3!\,4!}$$

where the sum is over all 5-tuples $(a,b,c,d,e) \equiv (1,2,3,4,5)$ (mod 5) obeying $a+b+c+d+e = 0$ and $a^2 + b^2 + c^2 + d^2 + e^2 = 10n$. Using this, an analogous formula can be found for η^{24}. Dyson knew that similar-looking formulas were also known for η^d for the values $d = 3, 8, 10, 14, 15, 21, 24, 26, 28, 35, 36, \ldots$.

What was ironic was that Dyson found that formula at the same time that Macdonald was finding the Macdonald identities. Both were at Princeton then, and would often chat a little when they bumped into each other after dropping off their daughters at school. But they never discussed work. Dyson didn't realise that his strange list of numbers has a simple interpretation: they are precisely the dimensions of the simple Lie algebras! $3 = \dim(A_1)$, $8 = \dim(A_2)$, $10 = \dim(C_2)$, $14 = \dim(G_2)$, etc. In fact these formulas for η^d are none other than (specialisations of) the Macdonald identities. For example, Dyson's formula is the denominator formula for $A_4^{(1)}$ ($24 = \dim(A_4)$). If they had spoken, they would probably have anticipated the affine denominator identity interpretation.

One curiousity apparently still has no algebraic interpretation: No simple Lie algebra has dimension 26, so the formula for η^{26} can't correspond to any Macdonald identity.

Macdonald didn't close the book on denominator identities. More recently Kac and Wakimoto [40] have used denominator identities for Lie superalgebras to obtain nice formulas for various generator functions involving sums of squares, sums of triangular numbers (triangular numbers are numbers of the form $\frac{1}{2}k(k+1)$), etc. For instance, the number of ways n can be written as a sum of 16 triangular numbers is

$$\frac{1}{3 \cdot 4^3} \sum ab\,(a^2 - b^2)^2$$

where the sum is over all odd positive integers a, b, r, s obeying $ar + bs = 2n + 4$ and $a > b$.

Another example of denominator identities is Borcherds' use of them in proving the Moonshine conjectures. In particular this motivated his introduction of the GKM algebras. The denominator identities for other GKM algebras were used by Borcherds to obtain results on the moduli spaces of e.g. families of K3 surfaces. They are also often turned-around now and used for learning about the positive roots in a given GKM.

2.9. Proof of the Moonshine Conjectures

The main Conway-Norton conjecture was proved almost immediately. Thompson showed that if $g \mapsto a_n(g)$ is a character for all $n \leq 1200$, then it will be for all n. He also showed that if certain congruence conditions hold for a certain number of $a_n(g)$ (all with $n \leq 100$), then all $g \mapsto a_n(g)$ will be *virtual* characters (i.e. a linear combination over $\mathbb{Z}$ of irreducible characters of $\mathbb{M}$; only if all coefficients are nonnegative will it be a true character). Atkin-Fong-Smith [54] used that to prove on a computer that indeed all were virtual characters. But their work didn't say anything about the underlying (possibly virtual) representation V. The real challenge was to construct (preferably in a natural manner) the representation which works. Frenkel-Lepowsky-Meurman [23] constructed a candidate for it (the Moonshine module $V^\natural$); it was Borcherds who finally proved $V^\natural$ obeyed the Conway-Norton conjecture. A good overview of Borcherds' work on Moonshine is provided in [32].

We want to show that the McKay-Thompson series $T_g(\tau) := q^{-1}\mathrm{tr}_{V^\natural}(gq^{L_0})$ of (2.1.4) equals the Hauptmodul $J_g(\tau)$ in (2.4.1) (the 'fudge factor' $q^{-1} = q^{-c/24}$ is familiar to e.g. KM algebras and CFT and was discussed at the end of §1.2). Borcherds' strategy was to bring in Lie theory and to use the corresponding denominator identity to provide useful combinatorial data. The first guess for this 'Monster Lie algebra' was the Kac-Moody algebra whose Dynkin diagram is essentially the Leech lattice (i.e. a node for each vector in Λ_{24}, and 2 nodes are connected by a number of edges depending on the value of a certain dot product). It was eventually discarded because some of the critical data (namely positive root multiplicities) needed in order to write down its denominator identity were too complicated. But looking at that failed attempt led Borcherds to a second candidate, now called the *fake Monster Lie algebra* $\mathfrak{g}'_M$. In order to *construct* it, he developed the theory of VOAs; and in order to *understand* it, he developed the theory of GKM algebras. We will define it shortly. $\mathfrak{g}'_M$ also turned out to be inadequate for proving the Moonshine conjectures; however it directly led him to the GKM algebra now called the true *Monster Lie algebra* $\mathfrak{g}_M$. And that directly led to the proof of Moonshine.

Step 1: Construct $\mathfrak{g}_M$ from $V^\natural = V_0^\natural \oplus V_1^\natural \oplus \cdots$. For later convenience, reparametrise these subspaces $V^i := V_{i+1}^\natural$. Recall the even indefinite lattice $II_{1,1}$ defined in (1.6.1). Of course the direct choice $\mathfrak{g}(V^\natural)$ is 0-dimensional because $V_1^\natural$ is trivial, so we must modify $V^\natural$ first.

The Monster Lie algebra is (essentially) defined to be the Lie algebra $\mathfrak{g}(V^\natural \otimes V_{II_{1,1}})$ associated to the vertex algebra $V^\natural \otimes V_{II_{1,1}}$ (strictly speaking, more of $\mathfrak{g}$ is quotiented away). By contrast, the fake Monster is the Lie algebra associated to the vertex algebra $V_{\Lambda_{24}} \otimes V_{II_{1,1}} \cong V_{II_{25,1}}$. Both of these are vertex algebras as opposed to VOAs, because of the presence of the indefinite lattices, but this isn't important here. $\mathfrak{g}_M$ inherits a $II_{1,1}$-grading from $V_{II_{1,1}}$: the piece of grading (m,n) is isomorphic (as a vector space) to V^{mn}, if $(m,n) \neq (0,0)$; the $(0,0)$ piece is isomorphic to $\mathbb{R}^2$. Borcherds uses the No-Ghost Theorem of string theory to show that the homogeneous pieces of $\mathfrak{g}_M$ are those of $V^\natural$.

Both $\mathfrak{g}_M$ and $\mathfrak{g}'_M$ are GKM algebras; for instance the $\mathbb{Z}$-grading of $\mathfrak{g}_M$ is given by $(\mathfrak{g}_M)_k = \oplus_{m+n=k} V^{mn}$ for $k \neq 0$, while the 0-part is $V^{-1} \oplus V^{-1}$. Although $\mathfrak{g}'_M$ is not used in the proof of the Monstrous Moonshine conjectures, it is related to some kind of Moonshine for the finite simple group .1, which is 'half' of the automorphism group .0 of the Leech lattice Λ_{24}. $\mathfrak{g}_M$ corresponds to the Cartan matrix

$$\begin{pmatrix} B_{-1,-1} & B_{-1,1} & B_{-1,2} & \cdots \\ B_{1,-1} & B_{1,1} & B_{1,2} & \cdots \\ B_{2,-1} & B_{2,1} & B_{2,2} & \cdots \\ \vdots & \vdots & \vdots & \ddots \end{pmatrix},$$

where $B_{i,j}$ for $i,j \in \{-1,1,2,3,\ldots\}$ is the $a_i \times a_j$ block with fixed entry $-i-j$ (the a_i as usual are the coefficients $\sum_n a_n q^n$ of $j - 744$).

2nd step: Compute the denominator identity of $\mathfrak{g}_M$: we get

$$p^{-1} \prod_{\substack{m>0 \\ n \in \mathbb{Z}}} (1 - p^m q^n)^{a_{mn}} = j(z) - j(\tau) \tag{2.9.1}$$

where $p = e^{2\pi i z}$. The result are various formulas involving the coefficients a_i, for instance $a_4 = a_3 + (a_1^2 - a_1)/2$. It turns out to be possible to 'twist' (2.9.1) by each $g \in \mathbb{M}$, obtaining

$$p^{-1} \exp[-\sum_{k>0} \sum_{\substack{m>0 \\ n \in \mathbb{Z}}} a_{mn}(g^k) \frac{p^{mk} q^{nk}}{k}] = T_g(z) - T_g(\tau) . \tag{2.9.2}$$

61

This looks a lot more complicated, but you can glimpse the Taylor expansion of $\ln(1-p^m q^n)$ there and in fact for $g = id$ (2.9.2) reduces to (2.9.1). This formula gives more generally identities like $a_4(g) = a_2(g) + (a_1(g)^2 - a_1(g^2))/2$, where $T_g(\tau) = \sum_i a_i(g)q^i$. These formulas involving the McKay-Thompson coefficients are equivalent to the replication formulae conjectured in §2.4.

3rd step: It was known earlier that all of the Hauptmoduls also obey those replication formula, and that anything obeying them will lie in a finite-dimensional manifold which we'll call R. In particular, if $B(q) = q^{-1} + \sum_{n>0} b_n q^n$ and $C(q) = q^{-1} + \sum_{n>0} c_n q^n$ both lie in R, and $b_n = c_n$ for $n \le 23$, then $B(q) = C(q)$. In fact, it turns out that if we verify for each conjugacy class $[g]$ of $\mathbb{M}$ that the first, second, third, fourth and sixth coefficients of the McKay-Thompson series T_g and the corresponding Hauptmodul J_g agree, then $T_g = J_g$, and we are done.

That is precisely what Borcherds then did: he compared finitely many coefficients, and as they all equalled what they should, this concluded the proof of Monstrous Moonshine!

However there was a disappointing side to his proof. While no one disputed its *logical* validity, it did seem to possess a disappointing *conceptual* gap. In particular, the Moonshine conjectures were made in the hope that proving them would help explain what the Monster had to do with the j-function and the other Hauptmoduls. A good proof says much more than 'True' or 'False'. The case-by-case verification occurred at the critical point where the McKay-Thompson series were being compared directly to the Hauptmoduls. The proof showed that indeed the Moonshine module establishes some sort of relation between T_g and J_g (namely, they must lie in the same finite-dimensional space), but why couldn't it be just a happy meaningless accident that they be equal? Of course we believe it's more than merely an accident, so our proof should reflect this: we want a more conceptual explanation.

This conceptual gap has since been filled [17] — i.e. the case-by-case verification has been replaced with a general theorem. It turns out that something obeying the replicable formulas will also obey something called *modular equations*. A modular equation for a function f is a polynomial identity obeyed by $f(x)$ and $f(nx)$. The simplest examples come from the exponential and cosine functions: note that for any $n > 0$, $\exp(nx) = (\exp(x))^n$ and $\cos(nx) = T_n(\cos(x))$ where T_n is a Tchebychev polynomial. A more interesting example of a modular equation is obeyed by $J(\tau) = j(\tau) - 744$: put $X = J(\tau)$ and $Y = J(2\tau)$, then

$$(X^2 - Y)(Y^2 - X) = 393768\,(X^2 + Y^2) + 42987520\,XY + 40491318744\,(X + Y)$$
$$- 120981708338256\,.$$

Finding modular equations (for various elliptic functions) was a passion of the great mathematician Ramanujan. His notebooks are filled with them. See e.g. [7] for an application of Ramanujan's modular equations to computing the first billion or so digits of π. Many modular equations are also studied in [10]. For more of their applications, see e.g. [16]. It can be shown that the only functions $f(\tau) = q^{-1} + a_1 q + \cdots$ which obey modular equations for all n, are $J(\tau)$ and the 'modular fictions' q^{-1} and $q^{-1} \pm q$ (which are essentially exp, cos, and sin).

It was proved in [17] that, roughly speaking, a function $B(\tau) = q^{-1} + \sum_{n>0} b_n q^n$ which obeys enough modular equations, will either be of the form $B(\tau) = q^{-1} + b_1 q$, or will necessarily be a Hauptmodul for a modular group containing some $\Gamma_0(N)$. The converse is also true: for instance, a modular equation for the Hauptmodul J_{25} of $\Gamma_0(25)$ given in (2.3.5) is

$$(X^2 - Y)(Y^2 - X) = -2(X^2 + Y^2) + 4(X + Y) - 4 \, ,$$

where $X = J_{25}(\tau)$ and $Y = J_{25}(2\tau)$. To eliminate the conceptual gap, this result should then replace step 3. Steps 1 and 2 are still required, however.

This conceptual gap should not take away from what was a remarkable accomplishment by Borcherds: not only the proof of the Monstrous Moonshine conjectures, but also the definition of two new and important algebraic structures. I hope the preceding sections give the reader some indication of why Borcherds was awarded one of the 1998 Fields Medals.

Another approach to the Hauptmodul property is by Tuite [57], who related it to the (conjectured) uniqueness of $V^\natural$. Norton has suggested that the reason $\mathbb{M}$ is associated to *genus-zero* modular functions could be what he calls its '6-transposition' property [47].

So has Moonshine been explained? According to Conway, McKay, and many others, it hasn't. They consider VOAs in general, and $V^\natural$ in particular, to be too complicated to be God-given. The progress, though impressive, has broadened not lessened the fundamental mystery, they would argue.

For what it's worth, I don't completely agree. Explaining away a mystery is a little like grasping a bar of soap in a bathtub, or quenching a child's curiosity. Only extreme measures like pulling the plug, or enrollment in school, ever really work. True progress means displacing the mystery, usually from the particular to the general. Why is the sky blue? Because of how light scatters in gases. Why are Hauptmoduls attached to each $g \in \mathbb{M}$? Because of $V^\natural$. Mystery exists wherever we can ask 'why' — like beauty it's in the beholder's eye.

Moonshine is now 'leaving the nest'. We are entering a consolation phase, tidying up, generalising, simplifying, clarifying, working out more examples. Important and interesting discoveries will be made in the next few years, and yes there still is mystery, but no longer does a Moonshiner feel like an illicit distiller: Moonshine is now a day-job!

Acknowledgments. I warmly thank the Feza Gursey Institute in Istanbul, and in particular Teoman Turgut, for their invitation to the workshop and hospitality during my month-long stay. These notes are based on 16 lectures I gave there in Summer 1998. I've also benefitted from numerous conversations with Y. Billig, A. Coste, C. Cummins, M. Gaberdiel, J. McKay, M. Tuite, and M. Walton — Mark Walton in particular made a very careful reading of the manuscript (and hence must share partial blame for any errors still remaining). My appreciation as well goes to J.-B. Zuber and P. Ruelle for sharing with me their personal stories behind the discoveries of, respectively, A-D-E in $A_1^{(1)}$ and Fermat in $A_2^{(1)}$. The research was supported in part by NSERC.

REFERENCES

1. V. I. Arnold, *Catastrophe Theory*, 2nd edn. (Springer, Berlin, 1997);
   M. Hazewinkel, W. Hesselink, D. Siersma, and F.D. Veldkamp, *Nieuw Arch. Wisk.* **25** (1977) 257;
   P. Slodowy, in: Lecture Notes in Math 1008, J. Dolgachev (ed.) (Springer, Berlin, 1983).
2. M. Bauer, A. Coste, C. Itzykson, and P. Ruelle, *J. Geom. Phys.* **22** (1997) 134.
3. D. Bernard, *Nucl. Phys.* **B288** (1987) 628.
4. J. Böckenhauer and D. E. Evans, *Commun. Math. Phys.* **200** (1999) 57; "Modular invariants, graphs and α-induction for nets of subfactors III", hep-th/9812110.
5. R. E. Borcherds, *Invent. math.* **109** (1992) 405.
6. R. E. Borcherds, "What is moonshine?", math.QA/9809110.
7. J. M. Borwein, P. B. Borwein and D. H. Bailey, *Amer. Math Monthly* **96** (1989) 201.
8. A. Cappelli, C. Itzykson, and J.-B. Zuber, *Commun. Math. Phys.* **113** (1987) 1.
9. R. Carter, G. Segal and I. M. Macdonald, *Lectures on Lie Groups and Lie algebras* (Cambridge University Press, Cambridge, 1995).
10. A. Cayley, *An Elementary Treatise on Elliptic Functions*, 2nd edn (Dover, New York, 1961).
11. J. H. Conway, *Math. Intelligencer* **2** (1980) 165.
12. J. H. Conway and S. P. Norton, *Bull. London Math. Soc.* **11** (1979) 308.
13. J. H. Conway and N. J. A. Sloane, *Sphere packings, lattices and groups*, 3rd edn (Springer, Berlin, 1999).
14. A. Coste and T. Gannon, *Phys. Lett.* **B323** (1994) 316.
15. R. Courant and D. Hilbert, *Methods of Mathematical Physics II* (Wiley, New York, 1989).
16. D. Cox, *Primes of the form $x^2 + ny^2$*, (Wiley, New York, 1989).
17. C. J. Cummins and T. Gannon, *Invent. math.* **129** (1997) 413.
18. P. Di Francesco, P. Mathieu and D. Sénéchal, *Conformal Field Theory* (Springer, New York, 1996).
19. L. Dixon, P. Ginsparg, and J. Harvey, *Commun. Math. Phys.* **119** (1988) 221.
20. F. Dyson, *Bull. Amer. Math. Soc.* **78** (1972) 635.
21. F. J. Dyson, *Math. Intelligencer* **5** (1983) 47.
22. D.E. Evans and Y. Kawahigashi, *Quantum symmetries on operator algebras* (Oxford University Press, Oxford, 1998).
23. I. Frenkel, J. Lepowsky, and A. Meurman, *Vertex operator algebras and the Monster* (Academic Press, San Diego, 1988).
24. J. Fuchs and C. Schweigert, *Symmetries, Lie algebras, and representations* (Cambridge University Press, Cambridge, 1997).
25. W. Fulton and J. Harris, *Representation Theory: A first course* (Springer, New York, 1996).
26. M. R. Gaberdiel and P. Goddard, "Axiomatic conformal field theory", hep-th/9810019.
27. M. R. Gaberdiel and P. Goddard, "An introduction to meromorphic conformal field theory and its representations", lecture notes in this volume.

28. D. Gaitsgory, "Notes on two dimensional conformal field theory and string theory", math.AG/9811061.
29. T. Gannon, "The Cappelli-Itzykson-Zuber A-D-E classification", math.QA/9902064.
30. K. Gawedzki, "Conformal field theory: a case study", lecture notes in this volume.
31. R. W. Gebert, *Internat. J. Mod. Phys.* **A8** (1993) 5441.
32. P. Goddard, "The work of R.E. Borcherds", math.QA/9808136.
33. D. Gorenstein, *Finite Simple Groups* (Plenum, New York, 1982).
34. D. Gorenstein, R. Lyons, and R. Solomon, *The Classification of the Finite Simple Groups* (AMS, Providence, 1994).
35. A. Hanany and Y.-H. He, "Non-abelian finite gauge theories", hep-th/9811183.
36. J. E. Humphreys, *Introduction to Lie algebras and representation theory* (Springer, New York, 1994).
37. V. G. Kac, In: Lecture Notes in Math 848 (Springer, New York, 1981).
38. V. G. Kac, *Infinite Dimensional Lie algebras*, 3rd edn (Cambridge University Press, Cambridge, 1990).
39. V. G. Kac, *Vertex Algebras for Beginners*, 2nd edn (AMS, Providence, 1998).
40. V. G. Kac and M. Wakimoto, In: *Lie Theory and Geometry in Honor of Bertram Kostant*, Progress in Math. **123** (Birkhäuser, Boston, 1994).
41. S. Kass, R. V. Moody, J. Patera and R. Slansky, *Affine Lie algebras, weight multiplicities, and branching rules*, Vol. 1 (University of California Press, Berkeley, 1990).
42. S. Lang, *Elliptic Functions*, 2nd edn (New York, Springer, 1997).
43. F. W. Lawvere and S. H. Schanuel, *Conceptual Mathematics: A first introduction to Categories* (Cambridge University Press, Cambridge, 1997).
44. H. Minc, *Nonnegative matrices* (Wiley, New York, 1988).
45. E.J. Mlawer, S.G. Naculich, H.A. Riggs, and H. J. Schnitzer, *Nucl. Phys.* **B352** (1991) 863.
46. W. Nahm, *Commun. Math. Phys.* **118** (1988) 171.
47. S. P. Norton, In: *Proc. Symp. Pure Math.* **47** (1987) 208.
48. A. Ocneanu, "Paths on Coxeter diagrams: From Platonic solids and singularities to minimal models and subfactors" (Lectures given at Fields Institute (1995), notes recorded by S. Goto).
49. L. Queen, *Math. of Comput.* **37** (1981) 547.
50. A.N. Schellekens and S. Yankielowicz, *Nucl. Phys.* **B327** (1989) 673.
51. M. Schottenloher, *A Mathematical Introduction to Conformal Field Theory* (Springer, Berlin, 1997).
52. G. Segal, In: *IXth Proc. Int. Congress Math. Phys.* (Hilger, 1989).
53. S. Singh, *Fermat's Enigma* (Penguin Books, London, 1997).
54. S. D. Smith, In: *Finite Groups – Coming of Age*, Contemp. Math. **193** (AMS, Providence, 1996).
55. W. Thurston, *Three-dimensional geometry and topology*, vol. 1 (Princeton, 1997).
56. L. Toti Rigatelli, *Evariste Galois* (Birkhäuser, Basel, 1996).
57. M. Tuite, *Commun. Math. Phys.* **166** (1995) 495.
58. V. G. Turaev, *Quantum invariants of knots and 3-manifolds* (de Gruyter, Berlin, 1994).

59. M. A. Walton, "Affine Kac-Moody algebras and the Wess-Zumino-Witten model", lecture notes in this volume.

60. M. Waldschmidt, P. Moussa, J.-M. Luck, and C. Itzykson (ed.), *From Number Theory to Physics* (Berlin, Springer, 1992).

61. D. Zagier, *Amer. Math. Monthly* **97** (1990) 144.

62. J.-B. Zuber, *Commun. Math. Phys.* **179** (1996) 265.

63. J.-B. Zuber, private communication, February 1999.

Affine Kac-Moody Algebras
and the
Wess-Zumino-Witten Model

Mark Walton

Physics Department, University of Lethbridge

Lethbridge, Alberta, Canada T1K 3M4

walton@uleth.ca

Abstract: These lecture notes are a brief introduction to Wess-Zumino-Witten models, and their current algebras, the affine Kac-Moody algebras. After reviewing the general background, we focus on the application of representation theory to the computation of 3-point functions and fusion rules.

1. Introduction

In 1984, Belavin, Polyakov and Zamolodchikov [1] showed how an infinite-dimensional field theory problem could effectively be reduced to a finite problem, by the presence of an infinite-dimensional symmetry. The symmetry algebra was the Virasoro algebra, or two-dimensional conformal algebra, and the field theories studied were examples of two-dimensional conformal field theories. The authors showed how to solve the minimal models of conformal field theory, so-called because they realise just the Virasoro algebra, and they do it in a minimal fashion. All fields in these models could be grouped into a discrete, finite set of conformal families, each associated with a representation of the Virasoro algebra.

This strategy has since been extended to a large class of conformal field theories with similar structure, the rational conformal field theories (RCFT's) [2]. The new feature is that the theories realise infinite-dimensional algebras that contain the Virasoro algebra as a subalgebra. The larger algebras are known as W-algebras [3] in the physics literature.

Thus the study of conformal field theory (in two dimensions) is intimately tied to infinite-dimensional algebras. The rigorous framework for such algebras is the subject of vertex (operator) algebras [4] [5]. A related, more physical approach is called meromorphic conformal field theory [6].

Special among these infinite-dimensional algebras are the affine Kac-Moody algebras (or their enveloping algebras), realised in the Wess-Zumino-Witten (WZW) models [7]. They

1

are the simplest infinite-dimensional extensions of ordinary semi-simple Lie algebras. Much is known about them, and so also about the WZW models. The affine Kac-Moody algebras are the subject of these lecture notes, as are their applications in conformal field theory. For brevity we restrict consideration to the WZW models; the goal will be to indicate how the affine Kac-Moody algebras allow the solution of WZW models, in the same way that the Virasoro algebra allows the solution of minimal models, and W-algebras the solution of other RCFT's. We will also give a couple of examples of remarkable mathematical properties that find an "explanation" in the WZW context.

One might think that focusing on the special examples of affine Kac-Moody algebras is too restrictive a strategy. There are good counter-arguments to this criticism. Affine Kac-Moody algebras can tell us about many other RCFT's: the coset construction [8] builds a large class of new theories as differences of WZW models, roughly speaking. Hamiltonian reduction [9] constructs W-algebras from the affine Kac-Moody algebras. In addition, many more conformal field theories can be constructed from WZW and coset models by the orbifold procedure [10] [11]. Incidentally, all three constructions can be understood in the context of gauged WZW models.

Along the same lines, the question "Why study two-dimensional conformal field theory?" arises. First, these field theories are solvable non-perturbatively, and so are toy models that hopefully prepare us to treat the non-perturbative regimes of physical field theories. Being conformal, they also describe statistical systems at criticality [12]. Conformal field theories have found application in condensed matter physics [13]. Furthermore, they are vital components of string theory [14], a candidate theory of quantum gravity, that also provides a consistent framework for unification of all the forces.

The basic subject of these lecture notes is close to that of [15]. It is hoped, however, that this contribution will complement that of Gawedzki, since our emphases are quite different.

The layout is as follows. Section 2 is a brief introduction to the WZW model, including its current algebra. Affine Kac-Moody algebras are reviewed in Section 3, where some background on simple Lie algebras is also provided. Both Sections 2 and 3 lay the foundation for Section 4: it discusses applications, especially 3-point functions and fusion rules. We indicate how *a priori* surprising mathematical properties of the algebras find a natural framework in WZW models, and their duality as rational conformal field theories.

2. Wess-Zumino-Witten Models

2.1. Action

Let G denote a compact connected Lie group, and g its simple Lie algebra[1]. Suppose γ is a G-valued field on the complex plane. The Wess-Zumino-Witten (WZW) action is written as [7][17]

$$S_k(\gamma) = -\frac{k}{8\pi} \int \mathcal{K}(\gamma^{-1}\partial^\mu\gamma, \gamma^{-1}\partial_\mu\gamma)\, d^2x \; + \; 2\pi k\, \tilde{S}(\gamma) \,, \qquad (2.1)$$

where $\partial_\mu = \partial/\partial x^\mu$, the summation convention is used with Euclidean metric, and $\mathcal{K}$ denotes the Killing form of g, which is nondegenerate for g simple,

$$\mathcal{K}(x,y) = \frac{\mathrm{Tr}(\mathrm{ad}\,_x \mathrm{ad}\,_y)}{2h^\vee} \;\;,\; x,y \in g \,. \qquad (2.2)$$

Here $h^\vee$ is an integer fixed by the algebra g, called the dual Coxeter number of g, and $\mathrm{ad}\,_x(z) := [x,z]$. The second term is the Wess-Zumino action. To describe it, imagine that the complex plane (plus the point at ∞) is a large 2-sphere S^2. γ then maps S^2 into the group manifold of G.

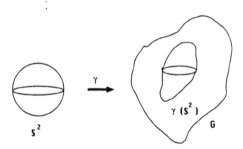

Figure 1. The map $\gamma:\; S^2 \to G$. B is the 3-dimensional solid ball with S^2 as its boundary, $\partial B = S^2$.

The homotopy groups $\pi_n(G)$ thus enter consideration (see [18], for example). The elements of $\pi_n(G)$ are the equivalence classes of continuous maps of the n-sphere S^n into (the group manifold of) G. Two such maps are equivalent if their images can be

---

[1] For an excellent, elementary introduction to Lie algebras, with physical motivation, see [16].

continuously deformed into each other. If all images of S^n in G are contractible to a point, then the n-th homotopy group of G is trivial, $\pi_n(G) = 0$. A non-trivial $\pi_n(G)$ indicates the presence of non-contractible n-cycles in G. (A cycle is a n-dimensional submanifold without boundary; a non-contractible one is also not a boundary itself.) So, homotopy is quite a fine measure of the topology of a group manifold G. For example, $\pi_n(G) = \mathbb{Z}$ implies there is a non-contractible n-cycle C_n in G that generates $\pi_n(G)$, and a map $S^n \to G$ can "wind around" this cycle any number ($\in \mathbb{Z}$) of times.

By a generalisation of Stokes' theorem, the existence of non-contractible cycles in G has to do with the existence of harmonic n-forms on G. A harmonic form h_n is a differential form that is closed ($dh_n = 0$), but not exact (no form p exists such that $h_n = dp$). Recall that if C_n is a non-contractible cycle, its boundary vanishes ($\partial C_n = 0$), and C_n is not a boundary itself. In the case $\pi_n(G) = \mathbb{Z}$ just mentioned, there exists a harmonic n-form h_n on G, that can be identified with the volume form on C_n, and can be normalised so that the volume of C_n computed with it is 1: $\int_{C_n} h_n = 1$.

Getting back to the Wess-Zumino term, since $\pi_2(G) = 0$ for any compact connected Lie group, γ can be extended to a map ($\tilde{\gamma}$ when we want to emphasise this) of B into G, where $\partial B = S^2$. The Wess-Zumino action can be written as

$$\tilde{S}(\gamma) = \frac{-1}{48\pi^2} \int_B \epsilon^{ijk} \mathcal{K}\left(\tilde{\gamma}^{-1}\frac{\partial\tilde{\gamma}}{\partial y^i}, \left[\tilde{\gamma}^{-1}\frac{\partial\tilde{\gamma}}{\partial y^j}, \tilde{\gamma}^{-1}\frac{\partial\tilde{\gamma}}{\partial y^k}\right]\right) d^3y \qquad (2.3)$$

where y^i ($i = 1, 2, 3$) denote the coordinates of B.

Let t^a denote the elements of a basis of g, i.e. $g = \mathrm{Span}\{t^a : a = 1, \ldots, \dim g\}$. For Hermitian t^a, $(t^a)^\dagger = t^a$, the commutation relations of g can be written as

$$[t^a, t^b] = \sum_c i f^{abc} t^c , \qquad (2.4)$$

where the structure constants f^{abc} are real. Normalising so that $\mathcal{K}(t^a, t^b) = \delta^{ab}$, we get

$$\mathcal{K}\left(t^a, [t^b, t^c]\right) = i f^{abc} . \qquad (2.5)$$

Since $\tilde{\gamma}^{-1}\frac{\partial\tilde{\gamma}}{\partial y^i}$ is an element of g, we see by (2.3) that the structure constants f^{abc} enter the Wess-Zumino action.

Now the totally antisymmetric structure constants f^{abc} of g define a harmonic 3-form h_3 on the group manifold of G. $\tilde{S}$ is an integral over the pull-back of this harmonic 3-form h_3 to the space B:

$$\tilde{S}(\gamma)_B := \int_B \tilde{\gamma}^* h_3 = \int_{\partial^{-1} S^2} \tilde{\gamma}^* h_3 . \qquad (2.6)$$

4

By the discussion of the previous paragraph, this points to a relation between the Wess-Zumino term and the homotopy of G. This will be made explicit soon.

If the WZW action is to describe a local theory on S^2, then the formal expression $B = \partial^{-1}S^2$ should indicate that the physics is independent of which 3-dimensional extension B of S^2 is used. Picture S^2 as a circle, in order to draw a simple diagram. $\gamma : S^2 \to G$ can be depicted as in Fig. 2.

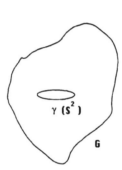

Figure 2. The map $\gamma : S^2 \to G$ depicted in one lower dimension (S^1 replaces S^2).

In Fig. 3, the images by γ of two different extensions B, B' of S^2 are pictured. In order that the physics described by B is equivalent to that described by B', we require

$$\exp\left[2\pi i k \tilde{S}(\gamma)_B\right] = \exp\left[2\pi i k \tilde{S}(\gamma)_{B'}\right] \tag{2.7}$$

or

$$\exp\left[2\pi i k \tilde{S}(\gamma)_{B'-B}\right] = 1 . \tag{2.8}$$

$\gamma(B' - B)$ is homotopically equivalent to S^3 (depicted as S^2 in Fig. 4).

Now to the homotopic significance of the WZ term: $\tilde{S}(\gamma)_{S^3} = N$ is the winding number of the map $\tilde{\gamma} : S^3 \to G$. Since $\pi_3(G) = \mathbb{Z}$ (for G any compact connected simple Lie group), we have $N \in \mathbb{Z}$. Therefore (2.8) requires $k \in \mathbb{Z}$, and since k and $-k$ yield indistinguishable physics, we use $k \in \mathbb{Z}_{\geq 0}$, which will be the so-called *level* of the affine Kac-Moody algebra realised by the WZW model.

The quantisation of the WZ term can also be understood by its relation to anomalies, which have topological significance (see [18], Chapter 13, for example). Consider first a

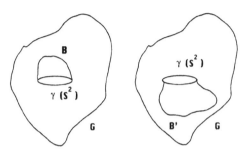

Figure 3. The γ-images of two different extensions B, B' of S^2.

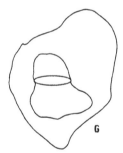

Figure 4. $\gamma(B - B')$ (see Fig. 3).

fermionic model with Lagrangian density

$$\mathcal{L} = \frac{1}{2}\bar{\Psi}\Gamma^\mu(\partial_\mu + \mathbb{A}_\mu)\Psi = \psi^\dagger(\partial_z + A)\psi + \bar{\psi}^\dagger(\partial_{\bar{z}} + \bar{A})\bar{\psi} . \qquad (2.9)$$

Here $z = x_1 + ix_2$, $\bar{z} = x_1 - ix_2$, and Γ^μ are the Dirac (gamma) matrices, with anti-commutation relations $\{\Gamma^\mu, \Gamma^\nu\} = 2\delta^{\mu,\nu}$. In the first expression, Ψ is a Dirac spinor, $\bar{\Psi} = \Psi^\dagger\Gamma^1$, and $\mathbb{A}_\mu$ is the gauge potential. In the second, the chiral components $\psi = (1+\Gamma)\Psi/2$, $\bar{\psi} = (1 - \Gamma)\Psi/2$ appear, where $\Gamma := i\Gamma^1\Gamma^2$ is the chirality operator, and $A = \mathbb{A}_1 - i\mathbb{A}_2$, $\bar{A} = \mathbb{A}_1 + i\mathbb{A}_2$. The Lagrangian is invariant under the gauge transformations

$$\begin{aligned} A &\to UAU^{-1} + U\partial_z U^{-1} , \quad \psi \to U\psi ; \\ \bar{A} &\to \bar{U}\bar{A}\bar{U}^{-1} + \bar{U}\partial_{\bar{z}}\bar{U}^{-1} , \quad \bar{\psi} \to \bar{U}\bar{\psi} , \end{aligned} \qquad (2.10)$$

where U and $\bar{U}$ are two independent elements of the gauge group G. This chiral $G \otimes G$ gauge invariance is the result of the vector and axial-vector gauge invariance of the Dirac Lagrangian: $U = \bar{U}$ specifies a vector gauge transformation, and when $U = \bar{U}^{-1}$, we get an axial-vector transformation.

In a spacetime of N dimensions, a gauge boson has $N - 2$ degrees of freedom. In $N = 2$, what this means is that all $\mathbb{A}_\mu$ can be obtained by applying gauge transformations to some fixed $\mathbb{A}_\mu$, 0, say. So we can parametrise

$$A = \alpha^{-1}\partial_z \alpha \ , \quad \bar{A} = \beta^{-1}\partial_{\bar{z}}\beta \ , \tag{2.11}$$

with $\alpha, \beta \in G$. Then, the gauge transformations (2.10) become

$$\alpha \ \rightarrow \ \alpha U \ , \quad \beta \ \rightarrow \ \beta \bar{U} \ . \tag{2.12}$$

Equivalently, we can say that the fields $A, \bar{A}$ are subsidiary fields that could have been eliminated in (2.9), giving a four-fermion interaction (and so a Thirring model).

The path integral

$$\int [d\psi][d\bar{\psi}] \exp\left(-\int \mathcal{L}\,d^2x\right) \ = \ e^{-S_{\text{eff}}} \tag{2.13}$$

defines an effective action S_{eff} from which the fermions have been eliminated. Because of the form of the Lagrangian (2.9) in (2.13), one often writes

$$S_{\text{eff}} \ = \ \log\det\left[\Gamma^\mu(\partial_\mu + \mathbb{A}_\mu)\right] \ . \tag{2.14}$$

In simple cases, these path integrals can be computed explicitly.

Suppose there are extra flavour indices for the fermions, suppressed in (2.9), running over a number N_L of (flavours of) left-handed fermions ψ, and a number N_R of right-handed fermions $\bar{\psi}$. If $N_L \neq N_R$, the axial-vector gauge invariance is destroyed when quantum corrections are taken into account. There is a chiral anomaly, proportional to $N_L - N_R$. In the path-integral formalism, this happens because the integration measure for the chiral fermions cannot be regularised in a way that preserves the invariance [19].

Following Polyakov and Wiegmann [20] (see also [21]), consider a fermionic model with N_L flavours of left-handed fermions ψ and similarly N_R right-handed fermions $\bar{\psi}$:

$$\mathcal{L} \ = \ \psi^\dagger(\partial_z + A)\psi + \bar{\psi}^\dagger(\partial_{\bar{z}} + \bar{A})\bar{\psi} + v\mathcal{K}(A, \bar{A}) \ , \tag{2.15}$$

where v is a constant.

First consider the case $N_L = N_R = 1$. In the gauge $\bar{A} = 0$, integrating out the fermions gives [20]

$$\log \det \left[\Gamma^\mu (\partial_\mu + \mathbb{A}_\mu) \right] = S_1(\alpha) . \qquad (2.16)$$

This contribution comes from the left-handed fermion ψ. In general gauge, one would expect terms $S_1(\alpha) + S_1(\beta^{-1})$, the second term coming from the left-handed fermion $\bar{\psi}$. This is not the complete answer, however, since by vector gauge invariance, we expect a result that depends only on $\alpha\beta^{-1}$. This is where the $\mathcal{K}(A, \bar{A})$ term comes in. One finds

$$\log \det \left[\Gamma^\mu (\partial_\mu + \mathbb{A}_\mu) \right] = S_1(\alpha\beta^{-1}) , \qquad (2.17)$$

and the constant v of (2.15) is adjusted so that

$$S_1(\alpha\beta^{-1}) = S_1(\alpha) + S_1(\beta^{-1}) - \frac{1}{4\pi} \int d^2x \, \mathcal{K}\left((\alpha^{-1}\partial_z\alpha), (\beta^{-1}\partial_{\bar{z}}\beta) \right) . \qquad (2.18)$$

This is the *Polyakov-Wiegmann identity*. As we'll see, the affine current algebra of the WZW model can be derived from it.

Now suppose $N_L \neq N_R$, so that the theory has a chiral anomaly. Eliminating fermions gives

$$\begin{aligned}
S_{\text{eff}}(A, \bar{A}) &= N_L S_1(\alpha) + N_R S_1(\beta^{-1}) - \frac{1}{8\pi f} \int d^2x \, \mathcal{K}(\gamma^{-1}\partial^\mu\gamma, \gamma^{-1}\partial_\mu\gamma) \\
&= \frac{1}{2}(N_L + N_R)\left[S_1(\alpha) + S_1(\beta^{-1}) \right] + \frac{1}{2}(N_L - N_R)\left[S_1(\alpha) - S_1(\beta^{-1}) \right] \qquad (2.19) \\
&\quad - \frac{1}{8\pi f} \int d^2x \, \mathcal{K}(\gamma^{-1}\partial^\mu\gamma, \gamma^{-1}\partial_\mu\gamma) ,
\end{aligned}$$

where f is a constant, related to v. It is not fixed by gauge invariance here. Now take the limit $N_L + N_R \to \infty$, with $N_L - N_R$ fixed. The term $\frac{1}{2}(N_L + N_R)\left[S_1(\alpha) + S_1(\beta^{-1}) \right]$ is forced to vanish. This implies pure gauge $A, \bar{A}$: $A = \gamma^{-1}\partial_z\gamma$, $\bar{A} = \gamma^{-1}\partial_{\bar{z}}\gamma$ (compare to (2.11)). So we have

$$S_{\text{eff}} = -\frac{1}{8\pi f} \int d^2x \, \mathcal{K}(\gamma^{-1}\partial_\mu\gamma, \gamma^{-1}\partial_\mu\gamma) + 2\pi k \tilde{S}(\gamma) , \qquad (2.20)$$

with $k = N_L - N_R$. From this point of view then, k is quantised because it is the difference in the number of left-handed and right-handed fermions.

This last action is not quite that of the WZW model, with f being an arbitrary constant. It's that of a (two-dimensional) principal chiral σ-model, with WZ term. Such a sigma model is asymptotically free, as is the σ-model without the WZ term. Without the WZ term, the sigma model is strongly interacting in the infrared. But with the WZ term present in the action, there is an infrared fixed point, at $\frac{1}{f} = k$. The WZW model describes the dynamics of this fixed point. We'll remain at $\frac{1}{f} = k$ henceforth.

2.2. Current algebra

Let's rewrite the Polyakov-Wiegmann identity (2.18) as

$$S_k(\gamma\varphi) \;=\; S_k(\gamma) \;+\; S_k(\varphi) \;+\; C_k(\gamma,\varphi) \;, \tag{2.21}$$

putting $\varphi = \beta^{-1}$, and using $S_k(\gamma) = kS_1(\gamma)$, and

$$C_k(\gamma,\varphi) \;=\; k\,C_1(\gamma,\varphi) \;=\; -\frac{k}{4\pi}\int d^2x\,\mathcal{K}\left((\gamma^{-1}\partial_z\gamma),(\varphi\partial_{\bar z}\varphi^{-1})\right) \;. \tag{2.22}$$

The term $C_k(\gamma,\varphi)$ is a cocycle:

$$C_k(\gamma\varphi,\sigma) \;+\; C_k(\gamma,\varphi) \;=\; C_k(\gamma,\varphi\sigma) \;+\; C_k(\varphi,\sigma) \;. \tag{2.23}$$

The presence of this cocycle indicates a projective representation, of the loop group LG of G [22]. Alternatively, we can say that the group $\widehat{LG}$, an extension of LG, is represented non-projectively. This extension $\widehat{LG}$ has as its Lie algebra the (untwisted) affine Kac-Moody algebra $\hat g$, the central extension of the loop algebra of g. We'll call $\hat g$ an affine algebra, for short. Let's see how the WZW model realises $\hat g \oplus \hat g$ as a current algebra [7][23]. Then conformal invariance can be established.

Since

$$C_k(\Omega,\bar\Omega^{-1}) \;=\; -\frac{k}{4\pi}\int d^2x\,\mathcal{K}(\Omega^{-1}\partial_{\bar z}\Omega,\bar\Omega^{-1}\partial_z\bar\Omega) \;, \tag{2.24}$$

if either $\partial_{\bar z}\Omega = 0$ or $\partial_z\bar\Omega = 0$, then $C_k(\Omega,\bar\Omega^{-1}) = 0$, and also $S_k(\Omega) = S_k(\bar\Omega) = 0$. (2.21) thus establishes the local $G \otimes G$ invariance of the WZW model:

$$S_k\left(\Omega(z)\gamma(z,\bar z)\bar\Omega^{-1}(\bar z)\right) \;=\; S_k\left(\gamma(z,\bar z)\right) \;, \tag{2.25}$$

sometimes called the "$G(z) \otimes G(\bar z)$ invariance".

For infinitesimal transformations $\Omega = \mathrm{id} + \omega(z)$, $\bar\Omega(\bar z) = \mathrm{id} + \bar\omega(\bar z)$, the WZW field γ transforms as

$$\delta_\omega\gamma \;=\; \omega\gamma, \quad \delta_{\bar\omega}\gamma \;=\; -\gamma\bar\omega \;. \tag{2.26}$$

With $\delta\gamma = \delta_\omega\gamma + \delta_{\bar\omega}\gamma$, we find

$$\delta S_k(\gamma) \;=\; -\frac{k}{\pi}\int d^2x\,\left\{ \mathcal{K}\left(\omega,\partial_{\bar z}(\partial_z\gamma\gamma^{-1})\right) \;-\; \mathcal{K}\left(\bar\omega,\partial_z(\gamma^{-1}\partial_{\bar z}\gamma)\right)\right\} \;. \tag{2.27}$$

The equations of motion of the WZW model are

$$\partial^\mu(\gamma^{-1}\partial_\mu\gamma) \;+\; i\epsilon_{\mu\nu}\partial^\mu(\gamma^{-1}\partial^\nu\gamma) \;=\; 0 \;. \tag{2.28}$$

9

Switching to the complex coordinates $z, \bar{z}$, and using $\partial^z = 2\partial_z$, $\epsilon_{z\bar{z}} = i/2$, etc., these give $\partial_z(\gamma^{-1}\partial_{\bar{z}}\gamma) = 0$, with hermitian conjugate $-\partial_{\bar{z}}(\partial_z\gamma\gamma^{-1}) = 0$. Defining

$$\boxed{J := -k\partial_z\gamma\gamma^{-1}, \quad \bar{J} := k\gamma^{-1}\partial_{\bar{z}}\gamma,} \qquad (2.29)$$

we have

$$\partial_{\bar{z}}J = 0, \quad \partial_z\bar{J} = 0. \qquad (2.30)$$

So the currents $J, \bar{J}$ are purely holomorphic, antiholomorphic, respectively; i.e. $J = J(z)$, $\bar{J} = \bar{J}(\bar{z})$. These currents will realise two copies of the affine algebra $\hat{g}$.

First we must explain the quantisation scheme. We consider the Euclidean time direction to be the radial direction, so that constant time surfaces are circles centred on the origin. More explicitly, the (conformal) transformation

$$z = e^{\tau+i\sigma}, \quad \bar{z} = e^{\tau-i\sigma} \qquad (2.31)$$

maps the complex plane (punctured at $z = 0, \infty$) to a cylinder, with Euclidean time coordinate $\tau \in \mathbb{R}$ running along its length, and a periodic space coordinate $\sigma \equiv \sigma + 2\pi$. The origin $z = 0$ then corresponds to the distant past $\tau = -\infty$, and the distant future $\tau = +\infty$ is at $|z| = \infty$.

This is called *radial quantisation*. In (3+1)-dimensional QFT the n-point functions are vacuum-expectation-values of time-ordered products of fields. Similarly, in radial quantisation one needs to consider radially-ordered products of fields:

$$R(A(z)B(w)) := \begin{cases} A(z)B(w), & |z| > |w|, \\ B(w)A(z), & |z| < |w|. \end{cases} \qquad (2.32)$$

Define the correlation functions as vacuum-expectation-values of radially ordered products of fields, i.e.

$$\langle A(z)B(w)\rangle := \langle 0|R(A(z)B(w))|0\rangle. \qquad (2.33)$$

We make the operator product expansion

$$R(A(z)B(w)) = \sum_{n=-n_0}^{\infty} (z-w)^n D_{(n)}(w), \quad (n_0 \geq 0). \qquad (2.34)$$

We are also assuming $n_0 < \infty$, i.e. that there is no essential singularity at $z = w$. Break this product up by defining the contraction

$$\overbrace{A(z)B(w)} := \sum_{n=-n_0}^{-1} (z-w)^n D_{(n)}(w) \qquad (2.35)$$

10

or singular part, and the normal-ordered product

$$N\left(A(z)B(w)\right) := \sum_{n\geq 0} (z-w)^n D_{(n)}(w) . \tag{2.36}$$

If we also define

$$N\left(AB\right)(w) := D_{(0)}(w) , \tag{2.37}$$

then we can write

$$R\left(A(z)B(w)\right) = \overline{A(z)B}(w) + N\left(A(z)B(w)\right)$$
$$= \overline{A(z)B}(w) + N\left(AB\right)(w) + O(z-w) . \tag{2.38}$$

Radial ordering will be assumed henceforth. Often it is only the singular parts of operator product expansions (OPE's) that are relevant. We write

$$A(z)B(w) \sim \sum_{n=-n_0}^{-1} (z-w)^n D_{(n)}(w) = \overline{A(z)B}(w) , \tag{2.39}$$

i.e. $\sim$ indicates that only the singular terms are written.

To show that the currents realise two copies of $\hat{g}$, we integrate the right hand side of (2.27) by parts, using counter-clockwise integration contours, to get

$$\frac{i}{4\pi} \oint_0 dz \, \mathcal{K}\left(\omega(z), J(z)\right) - \frac{i}{4\pi} \oint_0 d\bar{z} \, \mathcal{K}\left(\bar{\omega}(\bar{z}), \bar{J}(\bar{z})\right) . \tag{2.40}$$

($\oint_w dz$ will indicate integration around a contour enclosing the point $z = w$.) Expanding $\omega = \sum_a \omega^a t^a$, $J = \sum_a J^a t^a$, (and $\bar{\omega}$, $\bar{J}$ similarly), using $\mathcal{K}(t^a, t^b) = \delta^{ab}$, we get

$$\delta S_k(\gamma) = \frac{-1}{2\pi i} \oint_0 dz \sum_a \omega^a J^a + \frac{1}{2\pi i} \oint_0 d\bar{z} \sum_a \bar{\omega}^a \bar{J}^a . \tag{2.41}$$

This transformation rule leads to the $\hat{g} \oplus \hat{g}$ current algebra. In the Euclidean path integral formulation, a correlation function of the product X of fields is given by

$$\langle X \rangle = \frac{\int [d\Phi] \, X \, e^{-S[\Phi]}}{\int [d\Phi] \, e^{-S[\Phi]}} , \tag{2.42}$$

where $[d\Phi]$ indicates path integration over the fields Φ of the theory. If the action S transforms with $\delta S = -\oint_0 dz \, \delta s(z)$, then

$$\delta\langle X \rangle = -\oint_0 dz \, \langle (\delta s) X \rangle . \tag{2.43}$$

So (2.41) implies

$$\delta_\omega \langle X \rangle = \frac{1}{2\pi i} \oint_0 dz \sum_a \omega^a(z) \langle J^a(z) X \rangle , \qquad (2.44)$$

where we have put $\bar{\omega} = 0$ for simplicity. Put $X = J^b(w)$. With $J(w) = -k(\partial_z \gamma)\gamma^{-1}$ and $\delta_\omega \gamma = \omega \gamma$, we get

$$\delta_\omega J = [\omega, J] - k\partial_w \omega . \qquad (2.45)$$

More explicitly this is

$$\delta_\omega J^b(w) = \sum_{c,d} i f^{bcd} \omega^c(w) J^d(w) - k\partial_w \omega^b(w) . \qquad (2.46)$$

In (2.44) this gives

$$\frac{1}{2\pi i} \oint_w dw \, \omega^c(w) \langle i f^{cbd} \frac{J^d(w)}{z - w} + \frac{k\delta_{bc}}{(z - w)^2} \rangle$$
$$= \frac{1}{2\pi i} \oint_w dw \, \omega^a(w) \langle J^a(z) J^b(w) \rangle . \qquad (2.47)$$

This relation determines the singular part of the (radially-ordered) operator product of two currents:

$$\boxed{J^a(z) J^b(w) \sim \frac{k\delta_{ab}}{(z - w)^2} + \frac{i f^{abc} J^c(w)}{z - w} .} \qquad (2.48)$$

A similar OPE holds for the currents $\bar{J}^a(\bar{z})$. This OPE is equivalent to an affine algebra. The Laurent expansion of a current about $z = 0$ is $J^a(z) = \sum_{n \in \mathbb{Z}} J_n^a z^{-1-n}$, or equivalently, $J_n^a = (1/2\pi i) \oint_0 dz \, z^n J^a(z)$. We can translate this expansion, so that

$$J^a(z) = \sum_{n \in \mathbb{Z}} (z - w)^{-1-n} J_n^a(w) \qquad (2.49)$$

is the Laurent expansion about the point $z = w$, and $J_n^a(0) = J_n^a$. Of course, we also have

$$J_n^a(w) = \frac{1}{2\pi i} \oint_w dz \, (z - w)^n J^a(z) . \qquad (2.50)$$

This allows us to write

$$[J_m^a, J_n^b] = \frac{1}{2\pi i} \oint_0 dw \, w^n \frac{1}{2\pi i} \oint_{|z| > |w|} dz \, z^m J^a(z) J^b(w)$$
$$- \frac{1}{2\pi i} \oint_0 dw \, w^m \frac{1}{2\pi i} \oint_{|z| < |w|} dz \, z^m J^b(w) J^a(z) , \qquad (2.51)$$

12

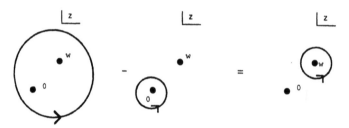

Figure 5. Subtraction of contours for (2.52).

where here radial ordering is not implicit in the operator products. Both operator products in the integrands are $R\left(J^a(z)J^b(w)\right)$, however. So, by subtraction of contours, we obtain

$$[J_m^a, J_n^b] \;=\; \frac{1}{2\pi i}\oint_0 dw\,\frac{1}{2\pi i}\oint_w dz\,z^m w^n\,R\left(J^a(z)J^b(w)\right)\;, \qquad (2.52)$$

as indicated in Fig. 5.

After substituting (2.48) into the last result, residue calculus then gives

$$[J_n^a, J_m^b] \;=\; \sum_c if^{abc}J_{n+m}^c \;+\; kn\delta^{ab}\delta_{m+n,0}\;. \qquad (2.53)$$

Identical commutation relations hold for the current modes $\bar{J}_m^a$.

These are the commutation relations of $\hat{g}\oplus\hat{g}$. It is easy to see that (2.53) is a central extension of the loop algebra of g. Consider $J^a\otimes s^n$, with s on the unit circle in the complex plane, and $n\in\mathbb{Z}$. The loop algebra of g is generated by the $J^a\otimes s^n$, since they are g-valued functions on S^1 (the loop). Now

$$[J^a\otimes s^m, J^b\otimes s^n] \;=\; [J^a, J^b]\otimes s^{m+n} \;=\; if^{abc}J^c\otimes s^{m+n}\;. \qquad (2.54)$$

So only the central extension term ($\propto k$) is missing.

The central extension term is known as a Schwinger term. (2.48) is not the usual form in quantum field theory, because radial quantisation is not typical. If we switch variables using $z=\exp(2\pi ix/L)$, then Laurent series become Fourier series, and we recover the more familiar form

$$[\tilde{J}^a(x), \tilde{J}^b(y)] \;=\; if^{abc}\tilde{J}^c(x)\delta(x-y) \;+\; \frac{1}{2\pi}\delta^{ab}k\delta'(x-y)\;, \qquad (2.55)$$

13

where we have put $\tilde{J}(x) := zJ^a(z)/L$. The Schwinger term is a quantum effect (as powers of $\hbar$ would show) and is related to chiral anomalies, as the presence of k suggests (recall that $k = N_L - N_R$ in the derivation of the WZW model as an effective theory with fermions integrated out).[2]

The conformal invariance of the model can now be established in a straightforward way. The Sugawara construction expresses the stress-energy tensor in terms of normal-ordered products of currents $J^a(z)$. The normal-ordered product (2.36) of two operators $X(w), Y(w)$ can be rewritten as

$$N(XY)(w) = \frac{1}{2\pi i} \oint_w \frac{dz}{z-w} X(z)Y(w) . \tag{2.56}$$

With this form of normal ordering, the appropriate version of Wick's theorem is

$$\overline{X(z)\, N(Y\, Z)}(w) = \frac{1}{2\pi i} \oint_w \frac{dx}{x-w} \Big\{ \overline{X(z)Y}(x)\, Z(w) \\ + Y(x)\, \overline{X(z)Z}(w) \Big\} . \tag{2.57}$$

Using this we calculate

$$\overline{J^a(z) \sum_b N(J^b\, J^b)}(w) = 2(k + h^\vee) \frac{J^a(w)}{(z-w)^2} , \tag{2.58}$$

using (2.48). Here $h^\vee = \sum_{a,b,c} f^{abc} f^{abc}/(2\dim g)$ is the dual Coxeter number of g (this is consistent with (2.2), (2.4), (2.5)). So

$$\sum_b \overline{N(J^b\, J^b)(z)\, J^a}(w) = 2(k + h^\vee) \left\{ \frac{J^a(w)}{(z-w)^2} + \frac{\partial J^a(w)}{z-w} \right\} . \tag{2.59}$$

The Sugawara stress-energy tensor is

$$T(z) = \frac{1}{2(k + h^\vee)} \sum_a N(J^a J^a)(z) . \tag{2.60}$$

Using (2.57) and (2.59) then gives

$$T(z)T(w) \sim \frac{c/2}{(z-w)^4} + \frac{2T(w)}{(z-w)^2} + \frac{\partial T(w)}{z-w} , \tag{2.61}$$

---

[2] For more detail on the relation between chiral anomalies and Schwinger terms, see [24], Chapter 5.

with the *central charge*

$$c =: c(g,k) = \frac{k \dim g}{k + h^\vee} .$$ (2.62)

This last result is the conformal algebra with central extension, or *Virasoro algebra*, in OPE form. So conformal invariance is established. Substituting $T(z) = \sum_{n \in \mathbb{Z}} z^{-2-n} L_n$ yields the usual form of Vir (the Virasoro algebra):

$$\boxed{[L_m, L_n] = (m-n)L_{m+n} + \frac{c}{12}(m^3 - m)\delta_{m+n,0} .}$$ (2.63)

For completeness, we also write

$$\boxed{T(z)J^a(w) \sim \frac{J^a(w)}{(z-w)^2} + \frac{\partial J^a(w)}{z - w} ,}$$ (2.64)

which corresponds to

$$\boxed{[L_m, J_n^a] = -nJ_{m+n}^a .}$$ (2.65)

This shows that $\hat{g}$ and Vir extend to a semi-direct product in the theory. Furthermore, the full *chiral algebra* of the WZW model is $Vir \ltimes \hat{g}$, with commutation relations (2.63),(2.53) and (2.65).

2.3. Factorisation and primary fields

A factorised form for $\gamma(z, \bar{z})$ solves the classical equation of motion:

$$\gamma(z, \bar{z}) = \gamma_L(z)\gamma_R(\bar{z}) \quad \Rightarrow \quad \partial_z(\gamma^{-1}\partial_{\bar{z}}\gamma) = 0 .$$ (2.66)

This factorisation survives in the following form in the quantum theory. As already mentioned, under an infinitesimal $G(z) \otimes G(z)$ transformation, we have

$$\delta_\omega \gamma = \omega\gamma , \qquad \delta_{\bar{\omega}}\gamma = -\gamma\bar{\omega} .$$ (2.67)

The currents $J^a(z), \bar{J}^a(\bar{z})$ generate the infinitesimal transformations of the fields, so we have

$$J^a(z)\gamma(w, \bar{w}) \sim \frac{-1}{z - w} t_\gamma^a \gamma(w, \bar{w}) ,$$
$$\bar{J}^a(z)\gamma(w, \bar{w}) \sim \frac{1}{\bar{z} - \bar{w}} \gamma(w, \bar{w})t_\gamma^a ,$$ (2.68)

where t_γ^a is the g-generator t^a in the representation appropriate to $\gamma(z, \bar{z})$.

15

The WZW model also contains other fields, besides $\gamma(z, \bar{z})$, that transform in similar fashion. These are the so-called *primary fields* $\Phi_{\lambda,\mu}(z, \bar{z})$:

$$
\begin{aligned}
J^a(z)\Phi_{\lambda,\mu}(w, \bar{w}) &\sim \frac{-1}{z - w} t_\lambda^a \Phi_{\lambda,\mu}(w, \bar{w}) , \\
\bar{J}^a(z)\Phi_{\lambda,\mu}(w, \bar{w}) &\sim \frac{1}{\bar{z} - \bar{w}} \Phi_{\lambda,\mu}(w, \bar{w}) t_\mu^a ,
\end{aligned}
\tag{2.69}
$$

Here λ, μ are two highest weights of integrable unitary irreducible representations $L(\lambda), L(\mu)$ of g, and t_λ^a, t_μ^a are the generators in those representations.

To find the action of the "current modes" J_n^a that generate $\hat{g}$, we use $J^a(z) = \sum_{n \in \mathbb{Z}} z^{-1-n} J_n^a$ in

$$
J^a(z)\Phi_{\lambda,\mu}(0, 0) \sim -\frac{1}{z} t_\lambda^a \Phi_{\lambda,\mu}(0, 0) \tag{2.70}
$$

to get

$$
\begin{aligned}
[J_0^a, \Phi_{\lambda,\mu}(0, 0)] &= -t_\lambda^a \Phi_{\lambda,\mu}(0, 0) , \\
[J_n^a, \Phi_{\lambda,\mu}(0, 0)] &= 0 , \text{ for } n > 0 .
\end{aligned}
\tag{2.71}
$$

This implies that the primary field $\Phi_{\lambda,\mu}$ transforms as a highest-weight representation $L(\hat{\lambda})$ of the affine algebra $\hat{g}$. Similar considerations work for the right action, so $\Phi_{\lambda,\mu}$ transforms as $L(\hat{\lambda}) \otimes L(\hat{\mu})$.[3]

For most purposes, it suffices to consider only the left *or* right action of $\hat{g}$. So we will write, instead of (2.69),

$$
\boxed{J^a(z)\phi_\lambda(w) \sim \frac{-t_\lambda^a \phi_\lambda(w)}{z - w} ,} \tag{2.72}
$$

and similarly for $\bar{J}^a(\bar{z})$ and $\bar{\phi}_\mu(\bar{z})$, if need be. We must emphasise, however, that $\phi_\lambda(z), \bar{\phi}_\mu(\bar{z})$ are *not* sensible local fields; they are only the holomorphic (left-moving) and antiholomorphic (right-moving) parts of the primary field $\Phi_{\lambda,\mu}(z, \bar{z})$. If you like, $\Phi_{\lambda,\mu}(z, \bar{z}) = \phi_\lambda(z)\bar{\phi}_\mu(\bar{z})$.

To see this, first note that the primary field $\Phi_{\lambda,\mu}(z, \bar{z})$ transforms nicely under conformal transformations. That's because of the Sugawara construction (2.60), expressing the stress-energy tensor as a normal-ordered product of the currents. In terms of modes, the Sugawara construction gives

$$
L_n = \frac{1}{2(k + h^\vee)} \sum_a \sum_{m \in \mathbb{Z}} N(J_{n-m}^a J_m^a) . \tag{2.73}
$$

where

$$
N(J_p^a J_q^b) = \begin{cases} J_p^a J_q^b , & p \le q ; \\ J_q^b J_p^a , & p > q . \end{cases} \tag{2.74}
$$

---

[3] As we'll see, a highest weight $\hat{\lambda}$ for $\hat{g}$ determines a highest weight λ for g.

We get

$$[L_n, \phi_\lambda(0)] = \begin{cases} 0, & n > 0 ; \\ h_\lambda \phi_\lambda(0) , & n = 0 . \end{cases} \tag{2.75}$$

where

$$h_\lambda = \frac{\sum_a \text{Tr}(t_\lambda^a t_\lambda^a)}{2(k + h^\vee) \, \text{Tr}(\text{id}_\lambda)} = \frac{(\lambda, \lambda + 2\rho)}{2(k + h^\vee)} \tag{2.76}$$

is the *conformal weight* of the "primary field" ϕ_λ, and ρ is the Weyl vector of g (the half-sum of the positive roots of g).

In OPE language, this is

$$\boxed{T(z)\phi_\lambda(0) \sim \frac{h_\lambda \phi_\lambda(0)}{z^2} + \frac{\partial \phi_\lambda(0)}{z} .} \tag{2.77}$$

Similarly,

$$\bar{T}(\bar{z})\bar{\phi}_\mu(0) \sim \frac{h_\mu \bar{\phi}_\mu(0)}{\bar{z}^2} + \frac{\bar{\partial}\bar{\phi}_\mu(0)}{\bar{z}} . \tag{2.78}$$

Now, the generator of infinitesimal scaling is $L_0 + \bar{L}_0$, as we'll show below. So $h_\lambda + h_\mu$ is the scaling dimension of the primary field $\Phi_{\lambda,\mu}$. (In radial quantisation, scaling = time-translation, so the Hamiltonian $\mathbf{H} = L_0 + \bar{L}_0$.) $L_0 - \bar{L}_0$ generates rotations, so that $h_\lambda - h_\mu$ is the *spin* of $\Phi_{\lambda,\mu}$. For a single-valued (local) field, we therefore require $h_\lambda - h_\mu \in \mathbb{Z}$. This is a highly nontrivial constraint on pairs (λ, μ), since $h_\lambda, h_\mu \in \mathbb{Q}$. It is in this sense that $\phi_\lambda(z)$ cannot be considered a sensible local field in its own right.

The fields $\Phi_{\lambda,\mu}$ are primary because all others are in the span of operator products of currents acting on them:

$$J^{a_1}(z_1) J^{a_2}(z_2) \cdots J^{a_n}(z_n) \bar{J}^{\bar{a}_1}(\bar{z}_1) J^{\bar{a}_2}(\bar{z}_2) \cdots J^{\bar{a}_{\bar{n}}}(\bar{z}_{\bar{n}}) \, \Phi_{\lambda,\mu}(z, \bar{z}) . \tag{2.79}$$

They are therefore called *descendant* fields. More usually, the basis elements are written as

$$J^{a_1}_{-n_1} \cdots J^{a_N}_{-n_N} \bar{J}^{\bar{a}_1}_{-\bar{n}_1} \cdots \bar{J}^{\bar{a}_{\bar{N}}}_{-\bar{n}_{\bar{N}}} \, \Phi_{\lambda,\mu}(z, \bar{z}) . \tag{2.80}$$

2.4. Field-state correspondence

$|0\rangle$ is the vacuum of the WZW model. $t_\lambda^a \phi_\lambda$ means $\sum_{v \in L(\lambda)} (t_\lambda^a)_{u,v} \phi_{\lambda,v}$. If $\phi_{\lambda,v} = \delta_{v,v_\lambda}$, where v_λ denotes the highest-weight vector of $L(\lambda)$, and $v \in L(\lambda)$, we have

$$\boxed{\phi_\lambda(0)|0\rangle = |v_\lambda\rangle .} \tag{2.81}$$

This is the basis of the field-state correspondence. More generally, defining

$$|\phi_\lambda\rangle := \sum_{v \in L(\lambda)} \phi_{\lambda,v}|v\rangle , \tag{2.82}$$

we can write

$$\phi_\lambda(0)|0\rangle \;=\; |\phi_\lambda\rangle \;. \qquad\qquad (2.83)$$

We can also consistently write

$$\phi_\lambda(z) \;=\; \sum_{u\in L(\lambda)} \phi_{\lambda,u}\, u(z)\;, \quad \text{with } u(0)\,|0\rangle \;=\; |u\rangle \;. \qquad (2.84)$$

In terms of $|v_\lambda\rangle$, the primary-field conditions read

$$J_0^a|v_\lambda\rangle \;=\; t_\lambda^a|v_\lambda\rangle\;, \quad J_n^a|v_\lambda\rangle \;=\; 0 \;(n>0)\;, \qquad (2.85)$$

in agreement with (2.75). The affine algebras are examples of triangularisable algebras (just like the simple Lie algebras) [25]. This means their generators can be written as a disjoint sum of three sets, with corresponding decomposition

$$\hat{g} \;=\; \hat{g}_- \;\oplus\; \hat{g}_0 \;\oplus\; \hat{g}_+ \;. \qquad\qquad (2.86)$$

$\hat{g}_0$ is the Cartan subalgebra, while $\hat{g}_\pm \oplus \hat{g}_0$ are Borel subalgebras. $\hat{g}_+(\hat{g}_-)$ correspond to positive (negative) roots, and so contain raising (lowering) operators. Now, in the basis used, $\hat{g}_+$ is generated by $\{J_{n>0}^a\} \oplus g_+$, where $g_+ \subset \{J_0^a\}$ contains the raising operators of $g \subset \hat{g}$. But since t_λ^a are the generators of g in a representation $L(\lambda)$ of highest weight λ, we know $g_+|v_\lambda\rangle = 0$. So by (2.85), $\hat{g}_+|v_\lambda\rangle = 0$, i.e. $|v_\lambda\rangle$ is the highest-weight state (highest state/vector) of the affine representation $L(\hat{\lambda})$ of $\hat{g}$.

The rest of the states in the representation $L(\hat{\lambda})$ can be obtained as descendant states, i.e. as linear combinations of states of the form

$$J_{-n_1}^{a_1} \cdots J_{-n_N}^{a_N} \phi_\lambda(0)|0\rangle \;. \qquad\qquad (2.87)$$

Now, there is still an infinite number of possible highest weights. But we'll find that for fixed $k \in \mathbb{Z}_{>0}$, only a finite number of inequivalent highest weights are possible. These are the (unitary) *integrable* highest weights; they generate representations of $\hat{g}$ that can be integrated to representations of $\widehat{LG}$. By the $G(z) \otimes G(\bar{z})$ invariance of the WZW model, these representations are precisely the relevant ones. We will arrive at this result from an algebraic perspective, however.

To do this, we first need to discuss $g, \hat{g}$ and their relation. This justifies an interesting digression on Kac-Moody algebras [26] [27] [28] [29].

3. Affine Kac-Moody Algebras

3.1. Kac-Moody algebras: simple Lie algebras

$g, \hat{g}$ are examples of Kac-Moody algebras, which can be presented in terms of a *Cartan matrix* $A = (A_{i,j})$, with integer entries (i.e. Kac-Moody algebras are generalised Cartan matrix Lie algebras). Let's first define g this way. If X is generated by x_1 and x_2, for example, we use the notation $X = \langle x_1, x_2 \rangle$.

Recall $g = g_+ \oplus g_0 \oplus g_-$. Now

$$g_+ = \langle e_i \; : \; i = 1, \ldots, r \rangle \; , \quad g_0 = \langle h_i \; : \; i = 1, \ldots, r \rangle \; , \quad g_- = \langle f_i \; : \; i = 1, \ldots, r \rangle \; , \quad (3.1)$$

where r is the rank of g, and $\{e_i, h_i, f_i \; : \; i = 1, \ldots, r\}$ are the *Chevalley generators* of g. The commutation relations of the generators can be expressed in terms of the Cartan matrix:

$$
\begin{aligned}
[h_i, h_j] &= 0 \\
[h_i, e_j] &= A_{j,i} e_j \\
[h_i, f_j] &= -A_{j,i} f_j \\
[e_i, f_j] &= \delta_{i,j} h_j \; .
\end{aligned}
\qquad (3.2)
$$

The Chevalley presentation of the algebra g is completed by the *Serre relations*:

$$
\begin{aligned}
[\text{ad}\,(e_i)]^{1-A_{j,i}} \; e_j &= 0 \; , \\
[\text{ad}\,(f_i)]^{1-A_{j,i}} \; f_j &= 0 \; .
\end{aligned}
\qquad (3.3)
$$

The $r \times r$ Cartan matrix has diagonal entries $A_{i,i} = 2$, so that $\langle e_i, h_i, f_i \rangle \cong s\ell(2)$ for all $i = 1, \ldots, r$. For simple g, $g \not\cong \oplus_{i=1}^{r} s\ell(2)$, so $A_{i,j} \neq 0$ for at least one pair $i \neq j$, if $r > 1$. For all Kac-Moody algebras (including semi-simple Lie, affine, hyperbolic, etc. algebras)[4], $A_{i,i} = 2 \; \forall i$, as just mentioned; $A_{i,j} \in -\mathbb{Z}_{\geq 0} \; \forall \; i \neq j$; and $A_{i,j} = 0 \Leftrightarrow A_{j,i} = 0$. In addition, the Cartan matrices are *symmetrisable*: there exist positive rational numbers q_j such that AD' is a symmetric matrix, where $D' = \text{diag}(q_j)$.

For g a semi-simple Lie algebra, $A_{i,j} \in \{0, -1, -2, -3\}$ for $i \neq j$, and most importantly, $\det A > 0$, i.e. the Cartan matrix is invertible. For simple g, A must be indecomposable.

The information contained in the Cartan matrix can be encoded in a so-called *(Coxeter-)Dynkin diagram*. r nodes are drawn, each associated with a row (or column) of A. Node i and node j ($j \neq i$) are joined by a number $A_{i,j} A_{j,i}$ of lines; and if $A_{i,j} \neq A_{j,i}$, so that

---

[4] For a discussion of generalised Kac-Moody algebras, or Borcherds-Kac-Moody algebras, see [5].

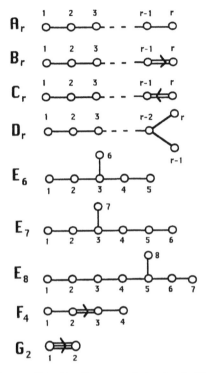

Figure 6. The (Coxeter-)Dynkin diagrams of the simple Lie algebras.

there are more than one lines, an arrow is drawn from node i to node j if $|A_{i,j}| > |A_{j,i}|$. The Coxeter-Dynkin diagrams of the simple Lie algebras are drawn in Figure 6.

A more direct significance, in terms of the roots of the algebra, can be given to the Cartan matrix and Coxeter-Dynkin diagrams. As we do this, we'll introduce another presentation of g, the *Cartan-Weyl presentation*.

First, find a maximal set of commuting Hermitian generators H^i, $(i = 1, \ldots, r)$:

$$\boxed{[H^i, H^j] = 0 \;\; (1 \le i, j \le r) \; .}$$
(3.4)

Any two such maximal Abelian subalgebras g_0 (Cartan subalgebras) are conjugate under the action of $\exp g$ (the covering group of a group with Lie algebra g). Fix a choice of Cartan subalgebra g_0.

Since the H^i mutually commute, they can be diagonalised simultaneously in any representation of g. The states of such a representation are then eigenstates of the H^i, $1 \le i \le r$.

20

We write

$$H^i \, |\mu; \, \ell\rangle \; = \; \mu^i \, |\mu; \, \ell\rangle \, . \tag{3.5}$$

Here μ is called a *weight vector* (or just a weight), with r components μ^i. The corresponding r-dimensional space is known as *weight space*. As we'll discuss shortly, it is dual to the Cartan subalgebra g_0. The ℓ of the kets $|\mu; \, \ell\rangle$ is meant to indicate any additional labels.

A basis for the whole of g can be constructed by appending $\{E^\alpha\}$, obeying

$$\boxed{[H^i, E^\alpha] \; = \; \alpha^i E^\alpha \; (1 \leq i \leq r) \, .} \tag{3.6}$$

α is a r-dimensional vector, called a *root*, and E^α is a step operator (raising or lowering operator, depending). For simple Lie algebras, E^α is determined by α, up to normalisation. If α is a non-zero root, the only multiple of α that is also a root is $-\alpha$, and we can take

$$E^{-\alpha} \; = \; \left(E^\alpha\right)^\dagger \, . \tag{3.7}$$

The set of roots of g will be denoted Δ.

To complete the presentation of g in the Cartan-Weyl basis, we also need:

$$[E^\alpha, E^\beta] \; = \; \begin{cases} N(\alpha, \beta) E^{\alpha+\beta} & \text{if } \alpha + \beta \in \Delta \\ 2\alpha \cdot H / \alpha^2 & \text{if } \alpha + \beta = 0 \\ 0 & \text{otherwise} \end{cases} \tag{3.8}$$

where $\alpha \cdot H$ means $\sum_{i=1}^{r} \alpha^i H^i =: H^\alpha$, $\alpha^2 := (\alpha, \alpha)$ (see below), and the $N(\alpha, \beta)$ are constants.[5] Here we use the scalar product (α, β), defined through the Killing form (2.2):

$$(\alpha, \beta) \; = \; \mathcal{K}(H^\alpha, H^\beta) \; = \; \text{Tr}\left(\text{ad}\,(H^\alpha), \text{ad}\,(H^\beta)\right) / 2h^\vee \, . \tag{3.9}$$

This completes the Cartan-Weyl presentation of g.

The Killing form also establishes an isomorphism between the Cartan subalgebra g_0 and its dual g_0^*, weight space: for every weight $\lambda \in g_0^*$, there corresponds an element $H^\lambda \in g_0$ by $\lambda(\cdot) = \mathcal{K}(H^\lambda, \cdot)$ (i.e. $\lambda(H^\beta) = \mathcal{K}(H^\lambda, H^\beta) = (\lambda, \beta)$ for $\beta \in \Delta$). This inner product can be extended, by symmetry, $(\alpha, \beta) = (\beta, \alpha)$, to all weights $\alpha, \beta \in g_0^*$. By (3.8), the rescaled root $2\alpha/\alpha^2$ has importance; it is called the *coroot* $\alpha^\vee$.

If we choose a fixed basis for the root lattice ($\subset$ weight lattice), we call α positive, $\alpha \in \Delta_+$, iff its first nonzero component in this basis is positive. Otherwise, $\alpha \in \Delta_-$, i.e. α is a negative root. E^α is considered a raising (lowering) operator if $\alpha \in \Delta_+$ ($\alpha \in \Delta_-$).

---

[5] (3.7) implies $N(\alpha, \beta) = -N(-\alpha, -\beta)$. Then for $\beta + \ell\alpha \in \Delta$ with $p \leq \ell \leq q$, we can set $N(\alpha, \beta)^2 = q(1 - p)(\alpha, \alpha)/2$.

A *simple root* is a positive root that cannot be written as a linear $\mathbb{Z}_{\geq 0}$-combination of other positive roots. The set of simple roots will be denoted $\Pi = \{\alpha_i \; : \; i = 1, \ldots, r\}$. The set of simple coroots is $\Pi^\vee = \{\alpha_i^\vee \; : \; i = 1, \ldots, r\}$. The basis dual to $\Pi^\vee$ is the Dynkin basis of *fundamental weights*:

$$\left(\Pi^\vee\right)^* \;=\; \{\omega^i \; : \; j = 1, \ldots, r \; \} \, . \tag{3.10}$$

That is, $(\omega^i, \alpha_j^\vee) = \delta_j^i$.

Let us now compare the Chevalley and Cartan-Weyl presentations of g. The Chevalley presentation emphasises the r subalgebras of type $s\ell(2) \cong A_1$ that are associated with each simple root (or fundamental weight). It is the more economical presentation, since it is written in terms of just $3r$ generators, those listed in (3.1). This economy allowed the discovery of the Kac-Moody algebras: it was natural to wonder whether loosening the constraints on the Cartan matrix would lead to other interesting types of algebras. The price to be paid is the imposition of the more complicated Serre relations (3.3). But these relations are what ensure that (among other things) a finite-dimensional algebra is generated.

In contrast, the Cartan-Weyl presentation makes use of the A_1-subalgebras associated with every positive root. For every positive root we get a raising and lowering operator, and the finite-dimensionality of the algebra is built in. Of course, the cost is the use of more generators, a total of $\dim g$ of them.

More concretely, it is not difficult to make the identifications

$$e_i \;=\; E^{\alpha_i} \, , \quad f_i \;=\; E^{-\alpha_i} \, , \quad h_i \;=\; \frac{2\alpha_i \cdot H}{\alpha_i^2} \;=\; \alpha_i^\vee \cdot H \, , \tag{3.11}$$

where $H = \sum_{i=1}^r \omega^i h_i$, and finally

$$\boxed{A_{i,j} \;=\; \frac{2\left(\alpha_i, \alpha_j\right)}{\alpha_j^2} \;=\; (\alpha_i, \alpha_j^\vee) \; .} \tag{3.12}$$

So, the Cartan matrix encodes the scalar products of simple roots with simple coroots.

Now, $\det A > 0$ guarantees that weight space is Euclidean. Consider the hyperplanes in weight space with normals α_i. The *primitive reflection* $r_{\alpha_i} = r_i$ of a weight $\lambda = \sum_{i=1}^r \lambda_i \omega^i$ across such a hyperplane is given by

$$r_{\alpha_i}\lambda \;=\; r_i\lambda \;=\; \lambda - (\lambda, \alpha_i^\vee)\alpha_i \, . \tag{3.13}$$

Being reflections, the r_i have order 2, and they generate a *Coxeter* group W, which can be presented as

$$W = \langle r_i : i = 1, \ldots, r \rangle , \tag{3.14}$$

with the relations

$$(r_i r_j)^{m_{ij}} = \text{id} . \tag{3.15}$$

Clearly, $m_{ii} = 1$ for all i, and it turns out that all $m_{ij} \in \{2, 3, 4, 6\}$, when $i \neq j$. This Coxeter presentation can be encoded in a *Coxeter diagram*: nodes are drawn for each primitive reflection, and $\{0, 1, 2, 3\}$ lines between nodes for $m_{ij} \in \{2, 3, 4, 6\}$, respectively ($i \neq j$). For simple g, we find the Coxeter diagrams are just the corresponding Dynkin diagrams (see Fig. 6), with the arrows omitted. In fact, the Coxeter group so obtained is the *Weyl group* of g.

The possible weights of integrable representations will lie on the *weight lattice* $P := \mathbb{Z}(\Pi^\vee)^*$, the points in weight space that are integer linear combinations of the fundamental weights. Of course, this lattice is periodic and "fills" weight space. So we can think of it as an infinite crystal. It has a point group isomorphic to the Weyl group, which explains the restriction $m_{ij} \in \{2, 3, 4, 6\}$, familiar from crystallography.

Still, it is remarkable that these Coxeter-Weyl groups almost determine the algebra g completely. More accurately (B_r and C_r have isomorphic Weyl groups), the simple Lie algebras are essentially those whose weight lattices can exist in a Euclidean weight space of dimension equal to the rank.

What is the geometry of the Weyl hyperplanes in weight space? There is a Weyl hyperplane for each root, not just for the simple roots, and they partition the r-dimensional weight space into a finite number of sectors. Each sector is of infinite hypervolume, and W acts simply transitively on them. The example of $g = A_2$ is pictured in Fig. 7.

We use the notation $\lambda = \sum_{i=1}^r \lambda_i \, \omega^i = (\lambda_1, \ldots, \lambda_r)$, and $L(\lambda_1, \ldots, \lambda_r) = L(\lambda)$. Fig. 8 is the weight diagram for the A_2 representation $L(2, 1)$. Notice it is symmetric under the action of the Weyl group $W \cong S_3$ for A_2. Let $\text{mult}(\lambda; \mu)$ denote the multiplicity of a weight μ in the representation $L(\lambda)$. Then this Weyl symmetry can be written as

$$\text{mult}(\lambda; \mu) = \text{mult}(\lambda; w\mu), \quad \forall w \in W . \tag{3.16}$$

The Weyl symmetry can also be expressed in terms of *characters*. Characters are to representations what weights are to states (vectors). They are simpler than the representations themselves, yet still contain sufficient information to be useful in many contexts. Precisely, the *formal character* of the g-representation $L(\lambda)$ is

$$\text{ch}_\lambda := \text{Tr}_{L(\lambda)} \, e^H . \tag{3.17}$$

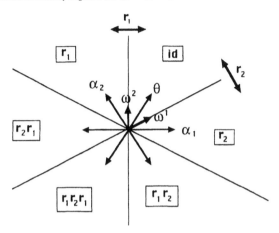

Figure 7. The Weyl sectors of A_2 weight space. Each sector is labelled by the Weyl element that maps it to the identity (id) sector. The identity sector is also known as the dominant sector. Also shown are the fundamental weights ω^1, ω^2, and the roots, including the simple roots α_1, α_2, and the highest root θ.

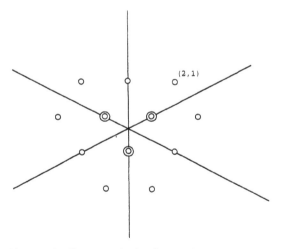

Figure 8. The weight diagram of $L(2,1)$, the A_2 representation of highest weight $(2,1)$.

Equivalently, if we define

$$P(\lambda) := \{\, \mu \in g_0^* \; : \; \text{mult}\,(\lambda;\mu) \neq 0 \,\}\,, \tag{3.18}$$

we can write

$$\text{ch}_\lambda \;=\; \sum_{\mu \in P(\lambda)} \text{mult}\,(\lambda;\mu)\,e^\mu \;. \tag{3.19}$$

In (3.17) and (3.19), e^H and e^μ are formal exponentials, with the additive property $e^\lambda e^\mu = e^{\lambda+\mu}$, for example.

The Weyl symmetry is made manifest in the celebrated *Weyl character formula*:

$$\text{ch}_\lambda \;=\; \frac{\sum_{w \in W} (\det w)\, e^{w.\lambda}}{\prod_{\alpha \in \Delta_+} (1 - e^{-\alpha})}\,, \tag{3.20}$$

where the *shifted Weyl action* is $w.\lambda := w(\lambda + \rho) - \rho$. Here $\det w = 1$ if w can be written as a composition of an even number of primitive reflections, and $\det w = -1$ for an odd number.

Since the character of the singlet representation $L(0)$ is $\text{ch}_0 = 1$, (3.20) gives the *denominator identity*

$$\prod_{\alpha \in \Delta_+} (1 - e^{-\alpha}) \;=\; \sum_{w \in W} (\det w)\, e^{w.0}\;. \tag{3.21}$$

So the Weyl character formula can also be written as

$$\text{ch}_\lambda \;=\; \frac{\sum_{w \in W} (\det w)\, e^{w(\lambda+\rho)}}{\sum_{w \in W} (\det w)\, e^{w\rho}}\;. \tag{3.22}$$

If we continue this relation to weights $\lambda \notin P_+$, we can also derive

$$\text{ch}_\lambda \;=\; (\det w)\,\text{ch}_{w.\lambda}\;. \tag{3.23}$$

This relation will be important later.

We can "informalise" the formal character in the following way:

$$\text{ch}_\lambda(\sigma) := \sum_{\mu \in P(\lambda)} \text{mult}\,(\lambda;\mu)\,e^{(\mu,\sigma)}\;. \tag{3.24}$$

The character $\text{ch}_\lambda(\sigma)$ is then a polynomial in the r indeterminates e^{σ_j}, $j = 1,\ldots,r$.

3.2. Kac-Moody algebras: affine algebras

As discussed above, the affine algebras $\hat{g}$ relevant to WZW models are central extensions of the loop algebras of g, for g semi-simple; we restrict to g simple here for simplicity. They are known as *untwisted affine* algebras. For such $\hat{g}$, $g \subset \hat{g}$ is known as the *horizontal subalgebra* of $\hat{g}$.

The Chevalley presentation for $\hat{g}$ is identical to that for g except that the $r \times r$ Cartan matrix $A = (A_{i,j})$ is replaced by the $(r+1) \times (r+1)$ Cartan matrix $\hat{A} = (\hat{A}_{i,j})_{i,j \in \{0,1,\dots,r\}}$, with $\hat{A}_{i,j} = A_{i,j}$ for $i, j \neq 0$. As for g, the elements of the Cartan matrix are determined by scalar products of simple roots and coroots:

$$\hat{A}_{i,j} = (\hat{\alpha}_i, \hat{\alpha}_j^\vee) \; . \tag{3.25}$$

Because of this structure, there is an intimate relation between the simple roots of g and those of $\hat{g}$.

An affine Kac-Moody Cartan matrix obeys all the conditions mentioned above that the simple Lie algebras obey, except that the det $A > 0$ condition is loosened. Let $\hat{A}_{(i)}$ denote the submatrix of $(r+1) \times (r+1)$ matrix $\hat{A}$ obtained by deleting the i-th row and column. Then we must have

$$\det \hat{A} = 0 \; , \quad \text{but} \quad \det \hat{A}_{(i)} > 0 \quad \forall i \in \{0, 1, \dots, r\} \; , \tag{3.26}$$

if $\hat{A}$ is to be an affine Cartan matrix. This means that the submatrices $\hat{A}_{(i)}$ must be Cartan matrices for semi-simple Lie algebras. Besides the untwisted affine algebras, twisted affine algebras also exist, but they are not so directly useful in conformal field theory.

(3.26) guarantees that $\hat{A}$ has no negative eigenvalues, and exactly one zero eigenvector. For all affine Cartan matrices $\hat{A}$, there exist positive integers $a_0, a_1, \dots, a_r$, called *marks*, such that $\sum_{i=0}^{r} a_i \hat{A}_{i,j} = 0$. If $\hat{A}$ is affine, meaning it is the Cartan matrix of an affine algebra, then so is $\hat{A}^T$ (their Dynkin diagrams are obtained from each other by reversing their arrows). Because of this, we also have $\sum_{j=0}^{r} \hat{A}_{i,j} a_j^\vee = 0$, where the $a_j^\vee$ are known as *co-marks* (notice this is consistent with the symmetrisability of $\hat{g}$). For untwisted $\hat{g}$, the marks and co-marks are determined by the *highest root* θ of g, which is its own co-root $\theta^\vee = \theta$ (here we use the normalisation convention $\alpha^2 = 2$ for the longest roots α). So we can expand

$$\theta = \sum_{i=1}^{r} a_i \alpha_i = \sum_{i=1}^{r} a_i^\vee \alpha_i^\vee \; , \tag{3.27}$$

with the expansion coefficients equalling the (co-)marks. $a_0 = a_0^\vee = 1$ completes their specification.

Equivalently, the Dynkin diagrams of the untwisted affine $\hat{g}$ are simply the *extended* Dynkin diagrams of the corresponding simple algebra g, obtained by augmenting the set of simple roots Π of g by $-\theta$. The Dynkin diagrams of the untwisted affine algebras are drawn in Fig. 9. So, the set of affine simple roots

$$\hat{\Pi} = \{ \hat{\alpha}_i \; : \; i \in \{0, 1, \ldots, r\} \} \tag{3.28}$$

is simply related to $\{-\theta, \alpha_1, \ldots, \alpha_r\}$.

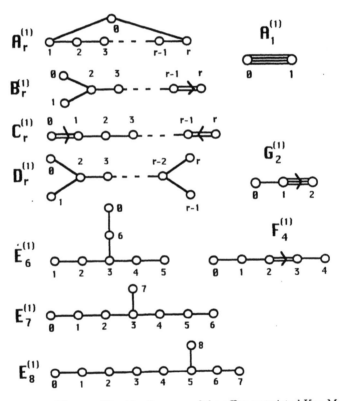

Figure 9. The (Coxeter-)Dynkin diagrams of the affine untwisted Kac-Moody algebras.

Notice that if any node of any of the Dynkin diagrams of Fig. 9 is omitted, one obtains the Dynkin diagram of a semi-simple Lie algebra. This is also true of the twisted affine

27

Dynkin diagrams (not drawn here), and is in agreement with the condition (3.26). It is only the untwisted Dynkin diagrams, however, that are isomorphic to the extended diagrams of simple Lie algebras.

Before making this precise, we must introduce another operator. For an arbitrary symmetrisable Kac-Moody algebra, one can define the inner product of simple roots directly from the Cartan matrix A by $(\alpha_i, \alpha_j) = A_{i,j} q_j$ (recall AD' is symmetric, with $D' = \text{diag}(q_i)$). This inner product is positive (semi-)definite iff A is of (affine) finite type. Now, many of the important results for $\hat{g}$ are obtained as straightforward generalisations of those for g, but these latter rely on having a non-degenerate bilinear form $(\cdot, \cdot)$. To make the bilinear form non-degenerate in the affine case, one needs to enlarge the Cartan subalgebra $\hat{g}_0$ of $\hat{g}$, to $\hat{g}_0^e$, by adding a derivation d. We will denote the *enlarged affine algebra* similarly, by $\hat{g}^e$.

The problem is the *canonical central element*

$$K = \sum_{i=0}^{r} a_i^\vee \hat{H}^i . \tag{3.29}$$

Clearly, $[K, \hat{H}^j] = 0$, and furthermore,

$$[K, E^{\pm \hat{\alpha}_i}] = \pm \sum_{j=0}^{r} a_j^\vee \hat{A}_{i,j} E^{\pm \hat{\alpha}_i} = 0 , \tag{3.30}$$

showing that K is indeed central. Actually, the coefficient k of the WZ term (2.3) in the WZW action (2.1) is to be identified with the eigenvalue of K, which is fixed in the current algebra of a given WZW model.

If we extend the bilinear form by choosing $\mathcal{K}(K, d) = 1$, $\mathcal{K}(K, K) = \mathcal{K}(d, d) = 0$, the resulting form is non-degenerate. The operator d is very natural in WZW models: $-d$ will be identified with the Virasoro zero mode L_0.

If a step operator $E^{\hat{\alpha}}$ is an element of $\hat{g}$, with

$$\begin{aligned}
[\hat{H}^i, E^{\hat{\alpha}}] &= \alpha^i E^{\hat{\alpha}} \quad (i \in \{1, \ldots, r\}) , \\
[K, E^{\hat{\alpha}}] &= 0 , \quad [d, E^{\hat{\alpha}}] = n ,
\end{aligned} \tag{3.31}$$

we denote $\hat{\alpha} = (\alpha, 0, n)$, and write $E^{\hat{\alpha}} =: E_n^\alpha$. So one can think of affine roots as vectors with $r + 2$ components, r of which describe a root of g, and the other two correspond to the elements $K, d \in \hat{g}^e$. The inner product on $\hat{g}_0^{e*}$ is

$$\boxed{(\hat{\alpha}, \hat{\beta}) = ((\alpha, k_\alpha, n_\alpha), (\beta, k_\beta, n_\beta)) = (\alpha, \beta) + k_\alpha n_\beta + n_\alpha k_\beta .} \tag{3.32}$$

It is determined by the symmetry and invariance:

$$\mathcal{K}(x,[z,y]) + \mathcal{K}([z,x],y) = 0 \quad (\forall x,y,z \in \hat{g}_0^e) , \qquad (3.33)$$

of the corresponding bilinear form on $\hat{g}_0^e$. Notice that k_α, n_α behave like light-cone coordinates $x_\pm = (t \pm x)/\sqrt{2}$ in a Minkowski metric, so the signature of the inner product on $\hat{g}_0^{e*}$ is Lorentzian.

Notice that $\delta = \sum_{i=0}^r a_i \hat{\alpha}_i = (0,0,1)$, so that δ is the root corresponding to $d = -L_0 \in \hat{g}_0^e$. The dual weight is denoted $\Lambda_0 = (0,1,0)$.

The affine simple roots are

$$\boxed{\hat{\alpha}_0 = (-\theta,0,1) , \quad \hat{\alpha}_{i\neq 0} = (\alpha_i,0,0) .} \qquad (3.34)$$

This explains why the extended Dynkin diagram of g is identical to the Dynkin diagram for $\hat{g}$.

The fundamental weights are

$$\boxed{\hat{\omega}^0 = (0,1,0) = (0,a_0^\vee,0) , \quad \hat{\omega}^{i\neq 0} = (\omega^{i\neq 0}, a_{i\neq 0}^\vee, 0) .} \qquad (3.35)$$

For an arbitrary affine weight $\hat{\lambda} = (\mu,\ell,n)$, ℓ is called the level of the weight, and n is called its grade. In the WZW context, the level of weight vectors will usually be fixed by the WZ coefficient k, and the grade is directly related to the eigenvalue of the Virasoro zero mode L_0. For $\hat{\lambda}$ as just written, we will adopt the notational convention that $\lambda = \mu$; if the "hat" is removed from an affine weight, the result is the horizontal projection, or "finite part" of it. This is consistent with (3.34) and (3.35), and also allows us to write $\omega^0 = 0$, $\alpha_0 = -\theta$, for examples. We also use ϕ_λ to denote $\phi_{\hat{\lambda}}$ (and have so already); so that $\phi_{k\hat{\omega}^0} = \phi_0$, for instance.

This notational convention also allows us to write

$$\hat{\lambda}_i = \lambda_i , \quad \forall\, i \in \{1,\ldots,r\} . \qquad (3.36)$$

These are the affine roots and weights. What about the affine Weyl group $\hat{W}$? It is also the Coxeter group associated with the corresponding Dynkin diagram, generated by the primitive Weyl reflections

$$\hat{r}_i \hat{\lambda} = \hat{\lambda} - (\hat{\lambda}, \hat{\alpha}_i^\vee) \alpha_i^\vee . \qquad (3.37)$$

Suppose $\hat{\lambda} = (\lambda, k, n)$, then this gives

$$\begin{aligned}
\hat{r}_i \hat{\lambda} &= (r_i\lambda, k, n) , \quad i \neq 0 ; \\
\hat{r}_0 \hat{\lambda} &= \hat{\lambda} - [k - (\lambda,\theta)] \hat{\alpha}_0 = (\lambda + [k - (\lambda,\theta)]\theta, k, n - [k - (\lambda,\theta)]) .
\end{aligned} \qquad (3.38)$$

Notice that $k - (\lambda, \theta)$ plays the role of $\hat{\lambda}_0$. This is justified by $(\delta, \hat{\lambda}) = \sum_{i=0}^{r} a_i^\vee \hat{\lambda}_i = \hat{\lambda}_0 + \sum_{i=1}^{r} a_i^\vee \lambda_i = \hat{\lambda}_0 + (\lambda, \theta)$, which should be the eigenvalue of K, i.e. the level.

Consequently, we sometimes use $\lambda_0 := k - (\lambda, \theta)$. So (3.36) can be extended to include $i = 0$, once the level k of an affine weight has been fixed, as it is in WZW models.

The relation between $\hat{W}$ and $W \subset \hat{W}$ is found by calculating $r_{\hat{\alpha}} \hat{\lambda}$ for $\hat{\alpha} = (\alpha, 0, m)$. One gets $r_{\hat{\alpha}} = r_\alpha (t_\alpha)^m$, where

$$t_\alpha \hat{\lambda} = \left(\lambda + k\alpha^\vee, k, n + \lambda^2 - (\lambda + k\alpha^\vee)^2/2k \right) . \tag{3.39}$$

$t_\alpha t_\beta = t_\beta t_\alpha$, so $\langle t_\alpha \rangle = T_{kQ^\vee}$, the translation group in the (scaled) co-root lattice $Q^\vee$ of g. Furthermore, $r_\beta t_\alpha r_\beta^{-1} = r_\beta t_\alpha r_\beta = t_{r_\beta(\alpha)}$, so $T_{kQ^\vee}$ is a normal subgroup of $\hat{W}$, and

$$\boxed{\hat{W} = W \ltimes T_{kQ^\vee} .} \tag{3.40}$$

This relation has important implications for the modular properties of affine characters, as we'll see.

The geometry of affine Weyl hyperplanes can be compared to that for the Weyl hyperplanes of g, at least after the horizontal projection $\hat{\lambda} = (\lambda, k, n) \mapsto \lambda$. The situation is analogous, with sectors of weight space labelled by elements of $\hat{W}$. But this time the sectors are of finite volume, and there is an infinite number of them, since $|\hat{W}| = \infty$. See Fig. 10 for a depiction of the case $g = A_2$.

This fact is highly suggestive: the integrable highest weights for g are those integral weights ($\lambda = \sum_{i=1}^{r} \lambda_i \omega^i$ with all $\lambda_i \in \mathbb{Z}$) contained in the sector labelled by id $\in W$, so we expect the integrable affine highest weights to be finite in number. How does this happen?

First, if $\hat{\lambda} = (\lambda, k, n)$ is to be a highest weight for an integrable representation of $\hat{g}$, then $g \subset \hat{g}$ implies λ must be one for g. So $\lambda \in P_+ = \{\mu = \sum_{i=1}^{r} \mu_i \omega^i : \mu_i \in \mathbb{Z}_{\geq 0}\}$, ensuring that each A_1 subalgebra $\langle e_i, h_i, f_i \rangle = \langle E_0^{\alpha_i}, H_0^i, E_0^{-\alpha_i} \rangle$ ($i \in \{1, \ldots, r\}$) is represented integrably, i.e. has $\lambda_i = 2j$ with "isospin" $j \in \mathbb{Z}_{\geq 0}/2$. The extra condition is simply that the A_1 subalgebra corresponding to the simple root α_0 be represented integrably. This just means $\lambda_0 \in \mathbb{Z}_{\geq 0}$ is required, i.e. $k - (\lambda, \theta) = k - \sum_{i=1}^{r} \lambda_i a_i^\vee \in \mathbb{Z}_{\geq 0}$. In other words,

$$\hat{\lambda} \in P_+^k = \left\{ \hat{\lambda} = \sum_{i=0}^{r} \lambda_i \hat{\omega}^i : \lambda_i \in \mathbb{Z}_{\geq 0}, \sum_{i=0}^{r} \lambda_i a_i^\vee = k \right\}, \tag{3.41}$$

explaining why there is a finite number of affine integrable highest weights at fixed level k. If the $r + 1$ simple-root A_1 subalgebras of $\hat{g}$ are all represented integrably, that turns

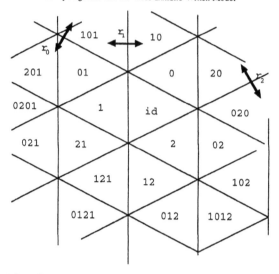

Figure 10. The affine Weyl sectors of the horizontal projection of $\hat{A}_2$ (affine A_2) weight space.

out to be sufficient to guarantee that the whole of $\hat{g}$ is so represented. We also write

$$\overline{P_+^k} \; = \; \left\{ \lambda = \sum_{i=1}^{r} \lambda_i \omega^i \; : \; \lambda_i \in \mathbb{Z}_{\geq 0}, \; \sum_{i=1}^{r} \lambda_i a_i^\vee \leq k \right\} \subset P_+ \tag{3.42}$$

for the set of horizontal projections of integrable affine highest weights at fixed level k.

Integrability is signalled by the presence of null vectors, vectors (states) of zero norm. For example, with $g = A_1$, and highest state $|v_\lambda\rangle$ with $\lambda = \lambda_1 \omega^1 = 2j\omega^1$, one finds the null states $e_1 |v_\lambda\rangle$ (from the highest-state condition) and $f_1^{\lambda_1+1} |v_\lambda\rangle = f_1^{2j+1} |v_\lambda\rangle$. See Fig. 11 for the example of $j = 3/2$, i.e. the A_1 representation $L(3)$. The existence of null vectors (and so integrability) goes hand-in-hand with the Weyl symmetry of representations.

For the integrable highest-weight representations of $\hat{g}$, with highest weight state satisfying $e_i |v_{\hat{\lambda}}\rangle = 0$ for all $i \in \{0, 1, \ldots, r\}$, there are $r+1$ *primitive null vectors* $|\eta_i\rangle := f_i^{1+\lambda_i} |v_{\hat{\lambda}}\rangle$. Primitive here means that all other null vectors (an infinite number of them) can be obtained as descendants of these ones. Because of (3.34), these are of exactly the same form as the primitive null vectors of the integrable g representation of highest weight λ, for $i \neq 0$. The additional ($i = 0$) primitive null vector has interesting consequences in the WZW model, as we'll see.

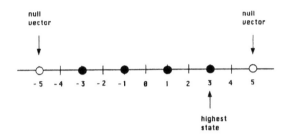

Figure 11. The weight diagram for the A_1 representation $L(3\omega^1)$, corresponding to angular momentum $j = 3/2$. The weights of the null vectors are shown.

The presence of the "extra" null vector is also consistent with the enlargement of the Weyl symmetry $W \to \hat{W}$. Consider the *affine formal character*

$$\begin{aligned} \mathrm{ch}_{\hat{\lambda}} &:= \mathrm{Tr}_{L(\hat{\lambda})}\left(e^{\hat{H}}\right) \\ &= \sum_{\hat{\mu} \in P(\hat{\lambda})} \mathrm{mult}\,(\hat{\lambda}; \hat{\mu})\, e^{\hat{\mu}} \quad, \end{aligned} \tag{3.43}$$

where, $\hat{H} = \sum_{i=0}^{r} \hat{\omega}^i h_i$, $\mathrm{mult}\,(\hat{\lambda}; \hat{\mu})$ denotes the multiplicity of the weight $\hat{\mu}$ in $L(\hat{\lambda})$, and we define

$$P(\hat{\lambda}) := \{\, \mu \in \hat{g}_0^{e*} \;:\; \mathrm{mult}\,(\hat{\lambda}; \hat{\mu}) \neq 0 \,\} \quad. \tag{3.44}$$

Then it is the *Weyl-Kac formula* that makes manifest the affine Weyl symmetry of affine characters:

$$\mathrm{ch}_{\hat{\lambda}} = \frac{\sum_{\hat{w} \in \hat{W}} (\det \hat{w})\, e^{\hat{w}.\hat{\lambda}}}{\prod_{\hat{\alpha} \in \hat{\Delta}_+} (1 - e^{-\hat{\alpha}})} \quad, \tag{3.45}$$

where $\hat{w}.\hat{\lambda}$ indicates the shifted action of $\hat{w}$: $\hat{w}.\hat{\lambda} := \hat{w}(\hat{\lambda} + \hat{\rho}) - \hat{\rho}$, with $\hat{\rho} = \sum_{j=0}^{r} \hat{\omega}^j$. $\hat{\Delta}_+$ indicates the set of positive roots of $\hat{g}$, to be specified shortly. $\mathrm{ch}_{\hat{0}} = 1$ leads to the affine denominator formula

$$\prod_{\hat{\alpha} \in \hat{\Delta}_+} (1 - e^{-\hat{\alpha}}) = \sum_{\hat{w} \in \hat{W}} (\det \hat{w})\, e^{\hat{w}\hat{\rho}} \quad, \tag{3.46}$$

so that (3.45) can also be written as

$$\mathrm{ch}_{\hat{\lambda}} = \frac{\sum_{\hat{w} \in \hat{W}} (\det \hat{w})\, e^{\hat{w}(\hat{\lambda} + \hat{\rho})}}{\sum_{\hat{w} \in \hat{W}} (\det \hat{w})\, e^{\hat{w}\hat{\rho}}} \quad. \tag{3.47}$$

One integrable affine highest weight is special: $\hat{\lambda} = k\hat{\omega}^0$ has horizontal projection $\lambda = 0$. This indicates that it corresponds to a state that is G-invariant; this state is the vacuum $|0\rangle$. The corresponding field $\phi_{k\hat{\omega}^0} = \phi_0$ is known as the *identity primary field*, because of its action on the vacuum:

$$\phi_{k\hat{\omega}^0}(0)\,|0\rangle \;=\; \phi_0(0)\,|0\rangle \;=\; |v_{k\hat{\omega}^0}\rangle \;=\; |v_0\rangle \;=\; |0\rangle \tag{3.48}$$

(see (2.81)). Now, more on the integrable highest-weight representations of $\hat{g}$ (they are sometimes called the *standard representations*, for short).

In terms of Chevalley generators, the highest state $|v_{\hat{\lambda}}\rangle$ is defined by $e_i|v_{\hat{\lambda}}\rangle = 0$, for all $i \in \{0, 1, \ldots, r\}$. But the highest state is annihilated by the raising operator corresponding to any positive root. So, using the Cartan-Weyl presentation, we have

$$E_0^\alpha|v_{\hat{\lambda}}\rangle \;=\; 0 \,,\; \forall\,\alpha \in \Delta_+ \;;\quad J_n^\alpha|v_{\hat{\lambda}}\rangle \;=\; 0 \,,\; \forall\,\alpha \in \Delta \,,\; n \in \mathbb{Z}_{>0} \,, \tag{3.49}$$

where $J_n^\alpha \in \{E_n^{\pm\alpha}, H_n^\alpha\}$. This then points to the appropriate choice of the set $\hat{\Delta}_+$ of positive roots of $\hat{g}$:

$$\hat{\Delta}_+ \;=\; \Delta_+ \,\cup\, \{n\delta \,:\, n \in \mathbb{Z}_{>0}\} \,\cup\, \{\alpha + n\delta \,:\, \alpha \in \Delta, n \in \mathbb{Z}_{>0}\} \,. \tag{3.50}$$

The full set of roots is $\hat{\Delta} = \hat{\Delta}_+ \cup \hat{\Delta}_- = \hat{\Delta}_+ \cup (-\hat{\Delta}_+)$. All roots except 0 and the *imaginary* ones $\{n\delta \,:\, 0 \neq n \in \mathbb{Z}\}$ have unit multiplicity, and each relates to a single element of $\hat{g}$: $E^{\alpha+n\delta} = E_n^\alpha$. $n\delta$ (including 0) has multiplicity r, relating to the existence of the r elements H_{-n}^i.

So, the generators of $\hat{g}$ can be written to emphasise their similarity with those for g: one simply adds "mode numbers" as subscripts to the symbols for the generators of g. Then their commutation relations also take a form that is simply related to those for g, written in (3.4),(3.6),(3.8):

$$
\begin{aligned}
[H_m^i, H_n^j] &= km\delta_{m+n,0}\delta^{i,j} \\
[H_m^i, E_n^\alpha] &= \alpha^i\,E_{m+n}^\alpha \\
[E_m^\alpha, E_n^\beta] &= \begin{cases} \alpha^\vee \cdot H_{m+n} + km(2/\alpha^2)\delta_{m+n,0} \,, & \alpha+\beta = 0 \\ N(\alpha,\beta)E_{m+n}^{\alpha+\beta} \,, & \alpha+\beta \in \Delta \\ 0 \,, & \alpha+\beta \notin \Delta \,. \end{cases}
\end{aligned}
\tag{3.51}
$$

Here of course, $\alpha, \beta \in \Delta$, and $m, n \in \mathbb{Z}$.

By the Sugawara construction, $|v_{\hat{\lambda}}\rangle$ is also the highest weight of a representation of Vir:

$$L_n|v_{\hat{\lambda}}\rangle \;=\; \sum_{m\in\mathbb{Z}}{}' N(J_m^\alpha J_{n-m}^\alpha)\,|v_{\hat{\lambda}}\rangle \;=\; 0 \,,\; \forall n \in \mathbb{Z}_{>0} \,. \tag{3.52}$$

We also have $L_0|v_{\hat{\lambda}}\rangle = h_\lambda |v_{\hat{\lambda}}\rangle$, as noted above, with conformal weight h_λ given in (2.76). Such an irreducible representation is not irreducible as a representation of Vir; rather, it decomposes into an infinite number of such representations.

The highest state is nevertheless the highest state of an irreducible representation of Vir. So, by the state-field correspondence, the $\hat{g}$-primary field also transforms as a Vir-primary field: under the conformal transformation $z \to w = w(z)$, an analytic function of z, and

$$\phi_\lambda(z) \to \phi_\lambda(w) = \left(\frac{dw}{dz}\right)^{-h_\lambda} \phi_\lambda(z) . \qquad (3.53)$$

So a Vir-primary field transforms in a tensorial way under conformal transformations.

Of particular use to us are the so-called *projective transformations*, where

$$w = \frac{az+b}{cz+d} , \quad \text{with } ad - bc = 1 . \qquad (3.54)$$

Writing a, b, c, d as the elements of a 2×2 matrix shows that these transformations form a group isomorphic to $PSL(2, \mathbb{C})$: P stands for projective, meaning the matrix and its negative describe equivalent transformations (3.54); S stands for special, i.e. the matrix has determinant one; and L means linear. The projective transformations are the only (invertible) conformal transformations that map the entire complex plane plus the point at ∞ to itself. They leave the vacuum invariant:

$$L_{\pm 1}|0\rangle = L_0|0\rangle = 0 , \qquad (3.55)$$

since the $L_{\pm 1}, L_0$ generate the $s\ell(2, \mathbb{C})$ algebra of the projective group.

For more details, consider infinitesimal conformal transformations, i.e. $w = z + \epsilon(z)$, with $|\epsilon(z)| \ll 1$. (3.53) then yields

$$\delta\phi_\lambda(z) = \left(\epsilon(z)\partial_z + h_\lambda\epsilon'(z)\right)\phi_\lambda(z) . \qquad (3.56)$$

If we don't restrict $\epsilon(z)$ further, we are considering general infinitesimal conformal transformations. ¿From their general form (3.54), one can see that infinitesimal projective transformations give

$$\epsilon(z) = c_{-1} + c_0 z + c_1 z^2 , \qquad (3.57)$$

where $c_{\pm 1}, c_0$ are constants. We write

$$\delta\phi_\lambda(z) = \sum_{m=-1}^{1} c_m [L_m, \phi_\lambda(z)] , \qquad (3.58)$$

34

and find

$$[L_m, \phi_\lambda(z)] = \left(z^{m+1}\partial_z + (m+1)h_\lambda z^m\right)\phi_\lambda(z) , \qquad (3.59)$$

with $m = \pm 1, 0$. This last formula is consistent with the commutation relations (2.63) of Vir, for the modes $L_{\pm 1,0}$, and shows they do generate $s\ell(2,\mathbb{C}) \subset Vir$.

Note that L_{-1} acts as the z-translation operator and L_0 as the generator of dilations:

$$e^{aL_{-1}} \phi_\lambda(z) e^{-aL_{-1}} = \phi_\lambda(z+a) , \quad e^{aL_0} \phi_\lambda(z) e^{-aL_0} = \phi_\lambda(e^a z) . \qquad (3.60)$$

Including both the holomorphic and antiholomorphic parts of the primary field, this last equation gives

$$e^{aL_0 + \bar{a}\bar{L}_0} \Phi_{\lambda,\mu}(z,\bar{z}) e^{-aL_0 - \bar{a}\bar{L}_0} = \Phi_{\lambda,\mu}(e^a z, e^{\bar{a}}\bar{z}) = e^{-ah_\lambda - \bar{a}h_\mu} \Phi_{\lambda,\mu}(z,\bar{z}) , \qquad (3.61)$$

using (3.53), where $\bar{a}$ denotes the complex conjugate of a. Putting $a = \alpha + i\theta$, with $\alpha, \theta \in \mathbb{R}$, we confirm that $L_0 + \bar{L}_0$ is the generator of dilations (radial Hamiltonian) and $L_0 - \bar{L}_0$ is the generator of rotations. Furthermore, $h_\lambda + h_\mu$ is the scaling dimension of $\Phi_{\lambda,\mu}$ and $h_\lambda - h_\mu$ is its spin.

The third generator L_1 generates what are known as special conformal transformations. All three types of transformations (translations, rotations and special conformal transformations) are conformal in any number N of dimensions. If we restrict the base field to $\mathbb{R}$, instead of $\mathbb{C}$, we have an algebra $s\ell(2,\mathbb{R})$. The antiholomorphic counterparts $\bar{L}_{\pm 1}, \bar{L}_0$ generate another copy of this algebra, and the direct sum of the two copies is isomorphic to a real form of $so(4)$. In N dimensions, the translations, rotations and special conformal transformations generate a real form of $so(N+2)$. In $N = 2$ the symmetry extends to an infinite-dimensional one, with infinite-dimensional algebra (2.54). After central extension, we find Vir.

A simple example of a standard representation of $\hat{g} = \hat{A}_1 = A_1^{(1)}$ is depicted in Fig. 12. There the weights in $\hat{g}_0^{e*}$ are drawn (except that the fixed eigenvalue $k = 2$ of K is not indicated as a coordinate). Note that the "horizontal" weight spaces are those for the simple Lie algebra $g = A_1$; in general, the horizontal subspaces of a representation $L(\hat{\lambda})$ of $\hat{g}$ will be (reducible) representations of g. (This is where the term horizontal subalgebra $g \subset \hat{g}$ comes from.) In particular, for the standard representation $L(\hat{\lambda})$, the horizontal representation of lowest L_0 eigenvalue is the irreducible representation $L(\lambda)$ of g. Notice also that the weight diagram is enclosed by a parabolic envelope: a parabola passes through all weights $\hat{\mu}$ such that $\hat{\mu} - \delta$ is not also in the diagram. Its curvature decreases with increasing level. The parabola becomes a paraboloid for higher rank algebras. The simple roots are indicated, as well as the weights of the primitive null vectors. The

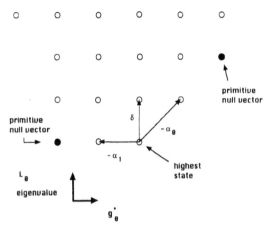

Figure 12. The weight diagram of the $\hat{A}_1$ representation $L(\hat{\omega}^0 + \hat{\omega}^1)$.

multiplicities of the weights rise rapidly with increasing L_0 eigenvalue n; asymptotically $\text{mult}\left(\lambda;\,(\mu, k, -n)\right) \sim n^{-3/4}\exp(\text{const}.n^{1/2})$.

Fig. 13 shows the standard representations for arbitrary affine algebras in very schematic fashion. There is a finite number (the case of card $P_+^k = 3$ is drawn) of such representations, each to be associated with a primary field in the corresponding WZW model. As mentioned above, the representation $L(k\hat{\omega}^0)$ is special among them: its lowest horizontal representation is $L(0)$, the scalar representation, and its lowest L_0 eigenvalue is the lowest of the low. That's because the single state in the representation $L(0)$ is to be identified with the vacuum of the WZW model. The corresponding primary field is called the *identity primary field*. The other standard representations have lowest horizontal representations of higher dimensions, and lowest L_0 eigenvalues that are higher than that of the vacuum; after all, $\mathbf{H} = L_0 + \bar{L}_0$, so the vacuum should have lowest energy. These last two effects go hand-in-hand, as the diagram is meant to indicate.

(2.65) shows that elements of g commute with L_0. Since $\mathbf{H} = L_0 + \bar{L}_0$, this means that $g \oplus g$ is a true symmetry algebra of the WZW model. The full affine algebra $\hat{g} \oplus \hat{g}$ plays the role of a spectrum-generating algebra in the theory, generating all the states in the towers corresponding to the primary fields.

The remarkable thing is that the states in these card P_+^k primary towers span the space of states of the WZW model! $\hat{g} \oplus \hat{g}$ generates the full spectrum of the model from the card P_+^k primary highest states. We can therefore say that the the infinite-dimensional

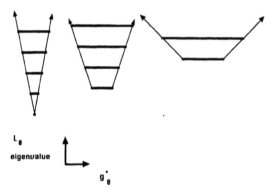

Figure 13. Schematic drawing indicating the standard affine representations that are relevant to a WZW model.

affine algebra effectively "finitises" the WZW field theory: one need only study the finite number of primary fields.

4. Affine Algebra Representations and WZW Models

In the last section we laid the basis for the application of the representation theory of untwisted affine algebras to WZW models. In this section we'll describe some specific results in detail.

4.1. Gepner-Witten equation

Null vectors constrain the possible couplings between WZW fields. Consider a primary field realising a standard representation of $\hat{g}$, with highest weight $\hat{\nu}$. That is, the primary field has holomorphic part $\phi_{\hat{\nu}}(z)$. This implies $f_i^{1+\nu_i}\phi_{\hat{\nu}} = 0$, for all $i \in \{0, 1, \ldots, r\}$. For $i = 0$, this can be rewritten as

$$\left(E^\theta_{-1}\right)^{1+\hat{\nu}_0} \phi_{\hat{\nu}}(z) = 0 \,. \tag{4.1}$$

Now, suppose this null field appears in a correlation function with the primary fields $\phi_1(z_1), \ldots, \phi_n(z_n)$; the correlator must then vanish:

$$\left\langle \left[\left(E^\theta_{-1}\right)^{1+\hat{\nu}_0}\phi_{\hat{\nu}}(z)\right]\phi_1(z_1)\cdots\phi_n(z_n)\right\rangle = 0 \,. \tag{4.2}$$

After using (2.50) to rewrite last equation as

$$\langle \frac{1}{2\pi i} \oint_z \frac{dw}{w-z} \left(E^\theta(w) \left[(E^\theta_{-1})^{\hat\nu_0} \phi_{\hat\nu}(z) \right] \right) \phi_1(z_1) \cdots \phi_n(z_n) \rangle = 0 , \qquad (4.3)$$

we can deform the contour of integration in the manner indicated in Fig. 14 to get

$$0 = \sum_{j=1}^n \frac{1}{2\pi i} \oint_{z_j} \frac{dw}{w-z} \langle \left[(E^\theta_{-1})^{\hat\nu_0} \phi_{\hat\nu}(z) \right] \phi_1(z_1) \cdots$$
$$\cdots \left[E^\theta(w) \phi_j(z_j) \right] \cdots \phi_n(z_n) \phi_n(z_n) \rangle . \qquad (4.4)$$

Now since the ϕ_j are primary, by (2.69) we can write

$$0 = \sum_{j=1}^n \frac{t^\theta_j}{z-z_j} \langle \left[(E^\theta_{-1})^{\hat\nu_0} \phi_{\hat\nu}(z) \right] \phi_1(z_1) \cdots \phi_n(z_n) \rangle , \qquad (4.5)$$

where the j on t^θ_j indicates that the generator should act on ϕ_j. If this process is repeated, we find

$$0 = \sum_{\substack{\{\ell_1,\ldots,\ell_n\} \\ \sum \ell_i = 1+\hat\nu_0}} \frac{(t^\theta_1)^{\ell_1}/\ell_1!}{(z-z_1)^{\ell_1}} \cdots \frac{(t^\theta_n)^{\ell_n}/\ell_n!}{(z-z_n)^{\ell_n}} \langle \phi_{\hat\nu}(z)\phi_1(z_1) \cdots \phi_n(z_n) \rangle . \qquad (4.6)$$

This is the *Gepner-Witten equation* [30]. Notice that it also holds if we replace $1+\hat\nu_0$ with any $p \geq 1 + \hat\nu_0$.

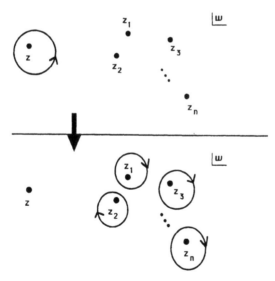

Figure 14. Contour deformation for (4.4).

38

The first consequence of (4.6) is that a non-integrable field has vanishing correlators with integrable fields, i.e. a non-integrable field decouples from integrable ones. To see this, let $\phi_{\hat{\nu}}$ be the identity field in (4.6), that is, set $\hat{\nu} = k\hat{\omega}^0$. Then

$$\phi_{\hat{\nu}}(z)|0\rangle = e^{zL_{-1}}\phi_{k\hat{\omega}^0}(0)e^{-zL_{-1}}|0\rangle = |0\rangle , \tag{4.7}$$

using (3.60),(3.55),(3.48). So we can remove $\phi_{\hat{\nu}}(z)$ from (4.6). Multiply by $(z - z_n)^{p-1}$ and integrate over z to get

$$
\begin{aligned}
0 &= \langle \phi_1(z_1) \cdots \phi_{n-1}(z_{n-1})(t_n^\theta)^p\phi_n(z_n) \rangle \\
&= \langle \phi_1(z_1) \cdots \phi_{n-1}(z_{n-1})\left[(E_0^\theta)^p\phi_n(z_n)\right] \rangle ,
\end{aligned}
\tag{4.8}
$$

for all $p \geq 1 + \hat{\nu}_0$. Now suppose ϕ_n is non-integrable, so that there is no $p \in \mathbb{Z}_{>0}$ such that $[(E_0^\theta)^p\phi_n] = 0$. Then (4.8) will only be satisfied if $\langle \phi_1(z_1) \cdots \phi_n(z_n) \rangle = 0$, i.e. if the non-integrable field ϕ_n decouples from the integrable ones $\phi_1, \ldots, \phi_{n-1}$.

4.2. 3-point functions

The second application of (4.6) will be to the 3-point correlation functions $\langle \phi_\lambda \phi_\mu \phi_\nu \rangle$ of primary fields. The 3-point functions encode the structure constants of the operator product algebra, the OPE coefficients (see below). So the 3-point functions are arguably the most important, since in principle, any n-point correlation function can be constructed using the operator product algebra.

The 3-point functions are highly constrained by global (z-independent) $G \otimes G$ invariance, and invariance under the projective transformations with $s\ell(2, \mathbb{C})$ algebra generated by $L_{\pm 1}, L_0$. We'll first examine these constraints, before applying the Gepner-Witten equation to arrive at a quite powerful result.

Let's work with the holomorphic parts of primary fields, for simplicity. Projective invariance yields

$$0 = \sum_{j=1}^{n} \left[z_j^{m+1}\frac{\partial}{\partial z_j} + (m+1)h_j z_j^m \right] \langle \phi_1(z_1) \cdots \phi_n(z_n) \rangle , \quad \text{for } m \in \{-1, 0, 1\} \tag{4.9}$$

(compare to (3.59)). The general solution to (4.9) is

$$\langle \phi_1(z_1) \cdots \phi_n(z_n) \rangle = F(\{z_{pqrs}\}) \prod_{\substack{i,j=1 \\ i<j}}^{n} (z_i - z_j)^{h_{ij}} , \tag{4.10}$$

where $h_{ij} = h_{ji}$, $\sum_{i \neq j} h_{ij} = 2h_j$, and F is an arbitrary function of the *anharmonic ratios*

$$z_{pqrs} := \frac{(z_p - z_q)(z_r - z_s)}{(z_q - z_r)(z_s - z_p)} . \tag{4.11}$$

These ratios are invariant under the projective transformations $z_i \to (az_i + b)/(cz_i + d)$, $ad - bc = 1$.

Only $n - 3$ of these anharmonic ratios are independent, so for the 3-point function, F is simply a constant. Before writing the explicit form, let us change notation somewhat. Replace ϕ_1, ϕ_2, ϕ_3 with $\phi_{\hat{\lambda}}, \phi_{\hat{\mu}}, \phi_{\hat{\nu}}$, and we'll drop the hats on the affine weights. We can then write

$$\langle \phi_\lambda(x)\phi_\mu(y)\phi_\nu(z) \rangle = (x - y)^{h_\nu - h_\lambda - h_\mu}(x - z)^{h_\mu - h_\nu - h_\lambda}(y - z)^{h_\lambda - h_\mu - h_\nu} \hat{C}_{\lambda,\mu,\nu} . \quad (4.12)$$

So the computation of the 3-point function boils down to a computation of a constant $\hat{C}_{\lambda,\mu,\nu}$. This constant is related to the *OPE of primary fields*

$$\phi_\lambda(z)\phi_\mu(0) \sim \frac{\hat{C}_{\lambda,\mu,\nu}\phi_{\nu^t}}{z^{h_\lambda + h_\mu - h_\nu}} , \quad (4.13)$$

and so is called an *operator product coefficient*. Here ν^t indicates the highest weight of the representation contragredient (charge-conjugate) to $L(\nu)$. (The corresponding field is the unique one with a non-vanishing 2-point function with the primary field ϕ_ν.)

The global G invariance of a correlation function of n primary fields imposes

$$0 = \sum_{j=1}^{n} t_j^a \langle \phi_1(z_1) \cdots \phi_n(z_n) \rangle . \quad (4.14)$$

That is, the n-point function must be a G-singlet. For this to be possible, the tensor product $L(\lambda) \stackrel{.}{\otimes} L(\mu) \otimes L(\nu)$ must contain the singlet representation $L(0)$. Alternatively, in the tensor-product decomposition

$$L(\lambda) \otimes L(\mu) = \sum_{\varphi \in P_+} T_{\lambda,\mu}^\varphi L(\varphi) , \quad (4.15)$$

we require that the *tensor-product coefficients* obey

$$T_{\lambda,\mu}^{\nu^t} = T_{\lambda,\mu,\nu} \neq 0 . \quad (4.16)$$

This implies that the corresponding *Clebsch-Gordan coefficient* $C_{\lambda,\mu,\nu} \neq 0$. In summary, then, the global G-invariance gives

$$\hat{C}_{\lambda,\mu,\nu} \neq 0 \quad \Rightarrow \quad C_{\lambda,\mu,\nu} \neq 0 . \quad (4.17)$$

We will henceforth concentrate on 3-point functions.

4.3. Depth rule

Let's apply the Gepner-Witten equation (4.6) to the 3-point function:

$$0 = \sum_{\substack{\ell_1,\ell_2 \\ \ell_1+\ell_2 \geq 1+\hat{\nu}_0}} \frac{(t_\mu^\theta)^{\ell_1}(t_\lambda^\theta)^{\ell_2}}{\ell_1!\ell_2!(z-z_1)^{\ell_1}(z-z_2)^{\ell_2}} \langle \phi_\nu(z)\phi_\mu(z_1)\phi_\lambda(z_2) \rangle . \tag{4.18}$$

Since the z, z_1, z_2 dependence of $\langle \phi_\nu(z)\phi_\mu(z_1)\phi_\lambda(z_2) \rangle$ is fixed, the terms in the summation are independent, and so

$$(t_\mu^\theta)^{\ell_1}(t_\lambda^\theta)^{\ell_2} \langle \phi_\nu(z)\phi_\mu(z_1)\phi_\lambda(z_2) \rangle = 0 \quad \forall\, \ell_1 + \ell_2 > \hat{\nu}_0 = k - (\nu,\theta) . \tag{4.19}$$

t^θ is the raising operator for the $A_1 \subset g$ subalgebra in the direction of the highest root θ of g. The maximum number of times it can be applied to a state (vector) v with non-vanishing result, is called the *depth* $d(\theta,v)$ of that state, or sometimes the θ-depth.

To understand the name, look at the example pictured in Fig. 8. There is drawn the weight diagram of the A_2-representation $L(2,1)$. Consider as the "upward" direction that of the highest root θ (see Figure 7). Since E^θ adds the root θ to the weight of a state, the depth of a state tells us how far "down" it is from its "top", roughly speaking. Concentrating on the 2-dimensional subspace of weight $(0,-1)$, we see that it breaks up into two one-dimensional subspaces of depths $d(\theta,v_1) = 1$ and $d(\theta,v_2) = 2$. An important point, however, is that $d(\theta, c_1v_1 + c_2v_2) = 2$, as long as $c_2 \neq 0$.

Now, if we write

$$\langle \phi_\nu(z)\phi_\mu(z_1)\phi_\lambda(z_2) \rangle = \\ \sum_{u\in L(\nu)} \sum_{v\in L(\mu)} \sum_{w\in L(\lambda)} \phi_{\nu,u}\phi_{\mu,v}\phi_{\lambda,w} \langle u(z)v(z_1)w(z_2) \rangle , \tag{4.20}$$

following (2.84), we can conclude from (4.19) that even if $\langle u|w \otimes v \rangle$ is non-zero, the 3-point function $\langle u(z)v(z_1)w(z_2) \rangle$ will vanish unless

$$\boxed{d(\theta,v) + d(\theta,w) < k+1 - (\theta,\nu) = 1 + \hat{\nu}_0.} \tag{4.21}$$

But if $\langle u(z)v(z_1)w(z_2) \rangle$ vanishes, then so does $\langle \phi_\nu(z)\phi_\mu(z_1)\phi_\lambda(z_2) \rangle$. Therefore, a necessary condition for $\langle \phi_\nu(z)\phi_\mu(z_1)\phi_\lambda(z_2) \rangle \neq 0$ is: whenever $\langle u|v \otimes w \rangle \neq 0$, (4.21) must be obeyed. This is the Gepner-Witten *depth rule*.

4.4. Tensor products and refined depth rule

Can we get additional constraints from other null vectors? No: the primitive null vectors are $|\eta_j\rangle = f_j^{1+\nu_j} |v_\nu\rangle$ for $i \in \{0, 1, \ldots, r\}$. We have just used the first ($j = 0$), and all others are present at the lowest grade in $L(\nu)$, and so are isomorphic to the primitive null vectors of the g-representation $L(\nu)$. These latter determine the Clebsch-Gordan coefficients for g. So, the depth rule should allow the determination of the operator product coefficients $\hat{C}_{\lambda,\mu,\nu}$ from a knowledge of the Clebsch-Gordan coefficients $C_{\lambda,\mu,\nu}$. This turns out to be less straightforward than one might hope, however.

To see why, we consider the simpler problem of computing the operator product multiplicities (called *fusion coefficients*) from the corresponding tensor-product multiplicities (we'll call them tensor-product coefficients). Being multiplicities, these coefficients are non-negative integers. A general tensor-product decomposition is written in (4.15); there the $T_{\lambda,\mu}^\varphi \in \mathbb{Z}_{\geq 0}$. The simplest example of such a decomposition is for $g = A_1$:

$$L(\lambda_1) \otimes L(\mu_1) = L(\lambda_1 + \mu_1) \oplus L(\lambda_1 + \mu_1 - 2) \oplus \cdots \oplus L(|\lambda_1 - \mu_1|) , \qquad (4.22)$$

where we write $L(\lambda_1)$ for $L(\lambda_1\omega^1)$, e.g. If we change notation using $\lambda_1 = 2j_1$, $\mu_1 = 2j_2$, and $\nu_1 = 2j$, we recognise the rule for the addition of quantum angular momenta.

Reasoning similar to that given above leads to the following rule. If $w, v, u \in L(\lambda), L(\mu), L(\nu)$, respectively, then $C_{\lambda,\mu}^\nu = 0$ (and so $T_{\lambda,\mu}^\nu = 0$) unless when $\langle u | w \otimes v \rangle \neq 0$, we have

$$\left(E^{-\alpha_i}\right)^{\ell_1} |w\rangle \otimes \left(E^{-\alpha_i}\right)^{\ell_2} |u\rangle = 0 , \qquad (4.23)$$

for all $\ell_1 + \ell_2 \geq 1 + \nu_i$, for $i \in \{1, \ldots, r\}$. If we generalise the definition of depth to:

$$d(\alpha, v) = \min\{ \ell \in \mathbb{Z}_{\geq 0} : (E^\alpha)^{\ell+1} v = 0 \} , \qquad (4.24)$$

then we require

$$d(-\alpha_i, v) + d(-\alpha_i, w) < 1 + \hat{\nu}_i , \quad \forall\, i \in \{1, \ldots, r\} . \qquad (4.25)$$

Compare this to (4.21).

Before writing a more useful version of this rule, let's look again at the example already mentioned above (4.20). Suppose we have a space V spanned by two independent states v_1, v_2, of depths $d(\theta, v_1) = 1$, $d(\theta, v_2) = 2$, respectively. Choosing a different basis, $(v_1 \pm v_2)/\sqrt{2}$, say, gives depths 2,2. The set of depths is a basis-dependent object. So, we should instead be concerned with the *dimensions* of the spaces $V_i := \{v \in V : (E^\theta)^{1+i} v = 0\}$, for $i = 1, 2$, which are 1 and 2 in this example, respectively. Of course, a particular choice of basis may help; $\{v_1, v_2\}$ is a basis of V that is good for the computation of the required dimensions, while the other basis is not.

Now, the highest-weight state $|v_\nu\rangle$ must appear in $L(\lambda) \otimes L(\mu)$ (as well as all others). So, we must be able to write

$$|v_\nu\rangle = \sum_{w \in L(\lambda)} \sum_{u \in L(\mu)} C^{v_\nu}_{w,u} |w\rangle \otimes |u\rangle , \qquad (4.26)$$

for some coefficients $C^{v_\lambda}_{w,u}$. It is not difficult to show that $|v_\nu\rangle$ must contain a non-zero component $\propto |w_\lambda\rangle \otimes |u\rangle$, where $|w_\lambda\rangle$ is the highest state of $L(\lambda)$, and $u \in L(\mu)$. Otherwise, $C^{v_\nu}_{w_\lambda,u} = 0$ would imply that $C^{v_\nu}_{w,u} = 0$ for all $w \in L(\lambda)$. Of course, so that the component $|w_\lambda\rangle \otimes |u\rangle$ has the correct weight ν, we must have

$$H|u\rangle = \left(\sum_{j=1}^{r} \omega^j h_i\right)|u\rangle = (\nu - \lambda)|u\rangle . \qquad (4.27)$$

Therefore, we can write (see [31], for example)

$$\boxed{\begin{aligned} T^\nu_{\lambda,\mu} &= \dim\left\{ u \in L(\mu; \nu - \lambda) \ : \ (E^{-\alpha_i})^{1+\nu_i} u = 0 \ \forall i = 1, \ldots, r \right\} \\ &=: \dim V^\nu_{\lambda,\mu} , \end{aligned}} \qquad (4.28)$$

where we have used

$$L(\mu; \nu - \lambda) := \left\{ u \in L(\mu) \ : \ Hu = (\nu - \lambda)u \right\} . \qquad (4.29)$$

For the operator product numbers, or fusion coefficients $^{(k)}N^\nu_{\lambda,\mu}$, we write a truncated tensor product, or *fusion product*:

$$L(\lambda) \otimes_k L(\mu) = \oplus_{\nu \in \overline{P^k_+}} \, ^{(k)}N^\nu_{\lambda,\mu} L(\nu) . \qquad (4.30)$$

Of course, since $\overline{P^k_+} \subset P_+$ (see (3.42)), $^{(k)}N^\nu_{\lambda,\mu}$ are undefined if any of $\lambda, \mu, \nu \in P_+$ are not in $\overline{P^k_+}$. But truncation here means that

$$^{(k)}N^\nu_{\lambda,\mu} \leq T^\nu_{\lambda,\mu} . \qquad (4.31)$$

An argument similar to that above leads to the following conjecture [32] (see also [33]):

$$\boxed{\begin{aligned} ^{(k)}N^\nu_{\lambda,\mu} &= \dim\left\{ u \in L(\mu; \nu - \lambda) \ : \ (E^{-\alpha_i})^{1+\nu_i} u = 0 \ \forall i = 0, \ldots, r \right\} \\ &=: \dim {}^{(k)}V^\nu_{\lambda,\mu} . \end{aligned}} \qquad (4.32)$$

Here, a special case of the "extra" conditions (4.21) has been incorporated into a formula similar to (4.28). (Recall that since $\hat{\alpha}_0 = (-\theta, 0, 1)$, $\alpha_0 = -\theta$.)

We'll call (4.32) the *refined depth rule*. Notice it explains (4.31). In fact, (4.32) implies the following stronger relations

$$^{(k)}N^{\nu}_{\lambda,\mu} \leq {}^{(k+1)}N^{\nu}_{\lambda,\mu}, \quad \lim_{k\to\infty} {}^{(k)}N^{\nu}_{\lambda,\mu} = T^{\nu}_{\lambda,\mu}. \tag{4.33}$$

All this can be encoded in the concept of a *threshold level* [34]. When $T^{\nu}_{\lambda,\mu} > 1$, we say that there are more than one different "couplings" $L(\nu) \subset L(\lambda) \otimes L(\mu)$. That is, there is more than one way to assemble the states of $L(\nu)$ in the tensor product $L(\lambda) \otimes L(\mu)$. If in addition, $L(\nu) \subset L(\lambda) \otimes_k L(\mu)$, we say that the coupling is also a "fusion coupling" at level k.

For each of the $T^{\nu}_{\lambda,\mu}$ couplings $L(\nu) \subset L(\lambda) \otimes L(\mu)$, there exists a threshold level k_t, such that the coupling is not a fusion coupling at levels $k < k_t$, and is for all levels $k \geq k_t$.

The threshold level allows a convenient notation: the fusion products for all levels can be written as the tensor product with threshold levels as subscripts:

$$L(\lambda) \otimes L(\mu) = \oplus_{k_t} \oplus_{\nu} {}^{(k_t)}n^{\nu}_{\lambda,\mu} L(\nu)_{k_t}. \tag{4.34}$$

Then

$$^{(k)}N^{\nu}_{\lambda,\mu} = \sum_{k_t}^{k} {}^{(k_t)}n^{\nu}_{\lambda,\mu}. \tag{4.35}$$

In (4.35) the sum is over all couplings, and any coupling with $k_t \leq k$ contributes once. The A_1 example (4.22) becomes

$$L(\lambda_1) \otimes L(\mu_1) = L(\lambda_1 + \mu_1)_{\lambda_1+\mu_1} \oplus L(\lambda_1 + \mu_1 - 2)_{\lambda_1+\mu_1-1}$$
$$\oplus \cdots \oplus L(|\lambda_1 - \mu_1|)_{\max(\lambda_1,\mu_1)}. \tag{4.36}$$

This can be derived from the depth rule, by considering the coupling $L(\nu_1) \subset L(\lambda_1) \otimes L(\mu_1)$ and corresponding $u \in L(\mu_1)$, with $Hu = [(\nu_1 - \lambda_1)\omega^1]u$. The depth is easily seen to be $d(\theta, u) = (\mu_1 - \nu_1 + \lambda_1)/2$, so that

$$k_t = (\mu_1 - \nu_1 + \lambda_1)/2 + (\nu, \theta) = (\lambda_1 + \mu_1 + \nu_1)/2. \tag{4.37}$$

Let us rewrite the A_1 fusion product one more way:

$$L(\lambda_1) \otimes L(\mu_1) = \oplus_{\nu_1=|\lambda_1-\mu_1|}^{\min(\lambda_1+\mu_1, 2k-\lambda_1-\mu_1)} L(\nu_1). \tag{4.38}$$

This the original form found by Gepner and Witten, and it makes clear the level-truncation of the tensor product.

An A_2 example is

$$L(1,1)^{\otimes 2} = L(0,0)_2 \oplus L(1,1)_2 \oplus$$
$$L(3,0)_3 \oplus L(1,1)_3 \oplus L(0,3)_3 \oplus \qquad (4.39)$$
$$L(2,2)_4 .$$

This is perhaps the simplest example with a $T^\nu_{\lambda,\mu} > 1$: we have $^{(2)}N^{(1,1)}_{(1,1),(1,1)} = 1$, $^{(k>2)}N^{(1,1)}_{(1,1),(1,1)} = 2$. This phenomenon occurs because algebras of rank greater than one (i.e. $g \neq A_1$) have most mult $(\mu; \varphi) := \dim L(\mu; \varphi) > 1$, the spaces of fixed weight φ in a representation $L(\mu)$ are typically not one-dimensional.[6]

4.5. Good bases and the Littlewood-Richardson rule

Now let's try to use the refined depth rule (4.32) to compute fusion products like (4.39). Since the level k will figure more prominently henceforth, we'll let g_k indicate the affine algebra $\hat{g}$ at fixed level k.

The problem is to find a good choice of basis of $L(\mu; \nu - \lambda)$ to simplify (as much as possible) the computation of $^{(k)}N^\nu_{\lambda,\mu} = \dim {}^{(k)}V^\nu_{\lambda,\mu} \subset V^\nu_{\lambda,\mu} \subset L(\mu; \nu - \lambda)$. Let's first back up and consider $T^\nu_{\lambda,\mu} = \dim V^\nu_{\lambda,\mu}$. After all, the problem of a good choice of basis already exists in the computation of $T^\nu_{\lambda,\mu}$ (for the α_i-depths, if not for the θ-depth). Suppose a basis $B(\mu; \nu - \lambda) = \{u^a : a = 1, \ldots, \dim L(\mu; \nu - \lambda)\}$ of $L(\mu; \nu - \lambda)$ were to be good; what should that mean? Ideally, one could test the u^a one-by-one, and those passing would form a basis of $V^\nu_{\lambda,\mu}$. That is, the subspace $V^\nu_{\lambda,\mu}$ would be spanned by the subset of $B(\mu; \nu - \lambda)$ that are elements of $V^\nu_{\lambda,\mu}$. It turns out that such *good bases* (or proper bases) $B^{(p)}(\mu; \nu - \lambda)$ exist [36]:

$$V^\nu_{\lambda,\mu} = \text{Span}\{ u^a \in B^{(p)}(\mu; \nu - \lambda) : u^a \in V^\nu_{\lambda,\mu} \}, \qquad (4.40)$$

for all possible highest weights $\lambda, \mu, \nu \in P_+$.

For the case $g = A_r$, the basis elements are indexed by *standard tableaux*, for example

$$\begin{array}{|c|c|c|c|}
\hline
1 & 2 & 2 & 4 \\
\hline
\end{array}$$

(4.41)

---

[6] Except for a G_2 exception, it turns out that mult $(\mu; \mu') = 1$ iff $\mu - \mu'$ has a unique expression as a $\mathbb{Z}_{\geq 0}$-linear combination of those positive roots α of g also obeying $(\mu, \alpha) > 0$ [35].

The tableaux with numbers removed are called *Young tableaux*, and their boxes are arranged in left-justified rows, of non-increasing length going from top to bottom. To obtain a standard tableau relevant to the A_r case, the numbers added must be from the set $\{1, 2, \ldots, r + 1\}$, and they must appear in non-decreasing order from left to right in the rows, and in increasing order from top to bottom in the columns. Notice this implies that a single column of a standard tableau can contain no more than $r + 1$ boxes, for the A_r case.

In combination with (4.28), the good bases (with elements indexed by standard tableaux) lead to a simple rule for the computation of tensor-product coefficients, the *Littlewood-Richardson rule*, to be explained below. Each box $\boxed{i}$ stands for a weight in the basic representation $L(\omega^1)$ of A_r, so the weight $wt(\boxed{i})$ of $\boxed{i}$ is

$$
wt(\boxed{i}) \;=\; \omega^1 - \alpha_1 - \alpha_2 - \ldots - \alpha_{i-1} \;=\;
\begin{cases}
\omega^1 \,, & i = 1 \\
-\omega^{i-1} + \omega^i \,; & i = 2, \ldots, r \\
-\omega^r \,, & i = r + 1
\end{cases}
\tag{4.42}
$$

Notice that $\sum_{i=1}^{r+1} wt(\boxed{i}) = 0$; therefore a column of length $r + 1$ can be dropped from a standard tableau. The *weight* $wt(T_\#)$ of a standard tableau $T_\#$ is just the sum of the weights of its component boxes. The *shape* $sh(T_\#)$ of a standard tableau $T_\#$ is the weight of the tableau obtained by replacing all the numbers in the i-th row with i's. Notice that the shape of a standard tableau will always be a dominant weight (i.e. an element of P_+); it will be the highest weight of the relevant representation. To restrict to those highest weights relevant to the A_r WZW model, the number of columns of length less than or equal to r must be less than or equal to k (see (3.42)). The weight of the Young tableau T obtained from the standard tableau $T_\#$ by removing its numbers, is defined by $wt(T) := sh(T_\#)$.

The elements of the good basis $B^{(p)}(\mu; \sigma)$ are indexed by the elements of

$$
T_\#(\mu; \sigma) \;=\; \{\, T_\# \in S_\# \;:\; sh(T_\#) = \mu, \; wt(T_\#) = \sigma \,\} \,,
\tag{4.43}
$$

where $S_\#$ just means the space of standard tableaux. We'll write $v(T_\#)$ for the state indexed by the standard tableau $T_\#$. The elements of these bases appropriate for the A_2 representation $L(1, 1)$ are shown in Fig. 15. There the tableaux are drawn, roughly at the positions of their weights in weight space.

The action of the step operators e_i and f_i ($i = 1, 2$), can be transcribed to an action on the standard tableaux. I will describe it later in the more convenient language of *paths*. For the moment, their actions (up to non-zero multiplicative factors) are indicated in Fig.

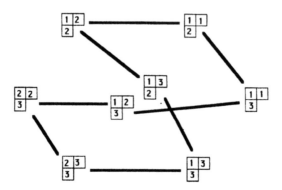

Figure 15. Shown are the standard tableaux that label the states of the A_2 representation $L(1, 1)$ in the good basis.

15 by the lines drawn. Notice that a single line specifies the action of both e_i and f_i (fixed i). That's because of the property

$$
\begin{aligned}
e_i T_\# \neq 0 &\Rightarrow f_i e_i T_\# \propto T_\# , \\
f_i T_\# \neq 0 &\Rightarrow e_i f_i T_\# \propto T_\# ,
\end{aligned}
\tag{4.44}
$$

which guarantees the "good"-ness property, as can be seen from the diagram.

Also, the depths $d(\alpha_i, T_\#) := d(\alpha_i, v(T_\#))$ of a standard tableau are easily found. Notice that the j right-most columns of a standard tableau form a standard tableau; call it $T_\#^{(j)}$, where j ranges from 0 to the full width of $T_\#$. It turns out that the weights of these sub-tableaux determine the depths $d(\alpha_i, T_\#)$ for all $i = 1, 2, \dots, r$. Precisely, we have

$$
d(\alpha_i, T_\#) = \max_j \left(wt(T_\#^{(j)}), -\alpha_i^\vee \right) .
\tag{4.45}
$$

We can also speak of the *height* $h(\alpha_i, T_\#)$. By $d(\alpha_i, T_\#)$ $(h(\alpha_i, T_\#))$ is meant the maximal non-negative integer ℓ such that $e_i^\ell T_\# \neq 0$ $(f_i^\ell T_\# \neq 0)$. So, $h(\alpha_i, T_\#) = d(-\alpha_i, T_\#)$, for example. They are also simply found, since for a state u in any $A_1 = \langle e_i, h_i, f_i \rangle \subset g$ representation, $h(\alpha_i, u) - d(\alpha_i, u) = (\sigma, \alpha_i^\vee)$, if $Hu = (\sigma, \alpha_i)u$. Now, notice that the tableaux drawn in Fig. 15, can be assigned to a single irreducible representation of $\langle e_i, h_i, f_i \rangle$ for each of $i = 1, 2$. This property generalises to all ranks r, and so

$$
h(\alpha_i, T_\#) = \left(wt(T_\#), \alpha_i^\vee \right) + d(\alpha_i, T_\#) .
\tag{4.46}
$$

47

The property just mentioned is another way of describing the "good"-ness of the required bases.

Consider the $g = A_2$ example $B^{(p)}((1,1);(0,0))$. Its elements are indexed by the tableaux

$$\boxed{\begin{array}{cc} 1 & 2 \\ \hline 3 \end{array}} \ , \ \boxed{\begin{array}{cc} 1 & 3 \\ \hline 2 \end{array}} \ . \tag{4.47}$$

Since

$$wt\left(\boxed{2}\right) \ = \ (-1,1) \ , \ wt\left(\boxed{\begin{array}{cc} 1 & 2 \\ \hline 3 \end{array}}\right) \ = \ (0,0) \ , \tag{4.48}$$

(4.45) gives

$$d\left(\alpha_1, \boxed{\begin{array}{cc} 1 & 2 \\ \hline 3 \end{array}}\right) \ = \ 1 \ \Rightarrow h\left(\alpha_1, \boxed{\begin{array}{cc} 1 & 2 \\ \hline 3 \end{array}}\right) \ = \ 1 \ ;$$

$$d\left(\alpha_2, \boxed{\begin{array}{cc} 1 & 2 \\ \hline 3 \end{array}}\right) \ = \ 0 \ \Rightarrow h\left(\alpha_2, \boxed{\begin{array}{cc} 1 & 2 \\ \hline 3 \end{array}}\right) \ = \ 0 \ . \tag{4.49}$$

Similarly,

$$wt\left(\boxed{3}\right) \ = \ (0,-1) \ , \ wt\left(\boxed{\begin{array}{cc} 1 & 3 \\ \hline 2 \end{array}}\right) \ = \ (0,0) \ , \tag{4.50}$$

implies

$$d\left(\alpha_1, \boxed{\begin{array}{cc} 1 & 3 \\ \hline 2 \end{array}}\right) \ = \ 0 \ \Rightarrow h\left(\alpha_1, \boxed{\begin{array}{cc} 1 & 3 \\ \hline 2 \end{array}}\right) \ = \ 0 \ ;$$

$$d\left(\alpha_2, \boxed{\begin{array}{cc} 1 & 3 \\ \hline 2 \end{array}}\right) \ = \ 1 \ \Rightarrow h\left(\alpha_2, \boxed{\begin{array}{cc} 1 & 3 \\ \hline 2 \end{array}}\right) \ = \ 1 \ . \tag{4.51}$$

A glance at Fig. 15 confirms these results.

When substituted into (4.28), the depths obtained by (4.45) yield the Littlewood-Richardson rule, a simple rule for the computation of the $g = A_r$ tensor-product coefficients $T^\nu_{\lambda,\mu}$:

$$\boxed{T^\nu_{\lambda,\mu} \ = \ \text{card}\{\, T_\# \in \mathcal{T}_\#(\mu; \nu - \lambda) \ : \ d(\alpha_i, T_\#) \le \nu_i \ \ (\forall i = 1, \ldots, r) \,\} \ .} \tag{4.52}$$

By (4.43), we have

$$\mathcal{T}_\#(\mu; \nu - \lambda) \ = \ \{\, T_\# \in \mathcal{S}_\# \ : \ sh(T_\#) = \mu, \ wt(T_\#) = \nu - \lambda \,\} \ . \tag{4.53}$$

We should mention that this is not precisely the form of the rule originally given by Littlewood and Richardson. It is, however, related to it by a simple transformation (see [37], for example).

To use the Littlewood-Richardson rule, one can first draw the Young tableau $\mathcal{T}_\lambda$ of shape λ. Consider a standard tableau $\mathcal{T}_\#$ of shape μ and weight $\nu - \lambda$. Add its sub-tableaux $\mathcal{T}^{(j)}_\#$

to $\mathcal{T}_\lambda$ by placing boxes $\boxed{\ell}$ to the right of the ℓ-th row (numbered from top to bottom)of $\mathcal{T}_\lambda$. An A_5 example, with $\lambda = (3,1,2,1)$, $\mu = (1,1,2,0)$ and $\nu = (2,0,3,2)$ is

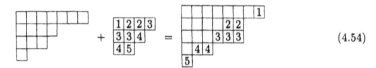

$$(4.54)$$

If the shape of the resulting *mixed tableau* (defined in the obvious way) is dominant (i.e. in P_+) for all $\mathcal{T}_\#^{(j)}$, then $\mathcal{T}_\#$ contributes 1 to $T_{\lambda,\mu}^\nu$. Also, the corresponding vector $v(\mathcal{T}_\#)$ is an element of the basis of $V_{\lambda,\mu}^\nu$.

To make things clear, consider a simple A_2 example: let's verify that $T_{\lambda,\mu}^\nu = T_{(1,0),(1,1)}^{(1,0)} = 1$. The tableaux of shape $\mu = (1,1)$ and weight $\nu - \lambda = (0,0)$ are those drawn in (4.47). Adding the sub-tableaux $\mathcal{T}_\#^{(j)}$ ($j = 1$ and 2) of the first one to $\mathcal{T}_\lambda = \boxed{}$ gives the mixed tableaux

$$\boxed{\begin{array}{c}\\ 2\end{array}} , \boxed{\begin{array}{c}1\\ 2\\ 3\end{array}} , \qquad\qquad (4.55)$$

so that the first tableau contributes 1 to the tensor-product coefficient. With the second, however, adding $\mathcal{T}_\#^{(1)} = \boxed{3}$ to $\mathcal{T}_\lambda$ gives

$$\boxed{\begin{array}{c}\\ 3\end{array}} . \qquad\qquad (4.56)$$

Therefore, the second tableau of (4.47) does not contribute, and we find $T_{(1,0),(1,1)}^{(1,0)} = 1$.

One nice feature of the Littlewood-Richardson rule is that one simply counts the number of standard tableaux of a certain type that pass a specific test. Furthermore, this test can be applied to the candidate tableaux one-by-one, without referring to the other candidates. For example, there are no redundancies or cancellations between candidates. For this reason, we'll call the Littlewood-Richardson rule a *combinatorial rule*.

The Littlewood-Richardson rule can be stated in many different ways. The version described is well suited to the application of (4.28), however, with its connection between standard tableaux and the vectors of $B^{(p)}(\mu; \nu - \lambda)$.

More importantly, the standard tableaux can be replaced by *universal* objects, that can be defined for any simple g in a uniform way. So the Littlewood-Richardson rule can be adapted from the case of $g = A_r$ to all simple g. It was Littelmann who completed this generalisation, first using sequences of Weyl group elements as the universal objects, and then using sequences of weights [38]. In the Littlewood-Richardson rule, both sequences have to do with the sub-tableaux $T_\#^{(j)}$ of a standard tableaux $T_\#$, or equivalently, the columns of $T_\#$, reading from right to left.

The more economical generalisation is the one using the sequences of weights. It is phrased in terms of *Littelmann paths* in weight space. We'll look at the $g = A_r$ case, but generalisation is straightforward. To each standard tableau, we can associate a piecewise linear path in weight space as follows: read the weights of the columns of the standard tableau from right to left, and associate a piece of the path to each. The vector position of the end of such a piece minus that of the beginning equals the weight of the corresponding column. So, by adding the pieces, in order (right to left on the tableau), we obtain the relevant path.

As an example, Fig. 16 shows the Littelmann paths for the outer weights of the A_2 representation $L(2,1)$.

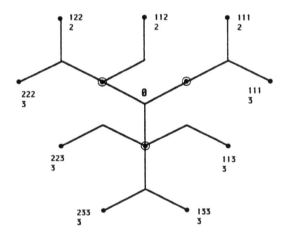

Figure 16. The Littelmann paths are drawn for the outermost weights of the representation $L(2,1)$ of A_2. For each such weight μ, the shortest route from 0 to μ is the relevant path.

As promised, we'll indicate the action of the step operators e_i, f_i ($i = 1, \ldots, r$) on the paths. By the equivalence of standard tableaux and Littelmann paths, this will also describe the actions on the standard tableaux. It is important to realise, however, that what we describe is only the action of the step operators up to normalisation. Let $v(\pi)$

denote the state labelled by the path π. e_i, f_i acting on a path π yield other paths, $e_i\pi$, $f_i\pi$. Then we have $e_i\,v(\pi) \propto v(e_i\pi)$ and $f_i\,v(\pi) \propto v(f_i\pi)$, with non-zero multiplicative factors left unspecified.

First consider e_i. Parametrise a path π with parameter $t \in [0,1]$. $wt\big(\pi(t)\big)$ is the weight of a point t of the path. $wt\big(\pi(0)\big) = 0$, and $wt\big(\pi(1)\big)$ is the weight of the path $wt(\pi)$, and the weight of the corresponding vector and standard tableau. Similarly, the shape of the path $sh(\pi)$ equals the shape of the corresponding Young tableau, or the weight of the highest vector. First find the minimum non-positive integer value of $\big(wt(\pi(t)), \alpha_i^\vee\big)$ for $t \in [0,1]$; call it M_i. Let t_2 be the minimum value of t where $\big(wt(\pi(t)), \alpha_i^\vee\big) = M_i$. If $M_i = 0$, then $e_i\pi = 0$. If $M_i \leq -1$, then find the maximum value of $t < t_2$ such that $\big(wt(\pi(t)), \alpha_i^\vee\big) = M_i + 1$, and call that t_1. Now break the path up into three pieces, corresponding to the intervals:

$$0 \leq t \leq t_1\ ,\quad t_1 \leq t \leq t_2\ ,\quad t_2 \leq t \leq 1\ . \tag{4.57}$$

Weyl reflect the middle piece across the hyperplane normal to α_i at $wt(\pi(t_1))$. Finally, re-attach the third piece (corresponding to $t_2 \leq t \leq 1$) at $\pi(t_2) + \alpha_i$, to obtain the path $e_i\pi$.

It is not hard to see that this action yields $d(\alpha_i, v(\pi)) = -M_i$ for the vector $v(\pi)$ that is indexed by the path π. A diagram sketching the action of e_i on a path π is given in Fig. 17.

The action of the lowering operator f_i is defined similarly; see Fig. 18. With M_i defined as above, consider $M_i' = wt(\pi) - M_i$. If $M_i' = 0$, then $f_i\pi = 0$. If $M_i' \geq 1$, first find the maximum value of t such that $\big(wt(\pi(t)), \alpha_i^\vee\big) = M_i$; call it t_1. Then obtain the minimum value of $t > t_1$ such that $\big(wt(\pi(t), \alpha_i^\vee\big) = M_i + 1$; call that t_2. Then the three intervals (4.57) are again relevant, and $f_i\pi$ is found by reflecting the middle piece across the hyperplane normal to α_i at $wt(\pi(t_1))$, and re-attaching the third piece at $\pi(t_2) - \alpha_i$.

It is not hard to see that M_i' so defined is the height $h(\alpha_i, v(\pi))$ for the vector $v(\pi)$ that is indexed by the path π.

With the actions so-defined, we've drawn in Fig. 19 the analogue of Fig. 15 for the A_2 representation $L(2,1)$. Again, the lines indicate the action of the step operators e_1, e_2, f_1, f_2.

Incidentally, the graphs of Figs. 15,19 are (essentially) examples of so-called *crystal graphs* [39][40]. These arise in the theory of *quantum groups* $U_q(g)$, the q-deformations of the universal enveloping algebra $U(g) = U_1(g)$ of g. Such quantum groups allow the construction of integrable lattice models in two-dimensions, in the spirit of the 2-dimensional Ising model. The graphs reflect the simplified representation theory of $U_q(g)$ at $q = 0$.

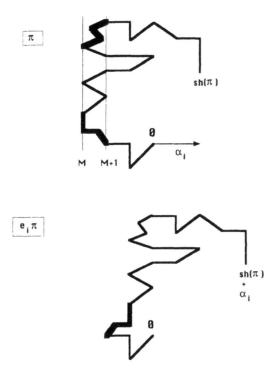

Figure 17. The action of e_i on a Littelmann path π. The thickened parts of the path π are those segments to be Weyl-reflected by e_i or f_i (see Fig. 18 for the latter).

(One can also say that the existence of the canonical bases at $q = 1$ is explained by the simplified representation theory at $q = 0$ [41].) But physically, $q = 0$ corresponds to absolute zero temperature, justifying the reference to crystals.

To make the actions of the ladder operators on paths completely clear, we'll do a couple of examples. First consider the path of weight (2,-2) drawn in Fig. 16. We indicate this path by the weights of its straight segments, written in sequence:

$$\pi \;=\; \{\,(0,-1),(1,0),(1,-1)\,\}\,. \tag{4.58}$$

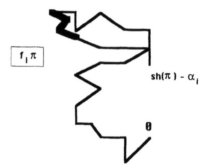

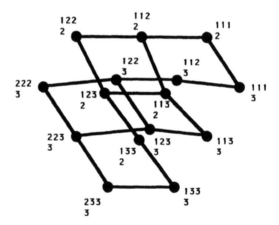

Figure 18. The action of f_i on the Littelmann path π of Fig. 17.

Figure 19. The standard tableaux of the A_2 representation $L(2,1)$ are drawn (roughly) at the positions of their weights. The lines indicate the action of the step operators e_1, e_2, f_1, f_2 on the good basis elements indexed by the tableaux (or by the corresponding Littelmann paths).

To find $e_2\pi$, we first need M_2, the minimum non-positive value of $\left(wt(\pi(t)), \alpha_2^\vee\right)$; here it is -2. We find $wt(\pi(t_1)) = (1,-1)$ and $wt(\pi(t_2)) = (2,-2)$. (Notice here that the "third piece" of (4.57) corresponds to the single value $t = 1$.) The "middle piece" consists of a single, straight segment, of weight $(1,-1)$, and so is reflected to a similar single, straight segment of weight $r_2(1, -1) = (0, 1)$. So we find

$$e_2\, \pi \;=\; \{\, (0, -1), (1, 0), (0, 1) \,\} \,, \tag{4.59}$$

a path that is not drawn in Fig. 16.

Now consider $f_1\pi'$, with π' the path

$$\pi' \;=\; \{\, (-1, 1), (1, 0), (0, 1) \,\} \,, \tag{4.60}$$

of weight $(0, 2)$. We have $M_1 = -1$, and so $M_1' = 0 - (-1) = 1$. $wt(\pi'(t_1)) = (-1, 1)$ and $wt(\pi'(t_2)) = (0, 1)$ indicate that the "middle piece" is a single, straight segment of weight $(1, 0)$, that must be replaced by one of weight $r_1(1, 0) = (-1, 1)$. As a result, we find

$$e_1\, \pi' \;=\; \{\, (-1, 1), (-1, 1), (0, 1) \,\} \,, \tag{4.61}$$

a path that is drawn in Fig. 16.

What is the form of the Littlewood-Richardson rule in terms of paths? To calculate $T_{\lambda,\nu}^\nu$, one considers all paths π of shape $sh(\pi) = \mu$ and weight $wt(\pi) = \nu - \lambda$. Such a path will contribute 1 to $T_{\lambda,\mu}^\nu$ iff $\lambda + wt(\pi(t))$ has non-negative Dynkin indices for all $t \in [0, 1]$. That is, we require $\left(\lambda + wt(\pi(t)), \alpha_i^\vee\right) \geq 0$, for all $i \in \{1, \ldots, r\}$, and all $0 \leq t \leq 1$. In other words, the λ-translated path $\lambda + wt\left(\pi(t)\right)$ must remain in the dominant sector; it must be a *dominant path*.

This form of the Littlewood-Richardson rule is very general, applicable to all symmetrisable Kac-Moody algebras. These paths also have some interesting invariances. For example, one can generate a valid set of paths for the vectors of $L(\mu)$ by acting successively with lowering operators on the "highest path", in all inequivalent ways. Furthermore, one can use any dominant path as the highest path [42], even the straight line from 0 to μ!

This is all interesting, but what we really want to know is how to adapt this machinery to the formula (4.32) for the computation of fusion coefficients $^{(k)}N_{\lambda,\mu}^\nu$. Can we find an analogue of the Littlewood-Richardson rule for fusions: a combinatorial rule that computes the fusion coefficients $^{(k)}N_{\lambda,\mu}^\nu$?

For that we also need to consider the action of E^θ. Restricting to $g = A_r$, we have $\theta = \alpha_1 + \alpha_2 + \ldots + \alpha_r$. By the Cartan-Weyl commutation relations (3.8),

$$E^\theta \;\propto\; \mathrm{ad}\,(e_1)\mathrm{ad}\,(e_2)\cdots \mathrm{ad}\,(e_r)\, e_r \,. \tag{4.62}$$

For example, if $r = 2$, $E^\theta \propto [e_1, e_2]$. Figures 15, 19 reveal that the basis $B^{(p)}(\mu; \sigma)$ that is good for the calculation of tensor-product coefficients, is *not* also good for the calculation of their truncated versions, the fusion coefficients [32][43].

For example, consider again $B^{(p)}((1,1);(0,0))$, and the relevant tableaux (4.47). We get

$$
\begin{aligned}
E^\theta \,\begin{array}{|c|c|}\hline 1 & 2 \\\hline 3 \\\cline{1-1}\end{array} &\propto [e_1,e_2]\,\begin{array}{|c|c|}\hline 1 & 2 \\\hline 3 \\\cline{1-1}\end{array} = -e_2 e_1 \begin{array}{|c|c|}\hline 1 & 2 \\\hline 3 \\\cline{1-1}\end{array} \propto \begin{array}{|c|c|}\hline 1 & 1 \\\hline 2 \\\cline{1-1}\end{array}\ , \\[2mm]
E^\theta \,\begin{array}{|c|c|}\hline 1 & 3 \\\hline 2 \\\cline{1-1}\end{array} &\propto [e_1,e_2]\,\begin{array}{|c|c|}\hline 1 & 3 \\\hline 2 \\\cline{1-1}\end{array} = e_1 e_2 \begin{array}{|c|c|}\hline 1 & 3 \\\hline 2 \\\cline{1-1}\end{array} \propto \begin{array}{|c|c|}\hline 1 & 1 \\\hline 2 \\\cline{1-1}\end{array}\ .
\end{aligned}
\tag{4.63}
$$

On the other hand, we know (by means to be discussed soon) that $^{(2)}N^{(1,1)}_{(1,1),(1,1)} = 1$, and $^{(\geq 3)}N^{(1,1)}_{(1,1),(1,1)} = 2$. This means that some linear combination v_0 of the vectors labelled by $\begin{array}{|c|c|}\hline 1 & 2 \\\hline 3 \\\cline{1-1}\end{array}$ and $\begin{array}{|c|c|}\hline 1 & 3 \\\hline 2 \\\cline{1-1}\end{array}$ has depth $d(\theta, v_0) = 0$ and an independent vector v_1 in their span has $d(\theta, v_1) = 1$.

Is there another basis that is good for $^{(k)}N^\nu_{\lambda,\mu}$? No. Such a basis must be good for the tensor-product coefficients $T^\nu_{\lambda,\mu}$, and in the case of $g = A_2$, such a basis is unique. So no such basis exists for A_2; neither does one for any $A_r \supset A_2$.

So, to find the fusion coefficients by this route, the appropriate linear combinations must be found. They would be useful for the computation of the OPE coefficients $\hat{C}^\nu_{\lambda,\mu}$. But for the fusion coefficients only, this is a bit much. Luckily, there is another way to find the fusion coefficients, due to Verlinde.

4.6. Affine characters and modular transformations

In order to motivate Verlinde's approach, let us first discuss some results on affine algebras that will turn out to be important for affine fusion [44]. Consider the affine character (3.43) made informal:

$$
ch_{\hat\lambda}(\hat\sigma) := \sum_{\hat\mu \in P(\hat\lambda)} \text{mult}\,(\hat\lambda; \hat\mu)\, e^{(\hat\mu, \hat\sigma)}\ .
\tag{4.64}
$$

We'll set

$$
\sigma = -2\pi i\left(\zeta := \sum_{j=1}^{r} z_j \alpha_j^\vee, \tau, 0\right),
\tag{4.65}
$$

and use (3.47) to bring in the affine Weyl group $\hat{W}$. Then one finds

$$
ch_{\hat\lambda}(\hat\sigma) = e^{-2\pi i \tau\left(h_\lambda - \frac{c(g,k)}{24}\right)} \frac{\sum_{w \in W} (\det w)\, \Theta^{(k+h^\vee)}_{w(\lambda+\rho)}(\tau, \{z\})}{\sum_{w \in W} (\det w)\, \Theta^{(h^\vee)}_{w\rho}(\tau, \{z\})}\ ,
\tag{4.66}
$$

where $\{z\}$ stands for $(z_1, z_2, \ldots, z_r)$, and the *theta functions* are

$$
\Theta^{(k)}_\lambda(\tau, \{z\}) := \sum_{\alpha \in Q^\vee} e^{-\pi i \left[2(\lambda + k\alpha, \zeta) - \frac{\tau}{k^2}|\lambda + k\alpha|^2\right]}\ .
\tag{4.67}
$$

Recall also that h_λ and $c(g,k)$ are the conformal weight of ϕ_λ, and the Virasoro central charge, respectively (see (2.76),(2.62)). It is convenient to define the *normalised character*

$$\chi_{\hat\lambda}(\tau, \{z\}) := e^{2\pi i \tau \left(h_\lambda - \frac{c(g,k)}{24}\right)} \, ch_{\hat\lambda}(\hat\sigma) \,. \tag{4.68}$$

Then we have the simple relation

$$\chi_{\hat\lambda}(\tau, \{z\}) = \frac{\sum_{w\in W} (\det w) \, \Theta^{(k+h^\vee)}_{w(\lambda+\rho)}(\tau, \{z\})}{\sum_{w\in W} (\det w) \, \Theta^{(h^\vee)}_{w\rho}(\tau, \{z\})} \,. \tag{4.69}$$

This last result is remarkable. First, notice that the sums are both over the finite Weyl group of the simple Lie algebra g. Consequently, there is a striking resemblance to the Weyl character formula (3.22) for g.

One can trace the appearance of the W-sums to the semi-direct product structure of $\hat W$, (3.40). When acting on a shifted weight, such as $\hat\lambda + \hat\rho$, $\hat\lambda \in P^k_+$, we have $\hat W = W \ltimes T_{(k+h^\vee)Q^\vee}$. The other factor $T_{(k+h^\vee)Q^\vee}$ is responsible for the presence of the theta functions. These latter are well-known to have remarkable transformation properties under the group $PSL(2,\mathbb{Z})$, the so-called *modular group* Γ. The modular group is generated by the elements $S : \tau \to -1/\tau$ and $T : \tau \to \tau + 1$, and the general modular transformation has the form

$$\tau \to \frac{a\tau + b}{c\tau + d} \,, \qquad a,b,c,d \in \mathbb{Z} \,, \qquad ad - bc = 1 \,. \tag{4.70}$$

So, this transformation can be encoded in a 2×2 integer matrix, of determinant one, or the negative of such a matrix; thus $\Gamma \cong PSL(2,\mathbb{Z})$.

As a consequence of the appearance of the theta functions in (4.69), we find that the normalised characters transform among themselves under modular transformations. Specifically, Kac and Peterson showed that

$$\begin{aligned}
\chi_{\hat\lambda}(-1/\tau, \{z/\tau\}) &= \sum_{\hat\mu \in P^k_+} S^{(k)}{}_{\lambda,\mu} \, \chi_{\hat\mu}(\tau, \{z\}) \,, \\
\chi_{\hat\lambda}(\tau + 1, \{z\}) &= \sum_{\hat\mu \in P^k_+} T^{(k)}{}_{\lambda,\mu} \, \chi_{\hat\mu}(\tau, \{z\}) \,.
\end{aligned} \tag{4.71}$$

Consider the card $P^k_+ \times$ card $P^k_+ =$ card $\overline{P^k_+} \times$ card $\overline{P^k_+}$ (see (3.42)) matrices $S^{(k)}$ and $T^{(k)}$, with elements $S^{(k)}{}_{\lambda,\mu}$ and $T^{(k)}{}_{\lambda,\mu}$, respectively; they turn out to be unitary, showing that the normalised characters form a unitary representation of (a subgroup of) the modular group Γ. The explicit forms of their elements are:

$$
\begin{aligned}
S^{(k)}{}_{\lambda,\mu} &= F(g,k) \sum_{w \in W} (\det w)\, e^{-\frac{2\pi i}{k+h^\vee}\left(\lambda+\rho,\, w(\mu+\rho)\right)} \ , \\
T^{(k)}{}_{\lambda,\mu} &= \delta_{\lambda,\mu}\, e^{-2\pi i\left(h_\lambda - \frac{c(g,k)}{24}\right)} \ ,
\end{aligned}
\tag{4.72}
$$

where $F(g,k)$ is a constant independent of λ, μ.

4.7. Kac-Peterson relation and Verlinde formula

The form of the $S^{(k)}{}_{\lambda,\mu}$ recalls the numerator of the Weyl character formula (3.22). To recover the full character, rather than just the numerator, we can take a ratio to find

$$
\frac{S^{(k)}{}_{\lambda,\mu}}{S^{(k)}{}_{0,\mu}} = \mathrm{ch}_\lambda\!\left(-2\pi i\,\frac{\mu+\rho}{k+h^\vee}\right) .
\tag{4.73}
$$

We'll call this important result the *Kac-Peterson* relation.

Why should there be such an intimate relation between the modular matrix $S^{(k)}$ of the affine characters and the characters of g? An answer is provided by conformal field theory, and more specifically here, by the WZW model.

First, recall one of the uses of characters in the representation theory of g. Consider the tensor-product decomposition $L(\lambda) \otimes L(\mu) = \sum_{\nu \in P_+} T^\nu_{\lambda,\mu} L(\nu)$, and the formal element e^H in it. Taking traces gives

$$
\mathrm{ch}_\lambda \mathrm{ch}_\mu = \sum_{\nu \in P_+} T^\nu_{\lambda,\mu} \mathrm{ch}_\nu \ .
\tag{4.74}
$$

One says that the simple Lie characters obey the tensor-product algebra.

It is therefore natural to wonder if the *Kac-Peterson ratios*

$$
\chi_\lambda^{(k)}(\mu) := \frac{S^{(k)}{}_{\lambda,\mu}}{S^{(k)}{}_{0,\mu}}
\tag{4.75}
$$

have interesting multiplicative properties. After all, (4.73) says that they are "discretised" simple Lie characters. One finds

$$
\chi_\lambda^{(k)}(\sigma)\, \chi_\mu^{(k)}(\sigma) = \sum_{\nu \in \overline{P_+^k}} {}^{(k)}N^\nu_{\lambda,\mu}\, \chi_\nu^{(k)}(\sigma) \ ,
\tag{4.76}
$$

valid $\forall \sigma \in \overline{P_+^k}$. That is, the Kac-Peterson ratios obey the WZW fusion algebra!

If we rewrite this in terms of the modular S matrix, $S^{(k)}$, we find

$$
^{(k)}N^\nu_{\lambda,\mu} \; = \; \sum_{\sigma \in \overline{P^k_+}} \frac{S^{(k)}_{\lambda,\sigma} \, S^{(k)}_{\mu,\sigma} \, S^{(k)}{}^*_{\nu,\sigma}}{S^{(k)}_{0,\sigma}} \tag{4.77}
$$

using the unitarity of $S^{(k)}$. This is the celebrated *Verlinde formula* [45]. It is valid for all RCFT's, when the corresponding modular S matrices are used.

4.8. Duality

The Verlinde formula, in the form (4.76), provides a rationale for the Kac-Peterson relation (4.73). As the depth rule makes clear, the WZW fusion products should be truncated versions of the g tensor products, because the horizontal subalgebra $g \subset \hat{g}$ is a true symmetry of the theory. It is perhaps not too surprising then that the quantities $\{\chi^{(k)}_\lambda(\sigma) \; : \; \lambda \in \overline{P^k_+}\}$ that obey the fusion rule algebra (4.76) turn out to be discretised characters of g. Roughly, that they are characters means that the fusion coefficients are intimately related to the corresponding tensor-product coefficients. Their "discretisation" is a consequence of constraints coming from the spectrum-generating algebra $\hat{g} \supset g$. It will imply that the fusion coefficients are bounded above by the tensor-product coefficients.

But that does not explain the Verlinde formula in any way: why should ratios of modular S matrix elements have anything to do with fusions, let alone represent the fusion algebra?

The answer arises from the powerful concept of *duality* in conformal field theory [46]. Consider an arbitrary correlation function in a conformal field theory, not necessarily a WZW model. One finds that such a correlation function factorises into holomorphic and antiholomorphic parts:

$$
\mathcal{C}(z_1, \ldots, z_n, \tau_1, \ldots, \tau_m \,;\, \bar{z}_1, \ldots, \bar{z}_n, \bar{\tau}_1, \ldots, \bar{\tau}_m) =
$$
$$
\sum_{I,J} \mathcal{C}_{I,J} \, \mathcal{B}_I(z_1, \ldots, z_n, \tau_1, \ldots, \tau_m) \, \bar{\mathcal{B}}_J(\bar{z}_1, \ldots, \bar{z}_n, \bar{\tau}_1, \ldots, \bar{\tau}_m) \; . \tag{4.78}
$$

Here $z_1, \ldots, z_n$ are meant to indicate the positions of the n fields (points) of the correlation function, and the τ_j are constants (sometimes moduli) set by the type of correlation function under consideration. For the general class of conformal field theories known as *rational conformal field theories (RCFT's)*, which includes the WZW models, the sums over $I, \bar{J}$ are (discrete and) finite. The functions $\mathcal{B}_I$, $\bar{\mathcal{B}}_J$ are known as *conformal blocks*.

The factorisation (4.78) is not unique, however. As was alluded to earlier, the conformal blocks can (in principle) be calculated using the operator product algebra. If we symbolise

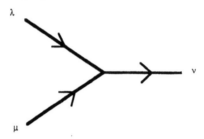

Figure 20. The graph labelling a conformal block for the 3-point function.

a non-zero operator product coefficient $\hat{C}^{\nu}_{\lambda,\mu}$ by the graph of Figure 20, we are associating that graph to the conformal block of a 3-point function. In a similar way, we can label a choice of conformal blocks by a trivalent graph. Part of the non-uniquenes of the conformal blocks comes from the non-uniqueness of the trivalent graph as label.

For example, consider a 4-point function. Its conformal blocks can be labelled by either of the two trivalent graphs of Fig. 21. The underlying assumption of duality is that the final physical correlation function must not depend on the choice of graph. Furthermore, duality states that there must be a linear relation between the conformal blocks associated with the two graphs, such that this is true. The matrix encoding the particular linear relation shown in the Figure, is called the fusing matrix F.

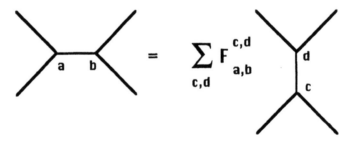

Figure 21. Duality for 4-point functions.

There is a another, related way of thinking of these trivalent graphs. When radial quantisation was mentioned earlier, the conformal transformation (2.31) from the cylinder

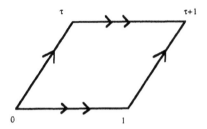

Figure 22. Torus, as parallelogram.

to the complex plane was discussed. One can think of the resulting plane as having two special, "marked" points, at the origin and at ∞. The inverse conformal transformation maps the plane with these two marked points (or punctures) to a Riemann surface, the cylinder. Letting the radius shrink to zero gives a particularly simple trivalent graph, a straight line.

If one considers a 4-point function, however, one has four marked points, and a conformal transformation can be found to map the plane to a Riemann surface. The Riemann surface will be topologically equivalent to a Riemann sphere with four marked points. One can then recover a trivalent graph by shrinking the sphere in different ways. In particular, one can recover the two different graphs of Fig. 21.

With this latter picture, duality can tell us why the modular group Γ enters consideration. The modular group is intimately connected with Riemann surfaces (with no marked points) of genus one, i.e. with tori. Represent a torus by a parallelogram with opposite sides identified, as in Fig. 22. We will consider a conformal field theory on such a torus. By conformal invariance, the overall scale doesn't matter, so we set the sides of the paralellogram as shown in the Figure. The conformal class of the torus can thus be specified by one complex number, its modulus τ. But this is still a redundant description. It is simple to see that $T : \tau \to \tau + 1$ does not change the underlying torus. Furthermore, after a rescaling, $S : \tau \to -1/\tau$ doesn't either. So the conformal class of the torus is invariant under the full modular group $\Gamma = \langle S, T \rangle$.

Now consider the "correlation function"

$$Z(\tau) = e^{2\pi i(\frac{c}{24}\tau - \frac{\bar{c}}{24}\bar{\tau})} \operatorname{Tr}_{\mathcal{H}} e^{-2\pi i\left(\mathbf{H}(i\operatorname{Im}\tau) + \mathbf{P}(\operatorname{Re}\tau)\right)}$$
$$= \operatorname{Tr}_{\mathcal{H}} e^{-2\pi i\left((L_0 - \frac{c}{24})\tau + (\bar{L}_0 - \frac{\bar{c}}{24})\bar{\tau}\right)} \,,$$

(4.79)

where $\bar{\tau}$ is the complex conjugate of τ, c and $\bar{c}$ are the holomorphic and antiholomorphic central charges, and $\mathcal{H}$ is the Hilbert space of the theory. We have used that the Hamiltonian and rotation operators are $\mathbf{H} = L_0 + \bar{L}_0$ and $\mathbf{P} = L_0 - \bar{L}_0$, respectively (see after (3.61)). (4.79) is known as the torus partition function.

Let us restrict once more to consideration of WZW models; what we'll say, however, can also be adapted straightforwardly to all RCFT's. Now, by the holomorphic-antiholomorphic factorisation of WZW models, the Hilbert space has the form

$$\mathcal{H} = \oplus_{\lambda,\mu\in\overline{P_+^k}} M_{\lambda,\mu}\,\mathcal{H}_\lambda \otimes \mathcal{H}_\mu \ , \tag{4.80}$$

with $M_{\lambda,\mu} \in \mathbb{Z}_{\geq 0}$. Here $\mathcal{H}_\lambda$ ($\lambda \in \overline{P_+^k}$) is the Hilbert space of states in the conformal tower of $L(\hat{\lambda})$ ($\hat{\lambda} \in P_+^k$). This factorisation is manifested in the partition function:

$$Z(\tau,\bar{\tau}) = \sum_{\lambda,\mu\in\overline{P_+^k}} M_{\lambda,\mu}\,\chi_{\hat{\lambda}}(\tau,\{0\})\chi_{\hat{\mu}}^*(\bar{\tau},\{0\}) \ . \tag{4.81}$$

That is, the conformal blocks for the torus partition function are the normalised characters. The corresponding trivalent graph is a loop, but the blocks are also labelled by the torus modulus τ. But τ, $\tau + 1$, $-1/\tau$, etc., are all different ways of labelling the same torus. So duality implies the modular covariance (4.71) of the normalised characters.

Summarising to this point: in a RCFT, (normalised) characters appear naturally as conformal blocks for the torus partition function, and duality implies that they must be modular covariant.

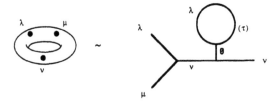

Figure 23. A depiction of a 3-point function on a torus, and one possible conformal block for it.

But this is just one simple consequence of duality; much more can be extracted, including the Verlinde formula. To derive it from duality, one needs to consider a correlation function that can involve both the 3-point functions (and so the fusion coefficients) and the torus (so that the modular transformations are involved). It is the 3-point functions on the torus that are pertinent in this context. Fig. 23 shows a torus with three marked points, and a choice of a trivalent graph to label the corresponding conformal block. That choice makes it clear how the 3-point functions and the characters appear. But there are many other choices of graphs, and one can also replace τ with $-1/\tau$, for example. These freedoms are not completely independent, however. They turn out to be sufficient to prove the Verlinde formula, but we will not provide the detailed argument here [46].

Recapping, we have seen that the remarkable modular properties of affine characters are accounted for in the physical context of WZW models. The underlying concept is duality, a property that extends to all RCFT's. It also implies many other important relations. The Verlinde formula is just one symptom of duality in conformal field theory.

4.9. Fusion coefficients as Weyl sums

To close this section, let us return to the problem of computing WZW fusion coefficients, and apply the Verlinde formula (4.77). The Kac-Peterson relation (4.73) in (4.74) implies

$$\chi_\lambda^{(k)}(\sigma)\,\chi_\mu^{(k)}(\sigma) \;=\; \sum_{\varphi \in P_+} T_{\lambda,\mu}^\varphi\, \chi_\varphi^{(k)}(\sigma)\,, \tag{4.82}$$

where we have used the notation (4.75). Compare this with (4.76). Both are valid $\forall \sigma \in P_+^k$. (4.76) led to (4.77) by the unitarity of the matrix $S^{(k)}$. A relation between the tensor-product coefficients $T_{\lambda,\mu}^\varphi$ and the fusion coefficients ${}^{(k)}N_{\lambda,\mu}^\nu$ should result from the same unitarity, except that the ranges are not identical. The sum over P_+ in (4.82) must be restricted to a sum over $\overline{P_+^k}$, as in (4.76).

To do so, we make use of the alternating Weyl symmetry (3.23), as applied to the Kac-Peterson ratios:

$$\chi_{w.\varphi}^{(k)}(\sigma) \;=\; (\det w)\,\chi_\varphi^{(k)}(\sigma)\,, \quad \forall\, w \in W\,. \tag{4.83}$$

This is not sufficient, however. To restrict the sum over $\varphi \in P_+$ to a sum over $\nu \in \overline{P_+^k}$, we need to use elements of the affine Weyl group $\hat{W}$, not just $W \subset \hat{W}$ (see Fig. 10, for example). More accurately, we need the action of the affine Weyl group projected onto the weight space of the horizontal subalgebra $g \subset \hat{g}$.

Consider the affine primitive reflection $\hat{r}_0$, with action on affine weights given in (3.38). Just as λ is used to denote the horizontal part of the affine weight $\hat{\lambda}$, $r_0.\lambda$ will indicate the horizontal part of $\hat{r}_0.\hat{\lambda}$. One finds

$$r_0.\lambda \;=\; r_\theta.\lambda \;+\; (k + h^\vee)\theta\,, \tag{4.84}$$

where $r_\theta \in W$ is the Weyl reflection across the hyperplane normal to the highest root θ. Then

$$e^{-\frac{2\pi i}{k+h^\vee}(r_0.\lambda,\sigma+\rho)} = e^{-\frac{2\pi i}{k+h^\vee}(r_\theta.\lambda,\sigma+\rho)} \, , \tag{4.85}$$

for $\sigma \in P_+$, since then $(\theta, \sigma + \rho) \in \mathbb{Z}$. Using this and the Weyl character formula (3.22), we find

$$\chi^{(k)}_{r_0.\varphi}(\sigma) = (\det r_\theta)\, \chi^{(k)}_\varphi(\sigma) = (\det r_0)\, \chi^{(k)}_\varphi(\sigma) \, . \tag{4.86}$$

Since the affine Weyl group $\hat{W}$ can be obtained from the Weyl group W of g by adjoining $\hat{r}_0$ as a generator, we have

$$\boxed{\chi^{(k)}_{w.\varphi}(\sigma) = (\det w)\, \chi^{(k)}_\varphi(\sigma) \, , \ \forall\, w \in \hat{W} \, .} \tag{4.87}$$

Using this in (4.82), and comparing with (4.76), we get

$$\sum_{\varphi \in \overline{P^k_+}} \sum_{w \in \hat{W}} (\det w)\, T^\varphi_{\lambda,\mu}\, \chi^{(k)}_\varphi(\sigma) = \sum_{\nu \in \overline{P^k_+}} {}^{(k)}N^\nu_{\lambda,\mu}\, \chi^{(k)}_\nu(\sigma) \, , \tag{4.88}$$

for all $\sigma \in \overline{P^k_+}$. The unitarity of $S^{(k)}$ means that the coefficients of $\chi^{(k)}_\nu(\sigma)$ on the left hand side of (4.88) can be equated with those on the right hand side, for all $\nu \in \overline{P^k_+}$. Therefore we find [26][47][48]

$$\boxed{{}^{(k)}N^\nu_{\lambda,\mu} = \sum_{w \in \hat{W}} (\det w)\, T^{w.\nu}_{\lambda,\mu} \, .} \tag{4.89}$$

The dependence of the right-hand side on the level k is implicit: for a fixed $\nu \in P_+$, $w.\nu$ can change with changing level, when $w \in \hat{W}$ (see (4.84)).

This last equation provides a fairly simple way of computing the fusion coefficients. For example, one can employ the Littlewood-Richardson rule (or Littelmann's generalisation) to first find the tensor-product coefficients $T^\nu_{\lambda,\mu}$, and then perform the alternating Weyl sum of (4.89). However, since it is an alternating sum (the signs are inherited from the Weyl character formula) cancellations occur, and the rule is not a combinatorial one.

Another expression can be given for the fusion coefficients as an alternating affine Weyl sum. This one will allow a connection with the refined depth rule (4.32). Using (3.19) brings in the weight multiplicities, and in (4.74) it leads to

$$T^\nu_{\lambda,\mu} = \sum_{w \in W} (\det w)\, \text{mult}\,(\mu; w.\nu - \lambda) \, . \tag{4.90}$$

In (4.89), this results in

$$
^{(k)}N_{\lambda,\mu}^{\nu} = \sum_{w \in \hat{W}} (\det w) \, \mathrm{mult}\,(\mu; w.\nu - \lambda) \,. \tag{4.91}
$$

This formula encodes a straightforward algorithm for the computation of the fusion coefficients. First, the weight $\lambda + \rho$ is added to all the weights of $P(\mu)$. the resulting weights are regarded as horizontal projections of affine weights, at level $k + h^{\vee}$. One then attempts to Weyl transform the resulting weights into the dominant ($w = \mathrm{id}$) sector, using the horizontal projection of the affine Weyl group. Some of the weights will be fixed by an affine Weyl reflection; that is, they can only lie on the boundary of the dominant sector after Weyl transformation. These should be ignored. The others will be transformed into the dominant sector, some by elements of $\hat{W}$ of determinant $+1$, and some by elements of determinant -1. The latter will always cancel some other dominant weights of the first kind. The final result will be a set of dominant weights $\{\nu + \rho\}$ with multiplicities $^{(k)}N_{\lambda,\mu}^{\nu}$, the quantities to be calculated.

The algorithm can be pictured in a weight diagram, for ranks ≤ 2. A simple example is drawn in Fig. 24. Recall now the refined depth rule (4.32). The conditions $(E^{-\alpha_i})^{1+\nu_i}u = 0$ can be interpreted as saying that the states u that count toward $^{(k)}N_{\lambda,\mu}^{\nu}$ are those for which the "cancelling states" $(E^{-\alpha_i})^{1+\nu_i}u$ do not exist. If one does, then the alternating Weyl sum formula (4.91) tells us that the cancelling state's weight will be reflected into the dominant sector, to cancel the weight ν, in the algorithm of the previous paragraph.

This concludes our discussion of the computation of fusion coefficients, as an example of an application of the representation theory of affine algebras to WZW models.

5. Conclusion

We hope these lecture notes have given some indication of the beauty of the subject of conformal field theory, and the associated infinite-dimensional algebras.

We'll close with a brief (and incomplete) guide to the review literature on conformal field theories, affine algebras, and their relation. The most comprehensive reference to date is [49]. Another monograph is [50].

For the beginner in conformal field theory, the reviews [51] [52] [53] and [54] (in French) are quite "user-friendly". So is Cardy's review [12], which emphasises the statistical mechanics applications of conformal field theory. For more applications, see [13].

For a review of rational conformal field theory, especially its duality, see [2].

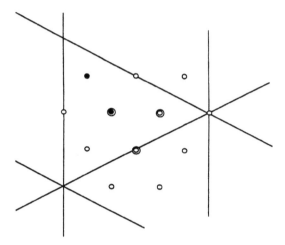

Figure 24. Weight diagram illustrating the computation by (4.91) of the A_2 fusion $L(2,0) \otimes_3 L(2,1) = L(0,3) \oplus L(1,1)$. Darkened circles indicate weights that correspond to the representations $L(0,3)$ and $L(1,1)$; the others are cancelled.

The ultimate reference for affine algebras is [26]. [27] is somewhat more accessible, however, but less comprehensive. Also accessible are [29] and [28]. [49] reviews the basic facts of affine algebras required in the study of conformal field theory.

For another treatment of affine algebras in conformal field theory, see [55]. [15] is an elegant exposition of functional integral methods applied to the WZW model.

Acknowledgements

It is a pleasure to thank the members of the Feza Gursey Institute for their warm hospitality. Thanks also go to the other lecturers and the participants in the summer school for enjoyable conversations. I am also grateful to Terry Gannon and Pierre Mathieu for helpful readings of the manuscript. Gannon gets additional thanks for encouragement, i.e. for gloating when he finished his lecture notes first.

The author acknowledges the support of a research grant from NSERC of Canada.

References

[1] A.A. Belavin, A.M. Polyakov, A.B. Zamolodchikov, Nucl. Phys. **B241** (1984) 333

[2] G. Moore, N. Seiberg, in the proceedings of the Trieste Spring School Superstrings, 1989, M. Green et al, eds. (World Scientific, 1990); and in the proceedings of the NATO Advanced Summer Institute and Banff Summer School, 1989, H.C. Lee, ed. (Plenum Press, 1990)

[3] P. Bouwknegt, K. Schoutens, Phys. Rep. **223** (1993) 183

[4] V. Kac, *Vertex Algebras for Beginners*, 2nd ed. (Amer. Math. Soc., 1998)

[5] T. Gannon, lecture notes, this volume

[6] M. Gaberdiel, P. Goddard, lecture notes, this volume

[7] E. Witten, Comm. Math. Phys. **92** (1984) 455

[8] P. Goddard, A. Kent, D. Olive, Phys. Lett. **152B** (1985) 88

[9] J. Balog, L. Feher, L. O'Raifeartaigh, P. Forgacs , A. Wipf, Annals Phys. **203** (1990) 76

[10] L. Dixon, J. Harvey, C. Vafa, E. Witten, Nucl. Phys. **B261** (1985) 620; **274** (1986) 285

[11] R. Dijkgraaf, C. Vafa, E. Verlinde, H. Verlinde, Commun. Math. Phys. **123** (1989) 485

[12] J. Cardy, in Les Houches session XILX, Fields, Strings, and Critical Phenomena, eds. E. Brézin, J. Zinn-Justin (Elsevier, 1989)

[13] I. Affleck, Acta Phys. Polon. **B26** (1995) 1869

[14] M. Green, J. Schwarz, E. Witten, *Superstring Theory, Vols. 1 & 2* (Cambridge U. Press, 1987)

[15] K. Gawedzki, lecture notes, this volume

[16] H. Georgi, Lie Algebras in Particle Physics (Benjamin/Cummings, 1982)

[17] S.P. Novikov, Usp. Mat. Nauk. **37** (1982) 3

[18] M. Nakahara, *Geometry, Topology and Physics* (IOP, 1990)

[19] K. Fujikawa, Phys. Rev. Lett. **42** (1979) 1195

[20] A.M. Polyakov, P.B. Wiegmann, Phys. Lett. **141B** (1984) 223; Phys. Lett. **131B** (1983) 121

[21] P. Di Vecchia, B. Durhuus, J.L. Petersen, Phys. Lett. **144B** (1984) 245

[22] A. Pressley, G. Segal, *Loop Groups* (Oxford U. Press, 1986)

[23] V.G. Knizhnik, A.B. Zamolodchikov, Nucl. Phys. **B247** (1984) 83

[24] J. Mickelsson, *Current Algebras and Groups* (Plenum, 1989)

[25] R.V. Moody, A. Pianzola, *Lie Algebras with Triangular Decomposition* (Wiley, 1995)

[26] V. Kac, *Infinite-dimensional Lie Algebras* (Cambridge U. Press, 1990)

[27] S. Kass, R.V. Moody, J. Patera, R. Slansky, *Affine Lie Algebras, Weight Multiplicities, and Branching Rules* (U. California Press, 1990)

[28] J. Fuchs, *Affine Lie Algebras and Quantum Groups* (Cambridge U. Press, 1992)

[29] P. Goddard, D. Olive, Int. J. Mod. Phys. **A1** (1986) 303

[30] D. Gepner, E. Witten, Nucl. Phys. **B278** (1986) 493

[31] D.P. Zelobenko, *Compact Lie Groups and Their Representations* (Am. Math. Soc., 1973)

[32] A.N. Kirillov, P. Mathieu, D. Sénéchal, M.A. Walton, Nucl. Phys. **B391** (1993) 651; p. 215, vol. 1 in M.A. del Olmo et al (editors), *Group Theoretical Methods in Physics, Proceedings of the XIXth International Colloquium, Salamanca, Spain, 1992* (CIEMAT, Madrid, 1993)

[33] M. A. Walton, Can. J. Phys. **72** (1994) 527

[34] C.J. Cummins, P. Mathieu, M.A. Walton, Phys. Lett. **254B** (1991) 386

[35] A.D. Berenstein, A.V. Zelevinsky, Funkt. Anal. Pril. **24** (1990) 1

[36] O. Mathieu, Geom. Ded. **36** (1990) 51;
I.M. Gelfand, A.V. Zelevinsky, Funkt. Anal. Pril. **19** (1985) 72

[37] J. Weyman, Contemp. Math. **88** (1989) 177

[38] P. Littelmann, J. Alg. **130** (1990) 328; Invent. Math. **116** (1994) 329

[39] M. Kashiwara, Commun. Math. Phys. **133** (1990) 249

[40] P. Littelmann, J. Alg. **175** (1995) 65

[41] G. Lusztig, J. Amer. Math. Soc. **3** (1990) 447; Prog. Theor. Phys. Suppl. **102** (1990) 175

[42] P. Littelmann, Ann. Math. **142** (1995) 499

[43] M.A. Walton, J. Math. Phys. **39** (1998) 665

[44] V.G. Kac, D. Peterson, Adv. Math. **53** (1984) 125

[45] E. Verlinde, Nucl. Phys. **B300** (1988) 360

[46] G. Moore, N. Seiberg, Phys. Lett. **212B** (1988) 451

[47] M.A. Walton, Phys. Lett. **241B** (1990) 365; Nucl. Phys. **B340** (1990) 777

[48] P. Furlan, A. Ganchev, V. Petkova, Nucl. Phys. **B343** (1990) 205

[49] P. Di Francesco, P. Mathieu, D. Sénéchal, *Conformal Field Theory* (Springer-Verlag, 1996)

[50] S.V. Ketov, *Conformal Field Theory* (World Scientific, 1995)

[51] J.-B. Zuber, Acta Phys. Polon. **B26** (1995) 1785

[52] P. Ginsparg, in Les Houches session XILX, Fields, Strings, and Critical Phenomena, eds. E. Brézin, J. Zinn-Justin (Elsevier, 1989)

[53] A.N. Schellekens, Fortschr. Phys. **44** (1996) 605

[54] Y. Saint-Aubin, Univ. de Montréal preprint, CRM-1472 (1987)

[55] J. Fuchs, lectures at the Graduate Course on Conformal Field Theory and Integrable Models (Budapest, August 1996), to appear in Springer Lecture Notes in Physics

INDEX

(NOTE: ROMAN NUMERALS I-VI INDICATE THE LECTURES IN THIS VOLUME.)

Milton Keynes UK
Ingram Content Group UK Ltd.
UKHW030902141024
449569UK00026B/1325

9 780813 342146